Understanding Mechanics

A.J. Sadler, B Sc

D.W.S. Thorning, B Sc, FIMA

Oxford University Press

Oxford University Press, Walton Street, Oxford OX2 6DP

Oxford New York Toronto
Delhi Bombay Calcutta Madras Karachi
Petaling Jaya Singapore Hong Kong Tokyo
Nairobi Dar es Salaam Cape Town
Melbourne Auckland

and associated companies in
Berlin Ibadan

Oxford is a trade mark of Oxford University Press

© A.J. Sadler, D.W.S. Thorning, 1983
Reprinted 1984, 1985, 1986, 1987, 1988 (twice), 1990 (twice), 1991

British Library Cataloguing in Publication Data

Sadler, A.J.
Understanding mechanics.
1. Mechanics, Analytic
I. Title II. Thorning, D.W.S.
531′.01′51 QA805

ISBN 0–19–914097–9

Typesetting and illustrations by Illustration Services, Oxford
Printed in Hong Kong

Preface

This book covers the mechanics required by a candidate taking the single subject GCE advanced level Mathematics and it makes a suitable first reading of the subject for those intending to go on to advanced level Applied Mathematics. The style and development of the topics should also appeal to teachers who are preparing candidates for the various alternative or additional mathematics examinations. The syllabuses at this level are varied, but by pursuing the relevant chapters to the level required, the teacher should find the book most suitable for such groups.

The principles of mechanics are not difficult to understand, but they need to be introduced with care and the student needs considerable practice in applying these principles to specific examples. To this end, this book presents topics in as simple and direct a manner as possible and gives ample opportunity for practice.

The theory sections in each chapter are followed by a number of worked examples which are typical of, and lead to the questions in the exercises. By reading the theory sections and following the worked examples, the reader should be able to make considerable progress with the exercise that follows.

Each exercise contains many carefully graded questions, beginning with those which give the student practice in applying the basic principles covered in the theory section, and progressing to more demanding questions.

Many pupils have difficulty in drawing the simple neat diagram that is the essential starting point to the solution of many applied mathematical problems; many of the exercises therefore start with questions in which the data are given on a simple diagram. In this way the student learns to start by drawing a diagram for each of the harder questions that occur later in the exercise.

Actual questions from past examination papers are included at the end of each chapter but the first. We are grateful to the following examination boards for permission to use these questions. The answers provided for these questions are the sole responsibility of the authors.

The Associated Examining Board
University of Cambridge Local Examinations Syndicate
Joint Matriculation Board
University of London School Examinations Council
Oxford Delegacy of Local Examinations
Southern Universities' Joint Board

A.J. Sadler
D.W.S. Thorning

Contents

1
Vectors

From simple arithmetic it is known that $3 + 2 = 5$.
It is possible however, in another context, to obtain a different answer when 3 and 2 are added.

Suppose a man walks 3 km due north and then 2 km due south. In order to find the total distance walked, the separate distances have to be added

i.e. 3 km + 2 km = 5 km ... [1]

This statement does not give the final position of the man at the end of his walk. In fact he is clearly 1 km due north of his starting point and this is his *displacement* from his original position.

i.e. 3 km due N + 2 km due S = 1 km due N ... [2]

These two statements give different information. The first statement adds *scalar* quantities, i.e. quantities which only have magnitude (or size). The answer gives the total distance travelled which is also a scalar.
The second statement deals with the addition of two *vector* quantities, i.e. quantities which have both magnitude and direction, and gives the vector displacement of the man at the end of his two walks.
The addition of vectors which have the same, or opposite, directions can be done quite simply. The addition of more general vectors requires a more sophisticated approach and can be done either by scale drawing or by calculation.

Example 1

A man walks 6 km south-west and then 4 km due west. How far, and in what direction, is he then from his starting point?

There are two ways of solving this problem:

(a) By scale drawing
Draw a sketch showing the two stages, OA and AB of his journey. From this sketch make a scale drawing with OA = 6 cm in a direction south-west of his starting-point O, and AB = 4 cm due west.

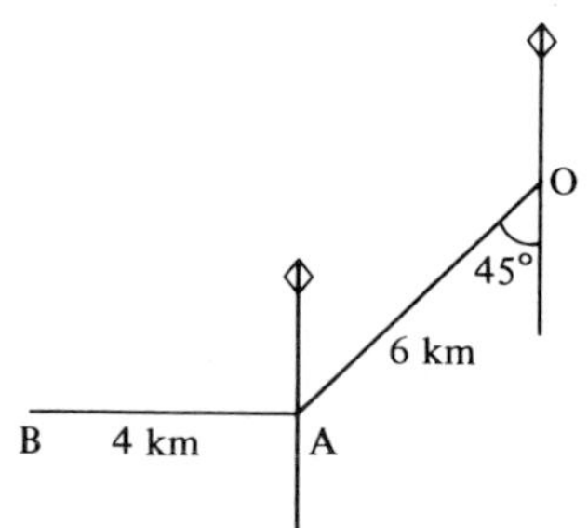

From the scale drawing, by measurement

$$OB = 9{\cdot}3 \text{ cm} \qquad A\hat{O}B \doteqdot 18^\circ$$

$$\therefore B\hat{O}S = 45^\circ + 18^\circ = 63^\circ$$

The man is 9·3 km from his starting point in a direction S 63° W.

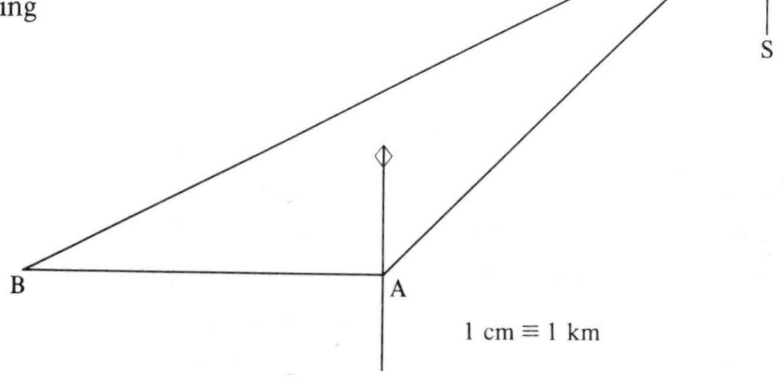

(b) By calculation

Draw a sketch showing the two stages, OA and AB, of his journey.

By the cosine rule

$$OB^2 = OA^2 + AB^2 - 2 \times OA \times AB \cos B\hat{A}O$$
$$= 6^2 + 4^2 - 2(6)(4)\cos 135$$

hence $OB = 9{\cdot}27$

Using the sine rule

$$\frac{OB}{\sin O\hat{A}B} = \frac{AB}{\sin B\hat{O}A}$$

thus $$\frac{9{\cdot}27}{\sin 135} = \frac{4}{\sin B\hat{O}A}$$

hence $B\hat{O}A = 17{\cdot}77^\circ$

and so $B\hat{O}S = 62{\cdot}77^\circ$

The man is 9·27 km from his starting point, in a direction S 62·77° W.

Exercise 1A

Solve each question by calculation or by scale drawing.

1. A woman cycles 5 km due east followed by 7 km due west. How far, and in what direction, is she then from her starting point?
2. A bird flies 40 km due south and then 30 km due east. Find the bird's distance and bearing from its original position.
3. A boat travels 6 km due east followed by 2·5 km due north. Find the distance the boat is then from its original position and the course it should set if it is to return by the shortest route.

4. A yacht sails 5 km in a direction N 30° E followed by 4 km due east. How far, and in what direction, is the yacht then from its original position?
   Would the yacht have reached the same position had it sailed 4 km due east followed by 5 km in a direction N 30° E?
5. I walk 800 m on a bearing 320° and then 500 m on a bearing 200°. Find how far I am then from my original position and the course I must set in order to return to my starting point by the shortest route.
6. An aeroplane flies from airport A to airport B 90 km away and on a bearing 070°. From B the aeroplane flies to airport C, 100 km from B on a bearing 210°. How far and on what course must the aeroplane now fly in order to return to A direct?
7. A ship travels 6 km north-east and then changes course and travels a further 3 km. If the ship is then 5 km from its original position find the two possible directions for the course set by the ship on the second stage of its journey.
8. A man walks 4 km due east, 3 km due north and then 3 km on a bearing S 60° E. By making an accurate scale drawing find the distance and bearing of the man's final position from his original position.

Resultant of vectors

In Example 1, the combined effect of two vectors was found by scale drawing and by calculation. This combined effect is said to be the *resultant* of the two vectors.
The resultant of two vectors is that single vector which could completely take the place of the two vectors, i.e. in Example 1, the man would have arrived at the same position had he walked 9·3 km in a direction S 63° W.

Example 2

Find the resultant of a vector of magnitude 5 units, direction 320°, and a vector of magnitude 8 units, direction 055°.

Make a scale drawing with
OA = 5 cm in direction 320°
AB = 8 cm in direction 055°

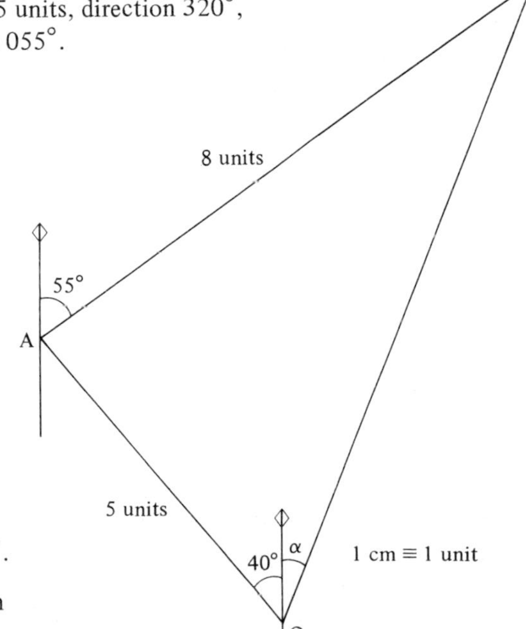

By measurement

$$OB = 9{\cdot}1 \text{ cm}$$
$$\alpha = 22^\circ$$

The resultant is 9·1 units in a direction 022°.

Alternatively, these answers could have been obtained by calculation.

Vector representation

In the solution of Example 2, a vector quantity was represented by a line segment of an appropriate length in a particular direction.
In order to distinguish between the distance OA and the vector OA, an arrow is placed over the letters of the vector. Thus $\overrightarrow{OA}$ represents the vector with magnitude and direction given by the line segment joining O to A.

Thus if a man walks from O to A and then from A to B, this could be written as a vector equation

$$\overrightarrow{OA} + \overrightarrow{AB} = \overrightarrow{OB},$$

since the vector represented by the line segment OB is clearly the resultant of $\overrightarrow{OA}$ and $\overrightarrow{AB}$.

Note carefully that the direction of the arrows on the diagram correspond to the order of the letters of the vectors which they represent.

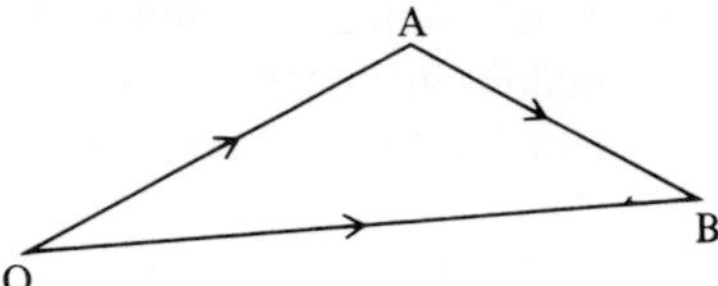

Vectors may also be written using single letters, and in this case heavy type is used,

thus $\quad \mathbf{x} + \mathbf{y} = \mathbf{z}$.

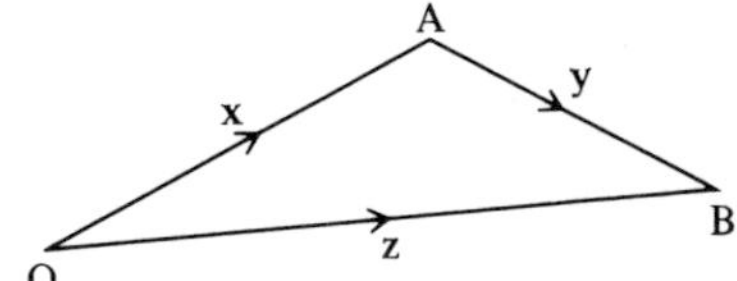

Since $\overrightarrow{OA} = \mathbf{x}$, it follows that $\overrightarrow{AO} = -\mathbf{x}$ because $\overrightarrow{AO}$ has the same length as $\overrightarrow{OA}$, but it is in the opposite direction.

Example 3

The diagram shows a parallelogram ABCD with $\overrightarrow{AB} = \mathbf{a}$ and $\overrightarrow{BC} = \mathbf{b}$. E is the mid-point of CD. Express the following vectors in terms of **a** and **b**.

(a) $\overrightarrow{AD}$, (b) $\overrightarrow{DC}$, (c) $\overrightarrow{CD}$, (d) $\overrightarrow{DE}$, (e) $\overrightarrow{AE}$.

(a) AD is the same length as BC and in the same direction

thus $\quad \overrightarrow{AD} = \overrightarrow{BC}$

$\therefore \quad \overrightarrow{AD} = \mathbf{b}$

(b) In a similar way,

$\overrightarrow{DC} = \mathbf{a}$

(c) CD is the same length as AB but in the opposite direction

$\therefore \quad \overrightarrow{CD} = -\mathbf{a}$

(d) $\quad \overrightarrow{DE} = \frac{1}{2}\overrightarrow{DC}$

$\therefore \quad \overrightarrow{DE} = \frac{1}{2}\mathbf{a}$

(e) $\quad \overrightarrow{AE} = \overrightarrow{AD} + \overrightarrow{DE}$

$\therefore \quad \overrightarrow{AE} = \mathbf{b} + \frac{1}{2}\mathbf{a}$

Exercise 1B

In questions **1** to **6**, the directions of the vectors are given as bearings, i.e. the angle the vector makes with the direction of north, measured clockwise from north.

1. Find the single vector that is the resultant of a vector of magnitude 7 units, direction 050°, and a vector of magnitude 4 units, direction 160°.
2. Find the single vector that is the resultant of two vectors of magnitude 6 units and 5 units and direction 240° and 260° respectively.
3. Find the resultant of a vector of magnitude 4 units, direction 040°, and a vector of magnitude 7 units, direction 130°.
4. Find the resultant of the vectors **a** and **b** if **a** has magnitude 6 units and direction 160° and **b** has magnitude 11 units and direction 320°.
5. By making a scale drawing, find the resultant of the vectors **a**, **b** and **c** given that **a** has magnitude 6 units and direction 060°, **b** has magnitude 7 units and direction 140° and **c** has magnitude 4 units and direction 020°.
6. By making a scale drawing, find the resultant of three vectors of magnitude 6 units, 9 units and 10 units and direction 330°, 200° and 080° respectively.

7. The diagram shows a parallelogram OABC with  $\overrightarrow{OA} = \mathbf{a}$ and $\overrightarrow{OC} = \mathbf{b}$. Express the following vectors in terms of **a** and **b**.

   (a) $\overrightarrow{AB}$ (b) $\overrightarrow{BA}$ (c) $\overrightarrow{CB}$ (d) $\overrightarrow{BC}$
   (e) $\overrightarrow{OB}$ (f) $\overrightarrow{BO}$ (g) $\overrightarrow{AC}$ (h) $\overrightarrow{CA}$

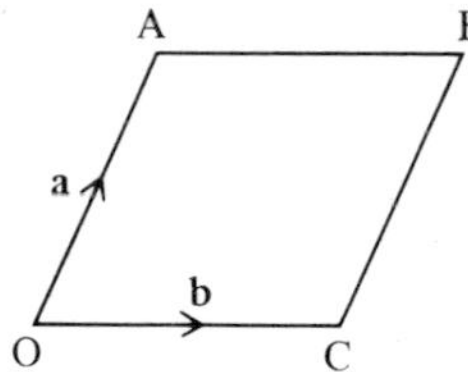

8. The diagram shows a triangle OAB with $\overrightarrow{OA} = \mathbf{a}$ and $\overrightarrow{OB} = \mathbf{b}$. C is the mid-point of AB. Express the following vectors in terms of **a** and **b**.

   (a) $\overrightarrow{AB}$ (b) $\overrightarrow{BA}$ (c) $\overrightarrow{AC}$ (d) $\overrightarrow{OC}$

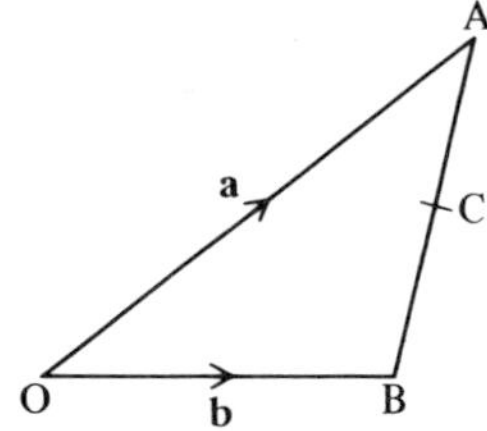

9. The diagram shows a trapezium OABC with $\overrightarrow{OA} = \mathbf{a}$ and $\overrightarrow{OC} = 2\mathbf{b}$. AB is parallel to and half as long as OC. Express the following vectors in terms of **a** and **b**.

   (a) $\overrightarrow{AB}$ (b) $\overrightarrow{OB}$ (c) $\overrightarrow{BC}$

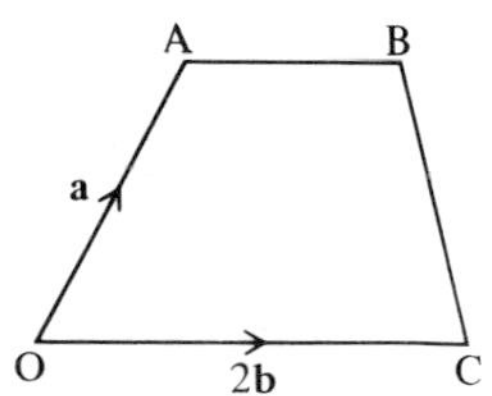

10. The diagram shows a triangle OAB with $\overrightarrow{OA} = \mathbf{a}$ and $\overrightarrow{OB} = \mathbf{b}$. C is a point on AB such that BC : CA = 1 : 2. Express the following vectors in terms of **a** and **b**.

    (a) $\overrightarrow{AB}$ (b) $\overrightarrow{AC}$ (c) $\overrightarrow{BC}$ (d) $\overrightarrow{OC}$.

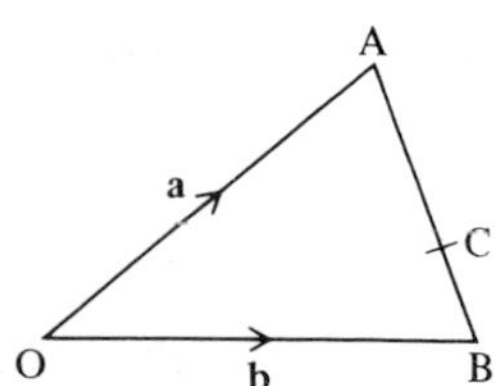

11. The diagram shows a parallelogram OABC with $\overrightarrow{OA} = \mathbf{a}$ and $\overrightarrow{OC} = \mathbf{c}$. E is a point on CB such that CE : EB = 1 : 3. D is a point on AB such that AD : DB = 1 : 2. Express the following vectors in terms of **a** and **c**.

(a) $\overrightarrow{AD}$ (b) $\overrightarrow{CE}$ (c) $\overrightarrow{OD}$ (d) $\overrightarrow{OE}$
(e) $\overrightarrow{AE}$ (f) $\overrightarrow{DE}$.

12. In a triangle OAB, the point C lies at the mid-point of OA and the point D lies on AB such that AD : DB = 3 : 1. If $\overrightarrow{OA} = \mathbf{a}$ and $\overrightarrow{OB} = \mathbf{b}$ express the following vectors in terms of **a** and **b**.

(a) $\overrightarrow{OC}$ (b) $\overrightarrow{AB}$ (c) $\overrightarrow{AD}$ (d) $\overrightarrow{OD}$ (e) $\overrightarrow{CB}$ (f) $\overrightarrow{CD}$.

Unit vectors

A unit vector is one with a magnitude of 1 unit. Unit vectors may be in any direction, but it is usual to denote a unit vector in the direction of the positive x-coordinate axis by **i**. A unit vector in the direction of the positive y-coordinate axis is denoted by **j**. A third axis, at right angles to the x and y axes, is referred to as the z-coordinate axis and a unit vector in this direction is denoted by **k**.
Almost without exception, the vectors used in this book will be in two dimensions only.

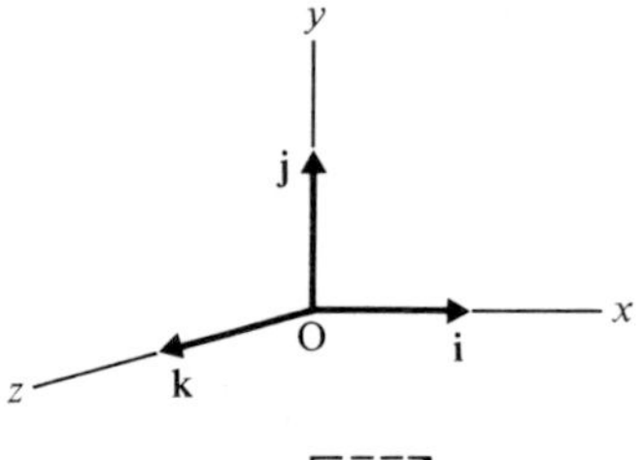

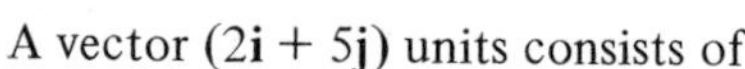

A vector (2**i** + 5**j**) units consists of

2 units in the direction of the unit vector **i**

and 5 units in the direction of the unit vector **j**.

These combine to give vector **R** shown in the diagram.

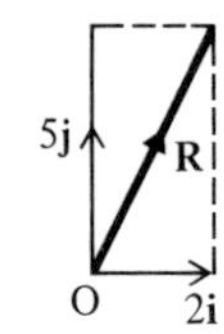

Addition of vectors

When vectors are given in terms of unit vectors, their addition is straightforward.

Example 4

Given $\mathbf{a} = (3\mathbf{i} + 2\mathbf{j})$ and $\mathbf{b} = (5\mathbf{i} - 6\mathbf{j})$, find the resultant of **a** and **b**.

$$\begin{aligned}\mathbf{a} + \mathbf{b} &= (3\mathbf{i} + 2\mathbf{j}) + (5\mathbf{i} - 6\mathbf{j}) \\ &= 8\mathbf{i} - 4\mathbf{j}\end{aligned}$$

The resultant is $(8\mathbf{i} - 4\mathbf{j})$

Since a vector may be expressed in different ways, it is necessary to be able to change from one form to another.

Example 5

Find the magnitude and the direction of the vector $(5\mathbf{i} + 12\mathbf{j})$ units.

Let the vector be represented by $\overrightarrow{OB}$ $(= \overrightarrow{OA} + \overrightarrow{OC})$

$$\text{magnitude of } \overrightarrow{OB} = OB$$

By Pythagoras,

$$OB^2 = OA^2 + AB^2 = 5^2 + 12^2$$
$$\therefore OB = 13 \text{ units}$$

direction of $\overrightarrow{OB}$ is given by $\quad \tan A\hat{O}B = \dfrac{AB}{OA} = \dfrac{12}{5}$

$$\text{or angle } A\hat{O}B = 67{\cdot}38^\circ$$

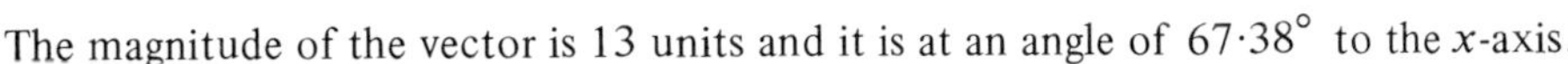

The magnitude of the vector is 13 units and it is at an angle of $67{\cdot}38^\circ$ to the x-axis.

Note: The magnitude of the vector $\overrightarrow{DE}$ is usually written $|\overrightarrow{DE}|$. Thus if $\overrightarrow{DE} = a\mathbf{i} + b\mathbf{j}$, $|\overrightarrow{DE}| = \sqrt{(a^2 + b^2)}$.

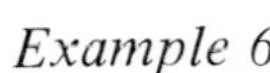

Example 6

A vector has a magnitude of 8 units on a bearing of 150°. Express this vector in the **i–j** form, where **i** is a unit vector due east and **j** is a unit vector due north.

Let the vector be represented by $\overrightarrow{OA}$
Complete the rectangle OEAS

thus $\quad \overrightarrow{OA} = \overrightarrow{OE} + \overrightarrow{EA}$

but $OE = OA \cos 60 \quad$ and $\quad EA = OA \sin 60$
$\qquad = 8 \cos 60 \qquad\qquad = 8 \sin 60$
$\qquad = 4 \qquad\qquad\qquad = 6{\cdot}93$

$\therefore \quad \overrightarrow{OE} = 4\mathbf{i} \qquad \therefore \quad \overrightarrow{EA} = -6{\cdot}93\mathbf{j}$

The vector $\overrightarrow{OA}$ is $(4\mathbf{i} - 6{\cdot}93\mathbf{j})$ units.

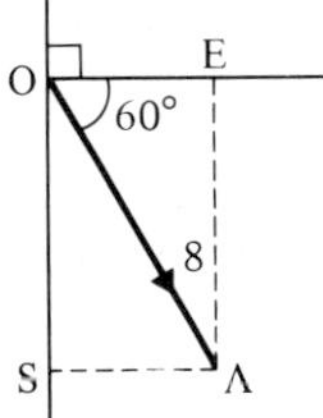

Example 7

Find the vector which has magnitude 20 units and is parallel to the vector $(8\mathbf{i} + 6\mathbf{j})$.

For the required vector to be parallel to the vector $(8\mathbf{i} + 6\mathbf{j})$, it must be some positive multiple of $(8\mathbf{i} + 6\mathbf{j})$.

$$\begin{aligned}
\text{Suppose the required vector } &= \lambda(8\mathbf{i} + 6\mathbf{j}) \\
&= 8\lambda\mathbf{i} + 6\lambda\mathbf{j} \\
\text{magnitude of this vector } &= \sqrt{[(8\lambda)^2 + (6\lambda)^2]} \\
&= \sqrt{(64\lambda^2 + 36\lambda^2)} \\
&= 10\lambda \\
\therefore\ 10\lambda &= 20 \\
\text{or} \quad \lambda &= 2
\end{aligned}$$

The required vector is $(16\mathbf{i} + 12\mathbf{j})$ units.

Exercise 1C

1. For each part of this question, (i) express the vector (shown as a heavy line) in the form $a\mathbf{i} + b\mathbf{j}$ where a and b are numbers, $\mathbf{i}$ and $\mathbf{j}$ are unit vectors in the directions Ox and Oy respectively, and the squares in each grid are of unit of length, (ii) find the magnitude of the vector, (iii) find the angle θ.

(a) 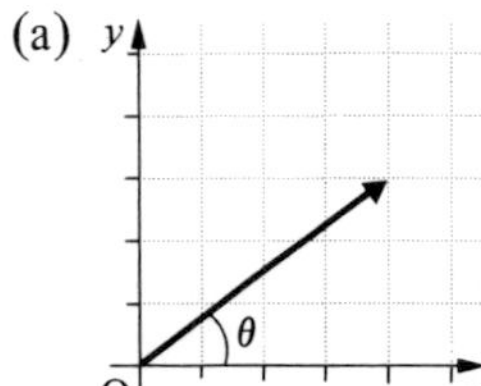

(b) 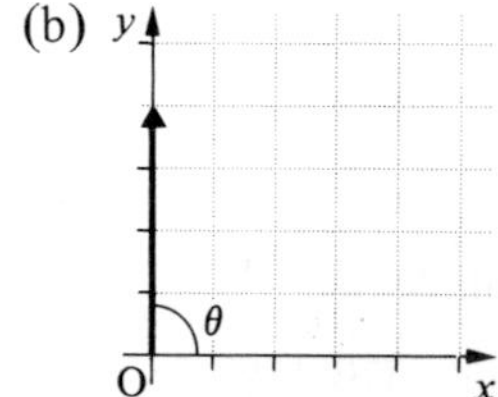

(c) 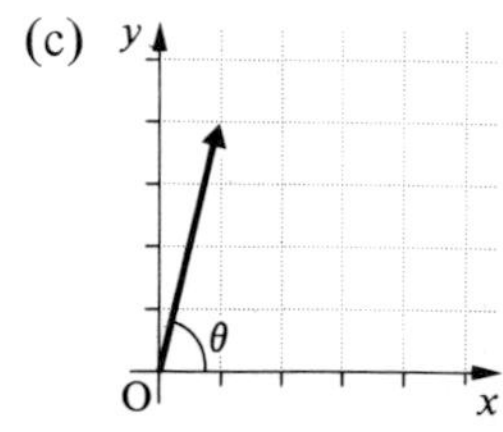

(d) 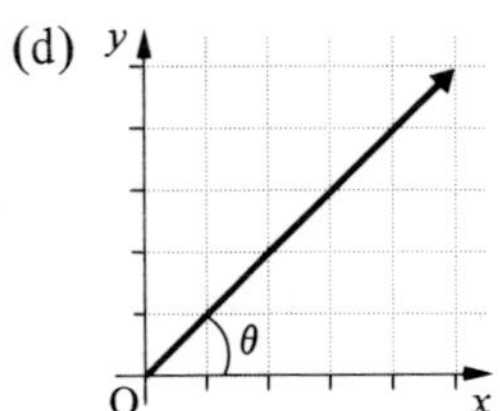

(e) 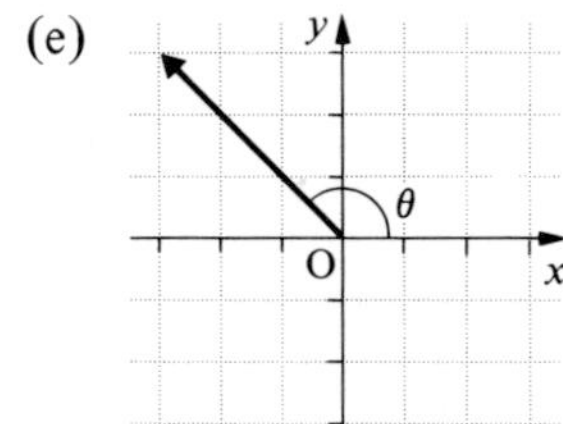

(f) 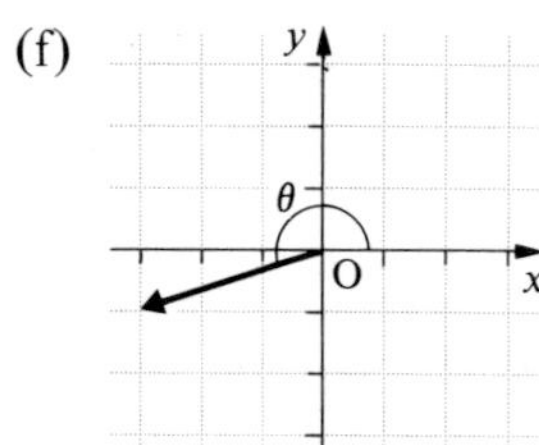

(g) 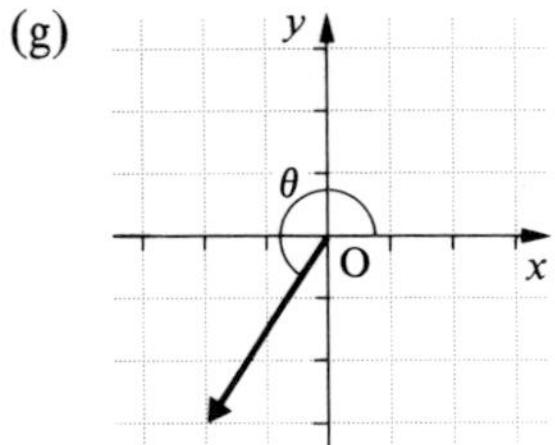

(h) 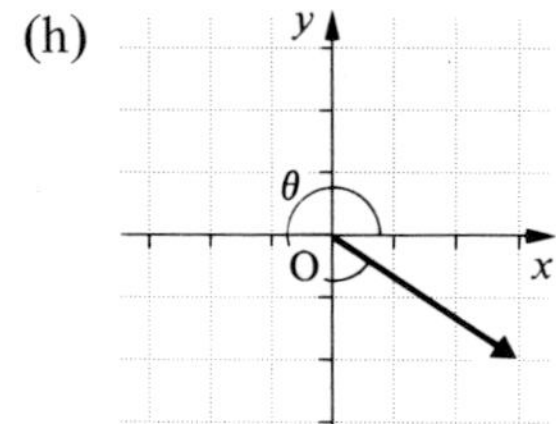

(i) 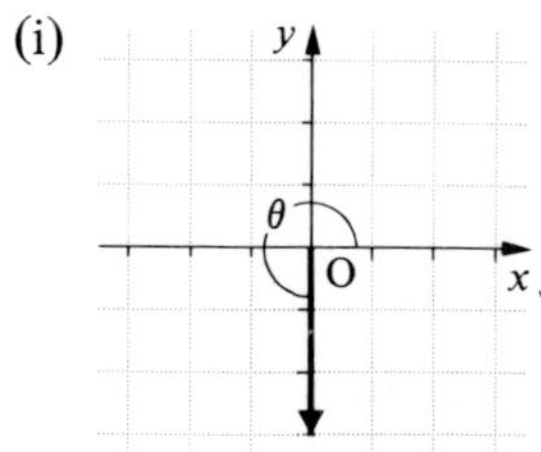

2. The following table gives the magnitude and direction of six vectors. Express each vector in the **i–j** form where **i** is a unit vector due east and **j** a unit vector due north.

vector	magnitude	direction (given as a bearing measured clockwise from north)
a	4 units	090°
b	7 units	180°
c	$5\sqrt{2}$ units	045°
d	10 units	060°
e	6 units	240°
f	10 units	335°

3. If $\mathbf{a} = 3\mathbf{i} + 4\mathbf{j}$, $\mathbf{b} = 7\mathbf{i} + 24\mathbf{j}$, $\mathbf{c} = 8\mathbf{i} - 15\mathbf{j}$ and $\mathbf{d} = -2\mathbf{i} + \mathbf{j}$ find
   (a) the resultant of **a** and **b**,
   (b) the resultant of **b** and **c**,
   (c) the resultant of **c** and **d**,
   (d) the magnitude of **a** (written $|\mathbf{a}|$),
   (e) $|\mathbf{b}|$,
   (f) $|\mathbf{c}|$,
   (g) $|\mathbf{a} + \mathbf{d}|$,
   (h) a vector that is parallel to **a** and three times as long as **a**,
   (i) a vector that is parallel to **b** and half the size of **b**,
   (j) a vector which has a magnitude of 15 units and is parallel to **a**,
   (k) a vector which has a magnitude of 100 units and is parallel to **b**,
   (l) a vector which has a magnitude of 10 units and is parallel to **d**,
   (m) a vector which is parallel to **a** and has the same magnitude as **b**,
   (n) a vector which has the same magnitude as **a** and is parallel to **d**,
   (o) a vector which has the same magnitude as **c** and is parallel to **a**.

2

Distance, velocity and acceleration

Constant speed and constant velocity

The statement that the *speed* of a car is 40 kilometres per hour (written 40 km/h or 40 km h^{-1}) means that, if the speed remains unchanged, the car will travel 40 km in each hour. The speed of the car is then said to be *uniform* or *constant*. At the same speed the car would travel 80 km in 2 hours, 120 km in 3 hours, etc.

$$\text{thus, distance travelled} = \text{speed} \times \text{time}$$
$$\text{or} \quad s = v \times t.$$

The *velocity* of a car is a measure of the speed at which it is travelling in a particular direction. If a car has constant, or uniform velocity then both the speed and the direction of motion of the car remain unchanged.
Thus the *velocity* of a car may be stated as 50 km/h due north and the *speed* of this car is then 50 km/h.
So it is seen that speed is a scalar quantity whereas velocity is a vector quantity and

$$\text{distance travelled in a particular direction} = \text{velocity} \times \text{time taken}$$
$$\text{or} \quad \mathbf{s} = \mathbf{v} \times t$$

The distance travelled in a particular direction may be referred to as the *displacement* of the body from some fixed position.
The letter v is used to denote both speed and velocity. This need cause no confusion provided that the difference between them is remembered, and it is clearly understood which is being used in a particular example.
In most cases, only linear motion will be considered, i.e. motion along a straight line. Therefore, the velocity can only be in one of two directions. The direction of the velocity can then be distinguished by the use of positive and negative.

For example: 5 m/s → denoted by velocity of 5 m/s

5 m/s ← denoted by velocity of −5 m/s

Change of units

The car which is travelling at 40 km/h is, of course, travelling a certain number of metres each second.

Example 1

Express a speed of 40 km/h in m/s.

$$40 \text{ km/h} = 40 \times 1000 \text{ m/h} = \frac{40 \times 1000}{60 \times 60} \text{ m/s} = 11\tfrac{1}{9} \text{ m/s}$$

A speed of 40 km/h is equivalent to a speed of $11\frac{1}{9}$ m/s.

Use of $s = vt$

When the relationship $s = vt$ is used, the units of the quantities involved must be consistent. If the speed is in km/h, the time must be in hours and the distance will then be in km.

Example 2

Find the distance travelled in 3 minutes by a body moving with a constant speed of 15 km/h. Find also the time taken by this body to travel 200 m at the same speed.

$$v = 15 \text{ km/h}$$
$$t = 3 \text{ minutes} = \tfrac{1}{20} \text{ h}$$

Using $\quad s = vt$

$$= 15 \times \tfrac{1}{20}$$
$$\therefore \quad s = \tfrac{3}{4} \text{ km or } 750 \text{ m}$$

The distance travelled is 750 m.

To find the time taken to travel 200 m

$$s = 200 \text{ m} \qquad v = 15 \text{ km/h}$$
$$= \frac{15 \times 1000}{60 \times 60} \text{ m/s} = \tfrac{25}{6} \text{ m/s}$$

Using $\quad s = vt$

$$200 = \tfrac{25}{6} \times t \text{ where } t \text{ is measured in seconds}$$
$$\therefore \quad t = 48 \text{ s}$$

The time taken to travel 200 m at 15 km/h is 48 s

Average speed

In practice the speed and velocity of a body are seldom constant. When a car travels 40 km in one hour, it is unlikely that its speed is constant. It is probable that for part of the time the car is travelling at more than 40 km/h, and for some of the time the car's speed is less than 40 km/h. Thus we often refer to the *average* speed, or the *average* velocity of a body.

$$\text{average speed} = \frac{\text{total distance travelled}}{\text{total time taken}}$$

and

$$\text{average velocity} = \frac{\text{total distance travelled in a particular direction}}{\text{total time taken}}$$

The distance travelled in a particular direction can more conveniently be referred to as the displacement from some fixed initial position.

Example 3

A, B and C are three points, in that order, on a straight road with AB = 40 km and BC = 90 km. A woman travels from A to B at 10 km/h and then from B to C at 15 km/h.
Calculate (a) the time taken to travel from A to B,
(b) the time taken to travel from B to C,
(c) the average speed of the woman for the journey from A to C.

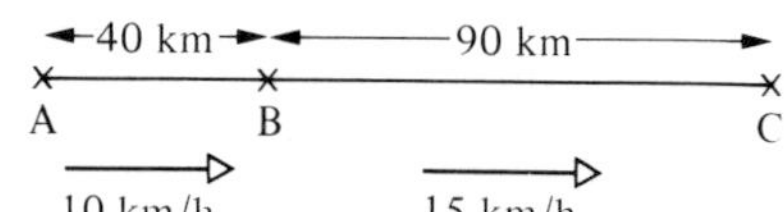

(a) using $s = vt$ for A to B,

$$40 = 10 \times t$$
$$\therefore \quad t = 4 \text{ h}$$

The time taken to travel from A to B is 4 h.

(b) using $s = vt$ for B to C,

$$90 = 15 \times t$$
$$\therefore \quad t = 6 \text{ h}$$

The time taken to travel from B to C is 6 h.

(c) using $\text{average speed} = \dfrac{\text{total distance travelled}}{\text{total time taken}}$

$$\text{for A to C, } v = \frac{40 + 90}{4 + 6}$$
$$\therefore \quad v = 13 \text{ km/h}$$

The average speed for the whole journey is 13 km/h.

Example 4

A man walks 400 m due east in a time of 190 s, and then 100 m due west in a time of 50 s. Calculate (a) his average speed, (b) his average velocity, for the whole journey.

(a) using $\text{average speed} = \dfrac{\text{total distance}}{\text{total time}}$

$$\text{average speed} = \frac{400 + 100}{190 + 50}$$
$$= \frac{500}{240} = 2\tfrac{1}{12} \text{ m/s}$$

The average speed is $2\frac{1}{12}$ m/s.

(b) using $\text{average velocity} = \dfrac{\text{displacement}}{\text{total time}}$

$$= \frac{400 \text{ m E} + 100 \text{ m W}}{240} = \frac{300 \text{ m E}}{240}$$
$$= 1\tfrac{1}{4} \text{ m/s E}$$

The average velocity is $1\frac{1}{4}$ m/s east.

Exercise 2A

1. Express a speed of 36 km/h in m/s.
2. Express a speed of 81 km/h in m/s.
3. Express a speed of 35 m/s in km/h.
4. Express a speed of 22 m/s in km/h.
5. Express a speed of 6 km/min in m/s.
6. A body travelling at a constant speed covers a distance of 200 m in 8 seconds. Find the speed of the body.
7. A body travelling at a constant speed covers a distance of 3 km in 2 minutes. Find the speed of the body.
8. Find the distance travelled in 5 seconds by a body moving with a constant speed of 3·2 m/s.
9. Find the distance travelled in 2 minutes by a body moving with a constant speed of 6 km/h.
10. Find the time taken by a body, moving with constant speed 3·5 m/s, to travel a distance of 21 m.
11. At time $t = 0$ a body passes through a point A and is moving with a constant velocity of 4 m/s. (a) Find how far the body is from A when $t = 3$ s.
    (b) What is the value of t when the body is 22 m from A?
12. The spacecraft Voyager II travels at a constant velocity of 80 000 km/h. Find the distance the spacecraft travels in (a) 1 hour (b) 15 minutes (c) 1 second.
13. The speed of sound is 340 m/s. Find the distance travelled in one minute by an aircraft flying at Mach 2 (i.e. twice the speed of sound).
14. The speed of light is 3×10^8 m/s. If the distance from the Sun to the Earth is $1{\cdot}5 \times 10^8$ km find how long it takes light from the Sun to reach the Earth.
15. If it takes 5 seconds for the sound of thunder to reach my ears, how far am I from he place that it actually occurred? (Speed of sound is 340 m/s.)
16. If an athlete runs a 1500 metre race in 3 minutes 33 seconds, find his average speed for the race.
17. A, B and C are three points lying in that order on a straight road with AB = 5 km and BC = 4 km. A man runs from A to B at 20 km/h and then walks from B to C at 8 km/h. Find (a) the total time taken to travel from A to C,
    (b) the average speed of the man for the journey from A to C.
18. A man walks 150 m due north, in a time of 70 s, and then 50 m due south, in a time of 30 s. Find (a) his average speed, (b) his average velocity.
19. A car is driven from Town A to Town B, 40 km away, at an average speed of 60 km/h. The car is at B for 10 minutes and is then driven back to A.
    (a) Find the average speed for the journey B → A if the average speed for the complete journey is 60 km/h.
    (b) What is the average velocity of the car for the complete journey?
20. A, B and C are three points lying in that order on a straight line with AB = 60 m and AC = 80 m. A body moves from A to B at an average speed of 10 m/s, then from B to C in a time of 4 s, and then returns to B. The average speed for the whole journey is 5 m/s. Find (a) the average speed of the body in the second stage of the motion (i.e. B → C),
    (b) the average speed of the body in moving from A to C,
    (c) the time taken for the third stage of the motion (i.e. C → B),
    (d) the average velocity for the complete motion.

i-j notation

In the first chapter, vectors were expressed using the **i**-**j** notation. Both the position and the velocity of a body can be given in this vector form.

Position vector

Using a suitable origin O, the position of a body at P may be given as the vector $\overrightarrow{OP}$ where

$$\overrightarrow{OP} = (a\mathbf{i} + b\mathbf{j})\text{ m}.$$

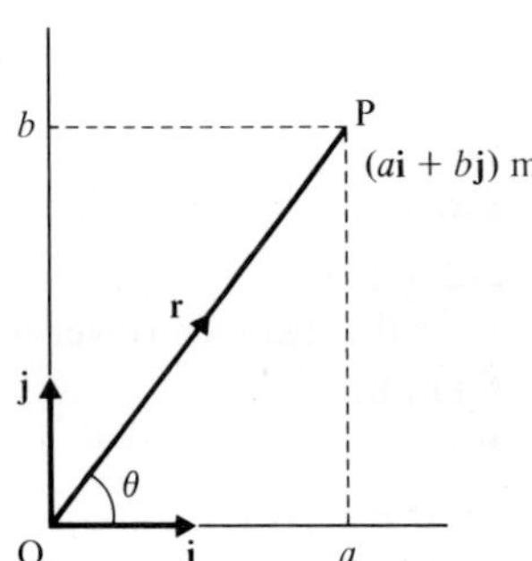

This is the position vector of the body.

As before, the vector may be denoted by a single letter, say, **r**

$$\text{i.e. } \mathbf{r} = \overrightarrow{OP} = (a\mathbf{i} + b\mathbf{j})\text{ m}.$$

The distance of P from the origin O and the direction of $\overrightarrow{OP}$ may then be found.

$$\text{distance OP} = |\overrightarrow{OP}| \text{ or } |\mathbf{r}|$$
$$= \sqrt{(a^2 + b^2)}$$

and the direction of $\overrightarrow{OP}$ is given by $\tan\theta = \dfrac{b}{a}$

Velocity vector

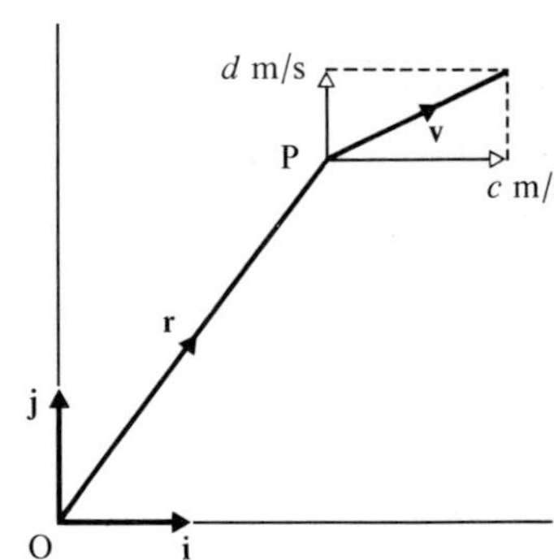

The velocity vector **v** can be expressed in the same way.
If the velocity vector of a body at P is given by

$$\mathbf{v} = (c\mathbf{i} + d\mathbf{j})\text{ m/s},$$

the body has a velocity of c m/s in the direction of the unit vector **i**, and d m/s in the direction of the unit vector **j**.

Example 5

The point O is the origin and the points P and Q have position vectors $(7\mathbf{i} - 24\mathbf{j})$ m and $(13\mathbf{i} - 16\mathbf{j})$ m respectively. Find (a) the distance OP, (b) the vector $\overrightarrow{PQ}$, (c) the distance PQ.

(a) $\overrightarrow{OP} = (7\mathbf{i} - 24\mathbf{j})$ m

$$\text{distance OP} = |\overrightarrow{OP}| = \sqrt{(7^2 + (-24)^2)} = 25\text{ m}$$

The distance OP is 25 m

(b) Since $\overrightarrow{OP} + \overrightarrow{PQ} = \overrightarrow{OQ}$

$$\overrightarrow{PQ} = \overrightarrow{OQ} - \overrightarrow{OP}$$
$$= (13\mathbf{i} - 16\mathbf{j}) - (7\mathbf{i} - 24\mathbf{j}) = (6\mathbf{i} + 8\mathbf{j})\text{ m}$$

The vector $\overrightarrow{PQ}$ is $(6\mathbf{i} + 8\mathbf{j})$ m

(c) From (b) $\overrightarrow{PQ} = (6\mathbf{i} + 8\mathbf{j})$ m

distance PQ $= |\overrightarrow{PQ}|$

$= \sqrt{(6^2 + 8^2)} = 10$ m

The distance PQ is 10 m

Example 6

A particle P has an initial position vector $(2\mathbf{i} + 9\mathbf{j})$ m. If the particle moves with a constant velocity of $(3\mathbf{i} - \mathbf{j})$ m/s, find its position vector after (a) 1 s, (b) 4 s.

(a) Given: $\overrightarrow{OP} = (2\mathbf{i} + 9\mathbf{j})$ m and $\mathbf{v} = (3\mathbf{i} - \mathbf{j})$ m/s

Let Q be the position after 1 s

then $\overrightarrow{OQ} = \overrightarrow{OP} + (1 \times \mathbf{v})$

$= (2\mathbf{i} + 9\mathbf{j}) + (1)(3\mathbf{i} - \mathbf{j}) = (5\mathbf{i} + 8\mathbf{j})$ m

The position vector after 1 s is $(5\mathbf{i} + 8\mathbf{j})$ m.

(b) Let S be the position after 4 s

then $\overrightarrow{OS} = \overrightarrow{OP} + (4 \times \mathbf{v})$

$= (2\mathbf{i} + 9\mathbf{j}) + 4(3\mathbf{i} - \mathbf{j}) = (14\mathbf{i} + 5\mathbf{j})$ m

The position vector after 4 s is $(14\mathbf{i} + 5\mathbf{j})$ m.

Exercise 2B

1. The point A has position vector $(7\mathbf{i} + 24\mathbf{j})$ m. Find how far A is from the origin.
2. The points B and C have position vectors $(8\mathbf{i} - 15\mathbf{j})$ m and $(5\mathbf{i} - 12\mathbf{j})$ m respectively.
   Find (a) how far B is from the origin,
   (b) how far C is from the origin,
   (c) $\overrightarrow{BC}$ in vector form (i.e. **i-j** notation),
   (d) $|\overrightarrow{BC}|$.
3. The point O is the origin and points A, B and C have position vectors $(3\mathbf{i} - 4\mathbf{j})$ m, $(8\mathbf{i} + 8\mathbf{j})$ m and $-7\mathbf{j}$ m respectively.
   Find (a) the distance OA, (b) the distance OB,
   (c) the distance OC, (d) the vector $\overrightarrow{AB}$,
   (e) the vector $\overrightarrow{BC}$, (f) the vector $\overrightarrow{CB}$,
   (g) the distance AB, (h) the distance BC.
4. Find the speed of a body moving with velocity $(6\mathbf{i} - 8\mathbf{j})$ m/s.
5. Find the speed of a body moving with velocity $(7\mathbf{i} - 24\mathbf{j})$ m/s.
6. Find the speed of a body moving with velocity $(-4\mathbf{i} + \mathbf{j})$ m/s.
7. Particle A has velocity $(5\mathbf{i} + 2\mathbf{j})$ m/s and particle B has velocity $(-4\mathbf{i} + 4\mathbf{j})$ m/s. Which particle has the greater speed?
8. A body moving with a velocity $(2\mathbf{i} + a\mathbf{j})$ m/s has a speed of 5·2 m/s. Find the value of a.
9. A body moving with a velocity of $(b\mathbf{i} + (b + 7)\mathbf{j})$ m/s has a speed of 17 m/s. Find the two possible values of b.

10. A particle has an initial position vector of $(5\mathbf{i} + 3\mathbf{j})$ m. If the particle moves with a constant velocity of $(2\mathbf{i} + 4\mathbf{j})$ m/s find its position vector after (a) 1 second, (b) 2 sceonds.

11. A particle has an initial position vector of $(5\mathbf{i} + 4\mathbf{j})$ m. If the particle moves with a constant velocity of $(2\mathbf{i} - \mathbf{j})$ m/s, find its position vector after (a) 3 seconds, (b) 5 seconds.

12. A particle has an initial position vector of $(7\mathbf{i} + 5\mathbf{j})$ m. The particle moves with a constant velocity of $(a\mathbf{i} + b\mathbf{j})$ m/s and after 3 seconds has a position vector of $(10\mathbf{i} - \mathbf{j})$ m. Find the values of a and b.

13. Find the speed of a body which is moving with a constant velocity of $(5\mathbf{i} - 12\mathbf{j})$ m/s. If the body is initially at a point with position vector $(\mathbf{i} + 6\mathbf{j})$ m, find the position vector of the body 3 seconds later and its distance from the origin at that time.

Uniform acceleration formulae

When the motion of a body is being considered, the letters u, v, a, t and s usually have the following meanings:

u = initial velocity v = final velocity
a = acceleration t = time interval or time taken
s = displacement

Consider a car travelling in a straight line. If initially its velocity is 5 m/s and 3 seconds later its velocity is 11 m/s, the car is said to be accelerating. Acceleration is a measure of the rate at which velocity is changing. In this example, the velocity increases by 6 m/s in 3 s. If the acceleration a is assumed to be uniform, then it is 6 m/s in 3 s, or 2 m/s each second, which is written 2 m/s^2.

$$\text{In general,} \quad \text{acceleration} = \frac{\text{change in velocity}}{\text{time interval}}$$

$$\therefore \quad a = \frac{v - u}{t}$$

$$\text{or} \quad at = v - u$$

$$\text{hence} \quad v = u + at \qquad \ldots [1]$$

If the acceleration is uniform, then the average velocity is the average of the initial and final velocities,

$$\text{i.e.} \quad \text{average velocity} = \frac{u + v}{2}$$

$$\text{but} \quad \text{average velocity} = \frac{\text{displacement}}{\text{time taken}} = \frac{s}{t}$$

$$\therefore \quad \frac{s}{t} = \frac{u + v}{2}$$

$$\text{or} \quad s = \frac{(u + v)t}{2} \qquad \ldots [2]$$

Substituting the value of v from equation [1] into equation [2]

$$s = \frac{(u + u + at)t}{2}$$

$$\therefore \quad s = ut + \tfrac{1}{2}at^2 \qquad \ldots [3]$$

Substituting for t from equation [1] into equation [2]

from equation [1] $$t = \frac{v-u}{a}$$

in equation [2] $$s = \frac{(u+v)}{2}\frac{(v-u)}{a}$$

$$\text{or} \quad 2as = v^2 - u^2$$

$$\therefore \quad v^2 = u^2 + 2as \qquad \ldots [4]$$

These four formulae are very important and should be committed to memory:

$$\left.\begin{aligned} v &= u + at \\ s &= \frac{(u+v)t}{2} \\ s &= ut + \tfrac{1}{2}at^2 \\ v^2 &= u^2 + 2as \end{aligned}\right\}$$

Remember these only apply to motion involving *uniform* acceleration

Distance and displacement

In the above formulae, s represents displacement. In practice, s is also used to denote distance because distance and displacement are often equal. There need be no confusion provided that care is taken in any particular question.

When the direction of motion of a body remains unchanged, then the distance travelled and the displacement are equal.

If the direction of motion changes part way through the motion, then the distance travelled and the displacement will not be equal.

Suppose a body moves 15 km due east and then 10 km due west:

distance moved = 15 km + 10 km = 25 km

displacement from initial position = 15 km E + 10 km W = 5 km E

Example 7

A body moves along a straight line from A to B with uniform acceleration $\frac{2}{3}$ m/s^2. The time taken is 12 s and the velocity at B is 25 m/s. Find (a) the velocity at A, (b) the distance AB.

Given: $a = \frac{2}{3}$ m/s^2 to find: (a) u

$t = 12$ s (b) s

$v = 25$ m/s

(a) using $$v = u + at$$
$$25 = u + (\tfrac{2}{3})(12)$$
$$\therefore \quad u = 17 \text{ m/s}$$

The velocity at A is 17 m/s.

(b) using
$$\begin{aligned} s &= ut + \tfrac{1}{2}at^2 \\ &= (17)(12) + \tfrac{1}{2}(\tfrac{2}{3})(12)^2 \\ &= 204 + 48 \\ s &= 252 \text{ m} \end{aligned}$$

or
$$\begin{aligned} v^2 &= u^2 + 2as \\ (25)^2 &= (17)^2 + 2(\tfrac{2}{3})s \\ 625 &= 289 + \tfrac{4}{3}s \\ s &= 252 \text{ m} \end{aligned}$$

or
$$\begin{aligned} s &= \frac{(u+v)t}{2} \\ &= \frac{(17+25)12}{2} \\ s &= 252 \text{ m} \end{aligned}$$

The distance AB is 252 m.

Example 8

A cyclist travelling downhill accelerates uniformly at $1\frac{1}{2}$ m/s^2. If his initial velocity at the top of the hill is 3 m/s, find (a) how far he travels in 8 s, (b) how far he travels before reaching a velocity of 7 m/s.

(a) Given: $a = 1{\cdot}5$ m/s^2 — to find: s
$u = 3$ m/s
$t = 8$ s

using
$$s = ut + \tfrac{1}{2}at^2$$
$$= 3(8) + \tfrac{1}{2}(1{\cdot}5)(8)^2$$
$$= 24 + 48$$
$$\therefore \quad s = 72 \text{ m}$$

In 8 s the cyclist travels 72 m.

(b) Given: $a = 1{\cdot}5$ m/s^2 — to find: s
$u = 3$ m/s
$v = 7$ m/s

using
$$v^2 = u^2 + 2as$$
$$(7)^2 = (3)^2 + 2(1{\cdot}5)s$$
or
$$49 - 9 = 3s$$
$$\therefore \quad s = 13\tfrac{1}{3} \text{ m}$$

The distance travelled is $13\frac{1}{3}$ m.

Retardation

If a body moving at 12 m/s is subsequently moving at 2 m/s, the body is said to be subject to a retardation, i.e. a negative acceleration. If the change in velocity takes place over a period of 4 s, the retardation is 10 m/s in 4 s or $2\frac{1}{2}$ m/s^2 and the acceleration is $-2\frac{1}{2}$ m/s^2.

Example 9

A stone slides in a straight line across a horizontal sheet of ice. It passes a point A with velocity 14 m/s, and point B $2\frac{1}{2}$ s later. Assuming the retardation is uniform and that AB = 30 m, find (a) the retardation, (b) the velocity at B, (c) how long after passing A the stone comes to rest.

(a) Given: $u = 14$ m/s — to find: retardation
$t = 2{\cdot}5$ s — let acceleration $= a$
$s = 30$ m

using
$$s = ut + \tfrac{1}{2}at^2$$
$$30 = (14)(2{\cdot}5) + \tfrac{1}{2}(a)(2{\cdot}5)^2$$
$$\therefore \quad 30 = 35 + \frac{25a}{8}$$
or
$$a = -1{\cdot}6 \text{ m/s}^2 \quad \text{i.e. a retardation}$$

The retardation is $1{\cdot}6$ m/s^2.

(b) let velocity at B $= v$

from part (a) retardation $= 1{\cdot}6$ m/s^2 or $a = -1{\cdot}6$ m/s^2

using
$$\begin{aligned} v^2 &= u^2 + 2as \\ &= (14)^2 + 2(-1{\cdot}6)(30) \\ &= 196 - 96 \\ \therefore \quad v &= 10 \text{ m/s} \end{aligned}$$

The velocity of the stone at B is 10 m/s.

(c) Given: $u = 14$ m/s to find: t
$a = -1{\cdot}6$ m/s^2
$v = 0$ when the stone comes to rest

using
$$\begin{aligned} v &= u + at \\ 0 &= 14 + (-1{\cdot}6)t \\ \therefore \quad t &= 8{\cdot}75 \text{ s} \end{aligned}$$

The stone is at rest 8·75 s after passing A.

Exercise 2C

Questions **1** to **20** involve a body moving with uniform acceleration along a straight line from point A to point B.

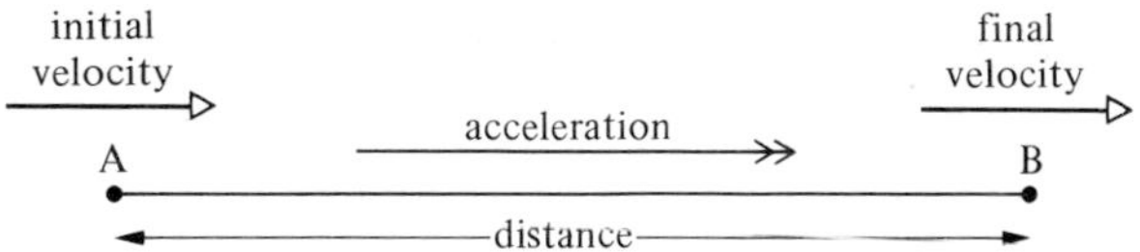

1. Initially at rest, acceleration = 4 m/s^2, time taken = 8 s. Find the distance.
2. Initial velocity = 3 m/s, acceleration = 2 m/s^2, time taken = 6 s. Find the final velocity.
3. Initially at rest, acceleration = 2 m/s^2, time taken = 4 s. Find the distance.
4. Initial velocity = 3 m/s, final velocity = 5 m/s, time taken = 10 s. Find the distance.
5. Initial velocity = 3 m/s, final velocity = 5 m/s, distance = 2 m. Find the acceleration.
6. Final velocity = 27 m/s, acceleration = 8 m/s^2, time taken = 2 s. Find the initial velocity.
7. Initial velocity = 7 m/s, final velocity = 3 m/s, distance = 5 m. Find the acceleration.
8. Distance = 28 m, acceleration = 1 m/s^2, time taken = 4 s. Find the initial velocity.
9. Distance = 20 m, initial velocity = 3 m/s, final velocity = 7 m/s. Find the time taken.
10. Initial velocity = 6 m/s, final velocity = 8 m/s, acceleration = 0·5 m/s^2. Find the distance.
11. Initial velocity = 2 m/s, final velocity = 50 m/s, time taken = 16 s. Find the acceleration.
12. Distance = 500 m, initial velocity = 1 m/s, time taken = 10 s. Find the acceleration.
13. Initial velocity = 10 m/s, final velocity = 2 m/s, acceleration = −4 m/s^2. Find the distance.
14. Initial velocity = 30 m/s, final velocity = 10 m/s, acceleration = −4 m/s^2. Find the time taken.
15. Initial velocity = 5 m/s, acceleration = 1 m/s^2, distance = 12 m. Find the time taken.
16. Distance = 60 m, final velocity = 8 m/s, time taken = 12 s. Find the initial velocity.
17. Initial velocity = 5 m/s, final velocity = 36 km/h, acceleration = $1\frac{1}{4}$ m/s^2. Find the distance.
18. Acceleration = 0·5 m/s^2, final velocity = 162 km/h, time taken = 1 minute. Find the initial velocity.
19. Acceleration = 2 m/s^2, final velocity = 10 m/s, time taken = 2 s. Find the distance.
20. Distance = 132 m, time taken = 12 s, acceleration = −1 m/s^2. Find the final velocity.

21. A train starts from rest and accelerates uniformly, at $1{\cdot}5$ m/s^2, until it attains a speed of 30 m/s. Find the distance the train travels during this motion and the time taken.

22. A cheetah can accelerate from rest to 30 m/s in a distance of 25 m. Find the acceleration (assumed constant).

23. The manufacturer of a new car claims that it can accelerate from rest to 90 km/h in 10 seconds. Find the acceleration (assumed constant).

24. In travelling the 70 cm along a rifle barrel, a bullet uniformly accelerates from its initial state of rest to a muzzle velocity of 210 m/s. Find the acceleration involved and the time for which the bullet is in the barrel.

25. According to the highway code, a car travelling at 20 m/s requires a minimum braking distance of 30 m. What retardation is this and what length of time will it take?

26. A car is initially at rest at a point O. The car moves away from O in a straight line, accelerating at 4 m/s^2. Find how far the car is from O after (a) 2 seconds, (b) 3 seconds. How far does the car travel in the third second?

27. A body moves along a straight line uniformly increasing its velocity from 2 m/s to 18 m/s in a time interval of 10 s. Find the acceleration of the body during this time and the distance travelled.

28. A particle is projected away from an origin O with initial velocity $0{\cdot}25$ m/s. The particle travels in a straight line and accelerates at $1{\cdot}5$ m/s^2.
Find (a) how far the particle is from O after 3 seconds,
(b) the distance travelled by the particle during the fourth second after projection.

29. When time $t = 0$, a body is projected from an origin O with an initial velocity of 10 m/s. The body moves along a straight line with a constant acceleration of -2 m/s^2.
(a) Find the displacement of the body from O when t equals 7 seconds.
(b) How far from O does the body come to instantaneous rest and what is the value of t then?
(c) Find the distance travelled by the body during the time interval $t = 0$ to $t = 7$ seconds.

30. A, B and C are three points lying in that order on a straight line. A body is projected from B towards A with speed 3 m/s. The body experiences an acceleration of 1 m/s^2 towards C. If BC = 20 m, find the time taken to reach C and the distance travelled by the body from the moment of projection until it reaches C.

31. A car is being driven along a road at a steady 25 m/s when the driver suddenly notices that there is a fallen tree blocking the road 65 metres ahead. The driver immediately applies the brakes giving the car a constant retardation of 5 m/s^2. How far in front of the tree does the car come to rest?
If the driver had not reacted immediately and the brakes were applied one second later with what speed would the car have hit the tree?

32. A train travels along a straight piece of track between two stations A and B. The train starts from rest at A and accelerates at $1{\cdot}25$ m/s^2 until it reaches a speed of 20 m/s. It then travels at this steady speed for a distance of $1{\cdot}56$ km and then decelerates at 2 m/s^2 to come to rest at B.
Find (a) the distance from A to B,
(b) the total time taken for the journey,
(c) the average speed for the journey.

33. A particle travels in a straight line with uniform acceleration. The particle passes through three points A, B and C lying in that order on the line, at times $t = 0$, $t = 2$ s and $t = 5$ s respectively. If BC = 30 m and the speed of the particle when at B is 7 m/s find the acceleration of the particle and its speed when at A.

34. A, B and C are three points which lie in that order on a straight road with AB = 95 m and BC = 80 m. A car is travelling along the road in the direction ABC with constant acceleration a m/s^2. The car passes through A with speed u m/s, reaches B five seconds later and C two seconds after that. Find the values of u and a.

35. A car A, travelling at a constant velocity of 25 m/s, overtakes a stationary car B. Two seconds later car B sets off in pursuit, accelerating at a uniform 6 m/s^2. How far does B travel before catching up with A?

Free fall under gravity

The uniform acceleration formulae developed in the last section may be used when considering the motion of a body falling under gravity. In such cases the acceleration of the body is 9·8 m/s^2 and this is commonly referred to as g, the acceleration due to gravity. If the motion is vertically upward, the body will be subject to a retardation of 9·8 m/s^2.

In fact the magnitude of g varies slightly at different places on the Earth's surface, but for our purposes it can be taken as having the constant value of 9·8 m/s^2.

Arrow convention

In any particular example, care is needed to ensure that the directions of the vectors involved are all the same.

$u = 25$ m/s $\uparrow$ implies that the initial velocity is 25 m/s upwards
$a = 9{\cdot}8$ m/s^2 $\downarrow$ implies a downward acceleration of 9·8 m/s^2
$= -9{\cdot}8$ m/s^2 $\uparrow$

Before substituting numerical values in the uniform acceleration formulae, the arrows of the vectors involved must all be in the same direction.

Example 10

A brick is thrown vertically downwards from the top of a building and has an initial velocity of 1·5 m/s. If the height of the building is $19\frac{2}{7}$ m, find
(a) the velocity with which the brick hits the ground,
(b) the time taken for the brick to fall.

(a) Given: $u = 1{\cdot}5$ m/s $\downarrow$ to find: (a) v
$s = 19\frac{2}{7}$ m $\downarrow$ (b) t
$a = 9{\cdot}8$ m/s^2 $\downarrow$

using
$$v^2 = u^2 + 2as$$
$$v^2 = (1{\cdot}5)^2 + 2(9{\cdot}8)\frac{(135)}{7}$$
$$= 380{\cdot}25$$
or $v = 19{\cdot}5$ m/s

The brick hits the ground with a velocity of 19·5 m/s.

(b) using
$$v = u + at$$
$$19{\cdot}5 = 1{\cdot}5 + 9{\cdot}8t$$
$$\therefore \quad t = \frac{18}{9{\cdot}8} = 1{\cdot}84 \text{ s}$$

The brick hits the ground after 1·84 s.

Example 11

A ball is thrown vertically upwards with a velocity of 14·7 m/s from a platform 19·6 m above ground level. Find (a) the time taken for the ball to reach the ground,
(b) the velocity of the ball when it hits the ground.

Given: $u = 14{\cdot}7 \text{ m/s} \uparrow$ to find: t

$a = 9{\cdot}8 \text{ m/s}^2 \downarrow = -9{\cdot}8 \text{ m/s}^2 \uparrow$

$s = 19{\cdot}6 \text{ m} \downarrow = -19{\cdot}6 \text{ m} \uparrow$

(a) using $s = ut + \frac{1}{2}at^2$

$$-19{\cdot}6 = 14{\cdot}7t + \tfrac{1}{2}(-9{\cdot}8)t^2$$

$$-4 = 3t - t^2$$

$$\therefore\ t^2 - 3t - 4 = 0$$

$$(t-4)(t+1) = 0 \quad \text{i.e. } t = 4 \text{ s or } -1 \text{ s}$$

The ball reaches the ground after 4 s.

(b) using $v = u + at$

$$v = 14{\cdot}7 + (-9{\cdot}8)4 = 14{\cdot}7 - 39{\cdot}2$$

$$v = -24{\cdot}5 \text{ m/s} \quad \text{i.e. } 24{\cdot}5 \text{ m/s}\downarrow$$

The ball hits the ground with a downward velocity of 24·5 m/s.

Example 12

A particle is projected vertically upwards with a velocity of 34·3 m/s. Find how long after projection the particle is at a height of 49 m above the point of projection for
(a) the first time, (b) the second time.

Given: $u = 34{\cdot}3 \text{ m/s} \uparrow$

$a = 9{\cdot}8 \text{ m/s}^2 \downarrow = -9{\cdot}8 \text{ m/s}^2 \uparrow$

$s = 49 \text{ m} \uparrow$

using $s = ut + \frac{1}{2}at^2$

$$49 = 34{\cdot}3t - \tfrac{1}{2}(9{\cdot}8)t^2$$

$$\therefore\ t^2 - 7t + 10 = 0$$

$$(t-5)(t-2) = 0 \quad \text{i.e. } t = 5 \text{ s or } 2 \text{ s}$$

The particle is 49 m above the point of projection (a) after 2 s, (b) after 5 s.

Exercise 2D

1. A book falls from a shelf 160 cm above the floor. Find the speed with which the book strikes the floor.
2. A stone is dropped from a position which is 40 metres above the ground. Find the time taken for the stone to reach the ground.
3. A stone is dropped from the top of a tower and falls to the ground below. If the stone hits the ground with a speed of 14 m/s, find the height of the tower.
4. A ball is thrown vertically downwards from the top of a tower and has an initial speed of 4 m/s. If the ball hits the ground 2 seconds later, find (a) the height of the tower, (b) the speed with which the ball strikes the ground.

5. A stone is projected vertically upwards from ground level with a speed of 21 m/s. Find the height of the stone above ground (a) 1 second after projection, (b) 2 seconds after projection, (c) 3 seconds after projection.
6. A ball is thrown vertically upwards with speed 28 m/s from a point which is 1 metre above ground level. Find (a) the speed the ball will have when it returns to the level from which it was projected, (b) the height above ground level of the highest point reached.
7. A ball is thrown vertically upwards from a point A, with initial speed of 21 m/s, and is later caught again at A. Find the length of time for which the ball was in the air.
8. A ball is kicked vertically upwards from ground level with an initial speed of 14 m/s. Find the height above ground level of the highest point reached and the total time for which the ball is in the air.
9. A stone is thrown vertically upwards with a speed of 20 m/s from a point at a height h metres above ground level. If the stone hits the ground 5 seconds later, find h.
10. A stone is projected vertically upwards from ground level at a speed of 24·5 m/s. Find how long after projection the stone is at a height of 19·6 m above ground (a) for the first time, (b) for the second time. For how long is the stone at least 19·6 m above ground level?
11. A ball is held 1·6 m above a concrete floor and released. The ball hits the floor and rebounds with half the speed it had just prior to impact. Find the greatest height the ball reaches after (a) the first bounce, (b) the second bounce.
12. A body is projected vertically upwards from ground level at a speed of 49 m/s. Find the length of time for which the body is at least 78·4 m above ground.
13. A bullet is fired vertically upwards at a speed of 147 m/s. Find the length of time for which the bullet is at least 980 m above the level of projection.
14. A body is projected vertically upwards with a speed of 14 m/s. Find the height of the body above the level of projection after (a) 1 second of motion, (b) 2 seconds of motion. Find the distance travelled by the body in the 2nd second of motion.
15. Two stones are thrown from the same point at the same time, one vertically upwards with speed 30 m/s, and the other vertically downwards at 30 m/s. Find how far apart the stones are after 3 seconds.

Graphical representation

Consider the motion of a body which accelerates uniformly from a speed u to a speed v in time t and then maintains constant speed v. Plotting velocity on the vertical axis and time on the horizontal axis, a velocity-time graph can be drawn.

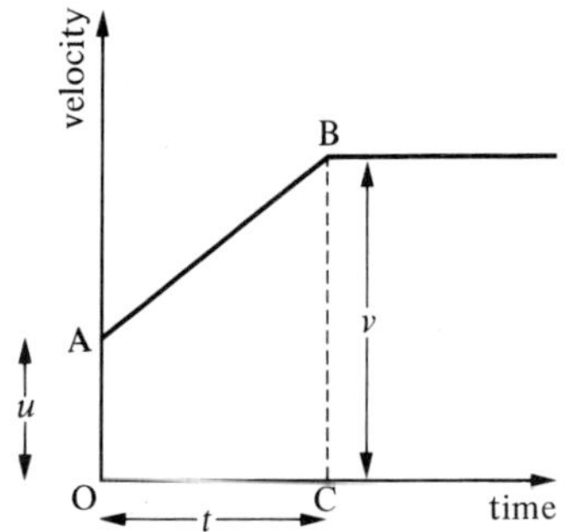

The acceleration of the body is defined as the rate of change of velocity, i.e. $a = \frac{v-u}{t}$ and so the acceleration during the time interval $0 \rightarrow t$ will be the gradient or slope of the line AB. From $s = \frac{(u+v)t}{2}$ it can be seen that the distance travelled by the body during the time interval $0 \rightarrow t$ is represented by the area OABC, i.e. the area 'under' the graph for that part of the motion.

Example 13

The velocity–time graph shown is for a body which starts from rest, accelerates uniformly to a velocity of 8 m/s in 2 seconds, maintains that velocity for a further 5 seconds and then retards uniformly to rest. The entire journey takes 11 seconds.

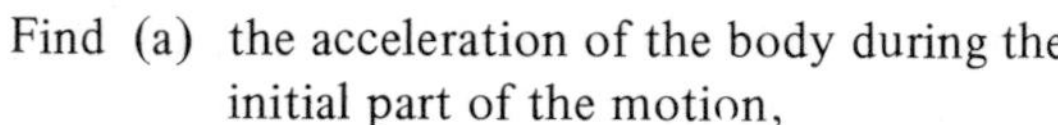

Find (a) the acceleration of the body during the initial part of the motion,
(b) the retardation of the body during the final part of the motion.
(c) the total distance travelled by the body.

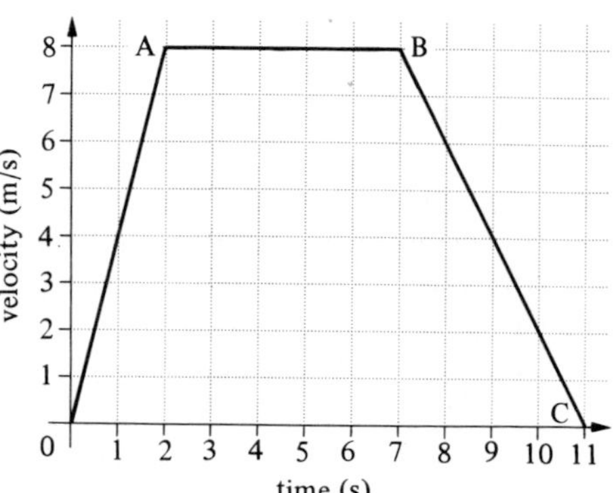

(a) The initial acceleration is given by the gradient of the line OA.

$$\text{gradient of OA} = \frac{\text{vertical increase from O to A}}{\text{horizontal increase from O to A}}$$

$$= \frac{8}{2} = 4$$

The initial acceleration is 4 m/s^2.

(b) The acceleration during the final part of the motion is given by the gradient of the line BC.

$$\text{gradient of BC} = \frac{\text{vertical increase from B to C}}{\text{horizontal increase from B to C}}$$

$$= \frac{-8}{4} = -2$$

The final retardation is 2 m/s^2.

(c) The total distance travelled is given by the area OABC. This is a trapezium and so:

$$\text{area OABC} = \frac{(5+11)8}{2}$$

$$= 64$$

The total distance travelled is 64 m.

Exercise 2E

1. Each of the following velocity–time graphs are for a body which accelerates uniformly for a time period of 4 seconds after which time it maintains its final velocity. In each case find
   (i) the initial velocity of the body, (ii) the final velocity of the body,
   (iii) the acceleration of the body during the 4 seconds,
   (iv) the distance travelled by the body during the 4 seconds.

(a)

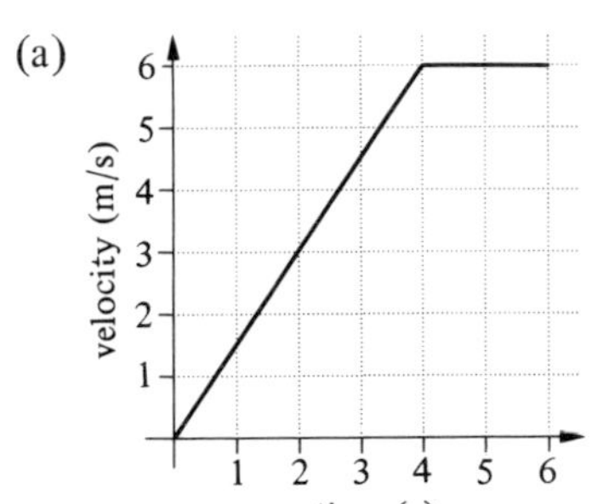

(b)

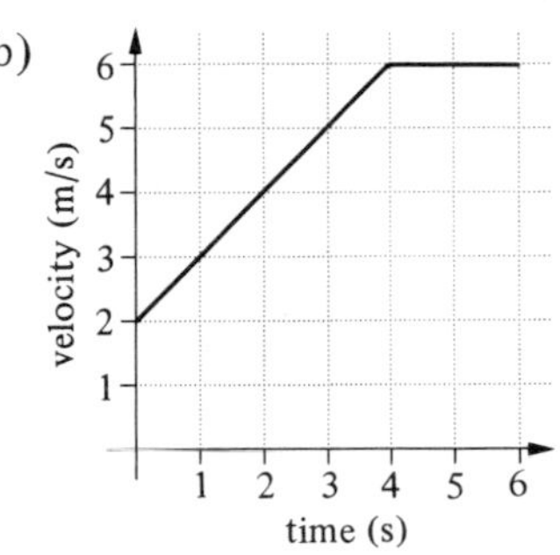

(c)

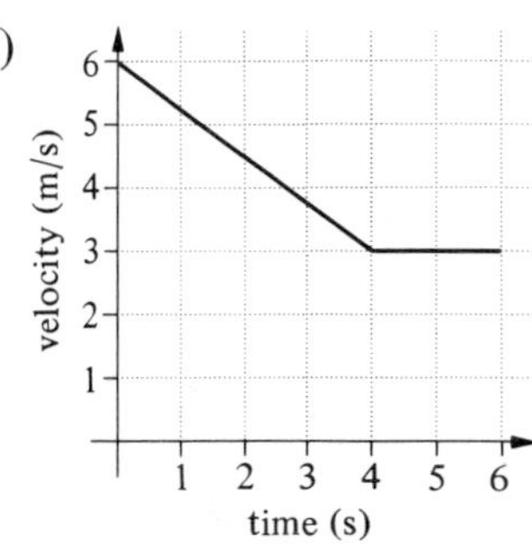

2. Each of the following velocity–time graphs are for a body which starts from rest, accelerates uniformly to a particular velocity, maintains that velocity for a period of time and then uniformly retards to rest. In each case find
   (i) the acceleration during the initial part of the motion,
   (ii) the retardation during the final part of the motion,
   (iii) the total distance travelled by the body during the motion.

(a)

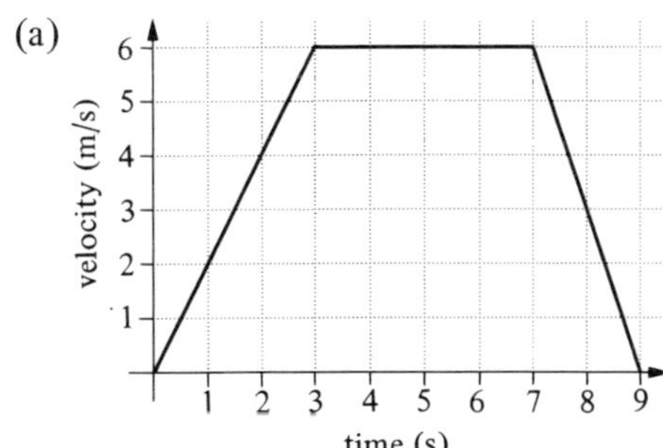

(b)

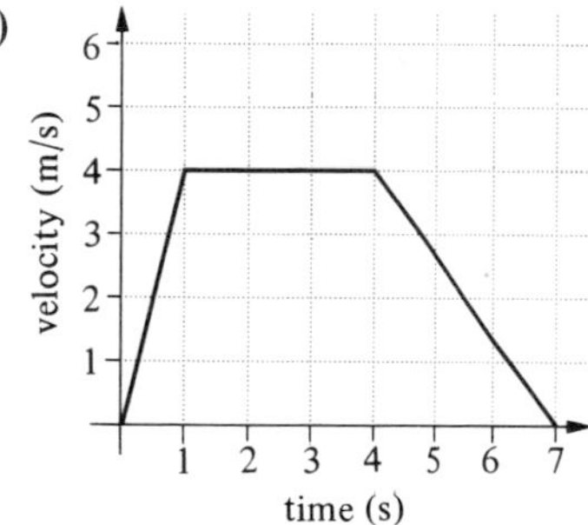

3. A cyclist rides along a straight road from a point A to a point B. He starts from rest at A and accelerates uniformly to reach a speed of 12 m/s in 8 seconds. He maintains this speed for a further 20 seconds and then uniformly retards to rest at B. If the whole journey takes 34 seconds, draw a velocity–time graph for the motion and from it find
   (a) his acceleration for the first part of the motion,
   (b) his retardation for the last part of the motion, (c) the total distance travelled.
4. A particle is initially at rest at a point A on a straight line ABCD. The particle moves from A to B with uniform acceleration, reaching B with a speed of 12 m/s after 2 seconds. The acceleration then alters to a constant 1 m/s^2 and 8 seconds after leaving B the particle reaches C. The particle then retards uniformly to come to rest at D after a further 10 seconds. Draw a velocity–time graph for the motion and from it find
   (a) the acceleration of the particle when travelling from A to B,
   (b) the speed of the particle on reaching C,
   (c) the retardation of the particle when travelling from C to D,
   (d) the total distance from A to D.

5. A and B are two points on a straight road. A car travelling along the road passes through A when $t = 0$ and maintains a constant speed until $t = 30$ seconds and in this time covers three-fifths of the distance from A to B. The car then retards uniformly to rest at B. Sketch a velocity–time graph for the motion and find the total time taken for the car to travel from A to B.
6. Two stations A and B are a distance of $6x$ m apart along a straight track. A train starts from rest at A and accelerates uniformly to a speed v m/s, covering a distance of x m. The train then maintains this speed until it has travelled a further $3x$ m, it then retards uniformly to rest at B. Make a sketch of the velocity–time graph for the motion and show that if T is the time taken for the train to travel from A to B then $T = \dfrac{9x}{v}$ seconds.

Exercise 2F Examination questions

1. The position of an aircraft flying on a straight course in a horizontal plane at a constant speed is plotted on a radar screen. At noon the position vector of this aircraft is $40\mathbf{i} + 16\mathbf{j}$ where $\mathbf{i}$ and $\mathbf{j}$ are unit vectors in directions east and north, respectively, referred to an origin O on the radar screen. Five minutes later the position of the aircraft on the screen is $33\mathbf{i} + 40\mathbf{j}$. Unit distance on the screen represents 1 km in the air. Calculate (i) the position vector of the aircraft at 12.15; and (ii) the velocity of the aircraft expressed as a vector in km h^{-1}. (S.U.J.B)

2. A train is uniformly retarded from 35 m/s to 21 m/s over a distance of 350 m. Calculate (a) the retardation, (b) the total time taken under this retardation to come to rest from a speed of 35 m/s. (London)

3. A particle moves with uniform acceleration $\frac{1}{2}$ m/s^2 in a horizontal line ABC. The speed of the particle at C is 80 m/s and the times taken from A to B and from B to C are 40 s and 30 s respectively. Calculate (a) the speed of the particle at A, (b) the distance BC. (London)

4. A particle P is at the origin and starts with velocity $2\mathbf{i} - 4\mathbf{j}$ and constant acceleration $3\mathbf{i} + 5\mathbf{j}$, where $\mathbf{i}$ and $\mathbf{j}$ are unit vectors in the directions East and North respectively and metres and seconds are the respective units for distance and time. After it has travelled for 2 seconds, find (i) its distance from the origin; (ii) its velocity, giving both magnitude and direction. (S.U.J.B)

5. A stone is thrown vertically upwards with initial speed 28 m/s. Find the time taken to reach the greatest height that it attains above the point of projection, and find this height. (London)

6. A ball A is thrown vertically upwards at 25 m/s from a point P. Three seconds later a second ball B is also thrown vertically upwards from the point P at 25 m/s. Taking the acceleration due to gravity as 10 m/s^2 calculate
   (i) how long A has been in motion when the balls meet,
   (ii) the height above P at which A and B meet. (A.E.B)

7. A body travelling in a straight line with a uniform acceleration of a m/s^2 passes a point O with a velocity of u m/s. During the first 5 seconds after passing O it travels 45 m and during the next second it travels a further 15 m. Calculate the value of a and of u. Calculate the velocity of the body when it is 140 m from O and the time taken to reach this point. (Cambridge)

8. A car moving along a straight line accelerates uniformly from rest until it has travelled x m. The car then moves for 50 s at a constant speed and travels a further x m. Finally the car uniformly decelerates and comes to rest after travelling a further $x/2$ m. Using a time-speed diagram, or otherwise, calculate the total time for the journey. (London)

9. A motorist starting a car from rest accelerates uniformly to a speed of v m/s in 9 seconds. He maintains this speed for another 50 seconds and then applies the brakes and decelerates uniformly to rest. His deceleration is numerically equal to three times his previous acceleration.
   (i) Sketch a velocity-time graph
   (ii) Calculate the time during which deceleration takes place.
   (iii) Given that the total distance moved is 840 m calculate the value of v.
   (iv) Calculate the initial acceleration. (Cambridge)

3
Force and Newton's laws

In the last chapter we considered bodies which changed their velocities. For this to occur, a *force* must act on the body.

Newton's First Law

A change in the state of motion of a body is caused by a force. The unit of force is the newton, abbreviated to N.

A body at rest

If forces act on a body and it does not move, the forces must balance. Hence, if a number of forces act on a body and it remains at rest, the resultant force in any direction must be zero.

Example 1

A body is at rest when subjected to the forces shown in the diagram. Find X and Y.

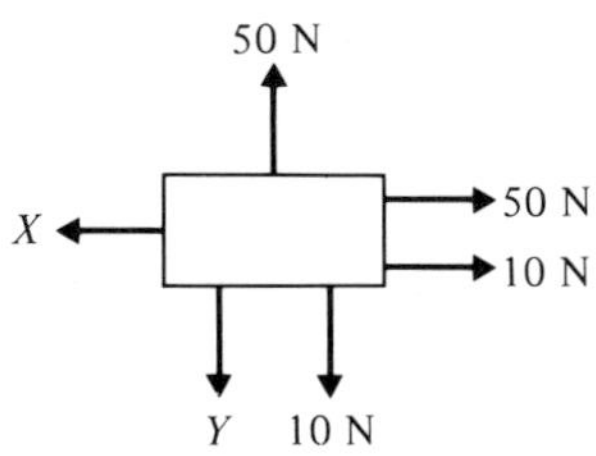

The horizontal forces balance

$$\therefore \quad X = 50 + 10$$
$$= 60 \text{ N}$$

The vertical forces balance

$$Y + 10 = 50$$
$$\therefore \quad Y = 40 \text{ N}$$

A body in motion

A body can only change its velocity, i.e. increase its speed, slow down or change direction, if a resultant force acts upon it. Thus, if a body is moving with constant velocity, there can be no resultant force acting on it.

Example 2

A body moves horizontally at a constant 5 m/s subject to the forces shown. Find P and S.

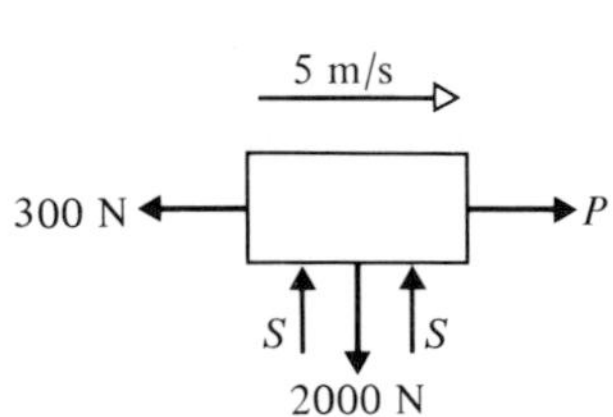

There is no vertical motion

$$\therefore \quad S + S = 2000$$
$$\therefore \quad S = 1000 \text{ N}$$

The horizontal velocity is constant

$$\therefore \quad P = 300 \text{ N}$$

Newton's First Law of Motion states that:

> A body will remain at rest, or will continue to move with constant velocity, unless external forces cause it to do otherwise.

Only when these external forces have a non-zero resultant will the body change from its previous state of rest or of constant velocity.

Newton's Second Law deals with such situations.

Newton's Second Law

When a resultant force acts on a body, it causes acceleration. The acceleration is proportional to the force. The same force will not produce the same acceleration in all bodies. The force which would give a cyclist, say, an acceleration of $\frac{1}{2}$ m/s^2, would, when applied to a car, produce a very much smaller acceleration.
The acceleration produced by a force depends upon the *mass* of the body on which it acts. The unit of mass is the kilogram, abbreviated to kg.
A force of 1 N produces an acceleration of 1 m/s^2 in a body of mass 1 kg. In general terms, a force of **F** newtons acting on a body of mass m kg produces an acceleration of **a** m/s^2 where

$$\mathbf{F} = m\mathbf{a}$$

This is a vector equation, so the acceleration produced is in the direction of the applied force, or of the resultant force if there is more than one force acting.

Newton's Second Law can be summarized by the equation $F = ma$ which is often referred to as the equation of motion.

Example 3

A body of mass 8 kg is acted upon by a force of 10 N.
Find the acceleration.

Using $F = ma$, the equation of motion is

$$10 = 8 \times a$$
$$\therefore \quad a = 1\tfrac{1}{4}$$

The acceleration is $1\frac{1}{4}$ m/s^2.

Example 4

Find the resultant force that would give a body of mass 200 g an acceleration of 10 m/s^2.

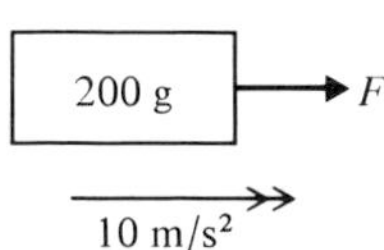

The mass is 200 g $= \frac{1}{5}$ kg.

Using $F = ma$, the equation of motion is

$$F = \tfrac{1}{5} \times 10 = 2$$

The force is 2 N.

Example 5

A body of mass 2 kg, subject to forces as shown in the diagram, accelerates uniformly in the direction indicated. Find the acceleration and the value of *P*.

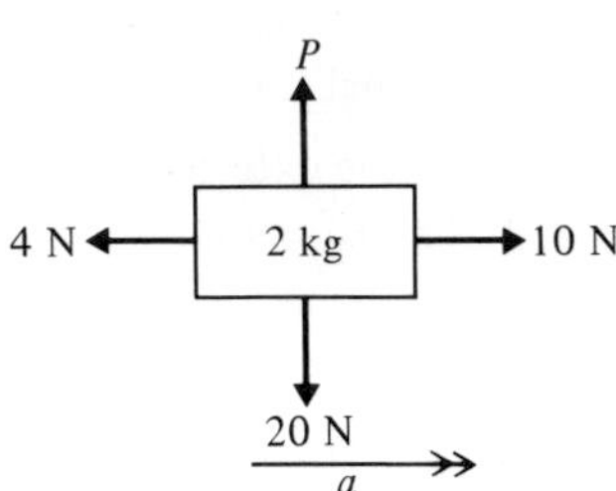

There is no vertical motion

$$\therefore \quad P = 20 \text{ N}$$

Horizontally, there are two forces acting, 10 N and 4 N, in opposite directions. Using $F = ma$, the equation of motion is

$$10 - 4 = 2 \times a$$
$$\therefore \quad a = 3$$

The horizontal acceleration is 3 m/s^2 and *P* is 20 N.

Example 6

Find, in vector form, the acceleration produced in a body of mass 5 kg subject to forces $(4\mathbf{i} + \mathbf{j})$ N and $(-\mathbf{i} + \mathbf{j})$ N. Also state the magnitude and the direction of the acceleration.

The resultant force acting is

$$(4\mathbf{i} + \mathbf{j}) + (-\mathbf{i} + \mathbf{j}) \text{ N}$$
$$= (3\mathbf{i} + 2\mathbf{j}) \text{ N}$$

Using $\mathbf{F} = m\mathbf{a}$, the equation of motion is

$$3\mathbf{i} + 2\mathbf{j} = 5 \times \mathbf{a}$$
$$\therefore \quad \mathbf{a} = \tfrac{3}{5}\mathbf{i} + \tfrac{2}{5}\mathbf{j}$$

The magnitude of $\mathbf{a} = |\mathbf{a}|$ $= \sqrt{((\tfrac{3}{5})^2 + (\tfrac{2}{5})^2)}$

$$= \tfrac{1}{5}\sqrt{(3^2 + 2^2)} = \tfrac{1}{5}\sqrt{13}$$
$$= 0{\cdot}72 \text{ m/s}^2$$

The acceleration produced is $(\tfrac{3}{5}\mathbf{i} + \tfrac{2}{5}\mathbf{j})$ m/s^2; its magnitude is 0·72 m/s^2 and its direction makes an angle $\tan^{-1}\tfrac{2}{3}$ with the unit vector **i**.

Exercise 3A

1. In each of the following situations a body is shown at rest under the action of certain forces. Find the magnitudes of the unknown forces *X* and *Y*.

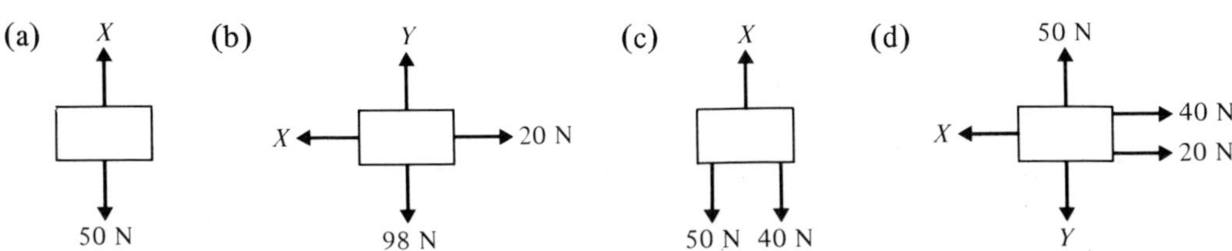

2. In each of the following situations a body is shown moving with constant velocity v under the action of certain forces. Find the magnitudes of the unknown forces X and Y.

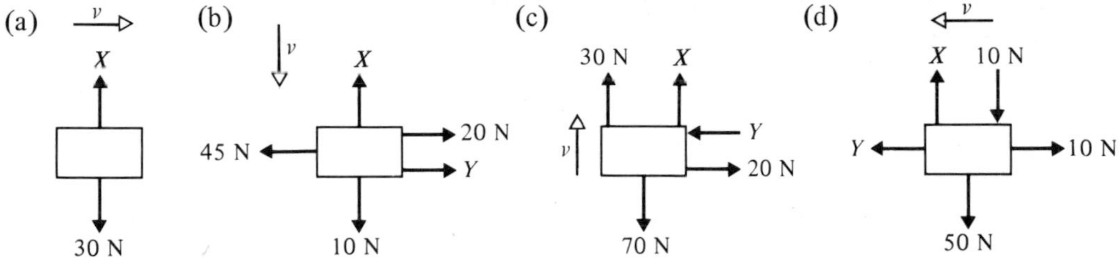

3. Find the acceleration produced when a body of mass 5 kg experiences a resultant force of 10 N.
4. Find the resultant force that would give a body of mass 3 kg an acceleration of 2 m/s^2.
5. A resultant force of 24 N causes a body to accelerate at 3 m/s^2. Find the mass of the body.
6. Find the acceleration produced when a body of mass 100 g experiences a resultant force of 5 N.
7. Find, in vector form, the resultant force required to make a body of mass 2 kg accelerate at $(5\mathbf{i} + 2\mathbf{j})$ m/s^2.
8. Find, in vector form, the acceleration produced in a body of mass 500 g subject to forces of $(4\mathbf{i} + 2\mathbf{j})$ N and $(-\mathbf{i} + \mathbf{j})$ N.
9. A car travels a distance of 24 m whilst uniformly accelerating from rest to 12 m/s. Find the acceleration of the car.
   If the car has a mass of 600 kg find the magnitude of the accelerating force.
10. A body of mass 500 g experiences a resultant force of 3 N.
    Find (a) the acceleration produced,
    (b) the distance travelled by the body whilst increasing its speed from 1 m/s to 7 m/s.
11. In each of the following situations the forces acting on the body cause it to accelerate as indicated. Find the magnitude of the unknown forces X and Y.

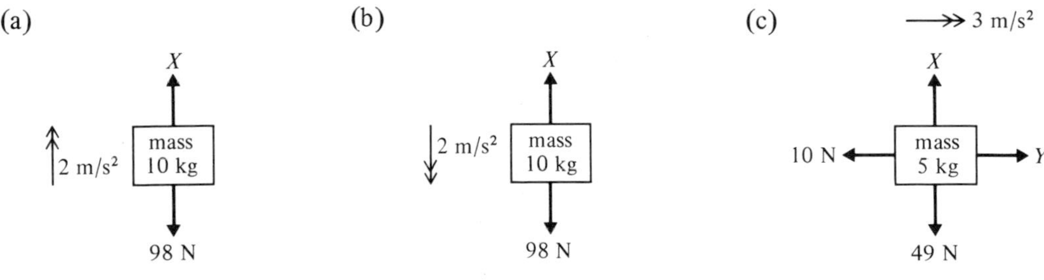

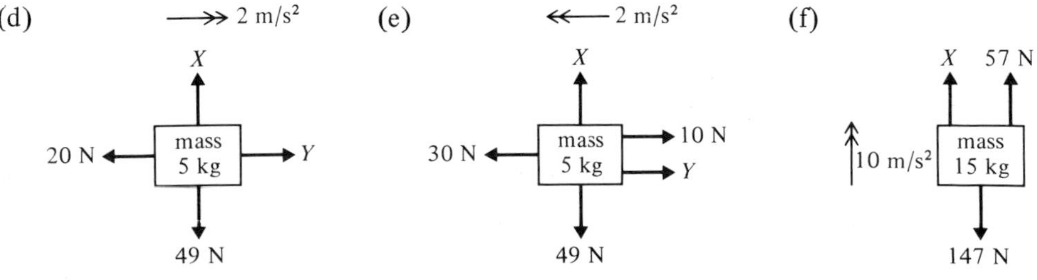

12. A car moves along a level road at a constant velocity of 22 m/s. If its engine is exerting a forward force of 500 N, what resistance is the car experiencing?
13. A car of mass 500 kg moves along a level road with an acceleration of 2 m/s^2. If its engine is exerting a forward force of 1100 N, what resistance is the car experiencing?
14. A van of mass 2 tonnes moves along a level road against resistances of 700 N. If its engine is exerting a forward force of 2200 N, find the acceleration of the van.
15. Find the magnitude of the resultant force required to give a body of mass 2 kg an acceleration of $(\mathbf{i} - 3\mathbf{j})$ m/s^2.
16. Find, in vector form, the acceleration produced in a body of mass 500 g when forces of $(5\mathbf{i} + 3\mathbf{j})$ N, $(6\mathbf{i} + 4\mathbf{j})$ N and $(-7\mathbf{i} - 7\mathbf{j})$ act on the body.
17. Forces of $(10\mathbf{i} + 2\mathbf{j})$ N and $(a\mathbf{i} + b\mathbf{j})$ N acting on a body of mass 500 g cause it to accelerate at $(24\mathbf{i} + 3\mathbf{j})$ m/s^2. Find the constants a and b.
18. Find the constant force necessary to accelerate a car of mass 600 kg from rest to 25 m/s in 12 s if the resistance to motion is (a) zero, (b) 350 N.
19. Find the constant force necessary to accelerate a car of mass 1000 kg from 15 m/s to 20 m/s in 10 s against resistances totalling 270 N.
20. A train of mass 60 tonnes is travelling at 40 m/s when the brakes are applied. If the resultant braking force is 40 kN, find the distance the train travels before coming to rest.
21. A train of mass 100 tonnes starts from rest at station A and accelerates uniformly at 1 m/s^2 until it attains a speed of 30 m/s. It maintains this speed for a further 90 s and then the brakes are applied, producing a resultant braking force of 50 kN. If the train comes to rest at station B, find the distance between the two stations.

Gravity and weight

As stated in chapter 2, a body falling freely under gravity experiences an acceleration of 9·8 m/s^2. From Newton's Laws it is clear that this acceleration must be caused by a force acting on the body. This force is called the *weight* of the body.

Consider a stone, of mass 2 kg, dropped from the top of a cliff.
It will fall with an acceleration of 9·8 m/s^2.
The force F which produces this acceleration is given by

$$\begin{aligned} F &= 2 \times 9{\cdot}8 \\ &= 19{\cdot}6 \text{ N} \end{aligned}$$

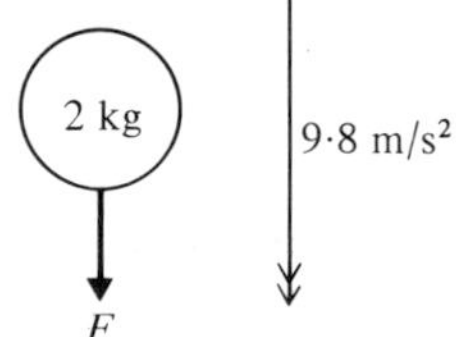

A body of mass m kg has a weight of mg N.

It should be remembered that, although the value of g (the acceleration due to gravity) has slightly different values at different places on the earth's surface, it should be taken as 9·8 m/s^2 unless stated otherwise.

Example 7

Find (a) the weight in newtons of a box of mass 5 kg,
(b) the mass of a stone of weight 294 N.

(a)

$$\begin{aligned} \text{mass} &= 5 \text{ kg} \\ \therefore \quad \text{weight} &= 5 \times 9{\cdot}8 \\ &= 49 \text{ N} \end{aligned}$$

The weight of the box is 49 N.

(b)

$$\text{weight} = 294 \text{ N}$$

$$\therefore \quad \text{mass} \times 9{\cdot}8 = 294 \qquad \therefore \quad \text{mass} = \frac{294}{9{\cdot}8} = 30 \text{ kg}$$

The mass of the stone is 30 kg.

The difference between the mass and the weight of a body is well illustrated by considering a 70 kg man on the Earth and the same man on the Moon. The acceleration due to gravity on the Moon is approximately 1·6 m/s^2.

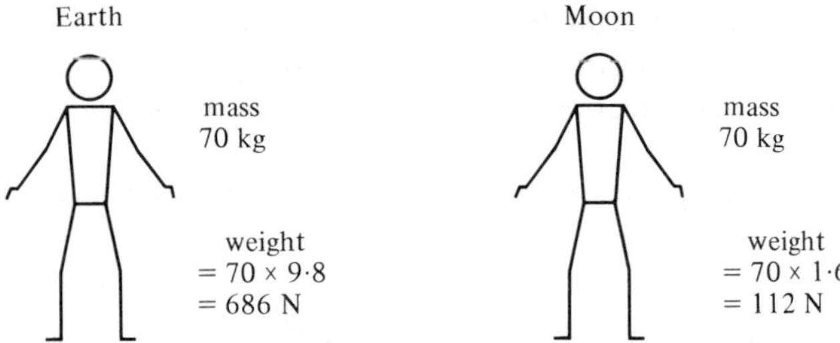

The man on the Moon has the same *mass* as he had on the Earth, but his *weight* is far less. He consequently feels lighter when on the Moon.

Example 8

A box of mass 5 kg is lowered vertically by a rope. Find the force in the rope when the box is lowered with an acceleration of 4 m/s^2.

$$\text{mass of box} = 5 \text{ kg}$$
$$\therefore \quad \text{weight of box} = 5g \text{ N}$$

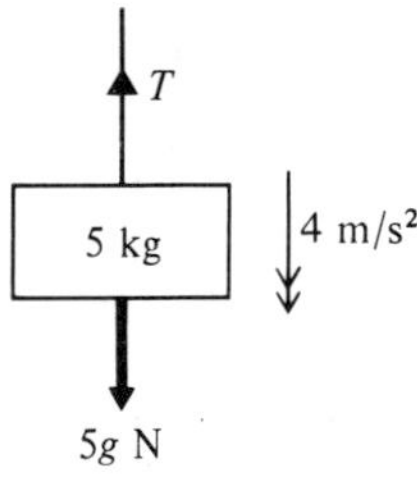

resultant vertical force on box is $(5g - T)$ N downwards.

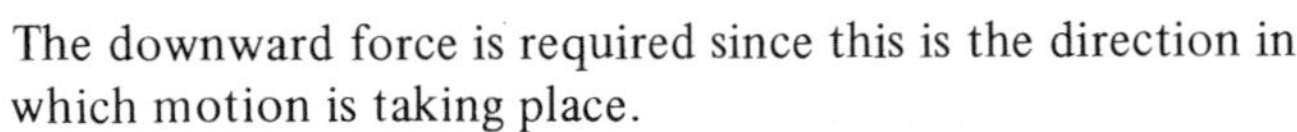

The downward force is required since this is the direction in which motion is taking place.

Using $F = ma$, the equation of motion is

$$5g - T = 5 \times 4$$
$$\therefore \quad T = 5g - 20$$
$$= 29$$

The force in the rope is 29 N.

Example 9

A pack of bricks of mass 100 kg is hoisted up the side of a house. Find the force in the lifting rope when the bricks are lifted with an acceleration of $\frac{1}{4}$ m/s^2.

$$\text{mass of bricks} = 100 \text{ kg}$$
$$\therefore \quad \text{weight of bricks} = 100g \text{ N}$$

resultant upward vertical force on the bricks is $(T - 100g)$ N.
(Upward force needed since motion is upward).

Using $F = ma$, the equation of motion is

$$T - 100g = 100 \times \tfrac{1}{4}$$
$$\therefore \quad T = 100g + 25$$
$$= 1005 \text{ N}$$

The force in the lifting rope is 1005 N.

Exercise 3B

Remember that g should be taken as $9{\cdot}8\ \text{m/s}^2$, unless otherwise stated.

1. Find the weight in newtons of a particle of mass 4 kg.
2. Find the mass of a car of weight 4900 N.
3. Find the weight, in newtons, of a particle of mass 100 g.
4. In each of the following situations, the forces acting on the body cause it to accelerate as indicated.
   In (a), (b) and (c), find the magnitude of the unknown forces X and Y.

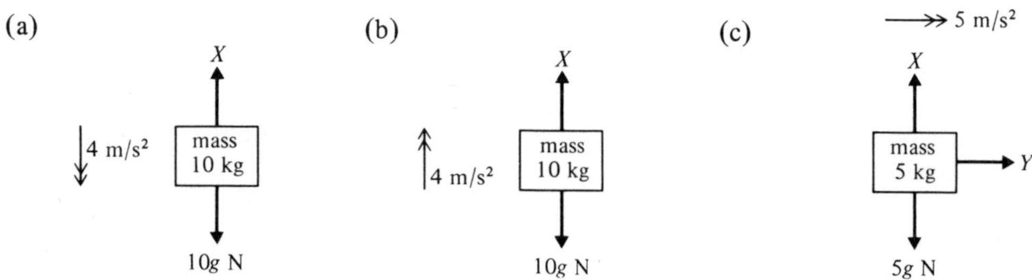

In (d), (e) and (f) find the mass m.

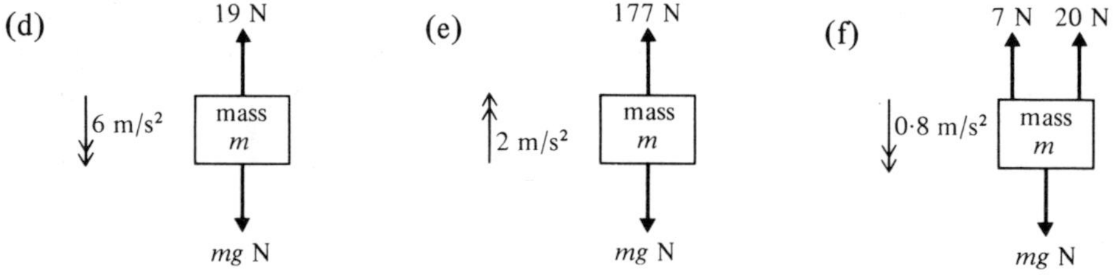

In (g), (h) and (i) find the magnitude of the acceleration a.

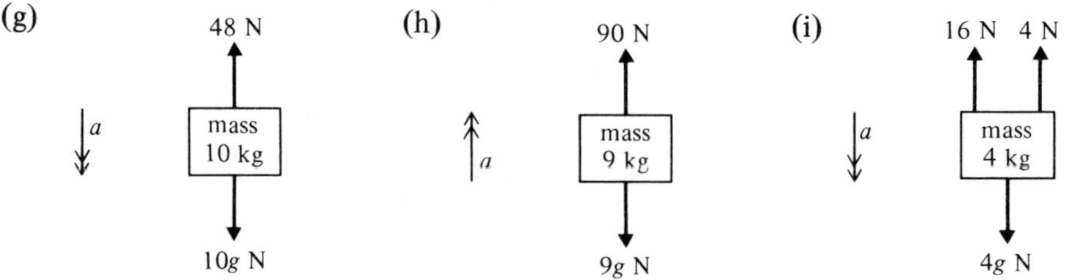

5. The diagram shows a body of mass 10 kg attached to a vertical string. Find the force T in the string in each of the following situations.
   (a) the string raises the body with an acceleration of $5\ \text{m/s}^2$,
   (b) the string lowers the body with an acceleration of $5\ \text{m/s}^2$,
   (c) the string raises the body at a constant velocity of 5 m/s,
   (d) the string lowers the body at a constant velocity of 5 m/s.

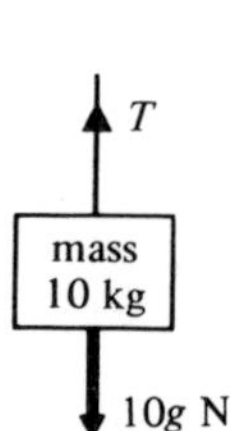

6. A particle of mass 100 g is attached to the lower end of a vertical string. Find the force in the string when it raises the particle with an acceleration of $1{\cdot}2$ m/s^2.
7. A concrete block of mass 50 kg is hoisted up the side of a building. Find the force in the lifting rope when the block is lifted with an acceleration of $\frac{1}{5}$ m/s^2.
8. A lift of mass 600 kg is raised or lowered by means of a cable attached to its top. When carrying passengers, whose total mass is 400 kg, the lift accelerates uniformly from rest to 2 m/s over a distance of 5 m. Find (a) the magnitude of the acceleration,
(b) the tension in the cable if the motion takes place vertically upwards,
(c) the tension in the cable if the motion takes place vertically downwards.

9. The hot air balloon shown in the diagram, rises from the ground with uniform acceleration. After 10 s the balloon has attained a height of 25 m. If the total mass of the balloon and basket is 250 kg, find the magnitude of the lifting force F.

10. A stone of mass 50 g is dropped into some liquid and falls vertically through it with an acceleration of $5{\cdot}8$ m/s^2. Find the force of resistance acting on the stone.
11. A tile of mass 2 kg falls from the roof of a building and hits the ground, $16{\cdot}6$ m below, 2 s later. Assuming the resistance experienced by the tile is constant throughout the motion, find this resistance.
12. A miners' cage of mass 420 kg contains three miners of total mass 280 kg. The cage is lowered from rest by a cable. For the first 10 seconds the cage accelerates uniformly and descends a distance of 75 m. Find the force in the cable during the first 10 seconds.
13. A bucket has a mass of 5 kg when empty and 15 kg when full of water. The empty bucket is lowered into a well, at a constant acceleration of 5 m/s^2, by means of a rope. When full of water the bucket is raised at a constant velocity of 2 m/s. Neglecting the weight of the rope, find the force in the rope (a) when lowering the empty bucket, (b) when raising the full bucket.

Newton's Third Law

This law states that: Action and Reaction are equal and opposite.

This means that if two bodies A and B are in contact and exert forces on each other, then the force exerted on B by A is equal in magnitude and opposite in direction to the force exerted on A by B. The following examples illustrate its application.

Example 10

A box of mass 5 kg rests on a horizontal floor. The box exerts a force on the floor and the floor 'reacts' by exerting an equal and opposite force on the box. As the box is at rest, this force of reaction R must equal the weight of the box, i.e. $R = 5g$ N.

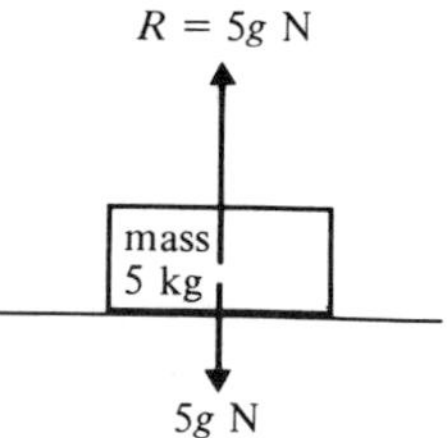

The force R is called the *normal reaction* as it acts at right angles to the surfaces in contact.

Example 11

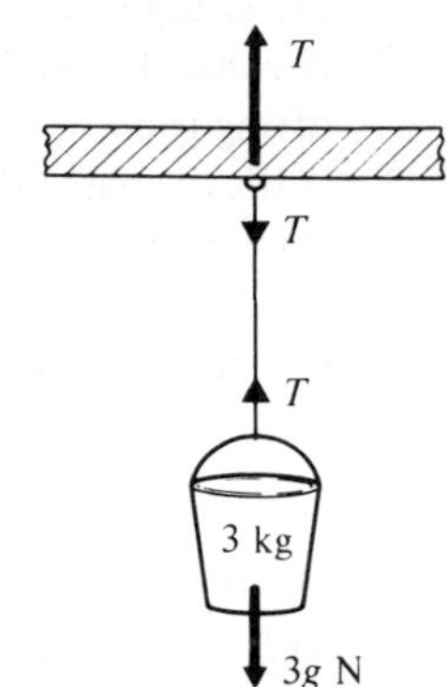

A bucket of mass 3 kg hangs on a vertical rope which is also attached to a beam. The bucket exerts a force on the rope, so the rope exerts an equal and opposite force on the bucket. As the bucket is at rest, this force T must equal the weight of the bucket, i.e. $T = 3g$ N.

At the point where the rope is attached to the beam, the rope exerts a downward force T on the beam. The beam exerts an equal and opposite force T on the rope. The rope therefore experiences a stretching force T at both ends.

The force T is called the tension in the rope.

Example 12

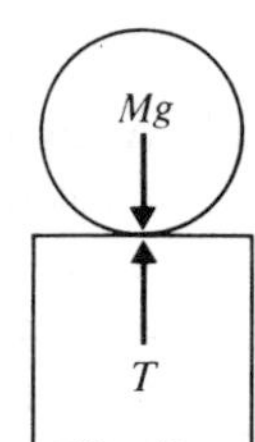

A granite sphere of mass M kg rests on top of a pillar. The sphere exerts a force on the pillar and the pillar exerts an equal and opposite force T on the sphere. As the sphere is at rest, T must equal the weight of the sphere, i.e. $T = Mg$ N.

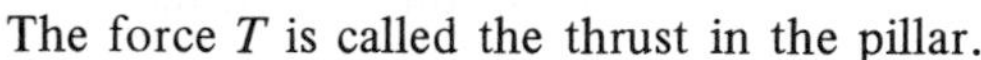

The force T is called the thrust in the pillar.

Example 13

A car pulls a trailer along a level road at a constant velocity v.
The trailer is pulled forwards by the tension T in the tow bar.
The trailer will exert an equal and opposite force T on the car.

If the car is accelerating, then there must be a force acting on the trailer to produce this acceleration, and this will be provided by the tension T in the tow bar.
On the other hand, when the car is slowing down, so also is the trailer. In the absence of brakes on the trailer, some force must act in the opposite direction to the motion of the car and trailer. In this case the tow bar will exert a *thrust* on both the car and the trailer.

Example 14

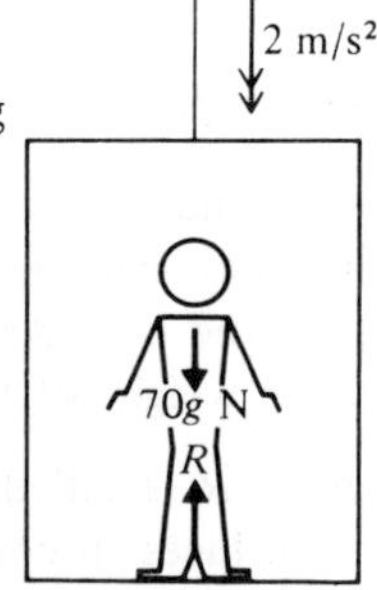

A man of mass 70 kg stands on the floor of a lift which is accelerating downwards at 2 m/s^2. The man exerts a force on the floor and the floor exerts an equal and opposite force R on the man. Thus the resultant downward vertical force on the man is $(70g - R)$ and the equation of motion for the man is

$$\begin{aligned} 70g - R &= 70 \times 2 \\ \therefore \quad R &= 686 - 140 \\ &= 546 \text{ N} \end{aligned}$$

It should be noted that in this case the reaction R is not equal to the weight of the man.

Connected particles

In the following examples the strings are all considered to be light and inextensible.

Note also that, when a surface is said to be smooth, it is to be assumed that the surface offers no resistance to the motion of a body across it.

Example 15

Consider a body of mass 3 kg at rest on a smooth horizontal table. This body is connected by a light string, which passes over a smooth pulley at the edge of the table, to another body of mass 2 kg hanging freely.
As the pulley is smooth, the tension in the string on both sides of the pulley will be the same.
The string is said to be light, so its weight can be ignored.
As the string is inextensible, when the system is released from rest the two bodies will have equal accelerations along the line of the string.

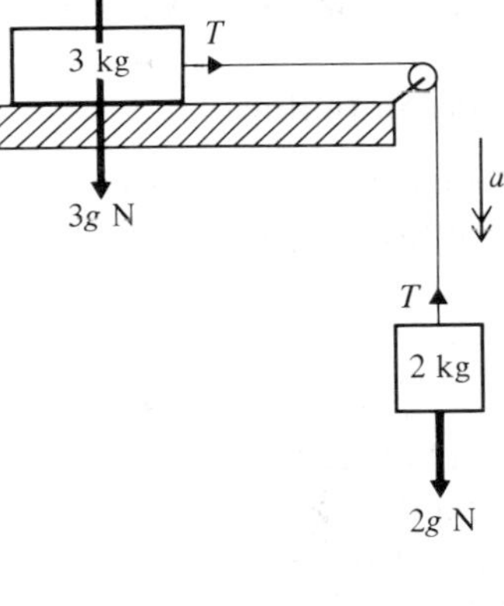

The 3 kg mass will not move in a vertical direction, so the vertical forces acting on it must balance.

$$\therefore \quad R = 3g$$

The horizontal force acting on the 3 kg mass is T.
Using $F = ma$, the equation of motion is

$$T = 3 \times a \qquad \ldots [1]$$

The 2 kg mass moves vertically downwards.
Using $F = ma$, the equation of motion is

$$2g - T = 2 \times a \qquad \ldots [2]$$

Solving equations [1] and [2] simultaneously

$$2g = 3a + 2a$$
$$\therefore \quad a = \tfrac{2}{5}g$$

The acceleration of both bodies is $\frac{2}{5}g$ m/s^2 along the line of the string; the tension in the string is $\frac{6}{5}g$ N, obtained by substituting for a in equation [1].

Example 16

Particles of mass 4 kg and 2 kg are connected by a light string passing over a smooth fixed pulley. The particles hang freely and are released from rest. Find the acceleration of the two particles and the tension in the string. Let the acceleration be a and the tension in the string be T.

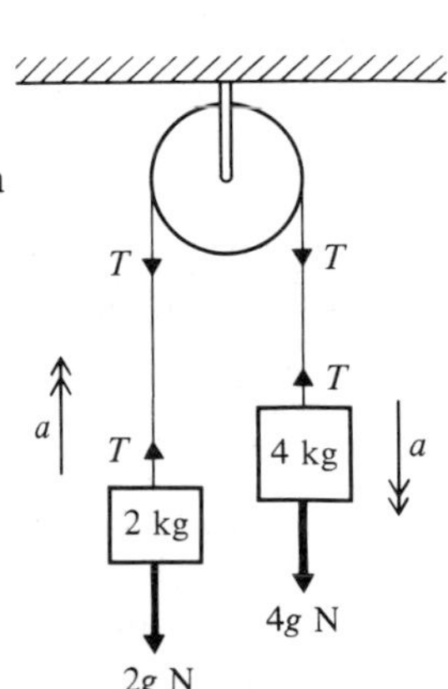

Using $F = ma$

for 2 kg mass: $\quad T - 2g = 2 \times a \qquad \ldots [1]$
for 4 kg mass: $\quad 4g - T = 4 \times a \qquad \ldots [2]$

Adding equations [1] and [2], $2g = 6a$

$$\therefore \; a = \tfrac{1}{3}g \;\text{ and }\; T = 2\tfrac{2}{3}g$$

Using $g = 9{\cdot}8$ m/s^2, the acceleration is $3{\cdot}27$ m/s^2 and the tension is $26{\cdot}1$ N.

Example 17

A body A rests on a smooth horizontal table. Two bodies of mass 2 kg and 10 kg, hanging freely, are attached to A by strings which pass over smooth pulleys at the edges of the table. The two strings are taut. When the system is released from rest, it accelerates at 2 m/s². Find the mass of A.

Let the mass of A be M kg. The tensions in the two strings will be different: let them be T_1 and T_2.

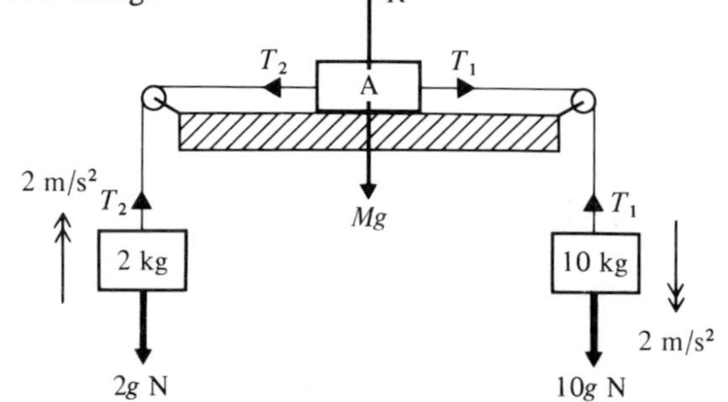

Using $F = ma$

for 2 kg mass: $T_2 - 2g = 2 \times 2$ . . . [1]
for A: $T_1 - T_2 = M \times 2$ . . . [2]
for 10 kg mass: $10g - T_1 = 10 \times 2$ . . . [3]

Adding equations [1], [2] and [3]

$$8g = 2M + 24$$

$$\therefore \quad M = 27{\cdot}2$$

The mass of the body A is 27·2 kg.

Force on pulley

It should be noted, in each of these examples, that there is a force acting on the fixed pulleys due to the tension in the string passing around the pulley. In Example 16 there is a downward force of $2T$ or 52·2 N acting on the fixed pulley, due to the string and the attached loads.

Exercise 3C

1. A box of mass 10 kg rests on a horizontal floor. What is the reaction that the floor exerts on the box?
2. A yo-yo of mass 200 g hangs at rest at the lower end of a vertical string. What is the tension in the string?
3. A cat of mass 4 kg sits on top of a vertical post. What is the thrust in the post?
4. The diagram shows a body of mass 5 kg hanging at rest at the end of a light vertical string. The other end of the string is attached to a mass of 2 kg which in turn hangs at the end of another light vertical string. Find the tension in each string.
5. A cube of mass 6 kg rests on top of a horizontal table. A smaller cube of mass 2 kg is placed on top of the 6 kg cube. Find the reaction between the two cubes and that between the larger cube and the table.
6. The diagram shows a car of mass m_1 pulling a trailer of mass m_2 along a level road. The engine of the car exerts a forward force F, the tension in the tow bar is T and the reactions at the ground for the car and the trailer are R_1 and R_2 respectively. If the acceleration of the car is a, write down the equation of motion for (a) the system as a whole, (b) the car, (c) the trailer.
   What can be said about R_1 and R_2?

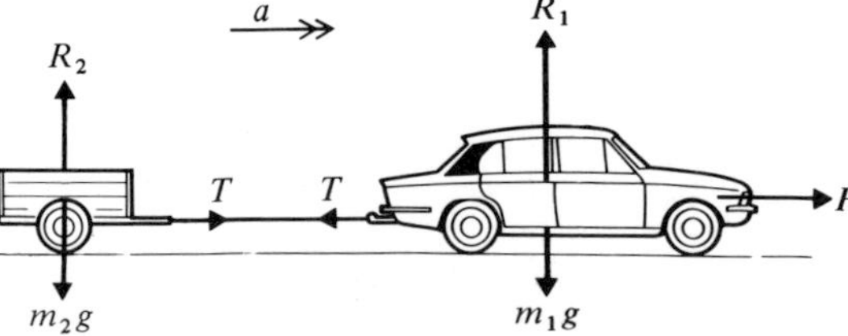

7. Masses m_1 and m_2 are connected by a light inextensible string passing over a smooth fixed pulley with $m_1 > m_2$. The masses move with acceleration a, as shown in the diagram. If the tension in the string is T, write down the equation of motion for (a) mass m_1, (b) mass m_2.

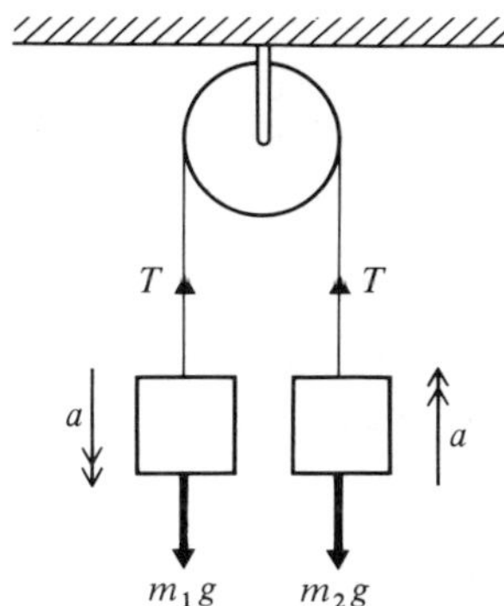

8. Mass m_1 lies on a smooth horizontal table and has one end of a light inextensible string attached to it. The string passes over a smooth fixed pulley at the edge of the table and carries a mass m_2 at its other end. T is the tension in the string, R is the reaction between m_1 and the table, and the acceleration a is as indicated in the diagram. Write down the equation of motion for (a) mass m_1, (b) mass m_2.

9. The diagram shows masses m_1, m_2 and m_3 connected by light inextensible strings such that m_1 and m_3 hang vertically and m_2 lies on a smooth horizontal surface. With $m_1 > m_3$, the forces and accelerations are as indicated in the diagram. Write down the equation of motion for
(a) mass m_1, (b) mass m_2,
(c) mass m_3.

10. A man of mass m is in a lift of mass M. The lift ascends with uniform acceleration a, and the tension in the cable is T. The force of reaction between the man and the floor of the lift is R. Write down the equation of motion for
(a) the system as a whole (Fig. 1),
(b) the lift (Fig. 2),
(c) the man (Fig. 3).

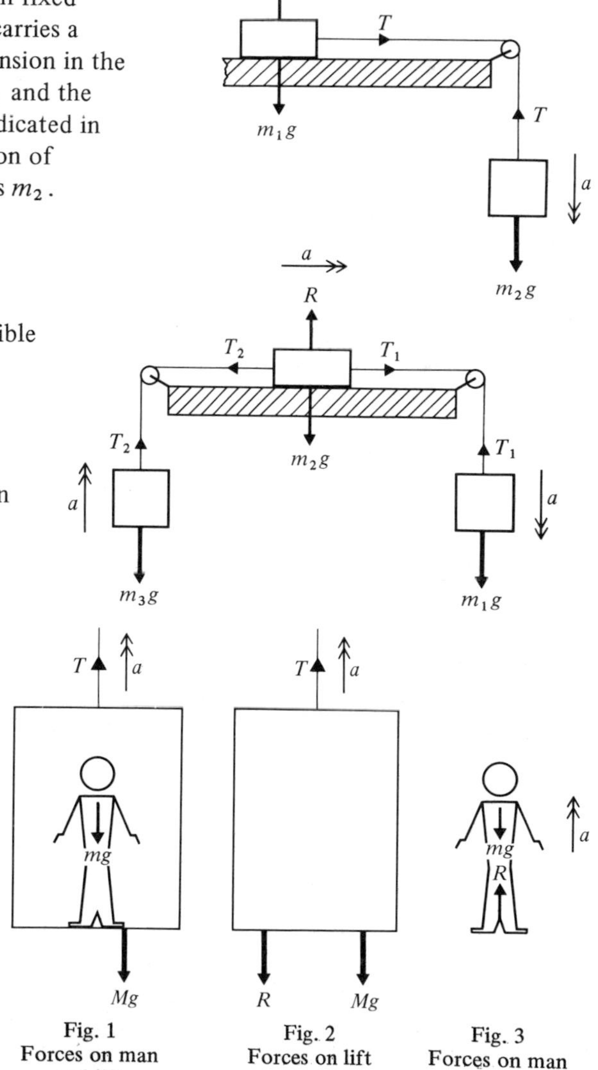

Fig. 1
Forces on man and lift

Fig. 2
Forces on lift

Fig. 3
Forces on man

11. A light inextensible string passes over a smooth fixed pulley and carries freely hanging masses of 6 kg and 4 kg at its ends. Find the acceleration of the system and the tension in the string.
12. Find the reaction between the floor of a lift and a passenger of mass 60 kg when the lift descends with constant acceleration of 1·3 m/s^2.
13. Find the reaction between the floor of a lift and a passenger of mass 60 kg when the lift ascends with constant acceleration of 1·2 m/s^2.
14. A light inextensible string passes over a smooth fixed pulley and carries freely hanging masses of 800 g and 600 g at its ends. Find the acceleration of the system and the force on the pulley.
15. A car of mass 900 kg tows a caravan of mass 700 kg along a level road. The engine of the car exerts a forward force of 2·4 kN and there is no resistance to motion. Find the acceleration produced and the tension in the tow bar.
16. Each of the following diagrams shows two freely hanging masses connected by a light inextensible string passing over a smooth fixed pulley. For each system find
(i) the acceleration of the masses, (ii) the magnitude of the tension T_1,
(iii) the magnitude of the tension T_2. (Assume the pulley to be light).

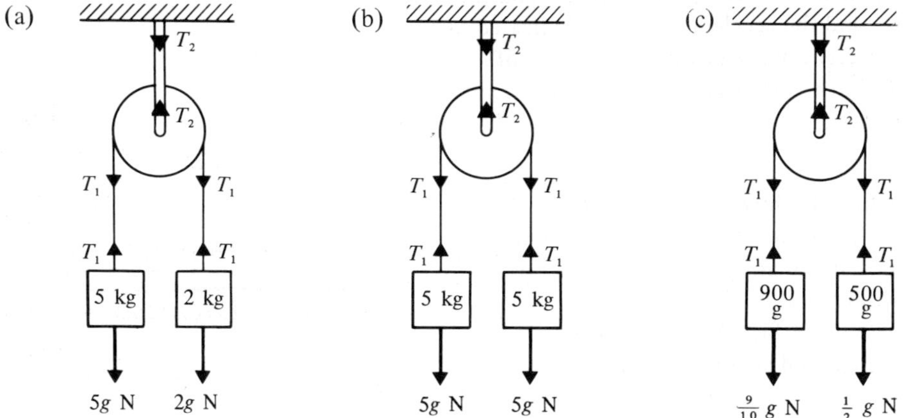

17. Bodies of mass 6 kg and 2 kg are connected by a light inextensible string which passes over a smooth fixed pulley. With the masses hanging vertically, the system is released from rest. Find the acceleration of the system and the distance moved by the 6 kg mass in the first 2 seconds of motion.

18. Each of the following diagrams shows three bodies connected by light inextensible strings passing over smooth pulleys. One mass lies on a smooth horizontal surface and the other two masses hang freely. In each case the masses, pulleys and strings all lie in the same vertical plane. With the strings taut, each system is released from rest. Find the ensuing accelerations and the tensions in the strings.

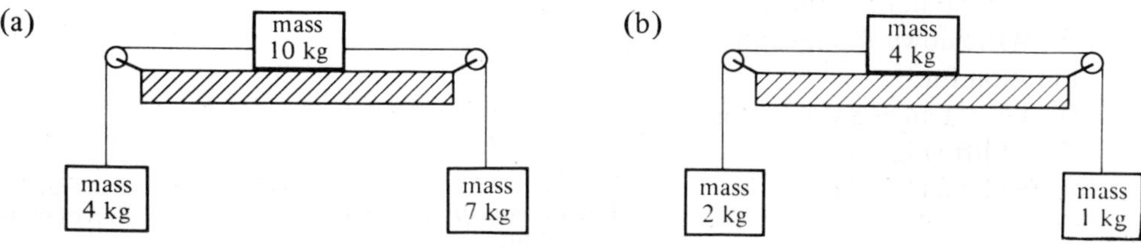

19. A body of mass 65 g lies on a smooth horizontal table. A light inextensible string runs from this body, over a smooth fixed pulley at the edge of the table, to a body of mass 5 g hanging freely. With the string taut, the system is released from rest. Find (a) the acceleration of the system, (b) the tension in the string, (c) the distance moved by the 5 g mass in the first 2 seconds of motion. (Assume that nothing impedes its motion in this time).
20. The motion of a lift, when ascending from rest, is in three stages. First, it accelerates at 1 m/s^2 until it reaches a certain velocity. It then maintains this velocity for a period of time after which it slows, with retardation 1·2 m/s^2, until it comes to rest. Find the reaction between the floor of the lift and a passenger, of mass 100 kg, during each of these three stages.
21. A car of mass 900 kg tows a trailer of mass 600 kg by means of a rigid tow bar. The car experiences a resistance of 200 N and the trailer a resistance of 300 N. If the car engine exerts a forward force of 3 kN, find the tension in the tow bar and the acceleration of the system.
If the engine is switched off and the brakes now apply a retarding force of 500 N, what will be the retardation of the system, assuming the same resistances apply? What will be the nature and magnitude of the force in the tow bar?

Exercise 3D. Harder questions

1. Particles of mass m_1 and m_2 ($m_2 > m_1$) are connected by a light inextensible string passing over a smooth fixed pulley. The particles hang vertically and are released from rest. Show that the acceleration of the system is $\frac{(m_2 - m_1)g}{m_1 + m_2}$ and the tension in the string is $\frac{2m_1 m_2 g}{m_1 + m_2}$.
2. A particle of mass m_1 lies on a smooth horizontal table and is connected to a freely hanging particle of mass m_2 by a light inextensible string passing over a smooth fixed pulley situated at the edge of the table. Initially the system is at rest with m_1 a distance d from the edge of the table. Show that the acceleration of the system is $\frac{m_2 g}{(m_1 + m_2)}$ and the time taken for m_1 to reach the edge of the table is $\sqrt{\frac{2d(m_1 + m_2)}{m_2 g}}$
3. The diagram shows the freely suspended particles A and C connected by means of light inextensible strings and smooth pulleys to particle B which lies on a smooth horizontal table. If the masses of A, B and C are $3m$, $3m$ and $4m$ respectively, find the acceleration of the system and the tensions in the strings.

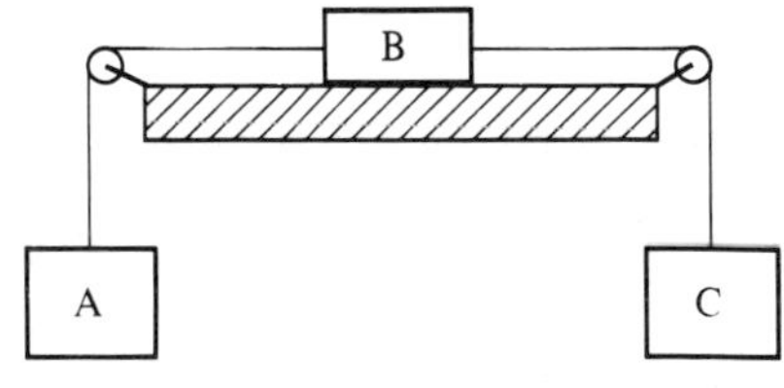

4. A car of mass 800 kg tows a trailer of mass 400 kg against resistances totalling 600 N. The separate resistances on the car and the trailer are proportional to their masses. If the car accelerates at 1·25 m/s^2 along a level road, find (a) the forward force exerted by the engine, (b) the tension in the tow bar.
5. Two particles A and B are connected by a light inextensible string passing over a smooth fixed pulley. The masses of A and B are $\frac{11}{2}m$ and $\frac{9}{2}m$ respectively. With A and B hanging vertically, the system is released from rest with particle A a distance d above the floor. If a time t elapses before A hits the floor, show that $20d = t^2 g$.

6. A lorry of mass 3 tonnes tows a trailer of mass 1 tonne along a level road and accelerates uniformly from rest to 18 m/s in 24 s. The resistances on the lorry and trailer are proportional to their masses and total 1200 N. Find (a) the driving force exerted by the engine of the lorry and (b) the tension in the tow bar.

7. The diagram shows a light inextensible string passing over a smooth fixed pulley, and carrying a particle A at one end and particles B and C at the other. The masses of A, B and C are $2m$, m and $2m$ respectively. Find the acceleration of the system when released from rest.
After C has travelled 50 cm it falls off and the system continues without it.
Find (a) the velocity of B at the instant C falls off,
(b) how much further B travels down before it starts to rise.

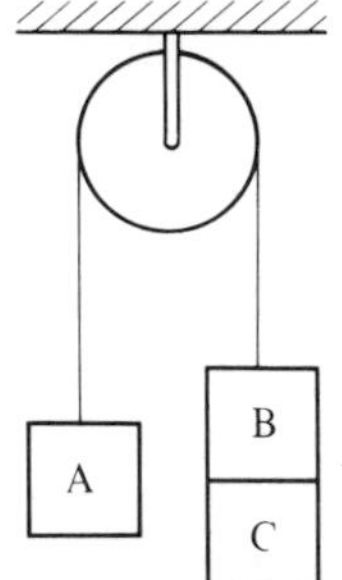

8. A car of mass 1 tonne exerts a driving force of 2·5 kN when pulling a trailer of mass 400 kg along a level road. The car and trailer start from rest and travel 18 m in the first 6 s of motion. If the resistances on the car and trailer are $1000x$ N and $400x$ N respectively, find the value of the constant x.

9. Particles of mass 600 g and 400 g are connected by a light inextensible string passing over a smooth fixed pulley. Initially both masses hang vertically, 30 cm above the ground. If the system is released from rest, find the greatest height reached above ground by the 400 g mass.

10. Particles of mass m_1 and m_2 ($m_2 > m_1$) are connected by a light inextensible string passing over a smooth fixed pulley. Initially both masses hang vertically with mass m_2 at a height x above the floor. Show that, if the system is released from rest, the mass m_2 will hit the floor with speed $\sqrt{\dfrac{2(m_2 - m_1)gx}{m_1 + m_2}}$ and the mass m_1 will rise a further distance $\dfrac{(m_2 - m_1)x}{m_1 + m_2}$ after this occurs.

Pulley systems

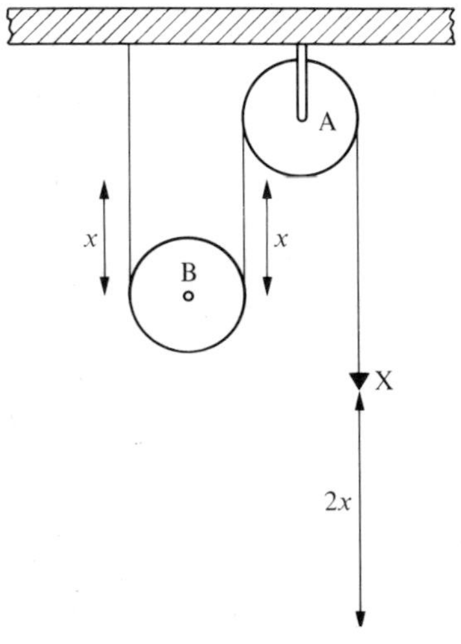

In the diagram, pulley A is fixed and pulley B may be raised by pulling down the end X of the string. All the parts of the string not in contact with the pulleys are vertical.
For B to move upwards a distance x, a length $2x$ of string must pass over the pulley A.
The distance between the pulley A and the end X of the string is therefore increased by $2x$.
Hence if B has an *upward* acceleration of a, then the end X of the string will have a *downward* acceleration of $2a$.

Example 18

In the pulley system shown, A is a fixed pulley and pulley B has a mass of 4 kg. A load of mass 5 kg is attached to the free end of the string. Assuming the pulleys to be smooth, the tension throughout the string will be T as shown. When the system is released from rest, let the upward acceleration of B be a. The downward acceleration of the 5 kg mass will then be $2a$.

Using $F = ma$

for 5 kg load: $5g - T = 5 \times 2a$

for pulley B: $2T - 4g = 4 \times a$

Solving these equations and substituting $g = 9{\cdot}8$ m/s^2, we obtain

$$a = 2{\cdot}45 \quad \text{and} \quad T = 24{\cdot}5$$

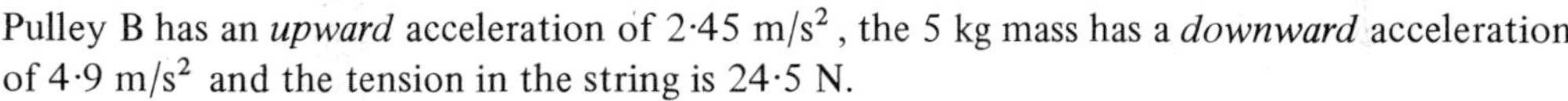

Pulley B has an *upward* acceleration of $2{\cdot}45$ m/s^2, the 5 kg mass has a *downward* acceleration of $4{\cdot}9$ m/s^2 and the tension in the string is $24{\cdot}5$ N.

In the pulley system shown, A and B are fixed pulleys and C is a moveable pulley. When the ends, X and Y, of the string move down distances x and y respectively, the length of the string between A and B is shortened by $(x + y)$. Pulley C will therefore move up a distance $\frac{1}{2}(x + y)$.
Hence, if the downward acceleration of X and Y are a_1 and a_2, the upward acceleration of C will be $\frac{1}{2}(a_1 + a_2)$.

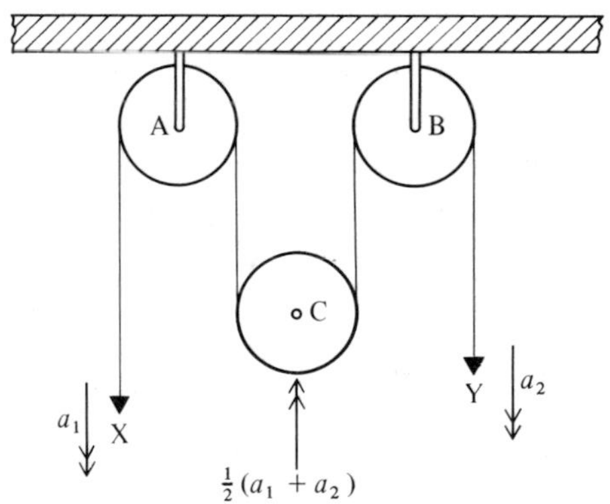

Example 19

A pulley system has loads of 6 kg and 3 kg at the ends of the string, and the moveable pulley has a mass of 2 kg as shown. Assuming the pulleys to be smooth, find the acceleration of pulley C.

Let the accelerations of the loads be a_1 and a_2. The acceleration of the pulley C will be $\frac{1}{2}(a_1 + a_2)$ in the opposite direction. The tension of the string will be T throughout.

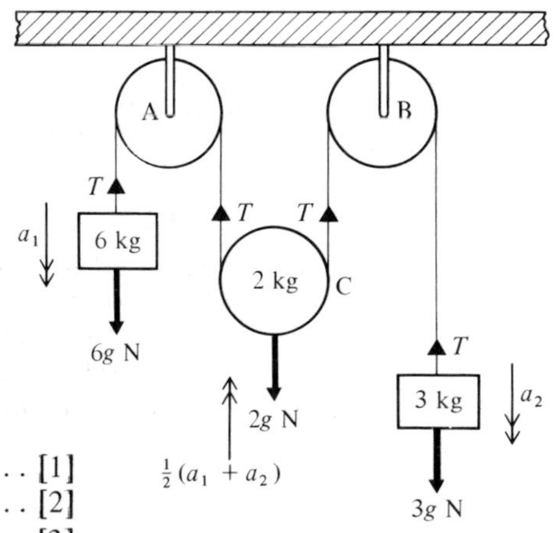

Using $F = ma$

for 6 kg load: $6g - T = 6 \times a_1$ ... [1]

for 3 kg load: $3g - T = 3 \times a_2$ ... [2]

for pulley C: $2T - 2g = 2 \times \frac{1}{2}(a_1 + a_2)$ ... [3]

From [1] and [2] $\quad 3g = 6a_1 - 3a_2$
and from [2] and [3] $\quad 4g = 7a_2 + a_1$

hence $a_1 = \frac{11}{15}g$ and $a_2 = \frac{7}{15}g$

The upward acceleration of C $= \frac{1}{2}(a_1 + a_2)$
$= \frac{3}{5}g$ m/s^2

Example 20

When a light pulley A is suspended from a fixed pulley, we have to consider relative accelerations. In the system shown, if the 3 kg load ascends with an acceleration a_1, the pulley A descends with an acceleration a_1.

Suppose the accelerations of the 2 kg and 6 kg loads, relative to pulley A, are a_2 upwards and a_2 downwards respectively.
The actual accelerations of these loads will then be $(a_2 - a_1)$ upwards and $(a_2 + a_1)$ downwards respectively. Find these accelerations.

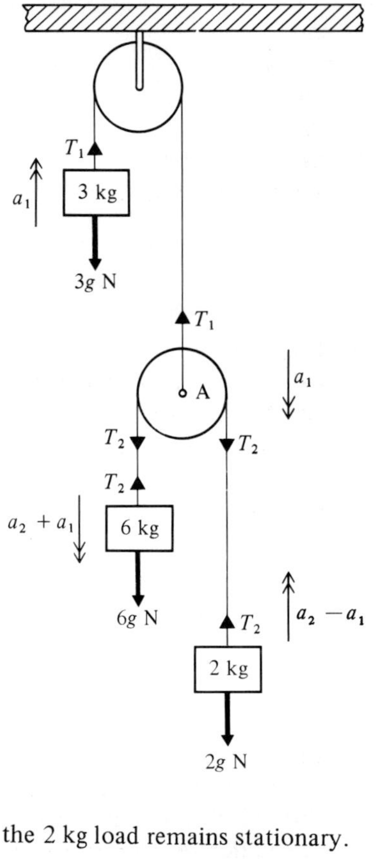

Let the tensions in the two strings be T_1 and T_2.
Assuming the pulley A to be weightless and using $F = ma$

for 3 kg load $\quad T_1 - 3g = 3a_1 \quad \ldots [1]$
for pulley A $\quad 2T_2 - T_1 = 0 \times a_1 \quad \ldots [2]$
for 6 kg load $\quad 6g - T_2 = 6 \times (a_2 + a_1) \quad \ldots [3]$
for 2 kg load $\quad T_2 - 2g = 2 \times (a_2 - a_1) \quad \ldots [4]$

Eliminating T_1 and T_2 from these equations

[3] and [4] gives $\quad 4g = 8a_2 + 4a_1$
[1] + [2] + 2[3] gives $\quad 9g = 12a_2 + 15a_1$

Solving simultaneously, $a_1 = \frac{1}{3}g$, $a_2 = \frac{1}{3}g$
The 3 kg load accelerates upwards at $\frac{1}{3}g$ m/s^2.

Acceleration of 6 kg load $= a_2 + a_1$
$= \frac{2}{3}g$ downward

Acceleration of 2 kg load $= a_2 - a_1$
$= 0$

The 6 kg load has a downward acceleration of $\frac{2}{3}g$ m/s^2 and the 2 kg load remains stationary.

Exercise 3E

In this exercise all pulleys are smooth, all strings are light and inextensible, and all those parts of the strings not in contact with the pulleys are vertical.

1. A string, with one end fixed, passes under a moveable pulley of mass 2 kg, over a fixed pulley and carries a 5 kg mass at its other end (see diagram). Find the acceleration of the moveable pulley and the tension in the string.

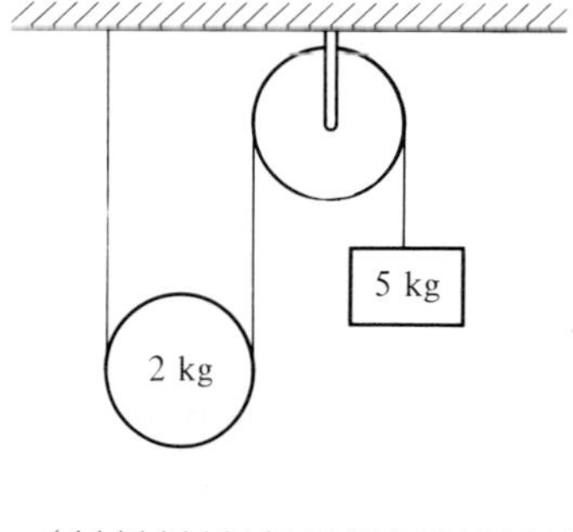

2. A string has a load of mass 2 kg attached to one end. The string passes over a fixed pulley, under a moveable pulley of mass 6 kg, over another fixed pulley and has a load of mass 3 kg attached to its other end.
   Find the acceleration of the moveable pulley and the tension in the string.

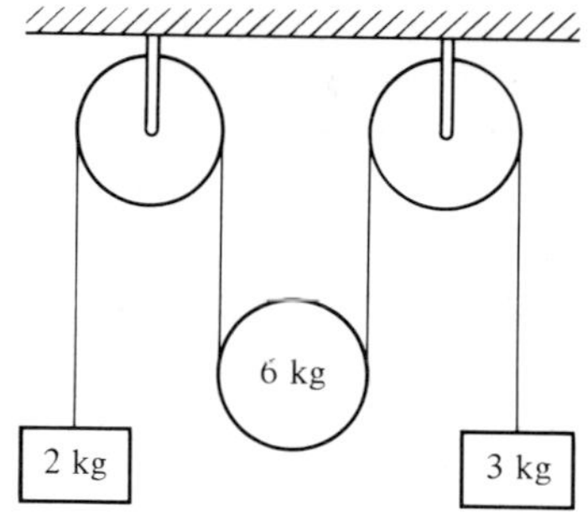

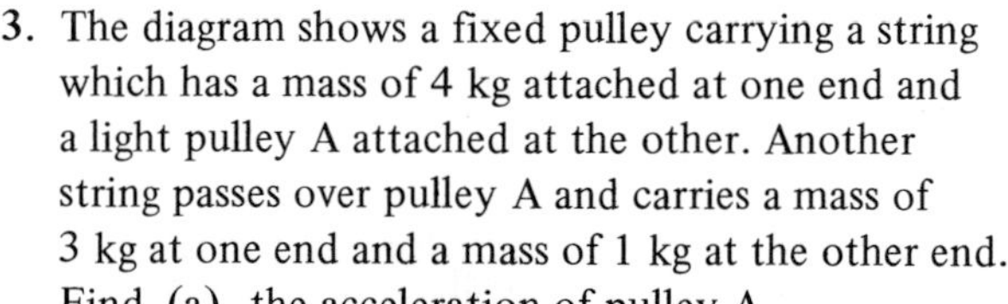

3. The diagram shows a fixed pulley carrying a string which has a mass of 4 kg attached at one end and a light pulley A attached at the other. Another string passes over pulley A and carries a mass of 3 kg at one end and a mass of 1 kg at the other end.
   Find (a) the acceleration of pulley A,
   (b) the acceleration of the 1 kg, 3 kg and 4 kg masses,
   (c) the tensions in the strings.

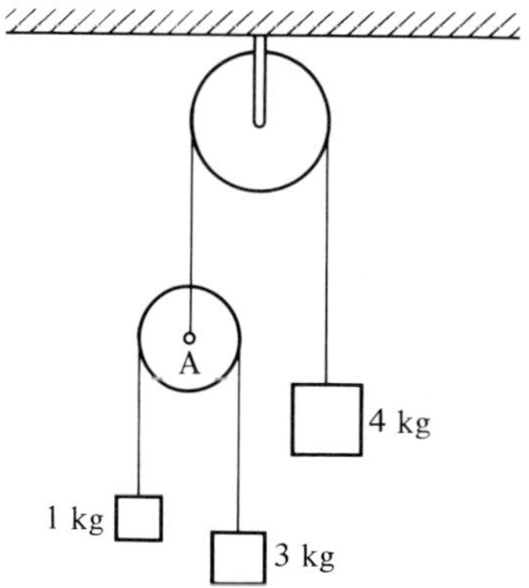

4. A fixed pulley carries a string which has a load of mass 7 kg attached to one end and a light pulley attached to the other end. This light pulley carries another string which has a load of mass 4 kg at one end, and another load of mass 2 kg at the other end. Find the acceleration of the 4 kg mass and the tensions in the strings.

5. A string, with one end fixed, passes under a moveable pulley of mass 8 kg, and over a fixed pulley; the string carries a 5 kg mass at its other end. Find the acceleration of the 5 kg mass and the tension in the string.
6. A string, carrying a particle A at one end, passes over a fixed pulley and has a light pulley attached to its other end. Over this light pulley runs another string carrying particle B at one end and particle C at the other. The masses of A, B and C are $3m$, $2m$ and m respectively. Find the acceleration of A and the tensions in the strings.
7. A string, with one end fixed, passes under a moveable pulley of mass m_1, over a fixed pulley and carries a mass m_2 at its other end. With the system released from rest, show that the tension in the string is $\dfrac{3m_1m_2g}{4m_2+m_1}$ and that, after time t, the moveable pulley has moved a distance $\dfrac{gt^2(2m_2-m_1)}{2(4m_2+m_1)}$.
8. A string, with a particle A attached to one end passes over a fixed pulley, under a moveable pulley B, over another fixed pulley and has a particle C attached to its other end. The masses of A, B and C are $3m$, $4m$ and $4m$ respectively. Find the acceleration of A and the tension in the string.
9. In the pulley system shown in the diagram, A is a heavy pulley which is free to move. Find the mass of pulley A if it does not move upwards or downwards when the system is released from rest.

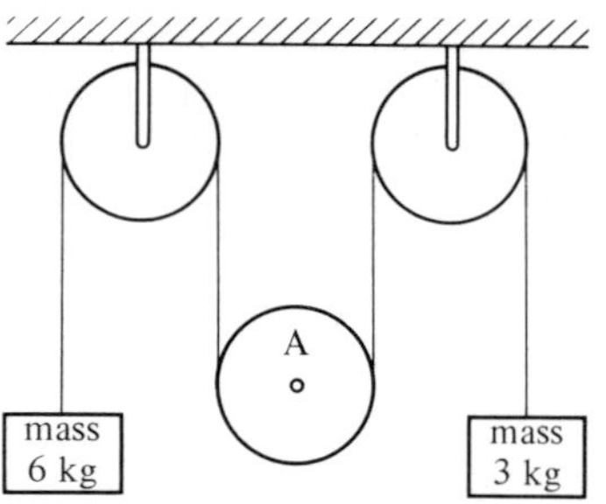

10. In the pulley system shown in the diagram, the pulley A is free to move.
Find the mass of the load B if, when the system is released from rest, pulley A does not move upwards or downwards.

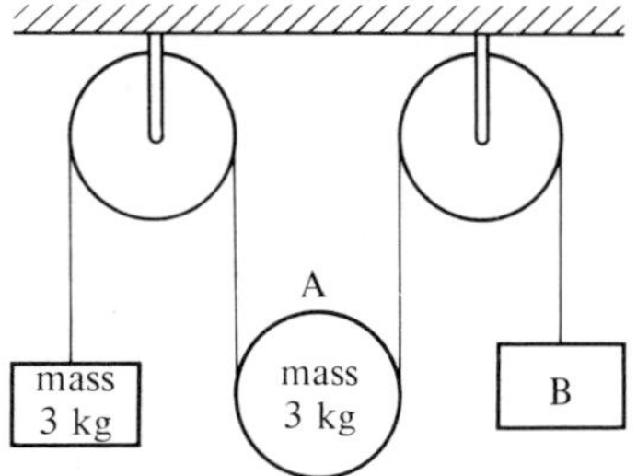

Exercise 3F Examination questions

1. Three coplanar forces act upon a mass of 5 kg. With the usual notation the forces, in newtons, are represented by the vectors $9\mathbf{i}-2\mathbf{j}$, $-3\mathbf{i}+10\mathbf{j}$ and $18\mathbf{i}-\mathbf{j}$. Calculate the resultant acceleration of the mass in magnitude and direction. (A.E.B)
2. A tug tows a barge of mass 10 000 kg with acceleration 0·1 m/s^2. The water resistance to the movement of the barge is 1000 N. Given that the towing rope is horizontal, calculate the tension in this rope.
[You may assume that the acceleration and velocity of the barge are in the direction of the tow rope.] (London)

3. A car of mass 800 kg is pulling a caravan of mass 600 kg by means of a tow-bar along a straight horizontal road. The resistive forces opposing the motions of the car and the caravan are 100 N and 60 N respectively. When the car is accelerating at $\frac{1}{2}$ m/s^2, calculate, in newtons, (a) the tension in the tow-bar, (b) the force being produced by the engine of the car. (London)

4. A particle A, of mass 0·5 kg, is placed on a smooth horizontal table and is connected by a light inextensible string passing over a small smooth pulley at the edge of the table to a particle B, of mass 0·2 kg, hanging freely. Each part of the string is taut and perpendicular to the edge of the table. The system is released from rest. Calculate the tension, in N, in the string when the system is in motion. (London)

5. A load of 400 kg is lifted by a cable through a vertical distance of 48 m. The load moves upwards from rest with a uniform acceleration of 0·5 m/s^2 over the first 36 m and then decelerates uniformly to rest. Take the acceleration due to gravity as 10 m/s^2 and calculate
   (i) the tension in the cable during acceleration,
   (ii) the maximum velocity attained by the load,
   (iii) the tension in the cable during deceleration. (Cambridge)

6. Particles of 30 g and 20 g are attached to the ends of a long light inextensible string which passes over a smooth fixed pulley. The system is released from rest with the particles hanging vertically. When the system has been in motion for 5 s the string is suddenly cut, with the 20 g particle at a point K. Calculate
   (i) the initial acceleration of the system,
   (ii) the velocities of the particles at the instant the string is cut,
   (iii) the height above K to which the 20 g particle rises,
   (iv) the time which elapses, after the string is cut, before the 20 g particle again passes through the point K. (A.E.B)

7. The vertical descent of a lift-cage is undertaken in three stages. During the first stage the lift uniformly accelerates from rest at $5k$ m/s^2, during the second stage it moves at a constant speed of 10 m/s and during the third stage it uniformly retards at $2k$ m/s^2 until it comes to rest.
   (i) Express, in terms of k, the times taken during the first and third stages of the descent.
   (ii) Given that the total distance covered by the lift during the descent is 350 m and that this distance is covered in 40 s, calculate the value of k.

   A tool-box of mass 5 kg was placed on the floor of the lift-cage before the descent.
   (iii) Find, in newtons, the force exerted by the floor of the lift-cage on the tool-box during each stage of the descent.

   (Take the acceleration due to gravity to be 10 m/s^2.) (A.E.B)

8. Two particles A and B, of masses $3m$ and m respectively, are connected by a light inextensible string which passes over a fixed light smooth pulley. The system is released from rest with the string taut and the straight parts of the string vertical. Find the acceleration of either particle and the tension in the string.
   The particles A and B start at the same level. The string breaks after time t_0 and the string is sufficiently long so that B does not hit the pulley. Show that the vertical distance between A and B when B reaches its highest point is $g{t_0}^2$. (London)

4

Resultants and components of forces

In the last chapter, we considered horizontal and vertical forces acting on bodies. However, not all forces act in these directions and we must therefore consider forces acting in any direction.

Resultant of two forces

The resultant R of two forces P and Q is that single force which could completely take the place of the two forces. The resultant R must have the same effect as the two forces P and Q.

When only parallel forces are involved, it is easy to find the resultant.
For example:

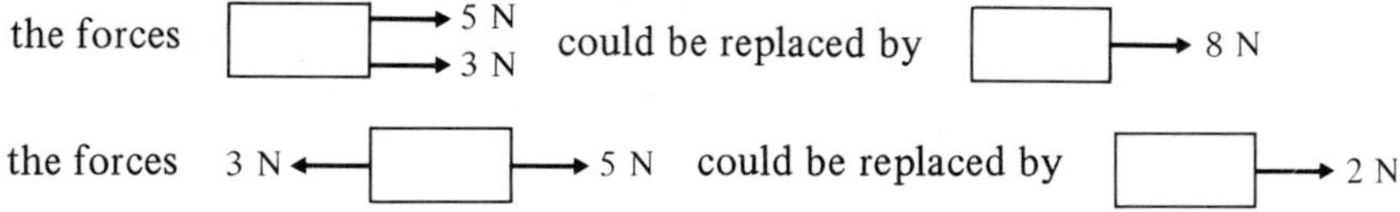

Parallelogram of forces

Two forces P and Q are represented by the line segments AB and AD.

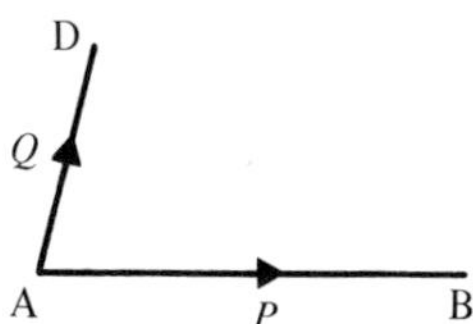

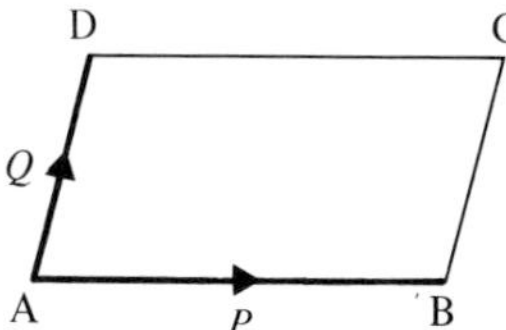

The parallelogram ABCD is completed by drawing BC and DC.

To find the resultant of the forces P and Q , we have to consider:

$$\overrightarrow{AB} + \overrightarrow{AD}$$

but $\overrightarrow{AD} = \overrightarrow{BC}$ as these are equivalent vectors

$$\therefore \quad \overrightarrow{AB} + \overrightarrow{AD} = \overrightarrow{AB} + \overrightarrow{BC}$$
$$= \overrightarrow{AC}$$

Hence the resultant of the two forces P and Q, which are represented by the line segments AB and AD, is fully represented by the line segment AC. This is the diagonal AC of the parallelogram ABCD, which is therefore referred to as a parallelogram of forces.

Example 1

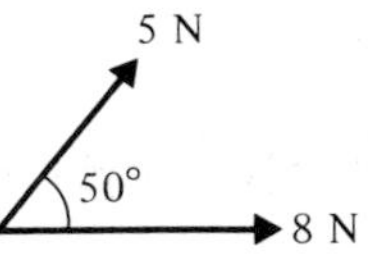

Find, by scale drawing, the magnitude of the resultant of the two forces shown in the sketch. Find also the angle that the resultant makes with the larger force.

Construct parallelogram ABCD with AB = 8 cm, AD = 5 cm and angle DAB = 50°.

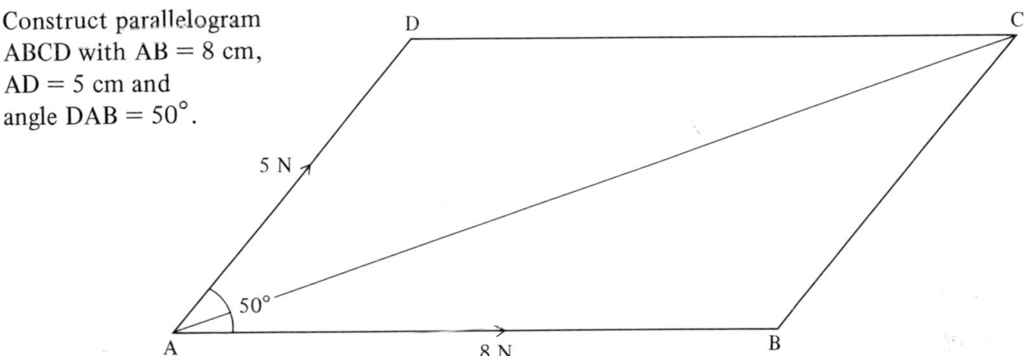

By measurement, AC = 11·9 cm and angle CAB = 19°
The resultant is 11·9 N and makes an angle of 19° with the larger force.

It should be noted that, in the last example, the magnitude and direction of the resultant could have been obtained by considering the triangle ABC, rather than the whole parallelogram. Thus the resultant of two forces, which are represented in magnitude and direction by the sides AB and BC of the triangle ABC, is fully represented by the side AC of the triangle.

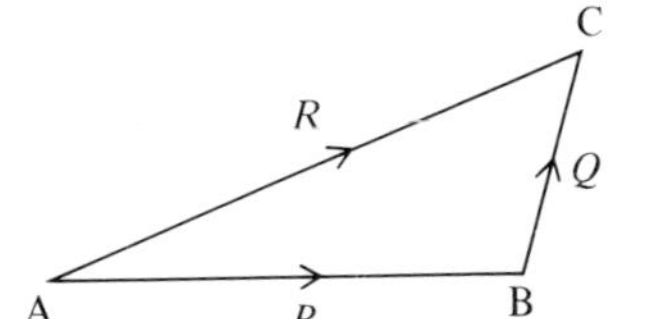

$\overrightarrow{AB} + \overrightarrow{BC} = \overrightarrow{AC}$

Note that the forces to be added are in the same sense around the triangle, in this case, anticlockwise.

From this triangle we can find the resultant, either by a scale drawing or by calculation.

Example 2

Find the magnitude of the resultant of the forces shown in the sketch, and the angle that the resultant makes with the larger force.

By scale drawing

Construct triangle ABC with AB = 7 cm, BC = 4 cm and angle ABC = 180° − 25° = 155°

By measurement, AC = 10·8 cm and angle CAB = 9°
The resultant is 10·8 N making an angle of 9° with the larger force.

By calculation

First, make a rough sketch.

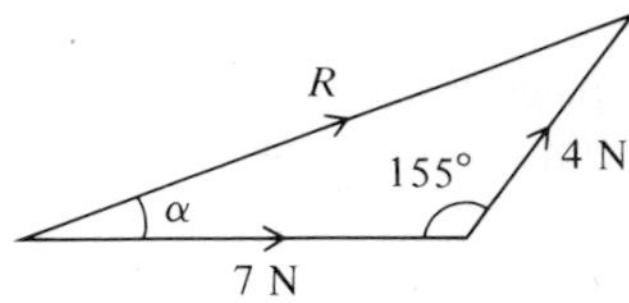

By the cosine rule $R^2 = 4^2 + 7^2 - 2 \times 4 \times 7 \cos 155$

$= 16 + 49 + 56 \cos 25$ since $\cos 155 = -\cos 25$

$= 65 + 50{\cdot}75$

$\therefore \quad R = 10{\cdot}75$ N

By the sine rule $\dfrac{4}{\sin \alpha} = \dfrac{10{\cdot}75}{\sin 155}$

$\therefore \quad \sin \alpha = \dfrac{4 \sin 155}{10{\cdot}75}$

$\therefore \quad \alpha = 9{\cdot}05°$

The resultant is 10·8 N making an angle of 9·05° with the larger force.

Angle between forces

In examples where a diagram is not given, it is necessary to interpret carefully the directions of the given forces.

If the angle between the forces is given as 35°, this should be interpreted as shown in the diagram.

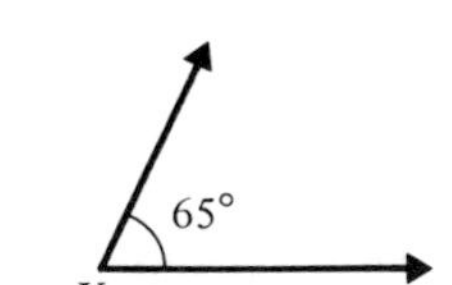

If it is stated that two forces act away from the point X and make an angle of 65° with each other, this should be interpreted as shown in the diagram.

65°
X

Example 3

Two forces of 7 N and 24 N act away from the point A and make an angle of 90° with each other. Find the magnitude and direction of their resultant.

First make a rough sketch.

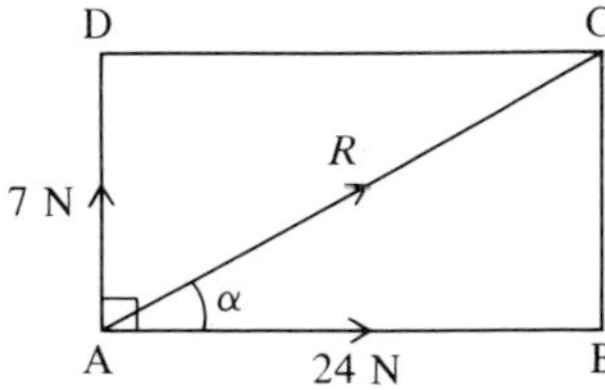

From triangle ABC: $R^2 = 7^2 + 24^2$ (Pythagoras)

$R^2 = 625$

$R = 25$ N

and $\tan \alpha = \dfrac{7}{24}$

$\therefore \quad \alpha = 16{\cdot}26°$

The resultant is 25 N making an angle of 16·26° with the 24 N force.

Example 4

Find the angle between a force of 7 N and a force of 4 N if their resultant has a magnitude of 9 N.

First make a rough sketch.

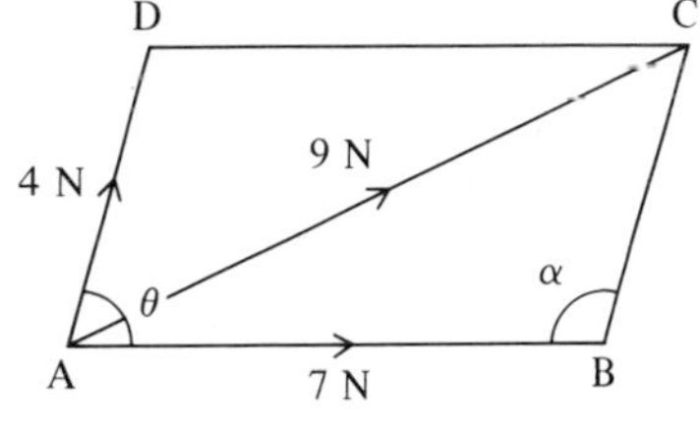

From triangle ABC, using the cosine rule:

$$9^2 = 7^2 + 4^2 - 2 \times 7 \times 4 \cos\alpha$$

$$\therefore \quad \cos\alpha = \frac{7^2 + 4^2 - 9^2}{2 \times 7 \times 4}$$

$$= -\frac{2}{7}$$

$$\text{hence} \quad \alpha = 106{\cdot}60^\circ$$

$$\therefore \quad \theta = 180^\circ - 106{\cdot}60^\circ$$

$$= 73{\cdot}40^\circ$$

The angle between the given forces is $73{\cdot}40^\circ$.

Resultant of any number of forces

It is now known that the resultant R of any two forces P and Q can be found by constructing the triangle shown:

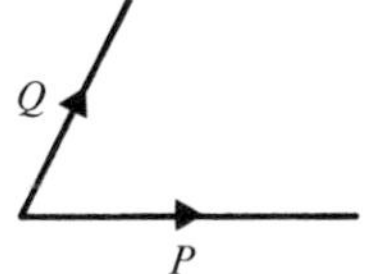

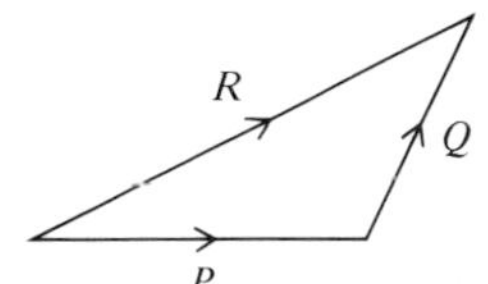

This method can be extended to find the resultant of any number of forces.
Consider the forces $P, Q, S, \ldots$
The forces P and Q can be combined and the resultant R_1 of these two forces can be then combined with the force S to find the resultant R_2 and so on.

Instead of drawing separate triangles, the forces can simply be added, paying due regard to their direction, and will form a polygon.

Thus to find the resultant R of the forces P, Q and S shown below, make an accurate scale drawing with the forces to be added following in the same sense around the figure:

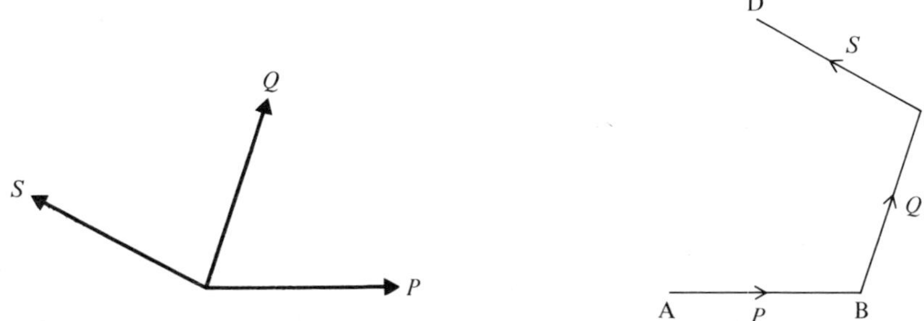

The line segment AD will then completely represent the resultant R.

Example 5

Forces of 6 N, 3 N and 4 N act as shown in the diagram. Find graphically the magnitude and the direction of the resultant of these forces.

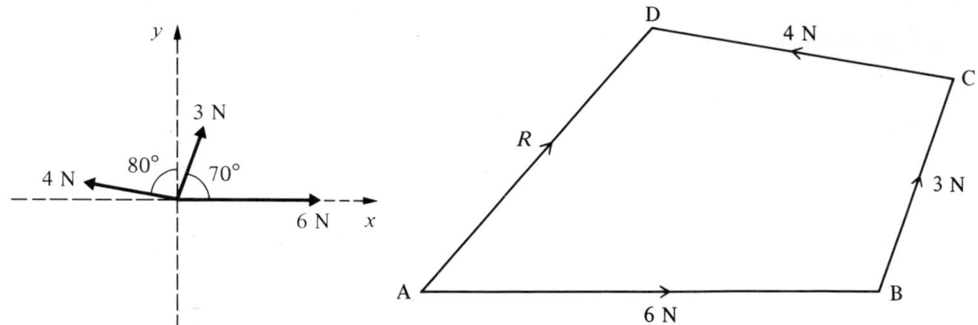

A line AB is drawn 6 cm in length and parallel to the force of 6 N.
A line BC is drawn 3 cm in length and parallel to the force of 3 N.
A line CD is drawn 4 cm in length and parallel to the force of 4 N.

Care is needed to ensure that each line is drawn in the correct direction, i.e. in the same direction as the force the line represents, and that the forces follow in the same sense around the polygon.

The polygon is completed by drawing the line AD which will represent the resultant in magnitude and direction.

By measurement, AD = 4·7 cm and angle DAB = 49°

The resultant is 4·7 N and makes an angle of 49° with the x-axis.

Example 6

Two forces of 5 N and 8 N act away from the point A and make an angle of 40° with each other. Find the angle which the resultant makes with the larger force.

Make a rough sketch and complete the parallelogram ABCD.

Note that, in this case, the magnitude of the resultant is not required.

Let N be the foot of the perpendicular from C to AB produced.

From the triangle BCN: $\quad BN = 5\cos 40 \qquad CN = 5\sin 40$

From the triangle ACN:

$$\tan\alpha = \frac{CN}{AN} = \frac{CN}{AB + BN} = \frac{5\sin 40}{8 + 5\cos 40}$$

$$\therefore \quad \alpha = 15{\cdot}20^\circ$$

The angle between the resultant and the larger force is 15·20°.
The angle α could have been found by first finding the magnitude of the resultant, and then using the sine rule in triangle ABC. The method of Example 6 avoids errors which could arise due to the incorrect determination of the resultant.

Exercise 4A

1. In each of the following diagrams, two forces are shown. Find, by scale drawing, the magnitude of their resultant and the angle it makes with the larger of the two forces.

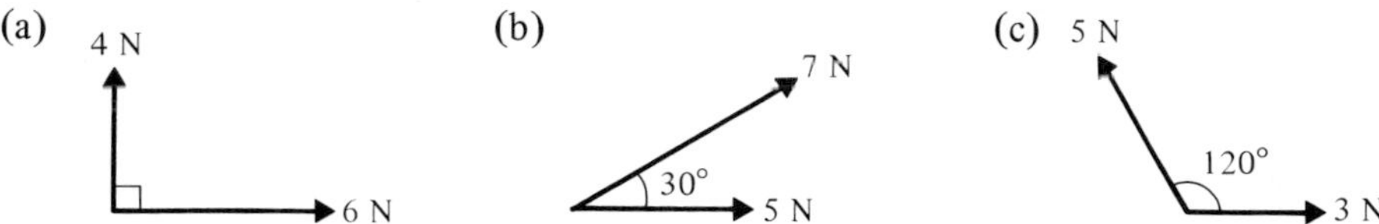

2. In each of the following diagrams, two forces are shown. Find, by calculation, the magnitude of their resultant and the angle it makes with the larger of the two forces.

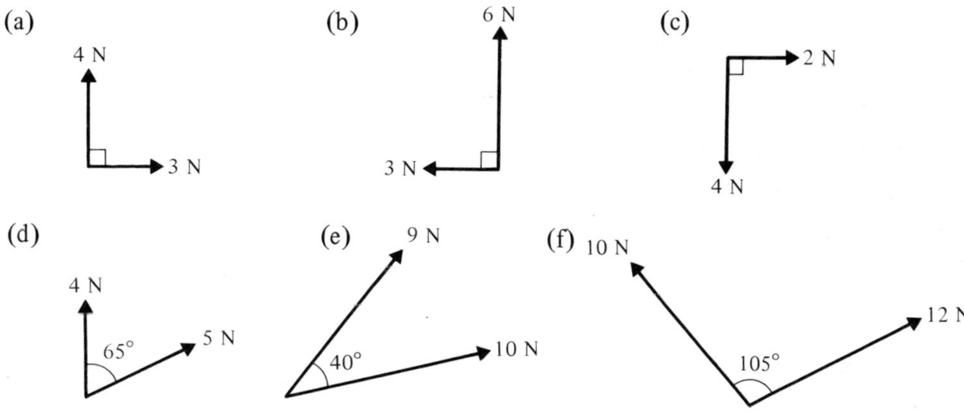

3. Find the magnitude and direction of the resultant of forces of 8 N and 3 N if the angle between the two forces is (a) 60° (b) 50° (c) 160°.
4. Forces of 3 N and 2 N act along OA and OB respectively, the direction of the forces being indicated by the order of the letters. If $A\hat{O}B = 150°$, find the magnitude of the resultant of the two forces and the angle it makes with OA.
5. Forces of 6 N and 4 N act along OA and BO respectively, the direction of the forces being indicated by the order of the letters. If $A\hat{O}B = 60°$, find the magnitude of the resultant of the two forces and the angle it makes with OA.
6. Find the angle between a force of 6 N and a force of 5 N given that their resultant has magnitude 9 N.
7. Find the angle between a force of 10 N and a force of 4 N given that their resultant has magnitude 8 N.
8. The angle between a force of 6 N and a force of X N is 90°. If the resultant of the two forces has magnitude 8 N find the value of X.
9. A force F N acts along $\overrightarrow{AB}$ and a force $2F$ N acts along $\overrightarrow{AC}$. If $B\hat{A}C = 60°$, find the magnitude of the resultant and the angle it makes with AB.
10. The angle between a force of P N and a force of 3 N is 120°. If the resultant of the two forces has magnitude 7 N, find the value of P.
11. The angle between a force of Q N and a force of 8 N is 45°. If the resultant of the two forces has magnitude 15 N find the value of Q.

12. Each of the following diagrams shows a number of forces. Find, by scale drawing, the magnitude of their resultant and the angle it makes with the x-axis.

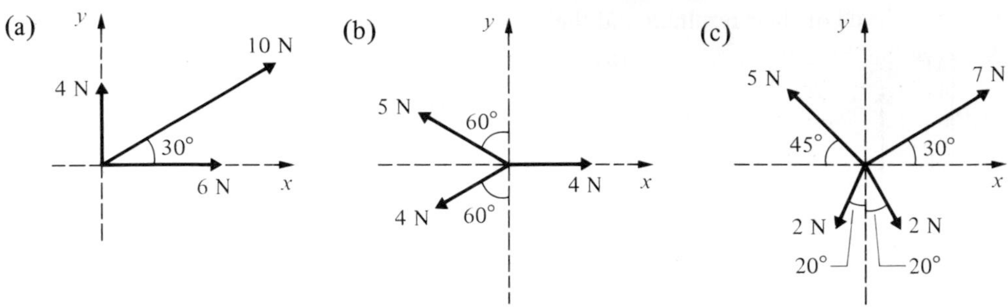

13. Find, by drawing, the magnitude and direction of the resultant of forces 5 N, 6 N, 3 N and 1 N in directions north, north-east, south-west and west respectively.
14. Find, by drawing, the magnitude and direction of the resultant of forces 5 N, 7 N, 6 N and 4 N acting in directions 050°, 100°, 200° and 310° respectively.
15. ABCD is a square. Forces of 4 N, 3 N, 2 N and 5 N act along the sides AB, BC, CD and AD respectively, in the directions indicated by the order of the letters. Find, by drawing, the magnitude of the resultant and the angle it makes with AB.
16. ABC is an equilateral triangle. Forces of 4 N, 4 N and 6 N act along the sides AB, BC and AC respectively, in the directions indicated by the order of the letters. Find, by drawing, the magnitude of the resultant and the angle it makes with AB.
17. A concrete block is pulled by two horizontal ropes. One rope has a tension of 500 N and is in a direction 050° and the other rope has a tension of 350 N and is in a direction 350°. Find the magnitude and direction of the resultant pull on the block.

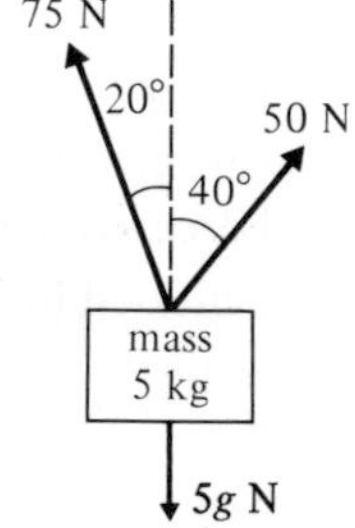

18. A body of mass 5 kg is being raised by forces of 75 N and 50 N as shown in the diagram. Find, by drawing, the magnitude of the resultant of the three forces acting on the body, and find the angle this resultant makes with the upward vertical.

19. Two forces have magnitudes P and Q and the angle between them is θ. If the resultant of these two forces has magnitude R, and makes an angle α with the force P, show that

(a) $R^2 = P^2 + Q^2 + 2PQ\cos\theta$ (b) $\tan\alpha = \dfrac{Q\sin\theta}{P + Q\cos\theta}$.

If $P = Q$ and $\theta = 40°$ find α.

Components

It has been seen that two forces can be combined into a single force which is called their resultant.

There is the reverse process which consists of expressing a single force in terms of its two *components*. These components are sometimes referred to as the *resolved parts* of the force.

It is particularly useful to find two mutually perpendicular components of a force.

The direction of the two components may, for example, be horizontal and vertical, or parallel and at right angles to the surface of an inclined plane.

Definition

The component of the force F in any given direction is a measure of the effect of the force F in that direction.

Suppose the force F acts at an angle θ to the x-axis as shown in the diagram. Let ON represent the force F and the angle NXO = 90°. Then OX and OY represent the horizontal and vertical components of F, along the x and y axes.

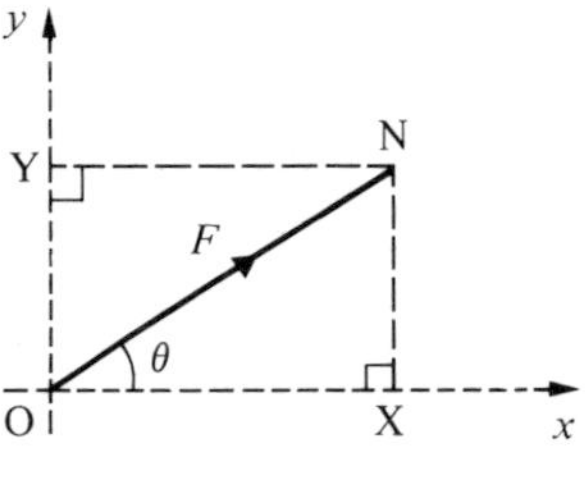

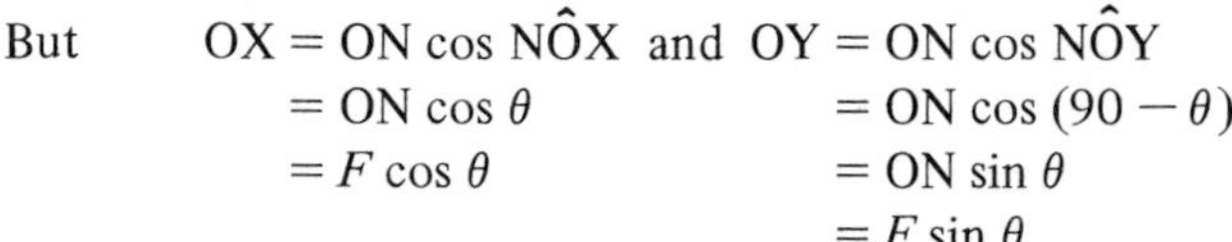

But $\quad OX = ON\cos N\hat{O}X$ and $OY = ON\cos N\hat{O}Y$

$$OX = ON\cos\theta = F\cos\theta$$

$$OY = ON\cos(90-\theta) = ON\sin\theta = F\sin\theta$$

Hence the components are $F\cos\theta$ and $F\sin\theta$ along the x and y axes respectively.
The rule for finding the components may be stated as:

> The component of a force in any direction is the product of the magnitude of the force and the cosine of the angle between the force and the required direction.

The components in two mutually perpendicular directions are then always $F\cos\theta$ and $F\cos(90-\theta)$ or as these are more usually written: $F\cos\theta$ and $F\sin\theta$.

It is important to remember that, when a force F has been resolved into its components in two mutually perpendicular directions, the force F is the resultant of these two components.

Example 7

Find the components of the given forces, in the direction of (i) the x-axis, (ii) the y-axis.

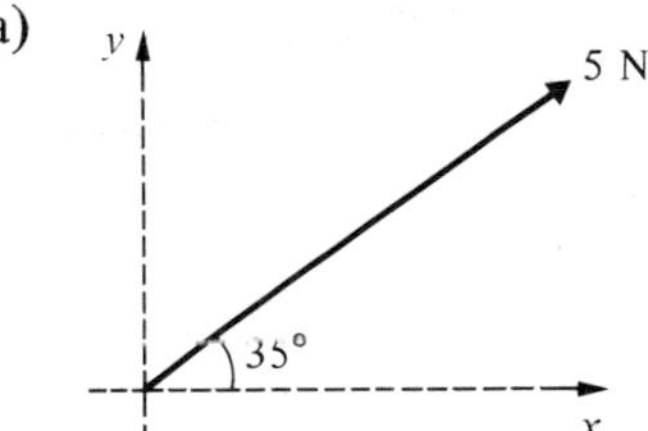

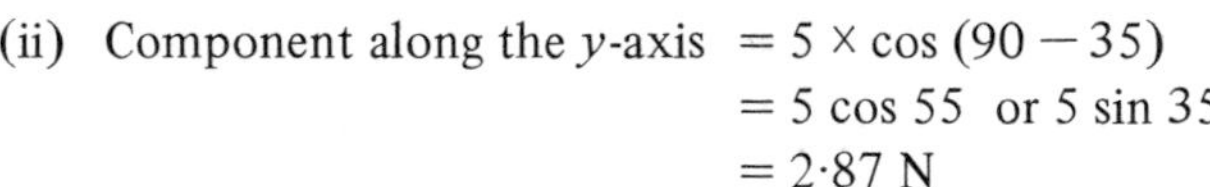

(i) Component along the x-axis $= 5\times\cos 35 = 4{\cdot}10$ N

(ii) Component along the y-axis $= 5\times\cos(90-35) = 5\cos 55$ or $5\sin 35 = 2{\cdot}87$ N

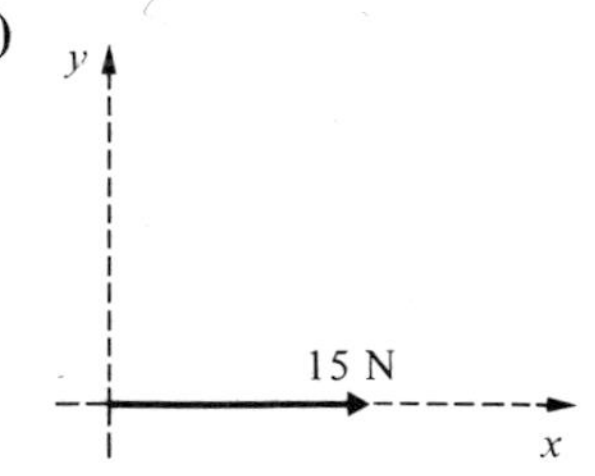

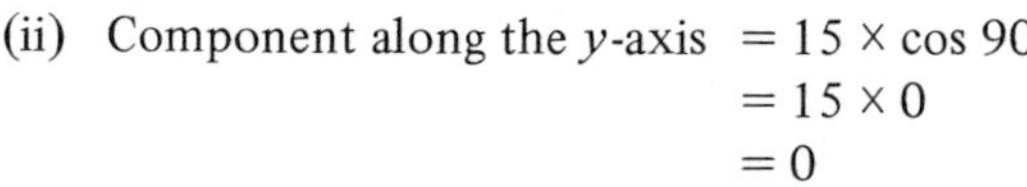

(i) Component along the x-axis $= 15\times\cos 0 = 15\times 1 = 15$ N

(ii) Component along the y-axis $= 15\times\cos 90 = 15\times 0 = 0$

It is seen in this last case that the component in the direction of the y-axis is zero. This agrees with our experience that a force has no effect in a direction at right angles to its line of action. Since the force acts along the x-axis, its component in that direction will be equal to the whole force, 15 N.

Example 8

Express each of the following forces in the form $(a\mathbf{i} + b\mathbf{j})$.

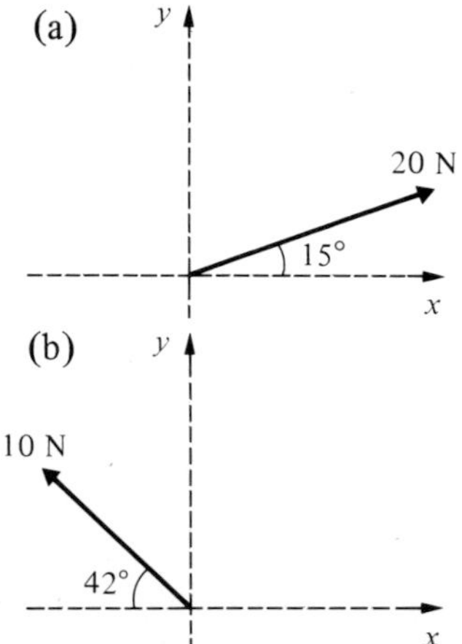

Remembering that **i** and **j** are the unit vectors in the directions of the x-axis and y-axis respectively:

(a) component along x-axis $= 20 \cos 15 = 19{\cdot}3$
component along y-axis $= 20 \sin 15 = 5{\cdot}18$

Form required: $(19{\cdot}3\mathbf{i} + 5{\cdot}18\mathbf{j})$ N.

(b) component along x-axis $= -10 \cos 42 = -7{\cdot}43$
component along y-axis $= \quad 10 \sin 42 = 6{\cdot}69$

Form required: $(-7{\cdot}43\mathbf{i} + 6{\cdot}69\mathbf{j})$ N.

Example 9

A body of mass 4 kg rests on an incline of 35°. Find the component of the weight of the body in each of the directions (i) down the plane, (ii) at right angles to the plane.

(i) Component down the plane

$= \text{force} \times \cos(\text{angle between force and plane})$
$= 4g \times \cos(90 - 35)$
$= 4g \times \sin 35$
$= 22{\cdot}5$ N

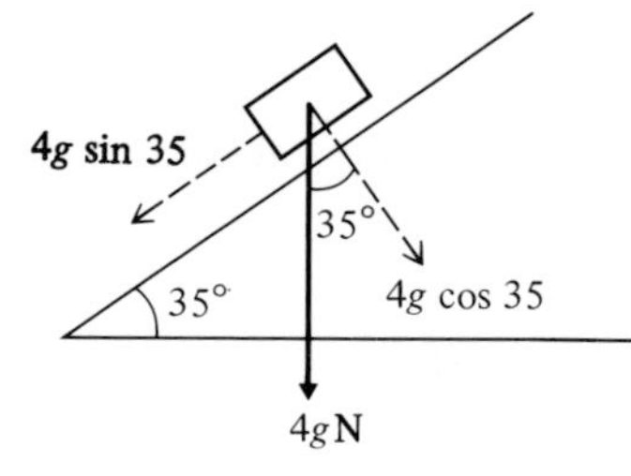

(ii) Component at right angles to plane

$= 4g \times \cos 35$
$= 4g \cos 35$ in the direction shown in the diagram
$= 32{\cdot}1$ N

Example 10

Find the sum of the components of the given forces in the direction of (i) x-axis, (ii) y-axis.

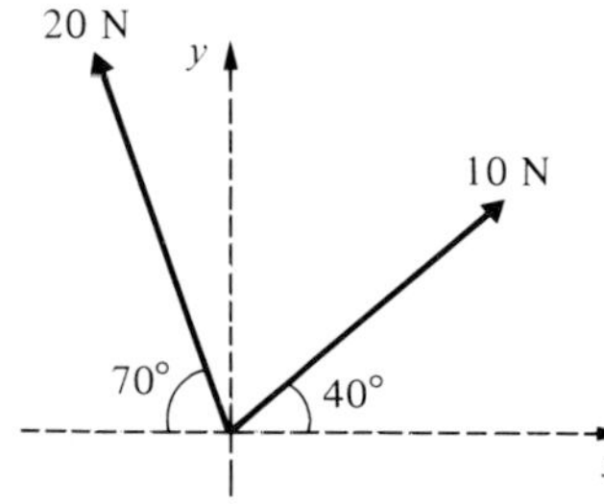

(i) resolving along the x-axis

$$10 \cos 40 - 20 \cos 70 = 7{\cdot}66 - 6{\cdot}84 = 0{\cdot}82 \text{ N}$$

(ii) resolving along the y-axis

$$10 \sin 40 + 20 \sin 70 = 6{\cdot}43 + 18{\cdot}79 = 25{\cdot}22 \text{ N}$$

This example illustrates that when there are a number of forces acting, their components in a particular direction can be added together, due regard being given to the directions of the components.

Exercise 4B

1. For each of the forces shown below, find the components in the direction of (i) the x-axis and (ii) the y-axis.

(a) y 8 N 30° x

(b)

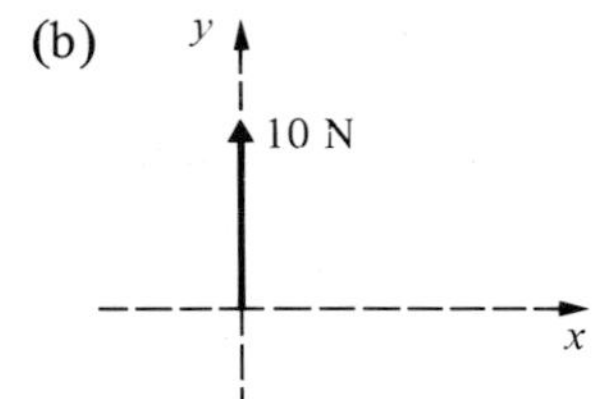

(c)

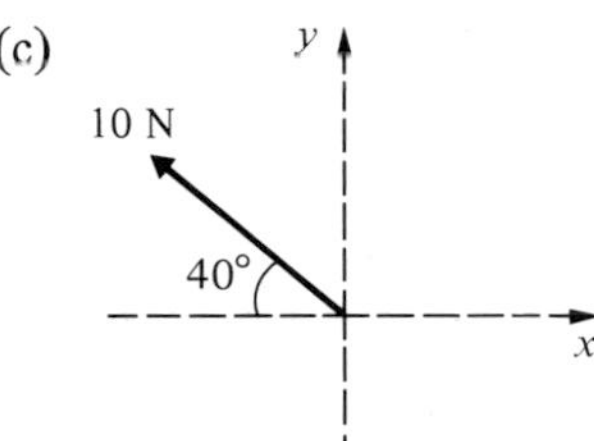

(d) y 45° x 4 N

(e)

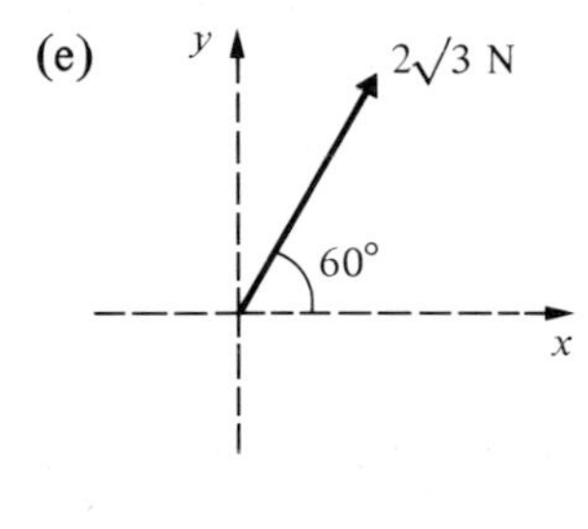

(f)

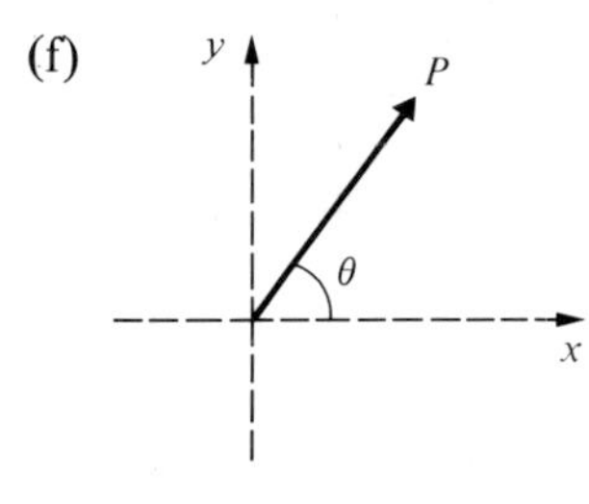

2. Express each of the following forces in the form $a\mathbf{i} + b\mathbf{j}$.

(a) y 16 N 60° x

(b)

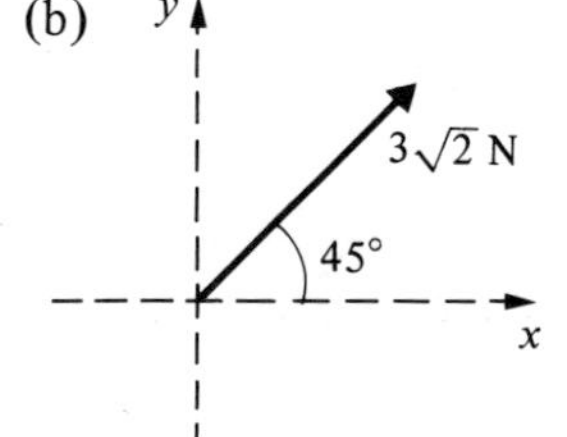

(c)

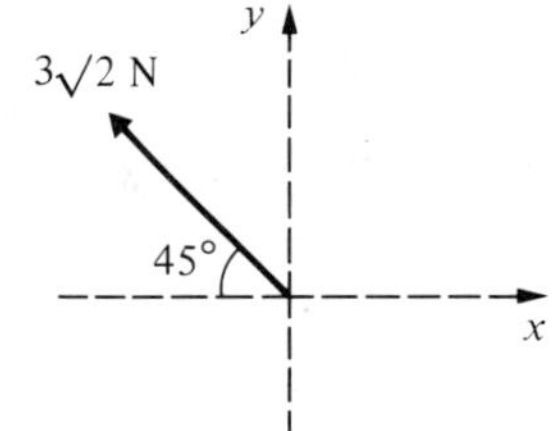

(d)

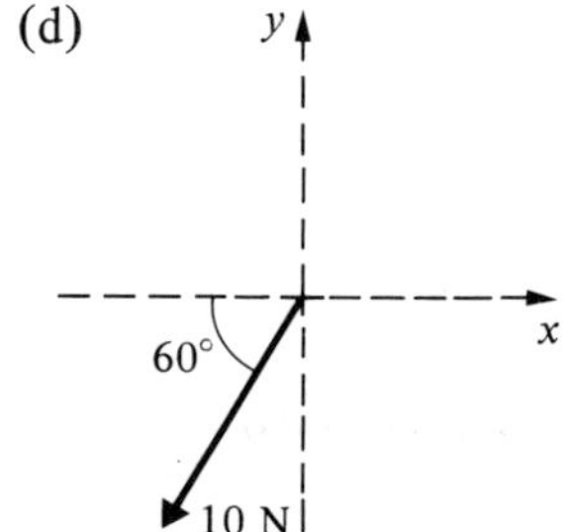

(e)

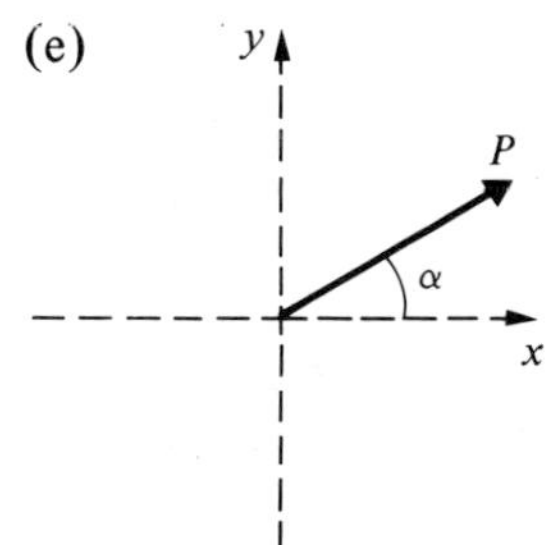

(f)

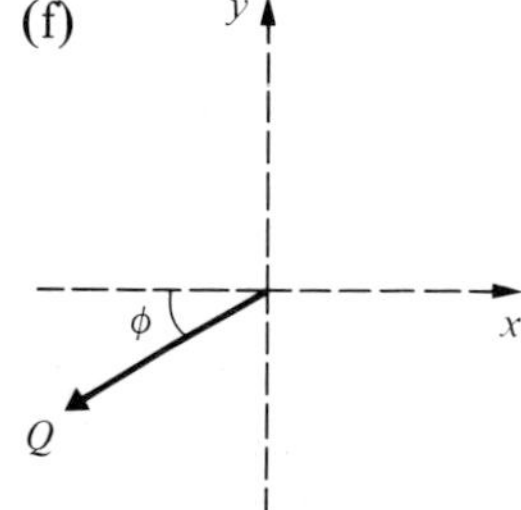

3. Each of the following diagrams shows a body of weight 10 N on an incline. In each case find the component of the weight of the body (i) in the Ox direction and (ii) in the Oy direction.

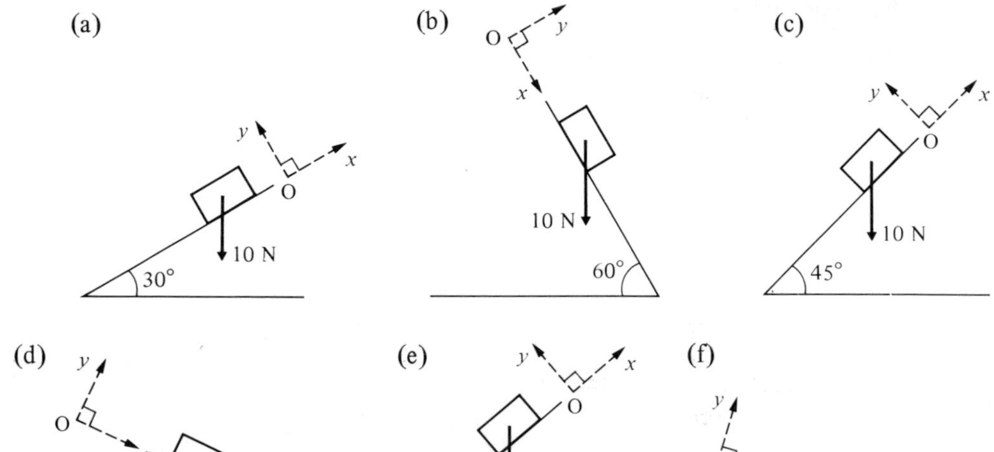

4. For each of the following systems of forces, find the sum of the components in the direction of (i) the x-axis and (ii) the y-axis.

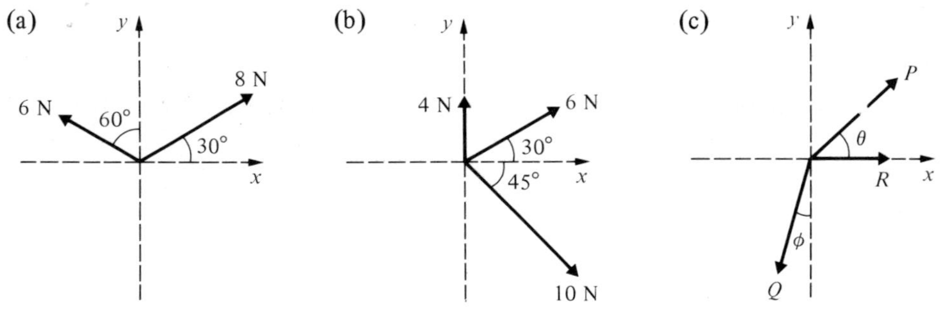

5. For each of the following systems of forces, find the sum of the components (i) in the Ox direction and (ii) in the Oy direction.

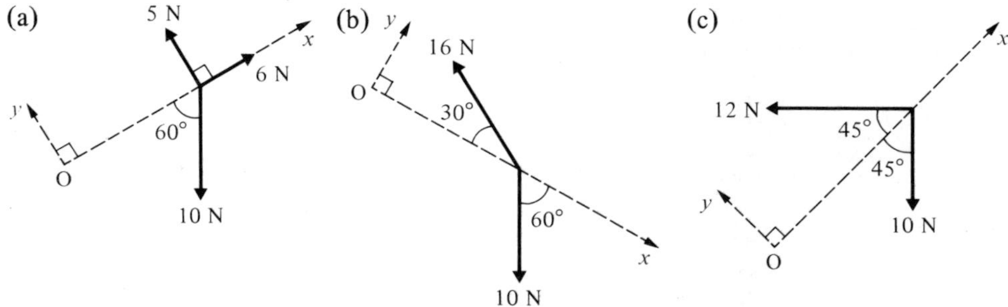

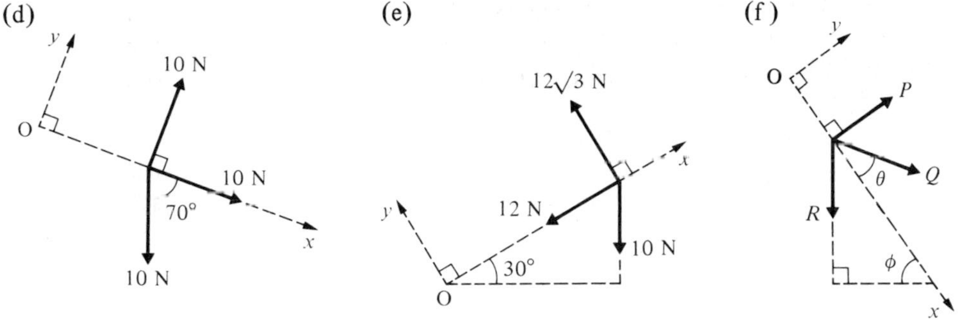

Resultant from sum of components

A number of forces, all of which lie in one plane, are said to be coplanar.

In a given system of coplanar forces, it is possible to choose two mutually perpendicular directions and find the components of all the forces in these two directions. By finding the algebraic sum of these components the resultant of the system can be found in both magnitude and direction.

If the forces are expressed in terms of the unit vectors **i** and **j**, the components of the forces are immediately known.

Example 11

Find the resultant of the following forces, giving the answer in the form $a\mathbf{i} + b\mathbf{j}$:

$(2\mathbf{i} + 4\mathbf{j})$ N, $(5\mathbf{i} - 7\mathbf{j})$ N and $(-2\mathbf{i} - \mathbf{j})$ N.

Resultant $= 2\mathbf{i} + 4\mathbf{j} + 5\mathbf{i} - 7\mathbf{j} - 2\mathbf{i} - \mathbf{j}$
$= (5\mathbf{i} - 4\mathbf{j})$ N

If the magnitude of the resultant is required:

magnitude $= \sqrt{(5^2 + 4^2)}$
$= \sqrt{41}$
$= 6{\cdot}40$ N

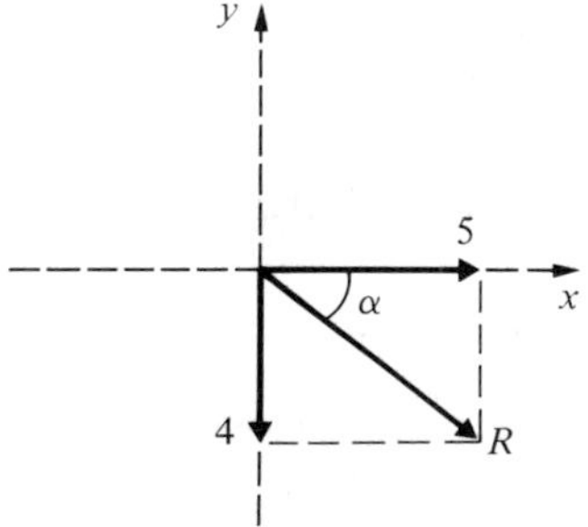

The direction is given by $\tan\alpha = \frac{4}{5}$ where α is the angle below the x-axis.

Hence $\alpha = 38{\cdot}66°$; the angle with the x-axis is $-38{\cdot}66°$

The resultant is $(5\mathbf{i} - 4\mathbf{j})$ N at an angle of $38{\cdot}66°$ below the x-axis.

Example 12

Find the resultant of the given forces, by finding the components of the forces in the direction of the x and y axes.

Components in direction of x-axis

$$8\sqrt{3}\cos 30 - 3\sqrt{2}\cos 45$$

$$= 8\sqrt{3} \times \frac{\sqrt{3}}{2} - 3\sqrt{2} \times \frac{1}{\sqrt{2}}$$

$$= 12 - 3$$

$$= 9 \text{ N}$$

Components in direction of y-axis

$$8\sqrt{3}\sin 30 + 3\sqrt{2}\sin 45 - 2\sqrt{3}$$

$$= 8\sqrt{3} \times \tfrac{1}{2} + 3\sqrt{2} \times \frac{1}{\sqrt{2}} - 2\sqrt{3}$$

$$= (2\sqrt{3} + 3) \text{ N}$$

$$\therefore \quad R^2 = 9^2 + (2\sqrt{3} + 3)^2$$

$\therefore \quad R = 11{\cdot}1$ N and the direction is at an angle α to the x-axis where

$$\tan\alpha = \frac{2\sqrt{3} + 3}{9}$$

$$\therefore \quad \alpha = 35{\cdot}69^\circ$$

The resultant is $11{\cdot}1$ N at an angle of $35{\cdot}69^\circ$ above the x-axis.

Example 13

ABCD is a rectangle. Forces of 9 N, 8 N and 3 N act along the lines DC, CB and BA respectively, in the directions indicated by the order of the letters.
Find the magnitude of the resultant and the angle it makes with DC.

Draw a diagram showing the forces

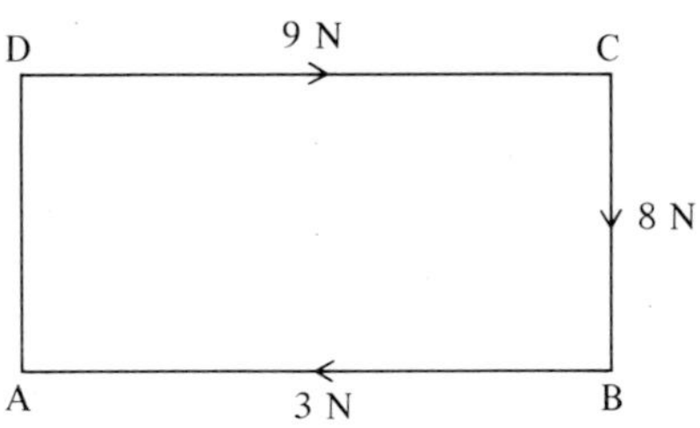

Resolving parallel to DC $\quad 9 - 3$

$$= 6 \text{ N}$$

Resolving parallel to CB $\quad$ 8 N

$$\therefore \quad R^2 = 6^2 + 8^2$$

$$= 36 + 64$$

$$\therefore \quad R = 10 \text{ N}$$

Drawing a diagram to show the two components the direction is seen to be given by:

$$\tan\alpha = \tfrac{8}{6}$$

$$\therefore \quad \alpha = 53{\cdot}13^\circ$$

The resultant is 10 N making an angle $53{\cdot}13^\circ$ with DC.

Exercise 4C

1. Find the resultant of each of the following sets of forces giving your answers in the form $a\mathbf{i} + b\mathbf{j}$.
   (a) $(2\mathbf{i} + 4\mathbf{j})$ N, $(3\mathbf{i} - 4\mathbf{j})$ N.
   (b) $(3\mathbf{i} - 5\mathbf{j})$ N, $(2\mathbf{i} - 5\mathbf{j})$ N, $(-\mathbf{i} + 7\mathbf{j})$ N.
   (c) $(6\mathbf{i} + 2\mathbf{j})$ N, $(-5\mathbf{i} + \mathbf{j})$ N, $(3\mathbf{i} - 3\mathbf{j})$ N.
   (d) $(2\mathbf{i} + 4\mathbf{j})$ N, $(3\mathbf{i} - 5\mathbf{j})$ N, $(6\mathbf{i} + 2\mathbf{j})$ N, $(-7\mathbf{i} - 7\mathbf{j})$ N.
2. The resultant of the forces $(5\mathbf{i} - 2\mathbf{j})$ N, $(7\mathbf{i} + 4\mathbf{j})$ N, $(a\mathbf{i} + b\mathbf{j})$ N and $(-3\mathbf{i} + 2\mathbf{j})$ N is a force $(5\mathbf{i} + 5\mathbf{j})$ N. Find a and b.
3. The resultant of the forces $(5\mathbf{i} + 7\mathbf{j})$ N, $(a\mathbf{i} + b\mathbf{j})$ N and $(b\mathbf{i} - a\mathbf{j})$ N is a force $(11\mathbf{i} + 5\mathbf{j})$ N. Find a and b.
4. Find the magnitude of the force $(4\mathbf{i} + 3\mathbf{j})$ N and the angle it makes with the direction of **i**.
5. Find the magnitude of the force $(-2\mathbf{i} + 4\mathbf{j})$ N and the angle it makes with the direction of **i**.
6. Find the magnitude of the resultant of each of the following sets of forces and state the angle that this resultant makes with the direction of **i**.
   (a) $(2\mathbf{i} + 3\mathbf{j})$ N, $(5\mathbf{i} - 2\mathbf{j})$ N, $(-3\mathbf{i} + 3\mathbf{j})$ N.
   (b) $(-2\mathbf{i} + 5\mathbf{j})$ N, $(\mathbf{i} + 2\mathbf{j})$ N.
   (c) $(4\mathbf{i} + 3\mathbf{j})$ N, $(-\mathbf{i} - 5\mathbf{j})$ N.
   (d) $(2\mathbf{i} + 4\mathbf{j})$ N, $(-6\mathbf{i} - 5\mathbf{j})$ N, $(2\mathbf{i} + \mathbf{j})$ N.
7. For each of the following systems of forces find the resultant in the form $a\mathbf{i} + b\mathbf{j}$. Hence find the magnitude of the resultant and the angle it makes with the x-axis.

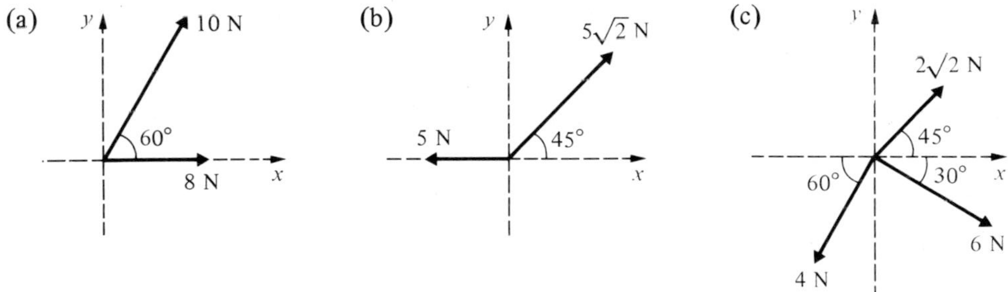

8. A sledge is being pulled across a horizontal surface by forces of $(6\mathbf{i} + 2\mathbf{j})$ N and $(4\mathbf{i} - 3\mathbf{j})$ N. What is the magnitude of the resultant pull on the sledge and what angle does this resultant make with the direction of **i**?
9. Find, by calculation, the resultant of forces of 5 N, 7 N, 8 N and 5 N acting in directions north, north-east, west and north-west respectively, giving your answer in the form $a\mathbf{i} + b\mathbf{j}$. (Take **i** as a unit vector due east and **j** as a unit vector due north).
10. Find, by calculation, the magnitude and direction of the resultant of forces of 10 N, 15 N and 8 N acting in directions 030°, 150° and 225° respectively.
11. ABCD is a rectangle. Forces of 3 N, 4 N and 1 N act along AB, BC and DC respectively, in the directions indicated by the order of the letters. By resolving in two mutually perpendicular directions find the magnitude of the resultant and the angle it makes with AB.
12. ABCD is a rectangle. Forces of $6\sqrt{3}$ N, 2 N and $4\sqrt{3}$ N act along AB, CB and CD respectively, in the directions indicated by the order of the letters. By resolving in two mutually perpendicular directions find the magnitude of the resultant and the angle it makes with AB.
13. ABCD is a rectangle. Forces of 8 N, 4 N, 10 N and 2 N act along AB, CB, CD and AD respectively, in the directions indicated by the order of the letters. Find the magnitude and direction of the resultant.

14. ABC is an equilateral triangle. Forces of 12 N, 10 N and 10 N act along AB, BC and CA respectively, the direction of the forces being indicated by the order of the letters. Find the magnitude and direction of the resultant.

15. ABCD is a rectangle with AB = 4 m and BC = 3 m. Forces of 3 N, 1 N and 10 N act along AB, DC and AC respectively, in the directions indicated by the order of the letters. Find the magnitude and direction of the resultant.

16. ABC is an equilateral triangle. Forces of 10 N act along AB, BC and AC in the directions indicated by the order of the letters. Find the magnitude of the resultant and the angle it makes with AB.

Exercise 4D Examination questions

1. **F** is a force acting at O of magnitude 10 N on a bearing of 060° while **G** is a force acting at O which has components 10 N due north and 5 N due east.
   Calculate the magnitude and direction of (a) **F** + **G** and (b) **F** − **G**. (Oxford)

2. Forces of 5 N acting due north, 6 N acting to the south-west and 8 N acting due west are represented by the vectors **p**, **q** and **r** respectively. Find, graphically, the magnitude and the direction of the resultant forces represented by (i) **p** + **q** + **r**, (ii) 2**p** + **q** − **r**. (A.E.B)

3. Two forces, of magnitude P and Q newtons, have a resultant of $2\sqrt{7}$ N when the angle between their lines of action is 90°. When the angle between the lines of action of the forces is 30° the resultant is of magnitude $2\sqrt{13}$ N. Calculate P and Q. (A.E.B)

4. Forces of 5 N, 9 N, 7 N act along the sides AB, BC, CA respectively of an equilateral triangle ABC in the directions indicated by the order of the letters. Find their resultant in magnitude and direction. (S.U.J.B)

5. The resultant of a force $2P$ N in a direction 060° and a force 10 N in a direction 180° is a force of $P\sqrt{3}$ N. Calculate the value of P and the direction of the resultant.
   A third force of 25 N, concurrent with the other two and in the same plane, is added so that the resultant of the system is in the direction 180°. Find the direction in which the third force is applied and find the magnitude of the resultant. (Cambridge)

6. In the diagram ABCD is a square of side 4 cm. The points E, F, G, H and J lie in the sides AB, BC, CD, CD and DA respectively in such a way that AE = BF = CG = HD = DJ = 1 cm.

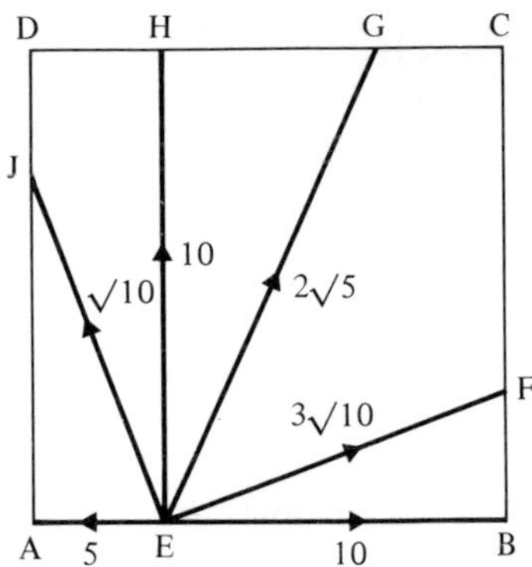

Forces of magnitude 10 N, $3\sqrt{10}$ N, $2\sqrt{5}$ N, 10 N, $\sqrt{10}$ N and 5 N act at the point E in the directions EB, EF, EG, EH, EJ and EA respectively. Calculate the magnitude of the resultant of these forces and show that its line of action makes an acute angle θ with EB where $\tan\theta = \frac{4}{3}$. (J.M.B)

7. (a) The position vectors of the points A, B and C are **a**, **b** and **c** respectively, referred to the point O as origin. Given that $3\mathbf{a} + \mathbf{b} = 4\mathbf{c}$, prove that the points A, B and C are collinear and find the ratio AB : AC.

   (b) Three forces $7\mathbf{i} + 5\mathbf{j}$, $2\mathbf{i} + 3\mathbf{j}$ and $\lambda\mathbf{i}$ act at the origin O, where **i** and **j** are unit vectors parallel to the x-axis and the y-axis respectively. The unit of force is the newton. If the magnitude of the resultant of the three forces is 17 N, calculate the two possible values of λ. Show that the two possible directions of the line of action of the resultant are equally inclined to Oy. (A.E.B)

5

Equilibrium and acceleration under concurrent forces

In Chapter 3 it was found that (i) if a body is not moving, then the resultant force acting in any direction must be zero,
(ii) if a body is accelerating, then the relationship $F = ma$ applies.

In this chapter these facts are now used, together with the skills of combining and resolving forces which were acquired in Chapter 4.

Terminology

Particle. A particle is that portion of matter which is so small in size that the distances between its extremities may be neglected.

Rigid body. A rigid body is, on the other hand, one in which the distances between its various parts are not negligible, and these distances remain fixed.

Equilibrium. A state of equilibrium is said to exist when two or more forces act upon a particle, or upon a rigid body, and motion does not take place.

Triangle of forces

It has already been seen in Chapter 4 that the resultant of two forces, acting at a point, can be found both graphically and by calculation.

The two forces P and Q are represented by the line segments AB and AD and the parallelogram ABCD is completed. The diagonal AC then fully represents, in magnitude and direction, the resultant R of the two forces P and Q.

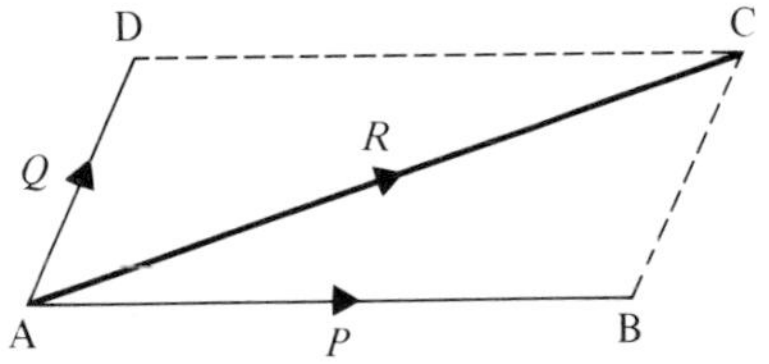

Thus $\overrightarrow{AB} + \overrightarrow{AD} = \overrightarrow{AC}$

$\therefore$ $\overrightarrow{AB} + \overrightarrow{BC} = \overrightarrow{AC}$ since $\overrightarrow{AD}$ and $\overrightarrow{BC}$ are equivalent vectors

$\therefore$ $\overrightarrow{AB} + \overrightarrow{BC} - \overrightarrow{AC} = 0$

i.e. $P + Q - R = 0$

Hence, if R is the resultant of the forces P and Q, then $-R$ added to the forces P and Q will produce equilibrium.

This means that, if three forces acting at a point can be represented by the sides of a triangle, and the forces all act in the same sense around the triangle, then these forces are in equilibrium.

The triangle DEF, shown below, is said to be a *Triangle of Forces* for the three forces X, Y and Z.

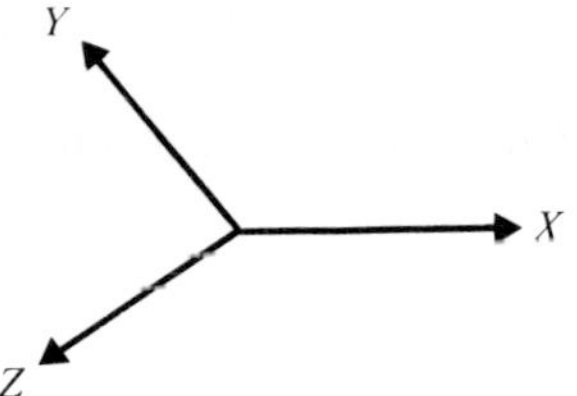

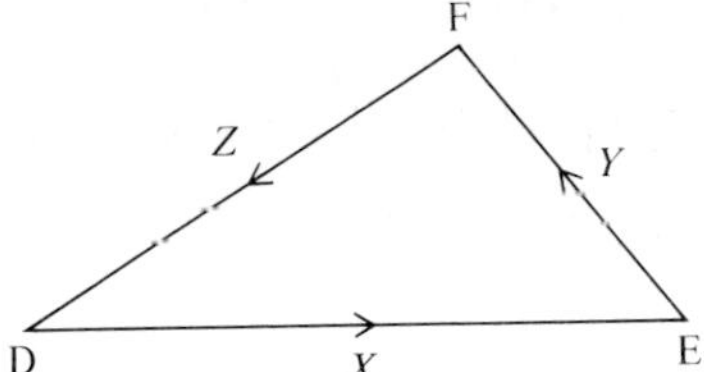

The significance of the different directions of the arrows on the side AC in the diagrams below should be carefully noted.

One force the resultant of two forces

$$\overrightarrow{AB} + \overrightarrow{BC} = \overrightarrow{AC}$$

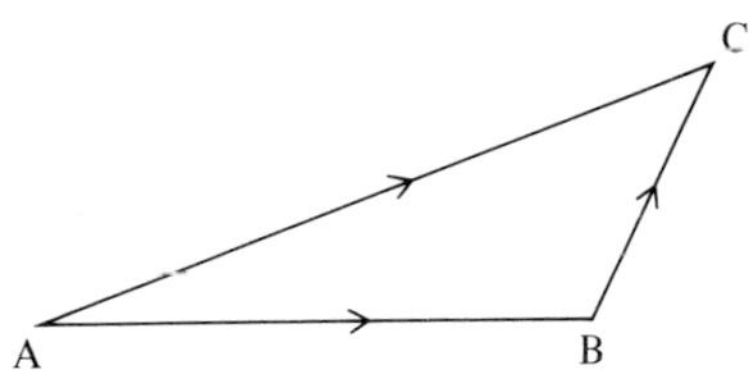

Three forces in equilibrium

$$\overrightarrow{AB} + \overrightarrow{BC} + \overrightarrow{CA} = 0$$

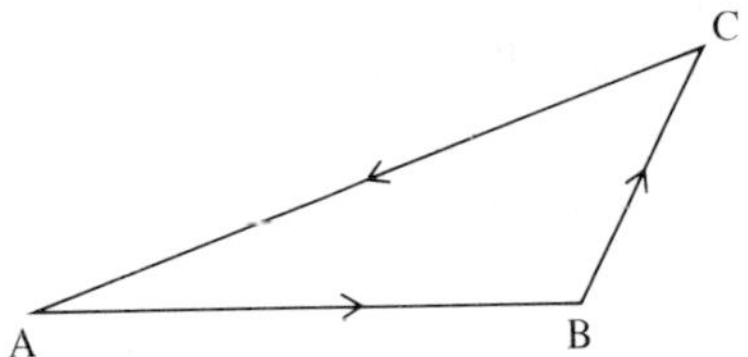

The converse of the Triangle of Forces is also true:

If three forces acting at a point are in equilibrium, they can be represented by the sides of a triangle.

It should be carefully noted that the directions of the forces must be parallel to the sides of the triangle, and such that the arrows on the sides of the triangle indicating the directions of the forces are all in the same sense.

Example 1

Given that the three forces shown in the diagram are in equilibrium, find, by scale drawing, the magnitude of S and θ.

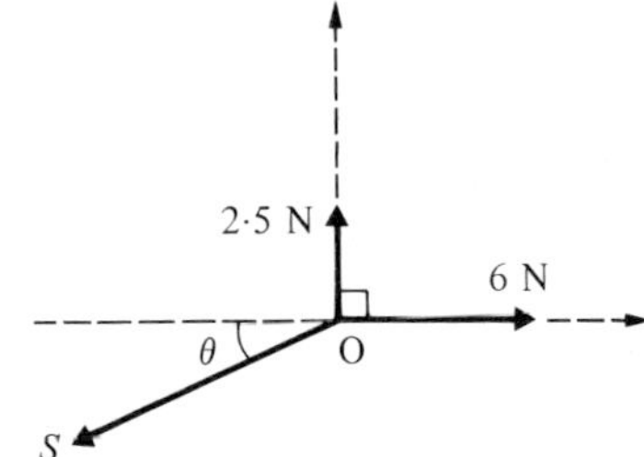

Draw a line OA, 6 cm in length, parallel to the 6 N force.
Draw a line AB, 2·5 cm in length, parallel to the 2·5 N force. Join BO.

By measurement BO = 6·5 cm and angle AOB = 23°

The force S is 6·5 N and $\theta = 23°$.

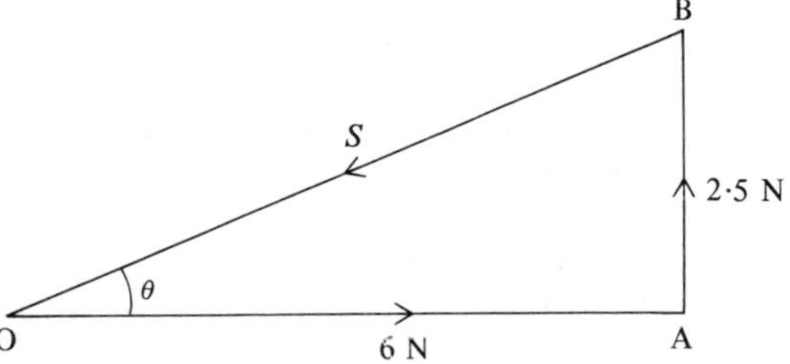

It will be noted that this involves the same process as was used in chapter 4 to find the *resultant* of two forces.

Polygon of forces

The resultant of a number of forces has already been found by extending the idea of drawing a triangle to that of drawing a polygon.
In the same way, given that a number of forces are in equilibrium, we can extend the idea of a triangle of forces to that of a polygon of forces.

Example 2

The forces shown in the diagram are known to be in equilibrium. By drawing a polygon of forces, find the magnitudes of S and θ.

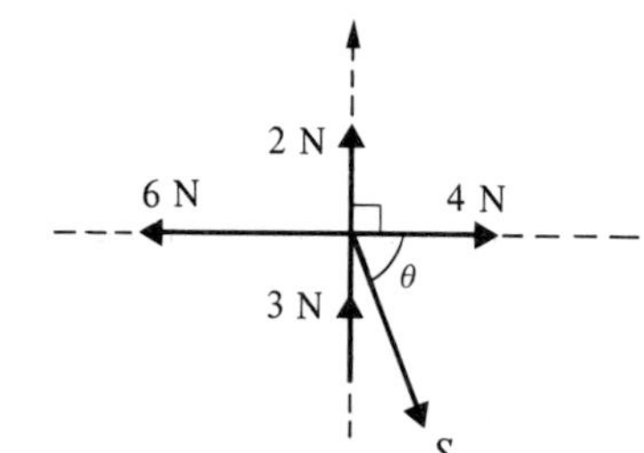

The polygon can be constructed in various ways.

The line AB is drawn parallel to the 2 N force and 2 units in length; BC is drawn 4 units in length and parallel to the 4 N force; CD is drawn 3 units in length and parallel to the 3 N force; DE is drawn 6 units in length and parallel to the 6 N force. The line EA then represents the force needed to produce equilibrium, i.e. the force S. Alternatively, a different polygon is obtained by drawing AW parallel to the 6 N force and 6 units in length, WX 3 units in length, XY 2 units in length and YE 4 units in length. The unknown force is again represented by the line EA required to complete the polygon.

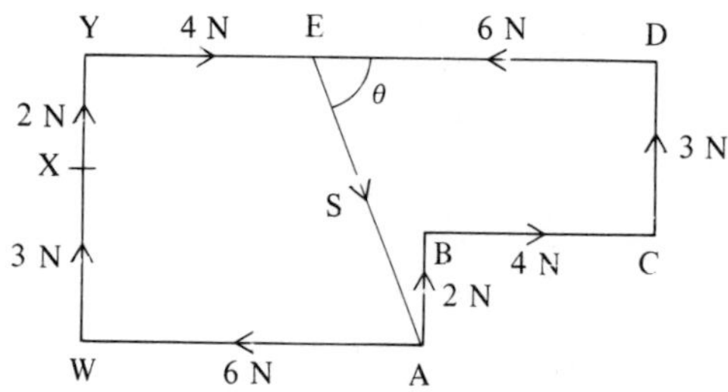

By measurement the force S is 5·4 N and the angle $\theta = 68°$.

Again, it should be carefully noted that the arrows on the sides of the polygon ABCDE are all in the same sense, as indeed they are in polygon AWXYE.

Three forces in equilibrium: solution by calculation

Examples 1 and 2 above were solved by graphical methods and, consequently, a high degree of accuracy is not easily obtained.

Given three forces in equilibrium, as in example 1, the triangle of forces can be sketched and then trigonometry can be used to calculate the unknown force and angle, as shown in examples 3 and 4.
Alternatively for some problems involving three forces in equilibrium, the theorem which follows gives a ready means of solution by calculation, as illustrated by examples 5 and 6.

Lami's Theorem

Suppose the forces X, Y and Z acting at a point are in equilibrium. The forces can therefore be represented by the sides of a triangle ABC.

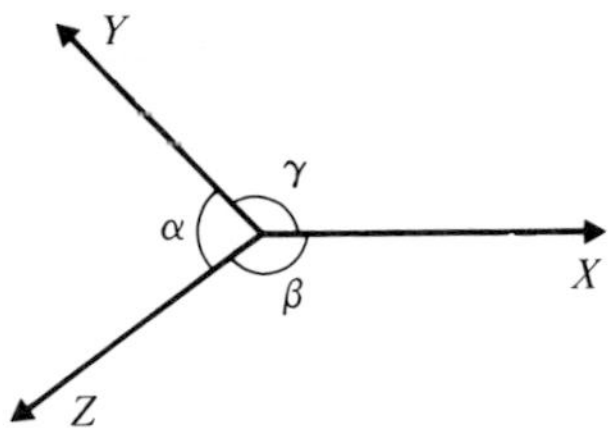

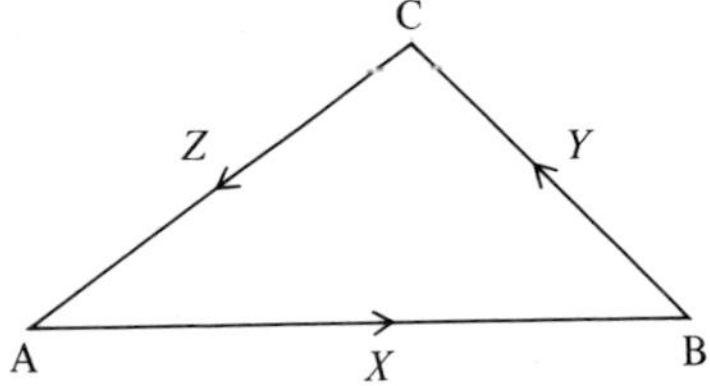

Applying the sine rule to the triangle ABC:

$$\frac{AB}{\sin B\hat{C}A} = \frac{BC}{\sin C\hat{A}B} = \frac{CA}{\sin A\hat{B}C}$$

But $B\hat{C}A = 180 - \alpha$, $C\hat{A}B = 180 - \beta$ and $A\hat{B}C = 180 - \gamma$

Hence $$\frac{AB}{\sin(180-\alpha)} = \frac{BC}{\sin(180-\beta)} = \frac{CA}{\sin(180-\gamma)}$$

$$\therefore \frac{AB}{\sin\alpha} = \frac{BC}{\sin\beta} = \frac{CA}{\sin\gamma}$$

or $$\frac{X}{\sin\alpha} = \frac{Y}{\sin\beta} = \frac{Z}{\sin\gamma}$$

This result, which only applies to three forces acting at a point, is known as Lami's Theorem.

Example 3

Given that the system of forces shown in the diagram is in equilibrium, sketch the triangle of forces and hence calculate the magnitude of the force X and the angle α.

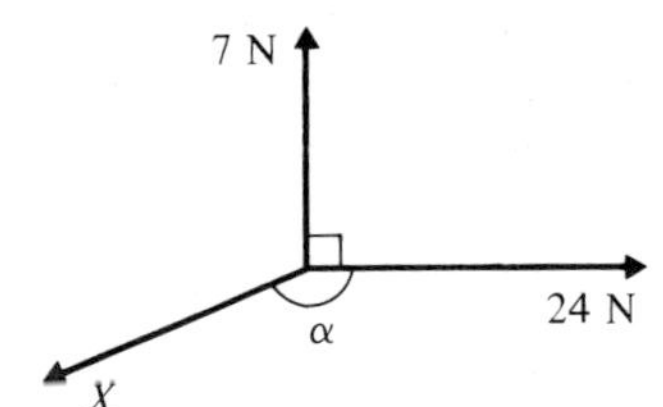

Sketch the triangle of forces.

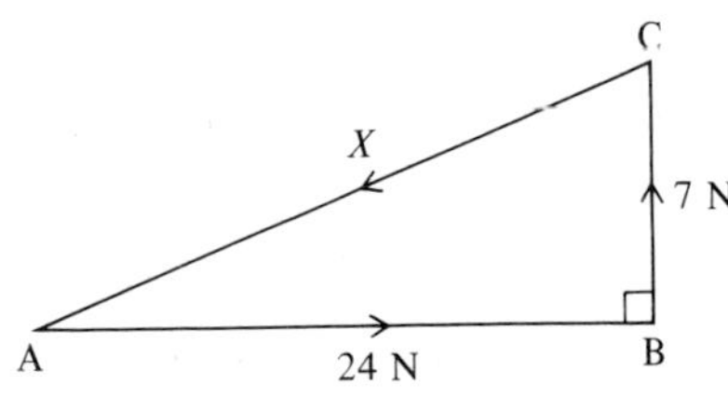

By Pythagoras: $AC^2 = AB^2 + BC^2$

$\therefore X^2 = 24^2 + 7^2$

$\therefore X = 25$ N

Also $\tan C\hat{A}B = \frac{7}{24}$

$\therefore C\hat{A}B = 16{\cdot}26^\circ$

But $\alpha = 180^\circ - C\hat{A}B = 163{\cdot}74^\circ$

The force X is 25 N and the angle α is $163{\cdot}74^\circ$.

Example 4

Sketch the triangle of forces for the given system of forces which is in equilibrium. Calculate the magnitude of P and θ.

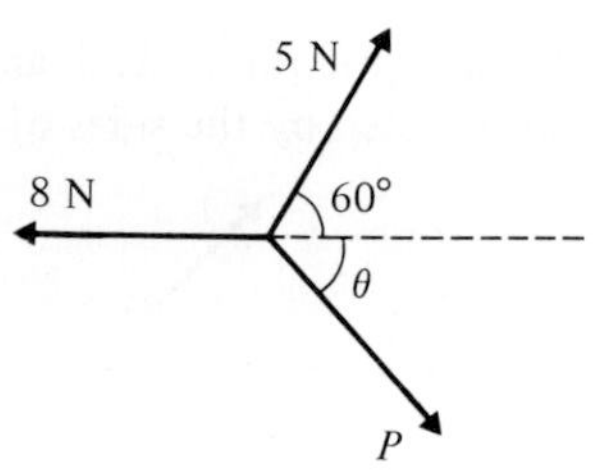

Sketch the triangle of forces

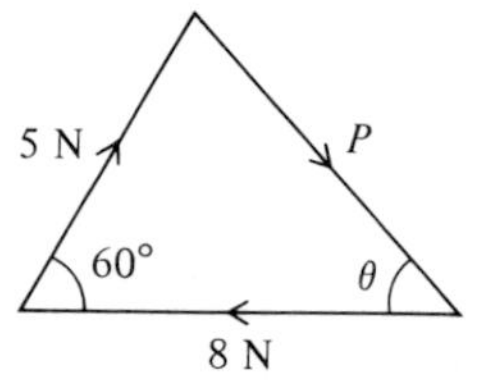

By the cosine rule from this triangle:

$$P^2 = 5^2 + 8^2 - 2 \times 5 \times 8 \times \cos 60$$
$$= 49$$
$$\therefore \quad P = 7 \text{ N}$$

By the sine rule:

$$\frac{7}{\sin 60} = \frac{5}{\sin \theta}$$
$$\therefore \quad \sin \theta = \frac{5 \sin 60}{7}$$
$$\therefore \quad \theta = 38{\cdot}21^\circ$$

The force P is 7·0 N and the angle θ is 38·21°.

It should be noted that various ways of determining the unknowns may be used, once the sketch of the triangle of forces has been made. Alternatively, Lami's Theorem may be applied directly as shown in the following example.

Example 5

The force system shown in the diagram is in equilibrium. Calculate P and Q.

By Lami's Theorem

$$\frac{8}{\sin 150} = \frac{P}{\sin 75} = \frac{Q}{\sin 135}$$

thus
$$\frac{8}{\sin 30} = \frac{P}{\sin 75}$$
$$\therefore \quad P = \frac{8 \sin 75}{\sin 30} = 15{\cdot}5$$

and
$$\frac{8}{\sin 30} = \frac{Q}{\sin 45}$$
$$\therefore \quad Q = \frac{8 \sin 45}{\sin 30} = 11{\cdot}3$$

The force P is 15·5 N and the force Q is 11·3 N.

Example 6

A mass of 5 kg is suspended, in equilibrium, by two light inextensible strings which make angles of 30° and 45° with the horizontal. Calculate the tensions in the strings.

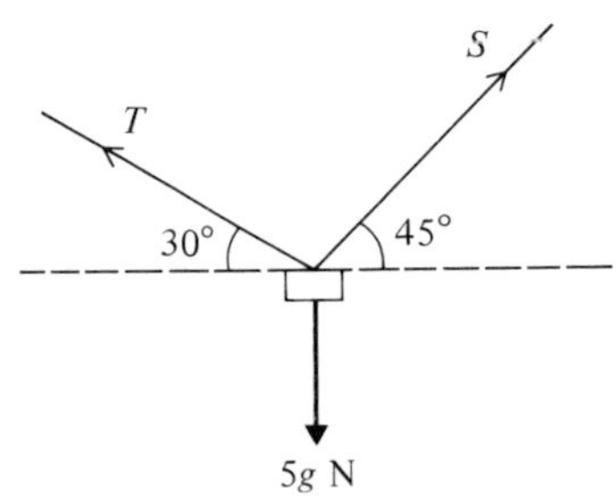

First, draw a diagram showing the position of equilibrium; let the tensions in the strings be S and T newtons. There are seen to be three forces, acting at a point, producing equilibrium, so by Lami's Theorem:

$$\frac{T}{\sin(90+45)} = \frac{5g}{\sin(180-30-45)}$$

$$\therefore \quad T = \frac{5g \sin 135}{\sin 105} = 35{\cdot}87$$

and

$$\frac{S}{\sin(90+30)} = \frac{5g}{\sin(180-30-45)}$$

$$\therefore \quad S = \frac{5g \sin 120}{\sin 105} = 43{\cdot}93$$

The tensions in the strings are 43·9 N and 35·9 N.

It should be noted that an equation involving T could have been used to determine the tension S, once the value of T had been found. It is better to avoid the use of a previously determined value if it is possible, since that value may be incorrect.

Exercise 5A

1. Each of the following systems of forces are in equilibrium. By making an accurate scale drawing of the triangle of forces, find the magnitude of forces P and Q and the size of angle θ.

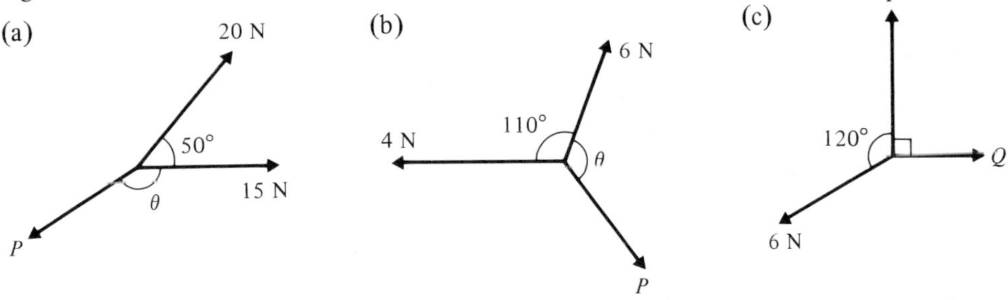

2. Each of the following systems of forces are in equilibrium. By making an accurate scale drawing of the polygon of forces, find the magnitude of forces P and Q and the size of angle θ.

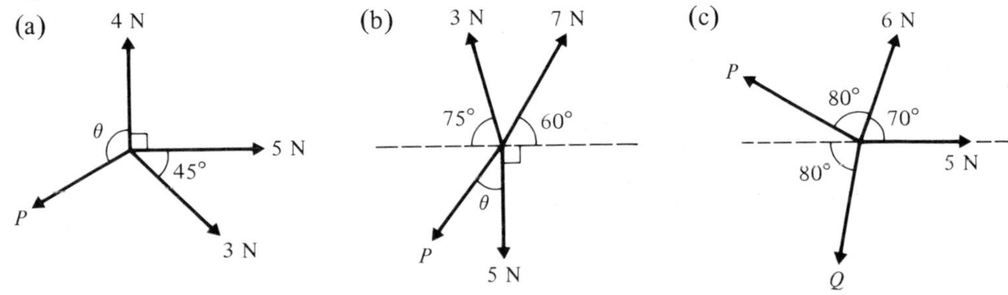

3. Each of the following systems of forces are in equilibrium. Make a sketch of the triangle of forces and hence calculate the magnitude of force P and the size of angle θ.

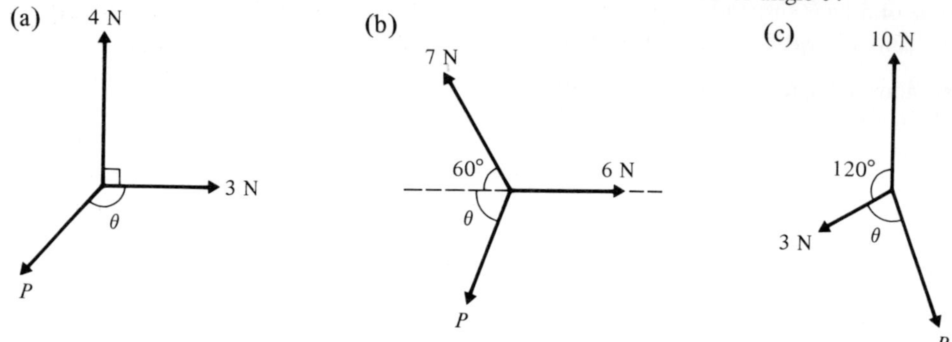

4. Each of the following systems of forces are in equilibrium. Use Lami's Theorem to find the magnitude of forces P and Q.

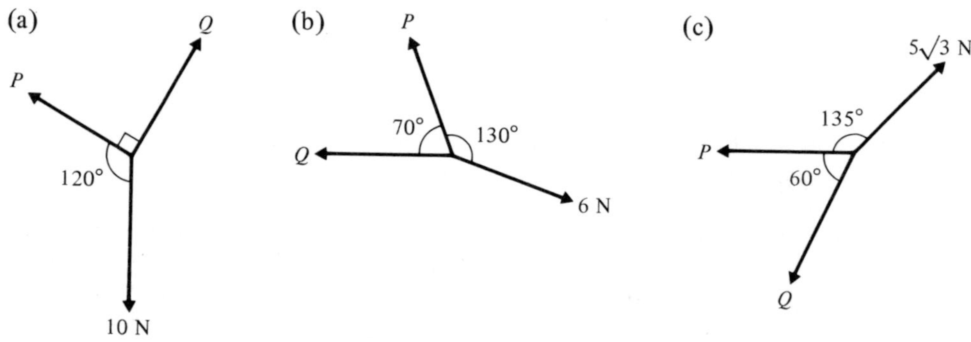

5. The diagram shows a body of weight 10 N supported in equilibrium by two light inextensible strings. The tensions in the strings are 7 N and T and the angles the strings make with the upward vertical are 60° and θ respectively. Using the triangle of forces, calculate T and θ.

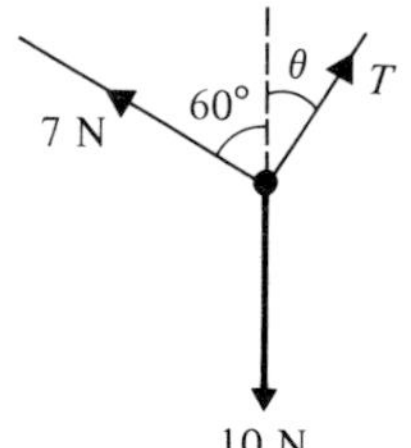

6. The diagram shows a light inextensible string with one end fixed at A and a mass of 5 kg suspended at the other end. The mass is held in equilibrium at an angle θ to the downward vertical by a horizontal force P. Find θ by trigonometry and then use Lami's Theorem to find the magnitude of the force P and the tension T.

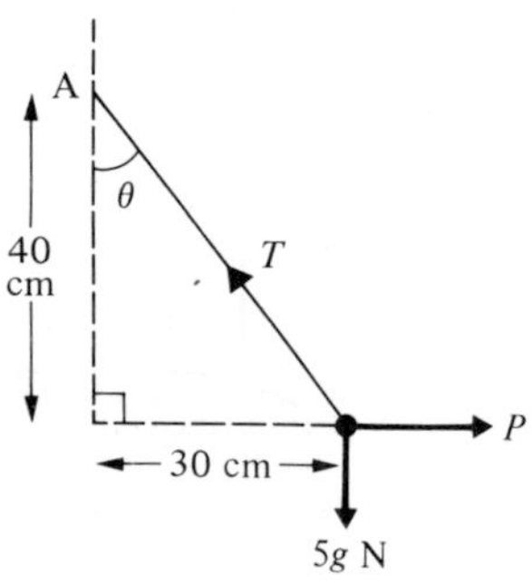

7. Four horizontal forces, all emanating from some point O, are in equilibrium. Three of the four forces have magnitudes 10 N, 20 N and 30 N in directions north, east and south-west respectively. Find the magnitude and direction of the fourth force by scale drawing.
8. A mass of 2 kg is suspended by two light inextensible strings, one making an angle of 60° with the upward vertical and the other 30° with the upward vertical. Find the tension in each string.
9. A light inextensible string of length 40 cm has its upper end fixed at a point A, and carries a mass of 2 kg at its lower end. A horizontal force applied to the mass keeps it in equilibrium, 20 cm from the vertical through A. Find the magnitude of this horizontal force and the tension in the string.
10. The diagram shows a body of mass 5 kg supported by two light inextensible strings, the other ends of which are attached to two points A and B on the same level as each other and 7 m apart. The body rests in equilibrium at C, 3 m vertically below AB. If $C\hat{B}A = 45°$ find T_1 and T_2, the tension in the strings.

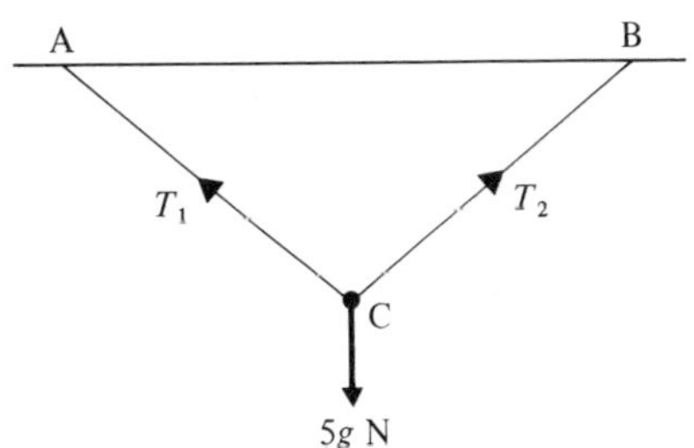

Particle in equilibrium under more than three forces: solution by calculation

When more than three forces act upon a particle and a state of equilibrium exists, it has been seen that a polygon of forces can be drawn and the magnitude and direction of an unknown force determined. Graphical methods such as this have only a limited degree of accuracy. Lami's Theorem cannot be used when there are more than three forces involved.

For a system of forces in equilibrium, the resultant force acting, in any direction, is zero. Thus the sum of the components (or resolved parts) of the forces in any and every direction must be zero. This result applies to a system of any number of forces which are in equilibrium, and gives a method of solving such problems.

For coplanar forces, we can choose two mutually perpendicular directions; by finding the components of all the forces, two equations will be obtained. Two unknown quantities can then be determined. Any other equation which may be obtained by resolving in some other direction will be a combination of the previous two equations, and it will not therefore enable more unknowns to be found.

Suppose the forces W, X, Y and Z act at the point O and are in equilibrium with the direction of the forces as shown in the diagram

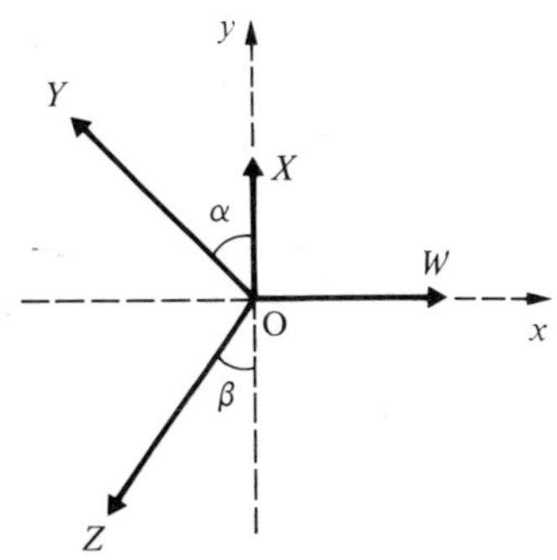

Since the resultant parallel to the x-axis is zero, the sum of the components of the forces in this direction is zero. Alternatively, the sum of the resolved parts in the direction of the positive x-axis must balance those in the opposite direction, and this will give the same equation.

Parallel to the x-axis

$$W - Y \sin\alpha - Z \sin\beta = 0 \quad \text{or} \quad W = Y \sin\alpha + Z \sin\beta \qquad \ldots [1]$$

Parallel to the y-axis

$$X + Y \cos\alpha - Z \cos\beta = 0 \quad \text{or} \quad X + Y \cos\alpha = Z \cos\beta \qquad \ldots [2]$$

Hence two equations are obtained and from these equations two unknowns can then be determined.

It should be noted that the alternative way of writing down the equations does give the same equations.

Example 7

The given forces act on a particle at O which is in equilibrium. By resolving in two directions, find P and S.

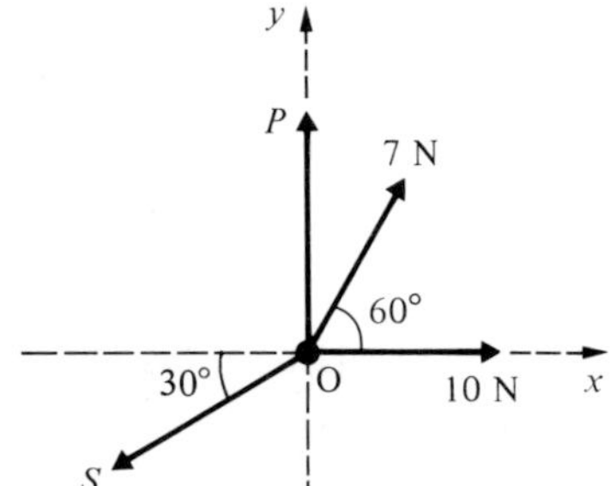

Resolving parallel to x-axis

$$10 + 7 \cos 60 = S \cos 30$$

$$\therefore \quad \frac{27}{2} = \frac{S}{2}\sqrt{3} \quad \text{or} \quad S = 9\sqrt{3}\ \text{N}$$

Resolving parallel to y-axis

$$P + 7 \cos 30 = S \cos 60$$

Substituting for S

$$\therefore \quad P + \tfrac{7}{2}\sqrt{3} = 9\sqrt{3} \times \tfrac{1}{2}$$

$$\therefore \quad P = \sqrt{3}\ \text{N}$$

The force P is $\sqrt{3}$ N and the force S is $9\sqrt{3}$ N.

Example 8

The forces $(3\mathbf{i} + 5\mathbf{j})$ N, $(a\mathbf{i} + b\mathbf{j})$ N, $(8\mathbf{i} - 6\mathbf{j})$ N and $(-4\mathbf{i} - 3\mathbf{j})$ N are in equilibrium. Find the values of a and b by calculation.

In this case a diagram is not necessary.

The resultant of these forces, in vector form, is

$$(3\mathbf{i} + 5\mathbf{j} + a\mathbf{i} + b\mathbf{j} + 8\mathbf{i} - 6\mathbf{j} - 4\mathbf{i} - 3\mathbf{j})$$
$$= (3 + a + 8 - 4)\mathbf{i} + (5 + b - 6 - 3)\mathbf{j}\ \text{N}$$
$$= (7 + a)\mathbf{i} + (b - 4)\mathbf{j}\ \text{N}$$

Since the forces are in equilibrium, the resultant is zero; therefore the sum of the components in any direction must also be zero.

Hence, parallel to the x-axis $\quad a + 7 = 0 \quad$ (in the direction of **i**)

$$\therefore \quad a = -7$$

and, parallel to the y-axis $\quad b - 4 = 0 \quad$ (in the direction of **j**)

$$\therefore \quad b = 4$$

The value of a is -7 and of b is $+4$.

Positions of equilibrium

If a position of equilibrium of a particle acted upon by a number of forces, is described, it is important to first draw a diagram showing all the forces acting on the particle. It can then be decided which are the best directions in which to resolve the forces.

Example 9

A particle of mass 2 kg is attached to the lower end of an inextensible string. The upper end of the string is fixed. A horizontal force of 21 N and an upward vertical force of $0{\cdot}5g$ N act upon the particle, which is in equilibrium with the string making an angle θ with the vertical. Calculate the tension in the string and the angle θ.

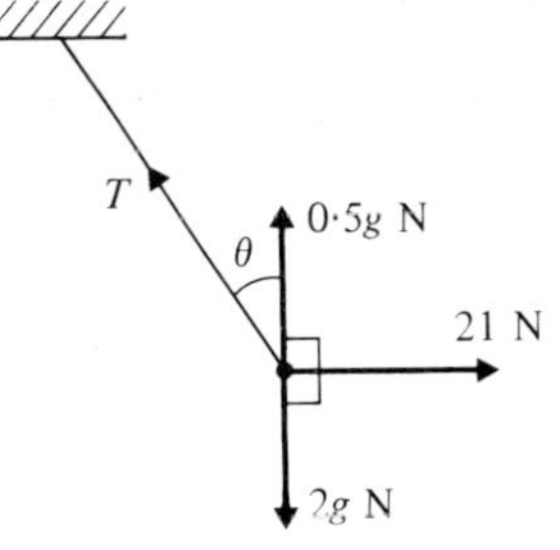

Draw a diagram showing the forces acting on the particle.
Let the tension in the string be T.

Resolving horizontally

$$21 = T \sin \theta \qquad \ldots [1]$$

Resolving vertically

$$2g = T \cos \theta + 0{\cdot}5g$$

$$\therefore \quad 14{\cdot}7 = T \cos \theta \qquad \ldots [2]$$

Dividing equation [1] by equation [2]

$$\tan \theta = \frac{21}{14{\cdot}7} = 1{\cdot}4286$$

$$\therefore \quad \theta = 55{\cdot}01^\circ$$

Substituting in equation [1]

$$T = \frac{21}{\sin 55{\cdot}01^\circ} = 25{\cdot}63 \text{ N}$$

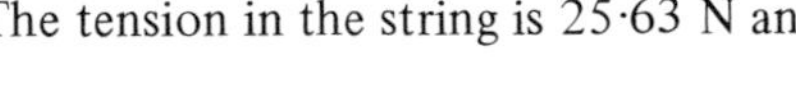

The tension in the string is $25{\cdot}63$ N and the angle θ is $55{\cdot}01^\circ$.

Components in other directions

Although resolving in a horizontal and vertical direction may frequently be convenient, this is not always so.
If it is possible to choose a direction, at right angles to which there is an unknown force, this may well be a sensible choice.

Inclined plane

When a mass is in equilibrium on an inclined plane, then it will usually be found expedient to resolve the forces parallel to, and at right angles to, the surface of the plane.

Example 10

A particle of mass 4 kg rests on the surface of a smooth plane which is inclined at an angle of 30° to the horizontal. When a force P acting up the plane and a horizontal force of $8\sqrt{3}$ N are applied to the particle, it rests in equilibrium. Calculate P and the normal reaction between the particle and the plane.

Draw a diagram showing the position of equilibrium and all the forces acting on the particle. Let the normal reaction be R.

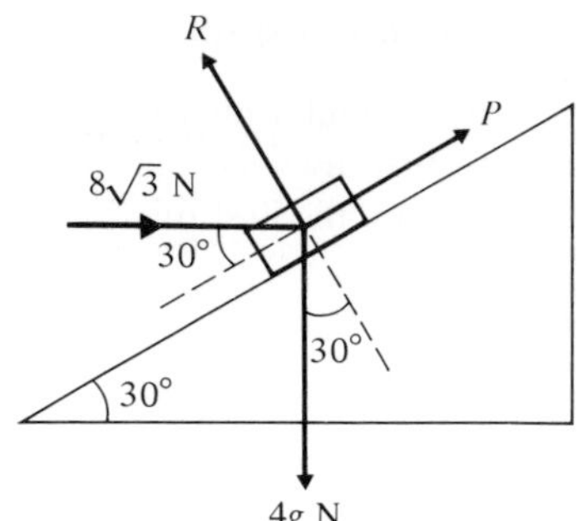

Resolving parallel to the surface of the plane

$$4g\cos 60 = 8\sqrt{3}\cos 30 + P$$

$$4g \times \tfrac{1}{2} = 8\sqrt{3} \times \frac{\sqrt{3}}{2} + P$$

$$\therefore \quad P = 7{\cdot}6 \text{ N}$$

Resolving at right angles to the surface of the plane

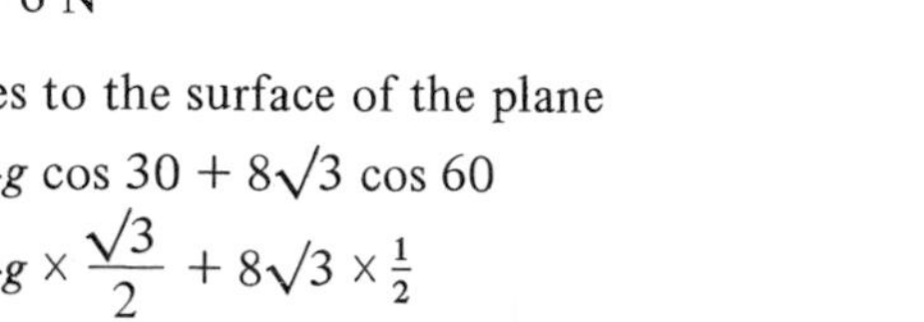

$$R = 4g\cos 30 + 8\sqrt{3}\cos 60$$

$$\therefore \quad R = 4g \times \frac{\sqrt{3}}{2} + 8\sqrt{3} \times \tfrac{1}{2}$$

$$\therefore \quad R = 40{\cdot}87 \text{ N}$$

The horizontal force is 7·6 N and the normal reaction of the plane is 40·9 N.

Systems involving more than one particle

In a more complicated system, there may be more than one particle involved in a state of equilibrium.
The diagram must show clearly all the forces acting on each particle. The equilibrium of each particle may then be considered separately. This means that the directions in which the forces are resolved for the two particles may not be the same.

Example 11

A light inextensible string passes over a smooth pulley fixed at the top of a smooth plane inclined at 30° to the horizontal. A particle of mass 2 kg is attached to one end of the string and hangs freely. A mass m is attached to the other end of the string and rests in equilibrium on the surface of the plane. Calculate the normal reaction between the mass m and the plane, the tension in the string and the value of m.

Draw a diagram showing the position of equilibrium and show the forces acting on each particle. Let the tension in the string be T.

R
T
T
30°
30°
2g N
mg

Consider the mass of 2 kg

Resolving vertically $\qquad T = 2g$

$$\therefore \quad T = 19{\cdot}6 \text{ N}$$

As no other forces act on this mass, there is only the one equation.

Consider the mass m

Resolving parallel to the surface of the plane

$$T = mg\cos 60$$

substituting for T $\qquad 19{\cdot}6 = mg \times \tfrac{1}{2}$

$$\therefore \quad m = 4 \text{ kg}$$

Resolving at right angles to the plane

$$R = mg\cos 30$$

substituting for m $\qquad \therefore \quad R = 33{\cdot}9 \text{ N}$

The normal reaction is 33·9 N, the tension in the string is 19·6 N and m is 4 kg.

Exercise 5B

Each of the diagrams in questions **1** to **9** shows a particle in equilibrium under the forces shown. In each case (a) obtain an equation by resolving in the direction Ox,
(b) obtain an equation by resolving in the direction Oy,
(c) use your equations for (a) and (b) to find the unknown forces and angles.

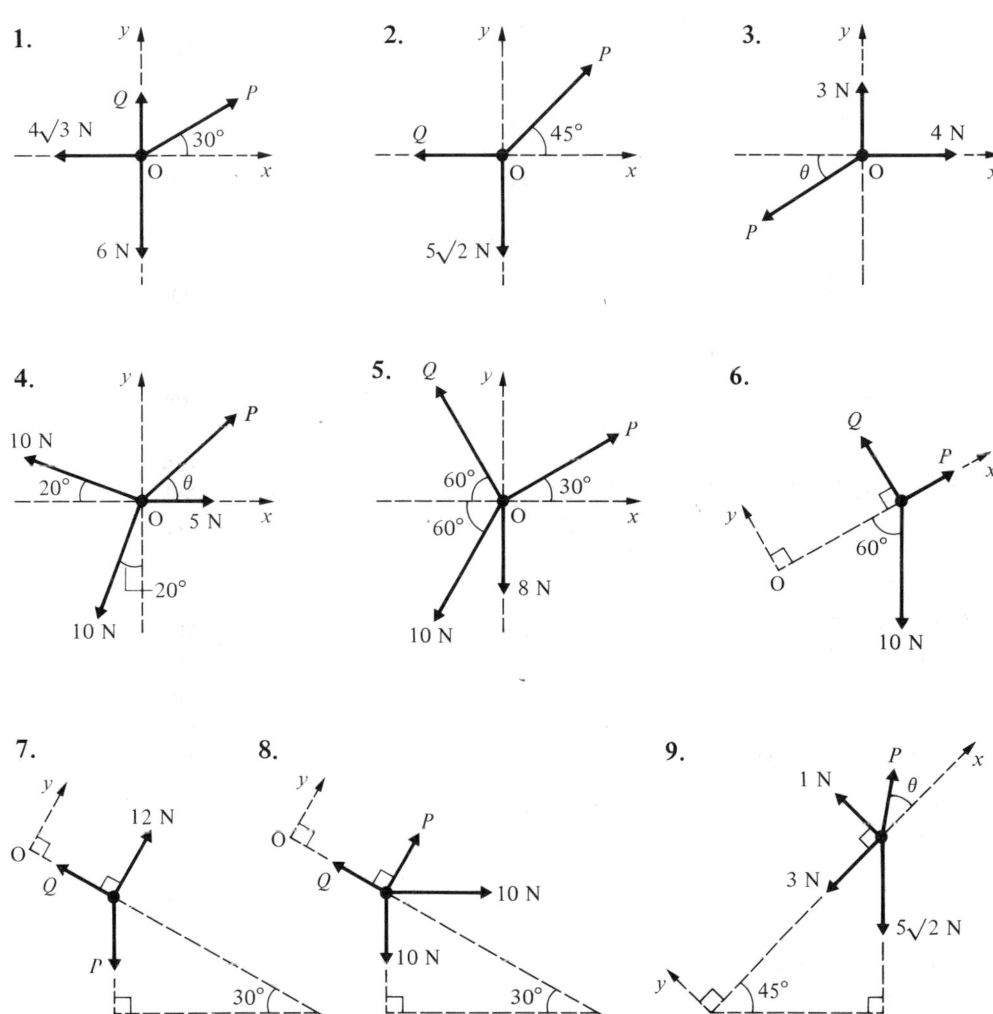

10. If each of the following sets of forces are in equilibrium, find the value of a and b in each case.

(a) $(6\mathbf{i} + 4\mathbf{j})$ N, $(-2\mathbf{i} - 5\mathbf{j})$ N, $(a\mathbf{i} + b\mathbf{j})$ N
(b) $(5\mathbf{i} + 4\mathbf{j})$ N, $(3\mathbf{i} + \mathbf{j})$ N, $(a\mathbf{i} + b\mathbf{j})$ N
(c) $(a\mathbf{i} + 3\mathbf{j})$ N, $(2\mathbf{i} - 5\mathbf{j})$ N, $(-7\mathbf{i} + b\mathbf{j})$ N
(d) $(a\mathbf{i} - 3b\mathbf{j})$ N, $(b\mathbf{i} - 2a\mathbf{j})$ N, $(-3\mathbf{i} + 8\mathbf{j})$ N
(e) $(-3\mathbf{i} + 2\mathbf{j})$ N, $(4\mathbf{i} + 7\mathbf{j})$ N, $(-8\mathbf{i} + 5\mathbf{j})$ N, $(a\mathbf{i} + b\mathbf{j})$ N

11. Each of the following diagrams shows a particle in equilibrium under the forces shown.

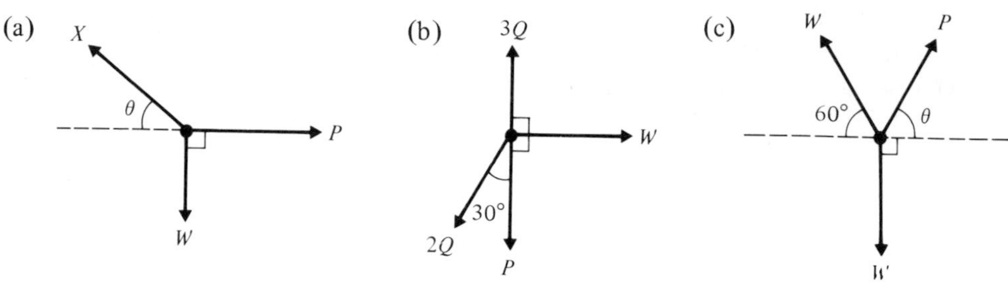

Prove: $P \tan\theta = W$ Prove: $P = W(3 - \sqrt{3})$ Prove: $\tan\theta = 2 - \sqrt{3}$

12. A light inextensible string of length 50 cm has its upper end fixed at point A and carries a particle of mass 8 kg at its lower end. A horizontal force P applied to the particle keeps it in equilibrium 30 cm from the vertical through A. By resolving vertically and horizontally, find the magnitude of P and the tension in the string.

13. A light inextensible string of length 26 cm has its upper end fixed at point A and carries a particle of mass m at its lower end. A force P at right angles to the string is applied to the particle and keeps it in equilibrium 10 cm from the vertical through A. By resolving vertically and horizontally find, in terms of m, the magnitude of P and the tension in the string.

14. A particle is in equilibrium under the action of forces 4 N due north, 8 N due west, $5\sqrt{2}$ N south-east and P. Find the magnitude and direction of P.

15. A force acting parallel to and up a line of greatest slope holds a particle of mass 10 kg in equilibrium on a smooth plane which is inclined at 30° to the horizontal. Find the magnitude of this force and of the normal reaction between the particle and the plane.

16. A horizontal force P holds a body of mass 10 kg in equilibrium on a smooth plane which is inclined at 30° to the horizontal. Find the magnitude of P and of the normal reaction between the particle and the plane.

17. A force P holds a particle of mass m in equilibrium on a smooth plane which is inclined at 30° to the horizontal. If P makes an angle ϕ with the plane, as shown in the diagram, find ϕ when R, the normal reaction between particle and plane, is $1{\cdot}5\ mg$.

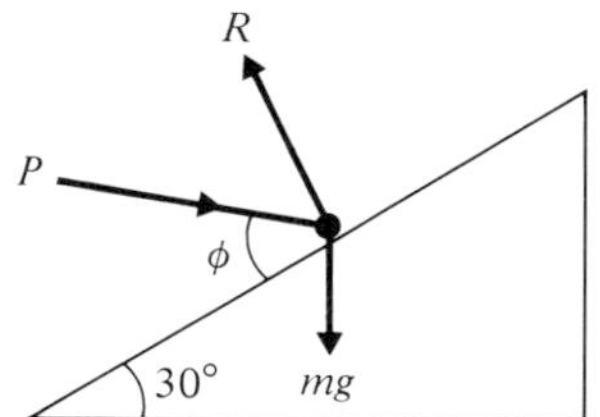

18. A particle of mass 3 kg lying on a smooth surface which is inclined at θ to the horizontal is attached to a light inextensible string which passes up the plane, along the line of greatest slope, over a smooth pulley at the top and carries a 1 kg mass freely suspended at its other end. If the system rests in equilibrium, find (a) the value of θ, (b) the tension in the string, (c) the normal reaction between the particle and the plane.

19. A 5 kg mass lies on a smooth horizontal table. A light inextensible string attached to this mass passes up over a smooth pulley and carries a freely suspended mass of 5 kg at its other end. The part of the string between the mass on the table and the pulley makes an angle of 25° with the horizontal. The system is kept in equilibrium by a horizontal force applied to the mass on the table. Find the magnitude of this horizontal force, the tension in the string and the normal reaction between the table and the mass resting on it.

20. The diagram shows masses of 8 kg and 6 kg lying on smooth planes of inclination θ and ϕ respectively. Light inextensible strings attached to these masses pass along the lines of greatest slope, over smooth pulleys and are connected to a 4 kg mass hanging freely. The strings both make an angle of $60°$ with the upward vertical as shown. If the system rests in equilibrium, find θ and ϕ.

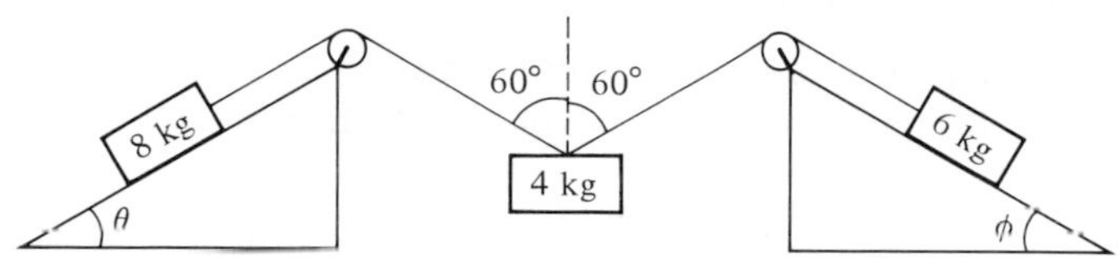

21. The diagram shows masses A and B each lying on smooth planes of inclination $30°$. Light inextensible strings attached to A and B pass along the lines of greatest slope, over smooth pulleys and are connected to a third mass C hanging freely. The strings make angles of ϕ and α with the upward vertical as shown. If A, B and C have masses $2m$, m and m respectively and the system rests in equilibrium, show that $\sin \alpha = 2 \sin \phi$ and $\cos \alpha + 2 \cos \phi = 2$. Hence find ϕ and α.

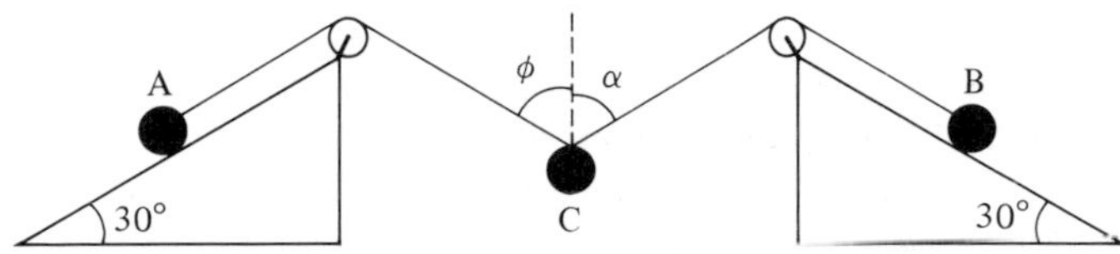

Motion of a particle on a plane

In the examples so far considered in this chapter, the particle has been in equilibrium under the action of a number of forces.

The motion of a particle (or a rigid body) can now be considered using Newton's Laws as in Chapter 3 and the idea of resolving a force in a particular direction.

Example 12

The body of mass 4 kg has an acceleration a when it is acted upon by a force of $25\sqrt{2}$ N which is inclined at $45°$ to the smooth horizontal surface on which the body rests, as shown. Resolve the forces acting on the body at right angles to the surface. Calculate the normal reaction between the body and the surface and the acceleration a of the body.

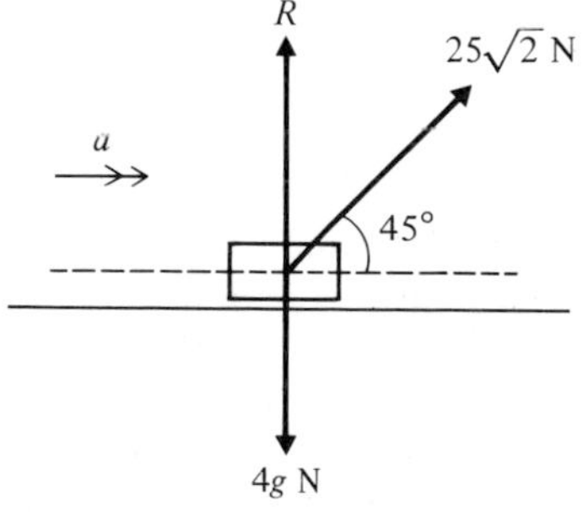

Resolving at right angles to the surface

$$R + 25\sqrt{2} \cos 45 = 4g$$

$$\therefore \quad R + 25\sqrt{2} \times \frac{1}{\sqrt{2}} = 39{\cdot}2$$

$$\therefore \quad R = 14{\cdot}2 \text{ N}$$

Resolving along the surface and applying $F = ma$,

$$25\sqrt{2} \cos 45 = 4 \times a$$

$$\therefore \quad 25 = 4a \text{ or } a = 6{\cdot}25 \text{ m/s}^2$$

The body has an acceleration of $6{\cdot}25$ m/s^2 along the surface, and the normal reaction between the body and the surface is $14{\cdot}2$ N.

Since the reaction R is found to be positive, i.e. $14{\cdot}2$ N, this implies that the body and the surface remain in contact.

Motion on an inclined plane

The same principles apply to motion on an inclined plane as were used in considering motion on a horizontal plane. The directions in which resolving takes place will necessarily be different.

Example 13

A body of mass $3\sqrt{3}$ kg on the surface of an inclined plane is acted upon by a horizontal force of $15g$ N, as shown in the diagram. Calculate the normal reaction of the plane on the body, and the acceleration of the body up the surface of the smooth inclined plane.

Let the acceleration of the body be a and the normal reaction R.

Resolving at right angles to the plane

$$R = 3\sqrt{3}g \cos 60 + 15g \cos 30$$
$$\therefore \quad R = 9\sqrt{3}g = 152{\cdot}8 \text{ N}$$

Resolving up the plane and applying $F = ma$

$$15g \cos 60 - 3\sqrt{3}g \cos 30 = 3\sqrt{3} \times a$$
$$\therefore \quad a = 5{\cdot}658 \text{ m/s}^2$$

The normal reaction is 153 N and the acceleration of the body is $5{\cdot}66$ m/s^2.

Connected particles

In a more complicated system involving the motion of more than one particle, the particles may again be considered separately. Care is needed to ensure that all the forces acting on each particle are considered. Any relationship which may exist between the accelerations of the various parts of the system must also be taken into account.
At this stage, strings will normally be inextensible and pulleys over which strings pass will be treated as smooth pulleys.

Example 14

The bodies shown are connected by a light string which passes over a smooth pulley. Calculate the tension T, the normal reaction R and the acceleration a.

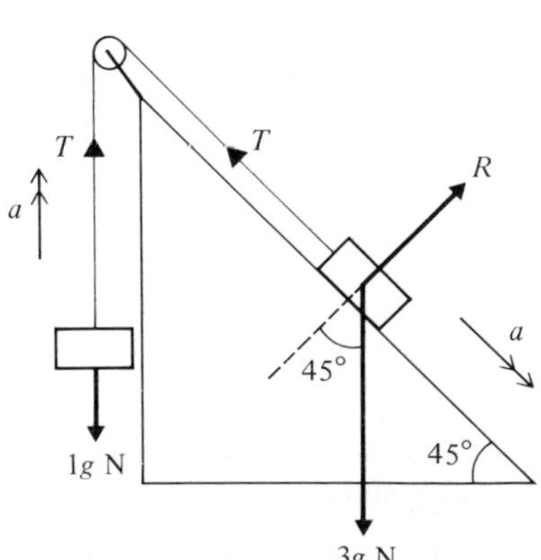

The diagram shows all the forces acting on the two bodies. Since the bodies are connected by an inextensible string, the accelerations must be equal in magnitude. If the 1 kg mass moves upwards, the 3 kg mass must move down the surface of the plane.

Applying $F = ma$ in a vertical direction for the 1 kg mass

$$T - 1g = 1 \times a \qquad \ldots [1]$$

Applying $F = ma$ down the plane for the 3 kg mass

$$3g \cos 45 - T = 3 \times a \qquad \ldots [2]$$

adding the equations $\frac{3}{2}g\sqrt{2} - g = 4a$

$$\therefore \quad a = 2{\cdot}747 \text{ m/s}^2$$

substituting in equation [1] $T = g + 2{\cdot}747$

$$= 12{\cdot}547 \text{ N}$$

Resolving at right angles to the surface of the plane, for the 3 kg mass (note that in this direction there is no acceleration)

$$R = 3g \cos 45$$
$$= 20{\cdot}79 \text{ N}$$

The tension in the string is 12·5 N, the normal reaction is 20·8 N and the acceleration of both particles is 2·75 m/s^2.

Rough surfaces

In the foregoing examples, motion has been taking place on smooth surfaces. In practice this does not happen; all surfaces tend to impede motion. The resistance to motion is an external force acting upon the body, parallel to the surfaces in contact. It will be considered to be a constant force.

Example 15

A body of mass 5 kg is released from rest on the surface of a rough plane which is inclined at 30° to the horizontal. If the body takes $2\frac{1}{2}$ seconds to acquire a speed of 4 m/s from rest, find the resistance to motion which the body must be experiencing.

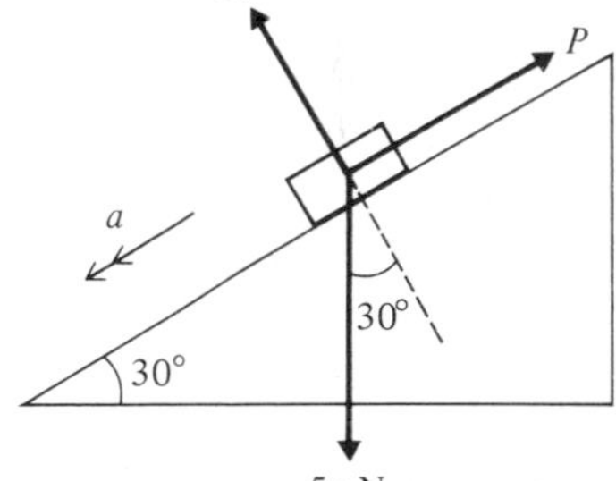

Assume that the force of resistance acting upon the body is P up the plane, and that the acceleration of the body is a down the plane.

Using $v = u + at$ for the motion of the body

$$4 = 0 + a \times 2\tfrac{1}{2}$$
$$\therefore \quad a = 1{\cdot}6 \text{ m/s}^2$$

Applying $F = ma$ to the motion of the body down the plane

$$5g \cos 60 - P = 5 \times a$$

substituting for a

$$5g \times \tfrac{1}{2} - P = 5 \times 1{\cdot}6$$
$$\therefore \quad P = 24{\cdot}5 - 8 = 16{\cdot}5 \text{ N}$$

The resistance to motion is a force of 16·5 N.

Inclination of a plane

In the examples so far considered, the inclination of the plane to the horizontal has been given in degrees.

The gradient of a hill or of a piece of railway track is usually relatively small. The inclination is frequently given in the form 1 in 8 (or $12\frac{1}{2}$%), and this means a rise of 1 unit for every 8 units measured horizontally.

Hence if the angle of inclination is θ, then

$$\tan\theta = \tfrac{1}{8} \text{ and } \therefore\ \theta = 7^\circ \text{ approximately.}$$

For angles of this size, the sine and tangent are equal within 1%, so it is usual to take the sine of the angle as the gradient.

For example: an incline of 1 in 80 is taken to mean that $\sin\theta = \frac{1}{80}$.

Example 16

A body of mass 8 kg is released from rest on the surface of a plane. If the resistance to motion is 1 N acting up the plane and the slope of the plane is 1 in 40, calculate the acceleration of the body down the plane and the speed acquired 6 seconds after release.

Draw a diagram showing the forces acting on the body.

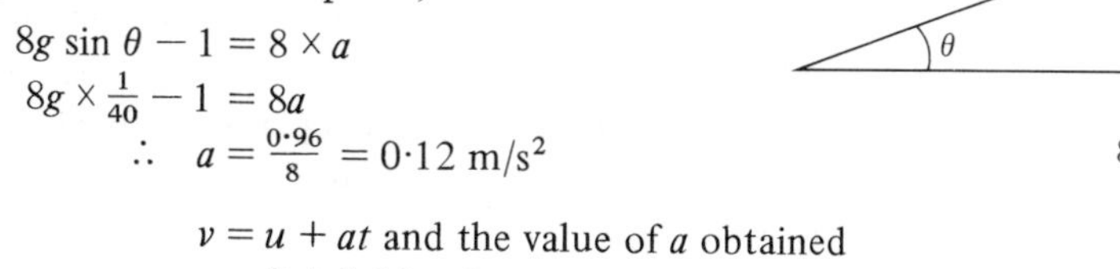

There is a component of the weight of the body which acts down the plane.

Applying $F = ma$ down the plane,

$$8g\sin\theta - 1 = 8 \times a$$
$$\therefore\ 8g \times \tfrac{1}{40} - 1 = 8a$$
$$\therefore\ a = \tfrac{0{\cdot}96}{8} = 0{\cdot}12 \text{ m/s}^2$$

using $v = u + at$ and the value of a obtained

$$v = 0 + 0{\cdot}12 \times 6$$
$$\therefore\ v = 0{\cdot}72 \text{ m/s}$$

The acceleration of the body is $0{\cdot}12$ m/s^2 down the plane and the speed acquired in 6 seconds is $0{\cdot}72$ m/s.

Exercise 5C

Each of the diagrams in questions **1** to **9** shows a body of mass 10 kg accelerating along a surface in the direction indicated. All of the forces acting are as shown. In each case

(a) obtain an equation by resolving perpendicular to the direction of motion,
(b) obtain an equation by applying $F = ma$ parallel to the direction of motion,
(c) use your equations to (a) and (b) to find the unknown forces, accelerations and angles.

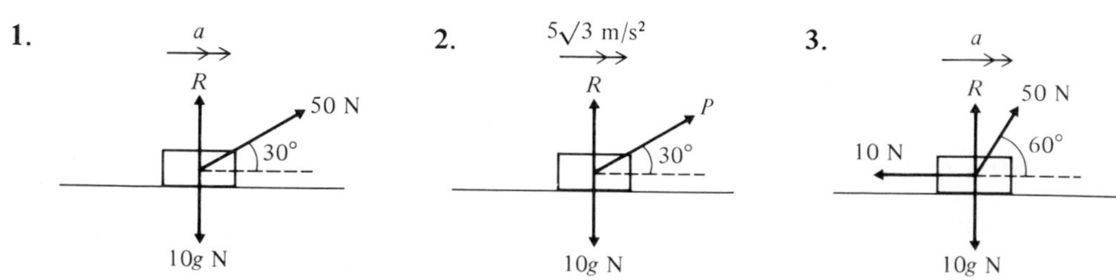

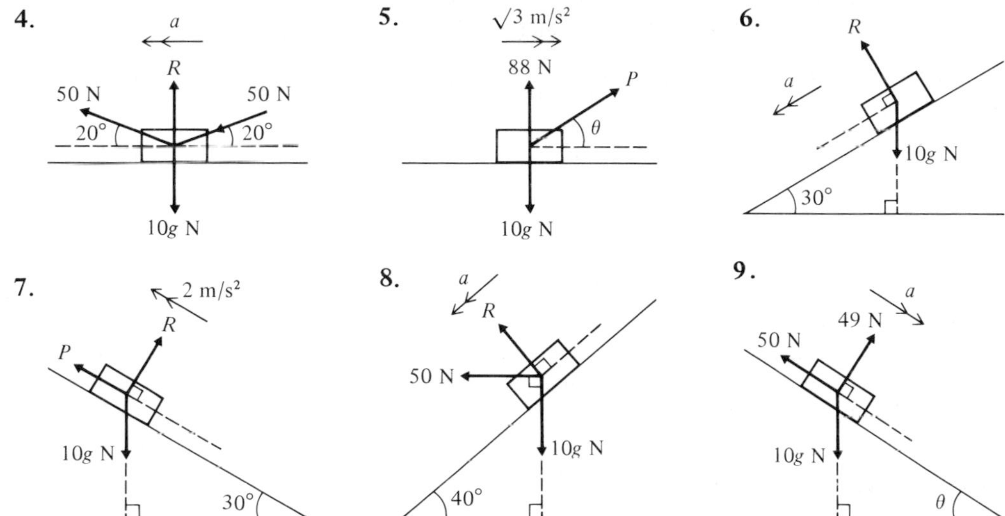

The diagrams for questions **10** and **11** show a body of mass 10 kg accelerating along an inclined plane in the direction indicated. In each case the 10 kg mass is connected to a freely hanging mass by a light inextensible string passing over a smooth pulley. All of the forces acting are as shown.

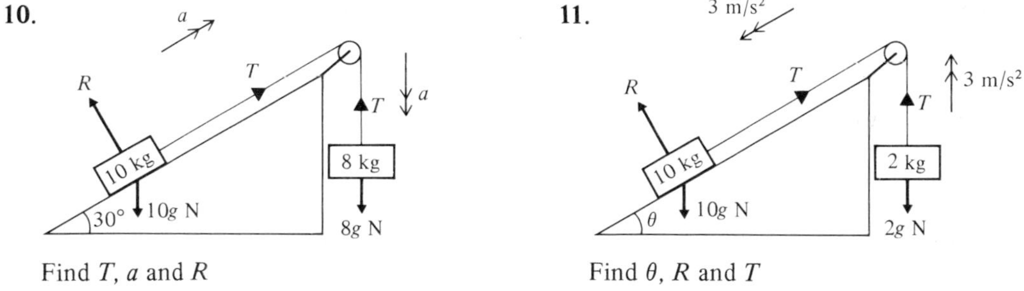

Find T, a and R

Find θ, R and T

12. A body of mass 10 kg is initially at rest on a rough horizontal surface. It is pulled along the surface by a constant force of 60 N inclined at 60° above the horizontal. If the resistance to motion totals 10 N, find the acceleration of the body and the distance travelled in the first 3 s.
13. A body of mass 5 kg, initially at rest on a smooth horizontal surface, is pulled along the surface by a constant force P inclined at 45° above the horizontal. In the first 5 seconds of motion the body moves a distance of 10 m along the surface. Find the acceleration of the body, the magnitude of P and the normal reaction between the body and the surface.
14. A mass of 5 kg is initially at rest at the bottom of a smooth slope which is inclined at $\sin^{-1}\frac{3}{5}$ to the horizontal. The mass is pushed up the slope by a horizontal force of 50 N. Find the normal reaction between the mass and the plane and the acceleration up the slope. How far up the slope will the mass travel in the first 4 s?
15. A body of mass 100 kg is released from rest at the top of a smooth plane which is inclined at 30° to the horizontal. Find the velocity of the body when it has travelled 20 m down the slope. What would your answer be if the mass had been 50 kg?

16. A body of mass 20 kg is released from rest at the top of a rough slope which is inclined at 30° to the horizontal. If the body accelerates down the slope at 3 m/s², find the resistance to motion experienced by the body. (Assume this resistance to be constant throughout).

17. A body of mass 20 kg is released from rest at the top of a rough slope which is inclined at 30° to the horizontal. Six seconds later the body has a velocity of 21 m/s down the slope. Find the resistance to motion experienced by the body. (Assume this resistance to be constant throughout).

18. Find the time interval between a particle reaching the bottom of a smooth slope of length 5 m and inclination 1 in 98, and another particle reaching the bottom of a smooth slope of length 6 m and incline 1 in 70. Both particles are released from rest at the top of their respective slopes at the same time.

19. A mass of 15 kg lies on a smooth plane of inclination 1 in 49. One end of a light inextensible string is attached to this mass and the string passes up the line of greatest slope, over a smooth pulley fixed at the top of the plane and has a freely suspended mass of 10 kg at its other end. If the system is released from rest, find the acceleration of the masses and the distance each travels in the first 2 s. Assume that nothing impedes the motion of either mass.

20. A mass of 2 kg lies on a rough plane which is inclined at 30° to the horizontal. One end of a light inextensible string is attached to this mass and the string passes up the line of greatest slope and over a smooth pulley fixed at the top of the slope; a freely suspended mass of 5 kg is attached to its other end. The system is released from rest; as the 2 kg mass accelerates up the slope, it experiences a constant resistance to motion of 14 N down the slope due to the rough nature of the surface. Find the tension in the string.

21. A mass of 10 kg lies on a smooth plane which is inclined at θ to the horizontal. The mass is 5 m from the top, measured along the plane. One end of a light inextensible string is attached to this mass; the string passes up the line of greatest slope and over a smooth pulley fixed at the top of the slope. The other end is attached to a freely suspended mass of 15 kg. This 15 kg mass is 4 m above the floor. The system is released from rest and the string first goes slack $1\frac{3}{7}$ s later. Find the value of θ.

22. One of two identical masses lies on a smooth plane, which is inclined at $\sin^{-1}\frac{1}{14}$ to the horizontal, and is 2 m from the top. A light inextensible string attached to this mass passes along the line of greatest slope, over a smooth pulley fixed at the top of the incline; the other end carries the other mass hanging freely 1 m above the floor. If the system is released from rest, find the time taken for the hanging mass to reach the floor.

Exercise 5D. Harder questions

Each of the diagrams in questions **1** to **9** shows a mass, or masses, accelerating in the directions indicated. In each case the forces acting are as shown and R is the normal reaction between the mass and the surface it is on.

1.

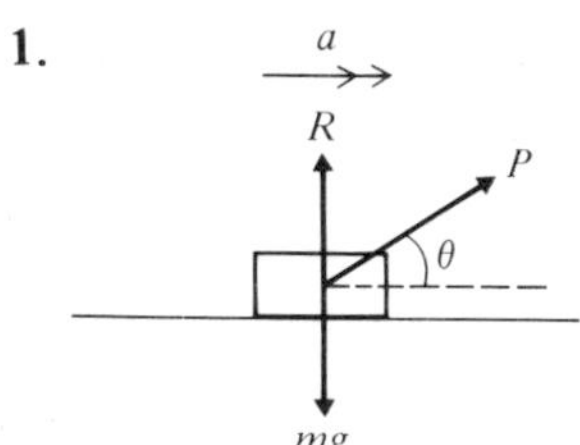

Prove: $\tan\theta = \dfrac{mg - R}{ma}$

2.

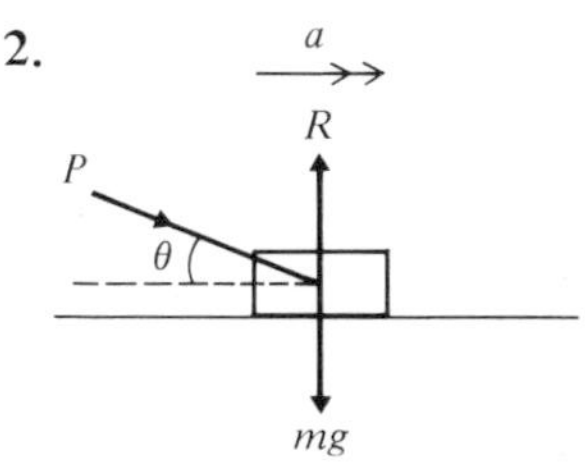

Prove: $\tan\theta = \dfrac{R - mg}{ma}$

3.

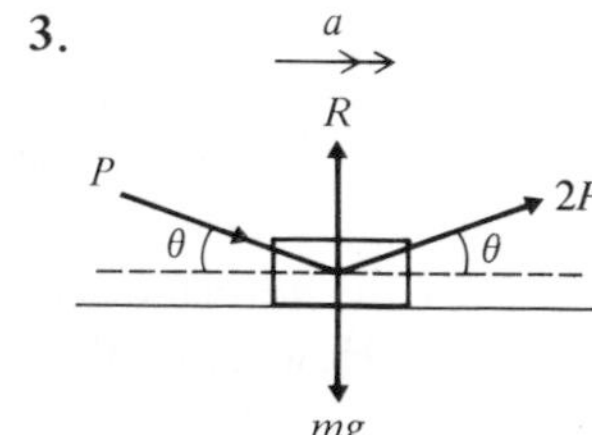

Prove: $3R = m(3g - a\tan\theta)$

4.

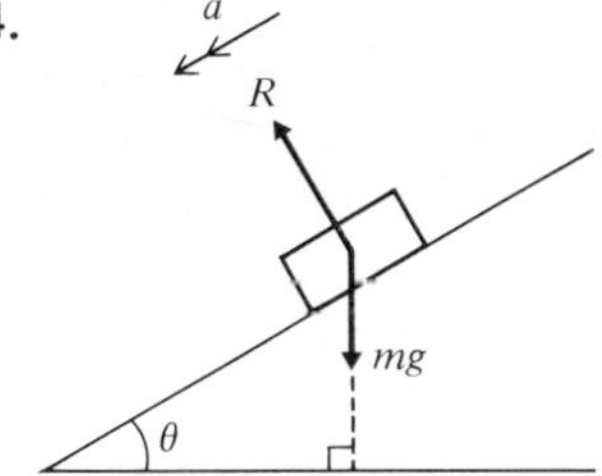

Prove: $R \tan \theta = ma$

5.

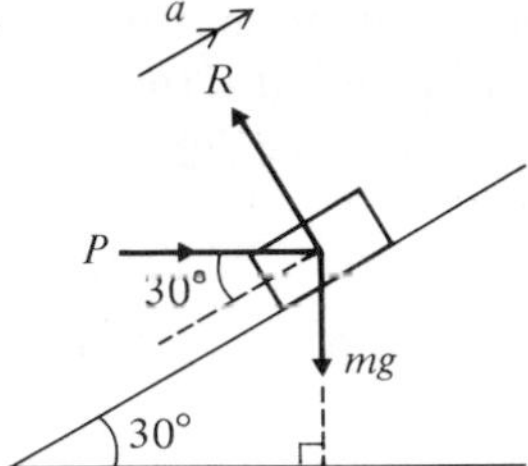

Prove: $\sqrt{3}\,P = m(2a + g)$

6.

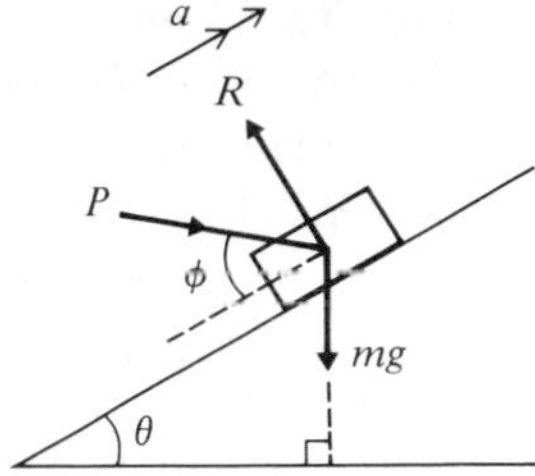

Prove: $R = m(g \cos \theta + a \tan \phi + g \tan \phi \sin \theta)$

7.

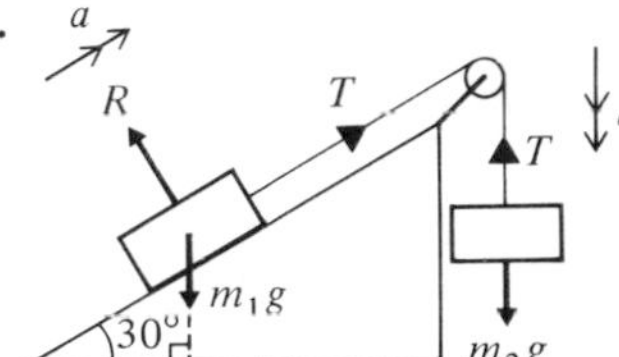

Prove: $a = \dfrac{(2m_2 - m_1)g}{2(m_1 + m_2)}$

8.

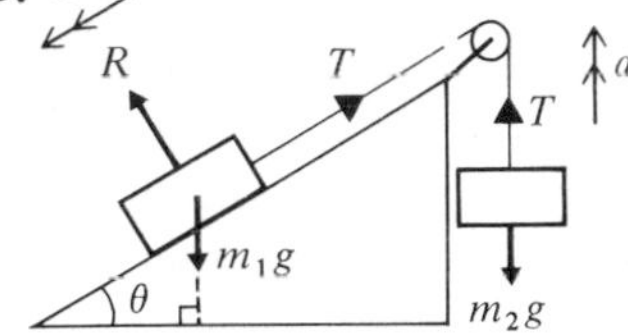

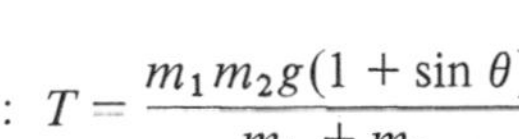

Prove: $T = \dfrac{m_1 m_2 g(1 + \sin \theta)}{m_1 + m_2}$

9.

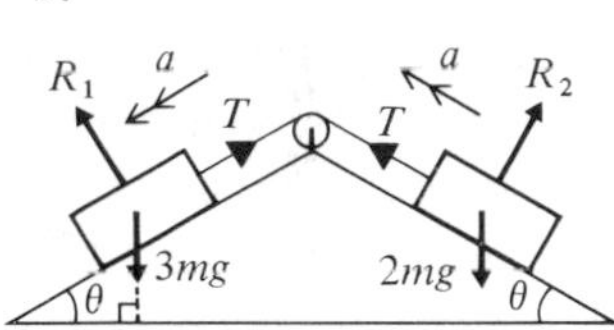

Prove: $g \sin \theta = 5a$

10. A body of mass m is pulled along a smooth horizontal surface by a force P inclined at θ above the horizontal. If the mass starts from rest, show that the distance moved in time t is given by $\dfrac{Pt^2 \cos \theta}{2m}$.

11. A body of mass m is pulled along a rough horizontal surface by a force P inclined at θ above the horizontal. If the mass accelerates from rest to velocity v in a distance d, show that the resistance to motion (assumed constant throughout) is $P \cos \theta - \dfrac{mv^2}{2d}$.

12. A mass of 5 kg is pulled along a rough horizontal surface by a force of 50 N inclined at 60° above the horizontal. The mass starts from rest and after 4 seconds the pulling force ceases. If the resistance to motion is 20 N throughout, find the total distance travelled before the mass comes to rest again.

13. A body of mass m is released from rest at the top of a smooth slope which is inclined at θ to the horizontal. Show that its velocity, when it has travelled a distance s down the slope, is given by $\sqrt{(2gs \sin \theta)}$.

14. A body of mass m is released from rest at the top of a rough plane which is inclined at θ to the horizontal. After time t the mass has travelled a distance d down the slope. Show that the resistance to motion experienced by the body is $\dfrac{m}{t^2}(gt^2 \sin \theta - 2d)$. (Assume the resistance to be constant throughout.)

Questions **15**, **16** and **17** refer to the situation shown. The body A lies on a smooth slope and body B is freely suspended. The pulley is smooth and the string light and inextensible.

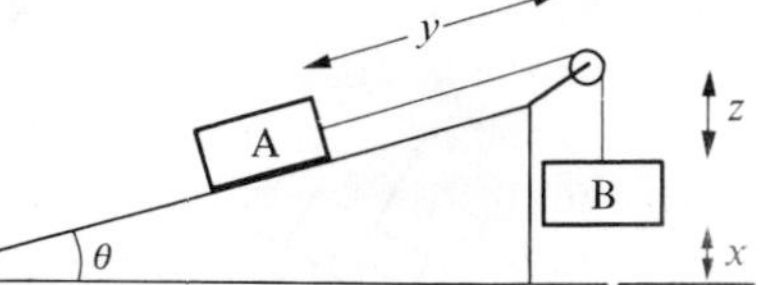

15. The mass of A is 4 kg and the mass of B is 3 kg. With $\theta = 30°$, body A will accelerate up the slope. If $y = 3$ m and $x = 2{\cdot}8$ m, find the velocity with which A hits the pulley.

16. If the mass of A is $2m$ and the mass of B is m, show that A will accelerate up the slope provided $\sin\theta < 0{\cdot}5$.
With this condition fulfilled and $y > x$, show that, if the system is released from rest, mass B hits the ground with velocity $\sqrt{\dfrac{2gx\,(1 - 2\sin\theta)}{3}}$ and that A reaches the pulley provided $x \geqslant \dfrac{3y\sin\theta}{1 + \sin\theta}$.

17. The mass of A is m_1 and the mass of B is m_2. Show that A will accelerate down the slope provided $m_1 \sin\theta > m_2$.
With this condition fulfilled, the system is released from rest and when A has travelled a distance d down the slope ($d < z$), the string connecting the two masses is cut. Show that the greatest height reached by B above the floor is $\dfrac{x(m_2 + m_1) + dm_1(1 + \sin\theta)}{(m_1 + m_2)}$.

18. Masses m_1 and m_2 are held at rest on inclined surfaces in the positions shown in the diagram ($s > d$). They are connected by a light inextensible string passing over a smooth pulley. Show that, when the system is released, m_1 will accelerate towards the pulley provided $\sqrt{3} > \dfrac{m_1}{m_2}$.

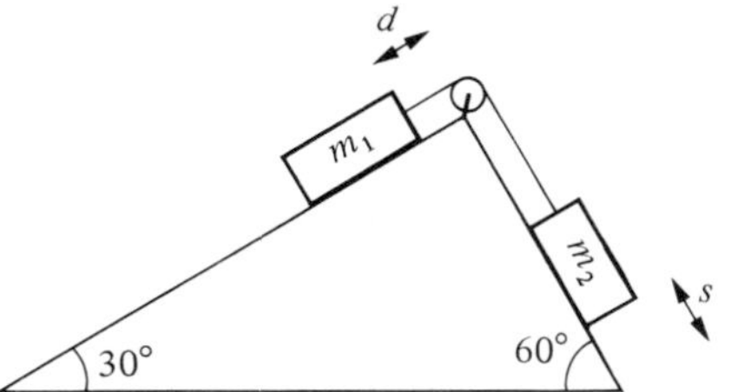

With this condition fulfilled show that m_1 hits the pulley with speed $\sqrt{\dfrac{dg(\sqrt{3}\,m_2 - m_1)}{m_1 + m_2}}$

Exercise 5E Examination questions

1. A particle is in equilibrium under the action of three forces **P**, **Q** and **R**. Given that $\mathbf{P} = (3\mathbf{i} + 5\mathbf{j})$, $\mathbf{Q} = (-2\mathbf{i} + 6\mathbf{j})$, calculate (a) the magnitude of **R**, (b) the tangent of the acute angle made by the line of action of **R** with the positive x-axis. (London)

2. The following horizontal forces pass through a point O: 5 N in a direction 000°, 1 N in a direction 090°, 4 N in a direction 225° and 6 N in a direction 315°. Find the magnitude and direction of their resultant.
Two further horizontal forces are introduced to act at O: P N in a direction 135° and Q N in a direction 225°. If the complete set of forces is now in equilibrium calculate the value of P and of Q. (Cambridge)

3. Four horizontal concurrent forces, 20 N acting in a direction 245°, 12 N in a direction 020°, P N in a direction 320° and Q N in a direction 110° are in equilibrium. Determine the value of P and of Q. (Cambridge)

4. The diagram represents a light inextensible string ABCDE in which AB = BC = CD = DE and to which are attached masses M, m and M at the points B, C and D respectively. The system hangs freely in equilibrium with the ends A and E of the string fixed in the same horizontal line. AB and DE each make an acute angle α with the vertical such that $\tan \alpha = \frac{3}{4}$. BC and CD each make an acute angle β with the vertical such that $\tan \beta = \frac{12}{5}$.

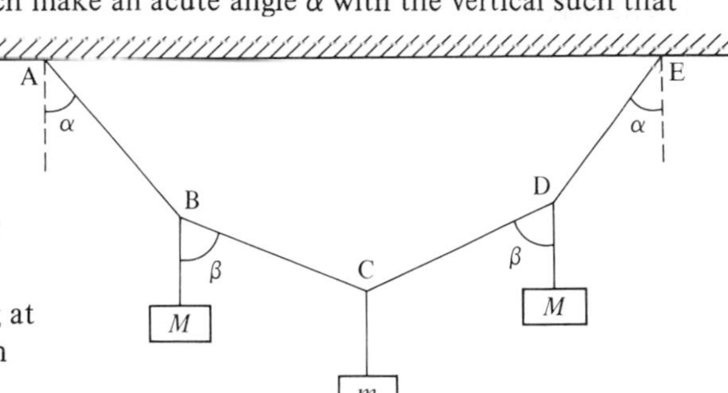

   (i) By considering the forces acting at C, calculate the tension in BC in terms of m and g.
   (ii) By considering the forces acting at B, calculate the tension in AB in terms of M and g.
   Show also that $10M = 11m$. (J.M.B)

5. A water skier of mass 95 kg is towed by a horizontal rope behind a boat. His body is straight, and the thrust of the water acts along the line of his body. When moving with uniform velocity, he is leaning back at 10° from the vertical. Find the tension in the rope. The boat begins to accelerate, and the skier leans back at 15°. The tension in the rope now becomes 500 N. Find the acceleration of the boat. (Oxford)

6. A children's slide consists of a sloping part 5 m long connected at the bottom of the slope to a horizontal part 5 m long. A boy of mass 30 kg starts from rest at the top of the slide and is subjected to a constant frictional resistance of 30 N while on either part of the slide. The component of his weight down the slide while he is on the sloping part is 120 N. Calculate
   (i) his acceleration down the sloping part;
   (ii) his speed on reaching the bottom of the slope;
   (iii) his retardation along the horizontal part;
   (iv) his final speed at the end of the horizontal part;
   (v) his total time on the whole slide. (Oxford)

7. Two particles A and B, of masses 0·4 kg and 0·3 kg respectively, are connected by a light inextensible string. The particle A is placed near the bottom of a smooth plane inclined at 30° to the horizontal. The string passes over a small smooth light pulley which is fixed at the top of the inclined plane and B hangs freely. The system is released from rest, with each portion of the string taut and in the same vertical plane as a line of greatest slope of the inclined plane. Calculate (a) the common acceleration, in m/s^2, of the two particles, (b) the tension, in N, in the string.
   Given that A has not reached the pulley, find (c) the time taken for B to fall 6·3 m from rest, (d) the speed that B has then acquired. (London)

8. With reference to Cartesian axes Ox, Oy, the vertices of a square are A(a, a), B($-a$, a), C($-a$, $-a$), D(a, $-a$). A particle P (x, y) is subject to forces acting along the lines joining P to the four vertices.
   (i) In the case when the forces are $k\overrightarrow{PA}$, $k\overrightarrow{BP}$, $k\overrightarrow{PC}$, $k\overrightarrow{DP}$ show that P is in equilibrium.
   (ii) In the case when the forces are $k\overrightarrow{PA}$, $k\overrightarrow{PB}$, $k\overrightarrow{PC}$, $k\overrightarrow{PD}$ show, by using components or otherwise, that the resultant force is $4k\overrightarrow{PO}$. (J.M.B)

6
Friction

Rough and smooth surfaces

A block of mass M kg rests on a horizontal table and a horizontal force of P newtons is applied to the block.
From Newton's Third law, it is known that equal and opposite forces act on the block and on the plane at right angles to the surfaces in contact.

$$\therefore \quad R = Mg$$

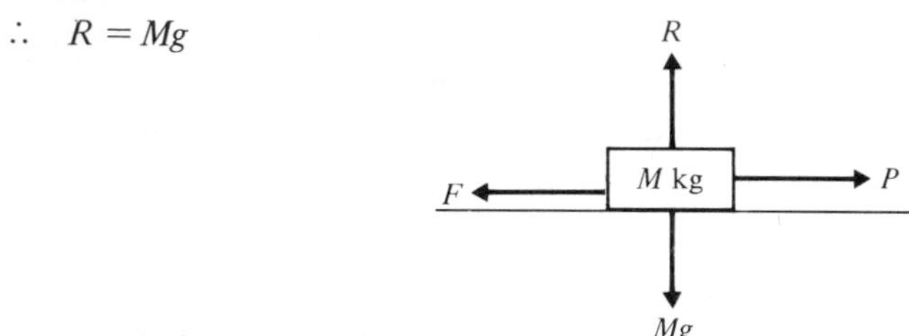

It is known from experience that, if the surfaces in contact are highly polished, it is easier to move the block than if both the underside of the block and the surface of the table are covered with sandpaper.
The force F which opposes the motion of the block, is called the frictional force and it acts in a direction to oppose the motion and is parallel to the surfaces in contact. If the surfaces were perfectly smooth, there would be no frictional force; hence $F = 0$ and motion would take place however small the applied force P might be. In practice, it is not possible to have perfectly smooth surfaces, although in particular instances we may consider the surfaces to be smooth.
When the surfaces are rough, the block will only move if P is greater than the frictional force F. The magnitude of the frictional force depends upon the roughness of the surfaces in contact and also upon the force P which is trying to move the block.

Limiting equilibrium

The frictional force F for a particular block and surface is not constant, but increases as the applied force P increases until the force F reaches a value F_{max} beyond which it cannot increase The block is then on the point of moving and is said to be in a state of *limiting equilibrium.*

$$\text{Suppose } F_{max} = P_1$$

If the applied force P is increased still further to a value P_2, the frictional force cannot increase as it has already reached its maximum value, and the block will therefore move.

Applying force = mass × acceleration, the equation of motion is:

$$P_2 - F_{max} = M \times a$$

Coefficient of friction

The magnitude of the maximum frictional force is a fraction of the normal reaction R. This fraction is called the coefficient of friction μ for the two surfaces in contact.

$$F_{max} = \mu R$$

For a perfectly smooth surface, $\mu = 0$.

It should be noted that the maximum frictional force will only act if (a) there is a state of limiting equilibrium, or (b) motion is taking place.

The frictional force F is only as large as is necessary to prevent motion.

Laws of friction

The laws governing the equilibrium of two bodies in contact and the motion of one body on another, may be summarized as follows.

The frictional force

(i) acts parallel to the surfaces in contact and in a direction so as to oppose the motion of one body across the other,

(ii) will not be larger than is necessary to prevent this motion,

(iii) has a maximum value μR, where R is the normal reaction between the surfaces in contact,

(iv) can be assumed to have its maximum value μR when motion occurs,

(v) depends upon the nature of the surfaces in contact and not upon the contact area.

Example 1

Calculate the maximum frictional force which can act when a block of mass 2 kg rests on a rough horizontal surface, the coefficient of friction between the surfaces being
(a) 0·7, (b) 0·2.

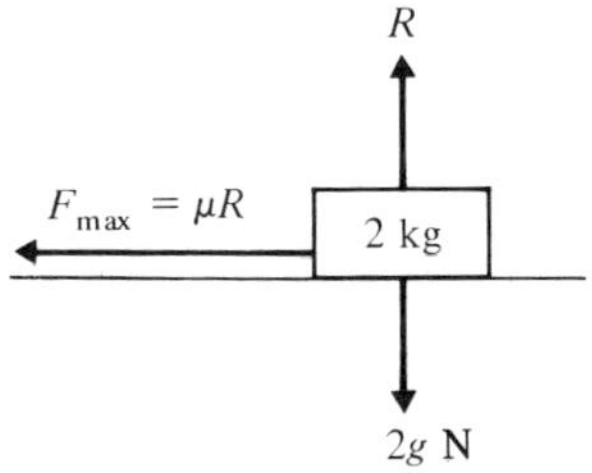

(a) There is no motion perpendicular to the plane

Resolving vertically $R = 2g$

$\therefore \quad R = 19{\cdot}6$ N

The maximum frictional force $F_{max} = \mu R$

$= (0{\cdot}7) \times 19{\cdot}6$

$= 13{\cdot}72$ N

(b) As before $R = 19{\cdot}6$ N

Maximum frictional force $= \mu R$

$= (0{\cdot}2) \times 19{\cdot}6$

$= 3{\cdot}92$ N

Example 2

A block of mass 5 kg rests on a rough horizontal plane, the coefficient of friction between the block and the plane being 0·6. Calculate the frictional force acting on the block when a horizontal force P is applied to the block and the magnitude of P is (a) 12 N, (b) 28 N, (c) 36 N. Also calculate the magnitude of any acceleration that may occur.

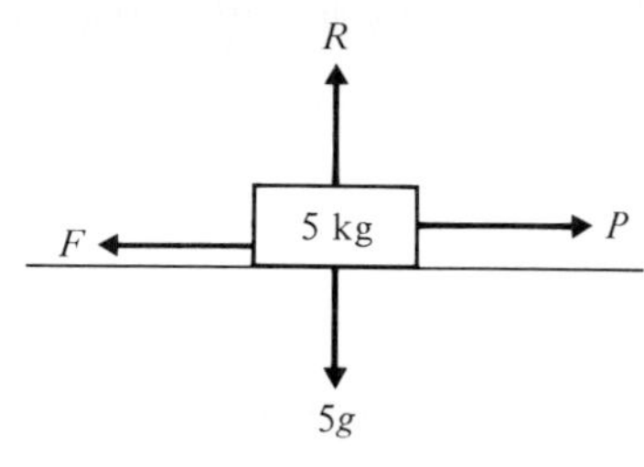

There is no motion perpendicular to the plane

Resolving vertically $R = 5g$

$\therefore \quad R = 49$ N

The frictional force will act in the direction opposite to that in which the force P acts. The maximum value of the frictional force is μR.

$$\begin{aligned}\mu R &= 0{\cdot}6 \times 49\\ &= 29{\cdot}4 \text{ N}\end{aligned}$$

(a) If $P = 12$ N, then P is less than μR, so there is no motion

frictional force $F = P$

$F = 12$ N

(b) If $P = 28$ N, then again P is less than μR and there is no motion

frictional force $F = P$

$F = 28$ N

(c) If $P = 36$ N, then P is greater than the maximum value of the frictional force, which is 29·4 N

frictional force acting = 29·4 N which does not prevent motion

The block will move and the maximum value μR of the frictional force will be maintained.

Using $F = ma$, the equation of motion is

$$\begin{aligned}P - \mu R &= m \times a\\ 36 - 29{\cdot}4 &= 5 \times a\\ \therefore \quad a &= 1{\cdot}32 \text{ m/s}^2\end{aligned}$$

Applied force not horizontal

When the force P acting on the block of mass M is inclined at an angle θ above the horizontal, this has two effects:

(i) the component of P in a vertical direction decreases the magnitude of the normal reaction R,

(ii) only the component of P in a horizontal direction is tending to move the block.

Hence the value of R is less than it would have been if the force P had been applied horizontally. The maximum frictional force μR is also therefore reduced. In addition, a smaller force is tending to cause motion, a component of P rather than P.

Example 3

A 10 kg trunk lies on a horizontal rough floor. The coefficient of friction between the trunk and the floor is $\frac{\sqrt{3}}{4}$. Calculate the magnitude of the force P which is necessary to pull the trunk horizontally if P is applied (a) horizontally, (b) at 30° above the horizontal.

(a) Resolving vertically $\qquad R = 10g$

In the position of limiting equilibrium

$$P_1 = \mu R$$
$$= \frac{(\sqrt{3})}{4} \times 10g$$
$$= 42{\cdot}43 \text{ N}$$

For motion to take place the applied force must exceed 42·43 N.

(b) Resolving vertically $\quad R + P_2 \cos 60 = 10g$

$$R = 98 - \frac{P_2}{2}$$

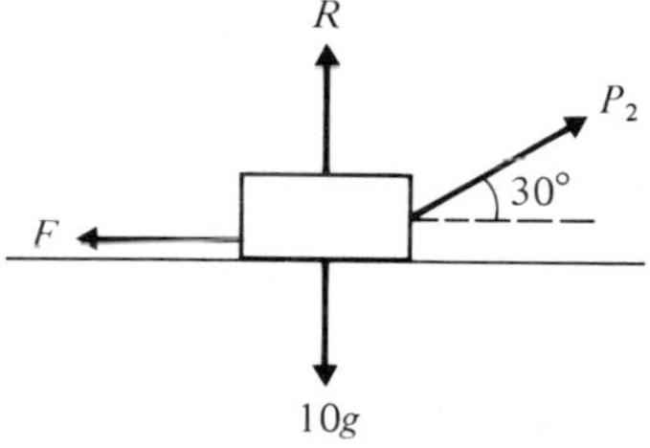

In the position of limiting equilibrium

$$P_2 \cos 30 = \mu R$$
$$\therefore \quad P_2 \cos 30 = \mu\left(98 - \frac{P_2}{2}\right)$$

Hence
$$P_2 \times \frac{\sqrt{3}}{2} + \frac{\sqrt{3}}{4} \times \frac{P_2}{2} = \frac{\sqrt{3}}{4} \times 98$$
$$\text{or} \quad P_2 = 39{\cdot}2 \text{ N}$$

For motion to take place the applied force must exceed 39·2 N.

Therefore it is easier to move such a trunk if the pulling force is inclined upwards as this reduces the frictional force opposing the motion.

Exercise 6A

1. Each of the following diagrams shows a body of mass 10 kg initially at rest on a rough horizontal plane. The coefficient of friction between the body and the plane is $\frac{1}{7}$. In each case, R is the normal reaction and F the frictional force exerted on the body, by the plane. Any other forces applied to the body are as shown. In each case, find the magnitude of F and state whether the body will remain at rest or will accelerate along the plane.

(a)

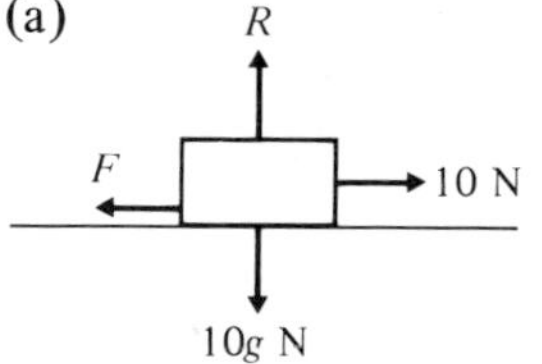

(b)

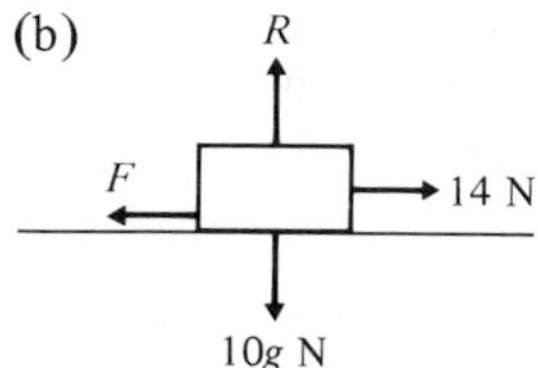

(c)

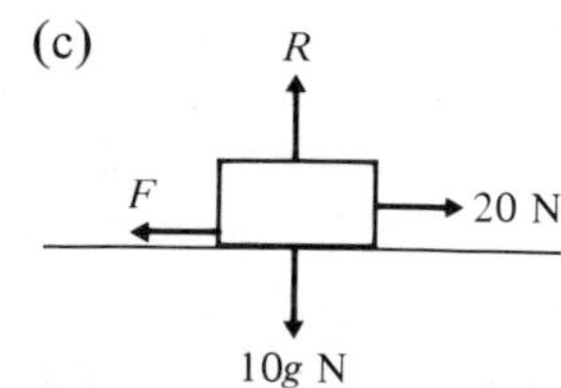

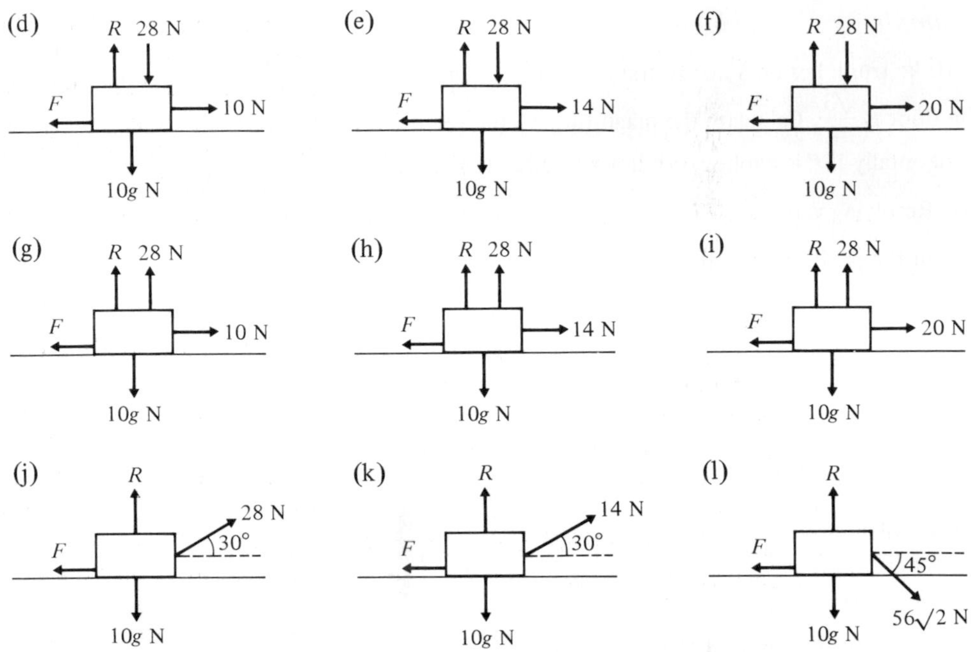

2. In each of the following situations, the forces shown cause the body of mass 5 kg to accelerate along the rough horizontal plane. The direction and magnitude of each acceleration is as indicated; R is the normal reaction and F the frictional force exerted on the body by the plane. For each case, find the coefficient of friction between the body and the plane.

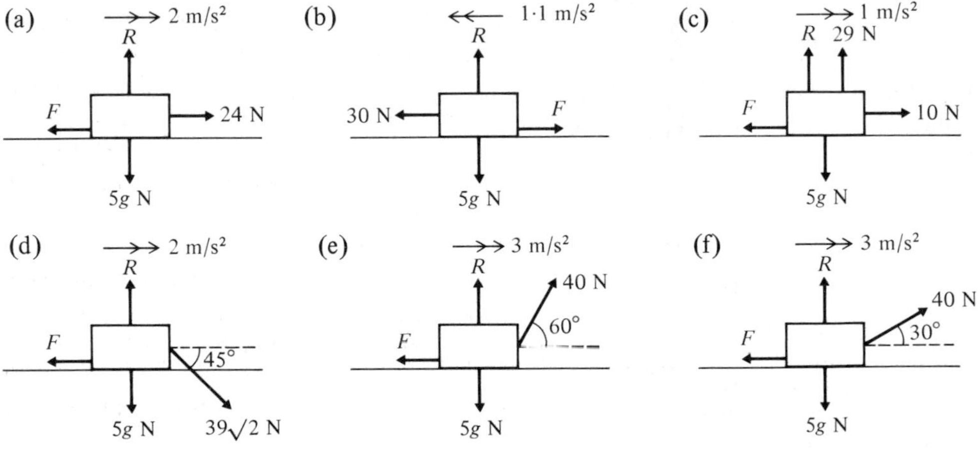

3. When a horizontal force of 28 N is applied to a body of mass 5 kg which is resting on a rough horizontal plane, the body is found to be in limiting equilibrium. Find the coefficient of friction between the body and the plane.
4. When a horizontal force of 0·245 N is applied to a body of mass 250 g which is resting on a rough horizontal plane, the body is found to be in limiting equilibrium. Find the coefficient of friction between the body and the plane.

5. A block of mass 20 kg rests on a rough horizontal plane. The coefficient of friction between the block and the plane is 0·25. Calculate the frictional force experienced by the block when a horizontal force of 50 N acts on the block. State whether the block will move and, if so, find its acceleration.
6. A block of mass 15 kg rests on a rough horizontal plane. The coefficient of friction between the block and the plane is 0·35. Calculate the frictional force experienced by the block when a horizontal force of 50 N acts on the block. State whether the block will move and, if so, find its acceleration.
7. A block of mass 500 g rests on a rough horizontal table. The coefficient of friction between the block and the table is 0·1. Calculate the frictional force experienced by the block when a horizontal force of 1 N acts on the block. State whether the block will move and, if so, find its acceleration.
8. A block of mass 2 kg is initially at rest on a rough horizontal table. The coefficient of friction between the block and the table is 0·5. Find the horizontal force that must be applied to the block to cause it to accelerate along the surface at (a) 5 m/s^2, (b) $0{\cdot}1 \text{ m/s}^2$.
9. When a horizontal force of 37 N is applied to a body of mass 10 kg which is resting on a rough horizontal surface, the body moves along the surface with an acceleration of $1{\cdot}25 \text{ m/s}^2$. Find μ, the coefficient of friction between the body and the surface.
10. A body of mass 2 kg is sliding along a smooth horizontal surface at a constant speed of 2 m/s when it encounters a rough horizontal surface, coefficient of friction 0·2. Find the distance that the body will move across the rough surface before it comes to rest.
11. A body of mass 1 kg is initially at rest on a rough horizontal surface, coefficient of friction 0·25. A constant horizontal force is applied to the body for 5 seconds and is then removed. Given that, when the force is removed, the body has a velocity of 3·5 m/s along the surface, find (a) the acceleration of the body when experiencing the applied force, (b) the magnitude of the applied force, (c) the retardation of the body when the force is removed, (d) the total distance travelled by the body.
12. A box of mass 2 kg lies on a rough horizontal floor, coefficient of friction 0·5. A light string is attached to the box in order to pull the box across the floor. If the tension in the string is T N, find the value that T must exceed for motion to occur if the string is (a) horizontal, (b) 30° above the horizontal, (c) 30° below the horizontal.
13. A box of mass 2 kg lies on a rough horizontal floor, coefficient of friction 0·2. A light string is attached to the box in order to pull the box across the floor. If the tension in the string is T N, find the value that T must exceed for motion to occur if the string is (a) horizontal, (b) 45° above the horizontal, (c) 45° below the horizontal.
14. A body of mass 100 g rests on a rough horizontal surface and has a light string, inclined at 20° above the horizontal, attached to it. When the tension in the string is 5×10^{-1} N, the body is found to be in limiting equilibrium. Find the coefficient of friction between the body and the surface.
    What would the tension in the string have to be for the body to accelerate along the surface at $1{\cdot}5 \text{ m/s}^2$?
15. A mass of 3 kg lies on a rough horizontal surface, coefficient of friction $\frac{1}{7}$. State whether or not the mass will slide along the surface when the surface is moved horizontally with an acceleration of (a) 1 m/s^2, (b) $1{\cdot}4 \text{ m/s}^2$, (c) 2 m/s^2.
16. A mass of 6 kg lies on a rough horizontal surface, coefficient of friction 0·25. State whether or not the mass will slide along the surface when the surface is moved horizontally with an acceleration of (a) $2{\cdot}4 \text{ m/s}^2$, (b) $2{\cdot}6 \text{ m/s}^2$, (c) 3 m/s^2.

17. A parcel is placed on the tail-board of a stationary lorry. The tail-board is horizontal and the coefficient of friction between the parcel and the tail-board is 0·2. State whether or not the parcel will slide along the surface of the tail-board when the lorry moves off horizontally with an acceleration of (a) 1 m/s^2, (b) 1·5 m/s^2, (c) 2·5 m/s^2.

18. A cake of mass 500 g lies on the horizontal surface of a plate. The coefficient of friction between the cake and the plate is 0·1. Will the cake slide across the plate when the plate is moved horizontally with an acceleration of 1·1 m/s^2?

Questions **19** to **22** refer to the system shown in the diagram. Body A lies on a rough horizontal table and is connected to the freely hanging body B by a light inextensible string passing over a smooth pulley.

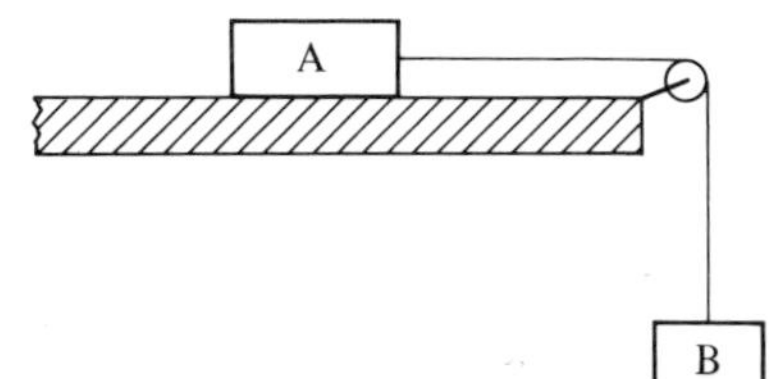

19. The masses of A and B are 6 kg and 1 kg respectively and the coefficient of friction between body A and the table is 0·2. If the system is released from rest, find the frictional force experienced by A and state whether motion will occur.

20. The masses of A and B are 90 g and 50 g respectively and the coefficient of friction between body A and the table is $\frac{1}{3}$. If the system is released from rest, find (a) the acceleration of the system, (b) the tension in the string, (c) the distance moved by A in the first second of motion, assuming that nothing impedes the motion of either mass.

21. The masses of A and B are 1 kg and 500 g respectively and the coefficient of friction between body A and the table is $\frac{1}{3}$. The system is released from rest with A 3 metres from the pulley and B 2·5 metres above the floor. Find (a) the initial acceleration of the system, (b) the speed with which B hits the floor, (c) the speed with which A hits the pulley.

22. The masses of A and B are m_1 and m_2 respectively and the coefficient of friction between body A and the table is μ. Show that, if the system is released from rest, motion will occur if $m_2 > m_1\mu$.
If this condition is fulfilled show that the resulting acceleration will be $\dfrac{g(m_2 - \mu m_1)}{m_1 + m_2}$.

Questions **23** to **25** refer to the system shown in the diagram. Body A lies on a rough horizontal table and is connected to freely hanging bodies B and C by light inextensible strings passing over smooth pulleys. The masses, pulleys and strings all lie in the same vertical plane.

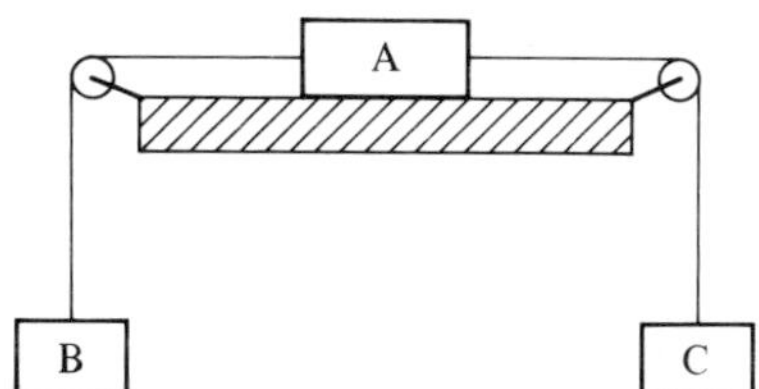

23. The masses of A, B and C are $4m$, m and $5m$ respectively and the coefficient of friction between body A and the table is $\frac{1}{4}$. Find the acceleration of the system when released from rest.

24. The masses of A, B and C are 5 kg, 3 kg and 2 kg respectively. When the system is released from rest, body B descends with an acceleration of 0·28 m/s^2. Find the coefficient of friction between body A and the table.

25. The masses of A, B and C are m_1, m_2 and m_3 respectively and the coefficient of friction between body A and the table is μ. Show that, if the system is released from rest, body B will move downwards, provided $m_2 > \mu m_1 + m_3$.

If this condition is fulfilled, show that the resulting acceleration will be $\dfrac{g(m_2 - \mu m_1 - m_3)}{m_1 + m_2 + m_3}$.

Rough inclined plane

A body of mass M kg rests on a plane which is inclined at θ to the horizontal.
The vertical force Mg can be resolved into two components, parallel to and perpendicular to the surface of the plane.
The plane exerts a normal reaction R on the body; since there is no motion at right angles to the plane, the normal reaction balances the component of Mg acting in this direction.

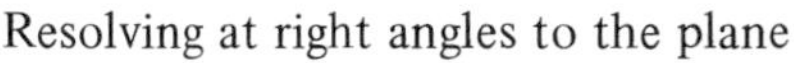

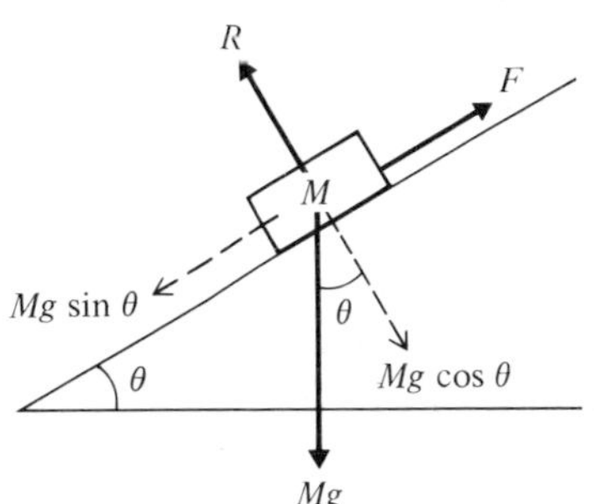

Resolving at right angles to the plane

$$R = Mg\cos\theta$$

The component $Mg\sin\theta$ acting down the plane will cause motion unless the frictional force F, acting up the plane, balances it.

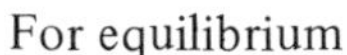

For equilibrium $\quad F = Mg\sin\theta$

The maximum value of F is, as before, μR

$$F_{max} = \mu R = \mu Mg\cos\theta$$

In the position of limiting equilibrium, the maximum frictional force must balance the force tending to produce motion.

$$\therefore \quad \mu Mg\cos\theta = Mg\sin\theta$$

For motion to take place down the plane, $Mg\sin\theta$ must exceed $\mu Mg\cos\theta$

$$\therefore \quad Mg\sin\theta > \mu Mg\cos\theta$$
$$\therefore \quad \tan\theta > \mu$$

Example 4

A mass of 6 kg rests in limiting equilibrium on a rough plane inclined at 30° to the horizontal. Find the coefficient of friction between the mass and the plane.

The mass is on the point of motion down the plane, so the frictional force F acts up the plane and has its maximum value μR.

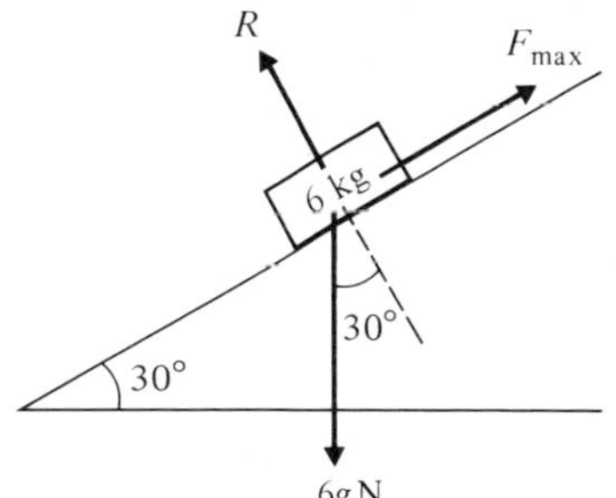

Resolving at right angles to the plane

$$R = 6g\cos 30°$$
$$\therefore \quad R = 3g\sqrt{3}$$

Resolving parallel to the surface of the plane

$$6g\sin 30 = \mu R$$
$$\therefore \quad 6g\sin 30 = \mu \times 3g\sqrt{3}$$
$$\therefore \quad \mu = \frac{1}{\sqrt{3}}$$

Example 5

A mass of 3 kg rests on a rough plane inclined at 60° to the horizontal and the coefficient of friction between the mass and the plane is $\frac{\sqrt{3}}{5}$. Find the force P, acting parallel to the plane, which must be applied to the mass in order to just prevent motion down the plane.
The frictional force F acts up the plane and together with the applied force P balances the component of the weight acting down the plane.

Resolving parallel to the surface of the plane

$$P + F = 3g \sin 60 \qquad \ldots [1]$$

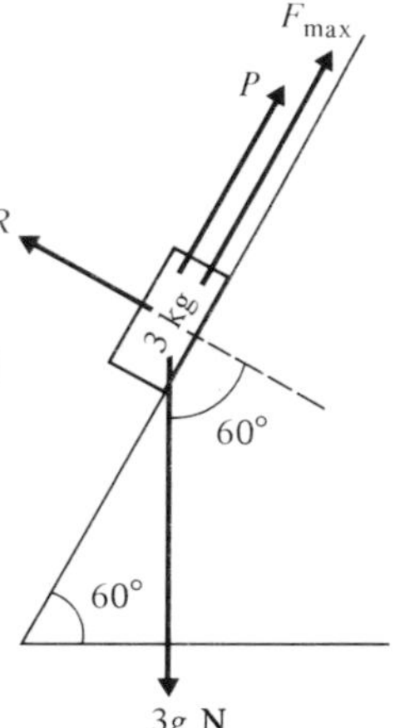

There is no motion perpendicular to the plane and so, resolving at right angles to the plane

$$R = 3g \cos 60$$

$$\therefore \quad R = \frac{3g}{2} \text{ N}$$

Since the mass is in limiting equilibrium (motion is *just* prevented)

$$F = \mu R$$

$$= \frac{\sqrt{3}}{5} \times \frac{3g}{2}$$

Substituting for F in equation [1]:

$$P + \frac{3\sqrt{3}}{10} g = \frac{3g\sqrt{3}}{2}$$

$$\therefore \quad P = \frac{6\sqrt{3}g}{5} \text{ N}$$

Motion up the plane

If the force P applied to the mass is larger than the component of Mg resolved down the plane, then the tendency will be for the mass to move *up* the plane. In this case the frictional force will act *down* the plane, opposing the motion of the mass.

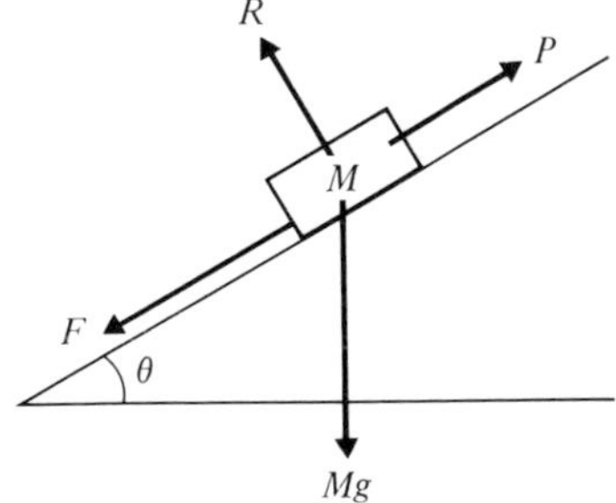

Motion will take place up the plane if

$$P > Mg \sin \theta + F_{max}$$

but

$$F_{max} = \mu R$$

$$= \mu Mg \cos \theta$$

So for motion up the plane $P > Mg \sin \theta + \mu Mg \cos \theta$

Example 6

A mass of 0·5 kg rests on a rough plane. The coefficient of friction between the mass and the plane is $\frac{1}{\sqrt{2}}$, and the plane is inclined at angle θ to the horizontal such that $\sin \theta = \frac{1}{3}$.

Investigate the motion of the mass when it experiences a force of 6 N applied up the plane along a line of greatest slope.

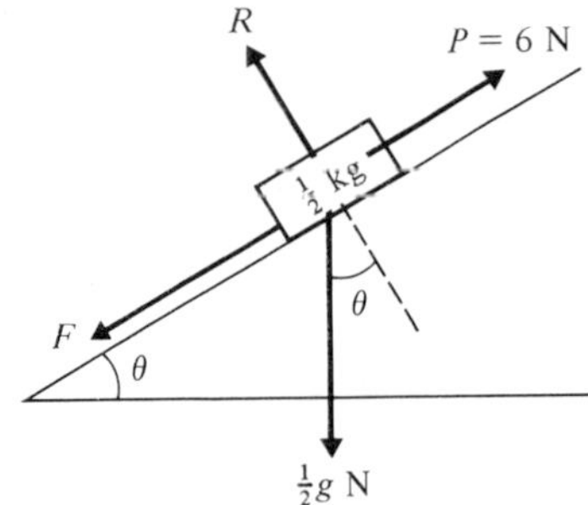

Since $\sin\theta = \frac{1}{3}$, $\cos\theta = \dfrac{2\sqrt{2}}{3}$

Resolving perpendicular to the plane

$$R = \tfrac{1}{2}g\cos\theta$$

$$\therefore\quad R = \tfrac{1}{2}g \times \frac{2\sqrt{2}}{3} = \frac{g\sqrt{2}}{3}$$

The forces acting down the plane are the component of the weight and the frictional force F if we assume the mass is tending to move up the plane. The component of the weight down the plane is $\frac{1}{2}g\sin\theta$ or $\frac{1}{2}g \times \frac{1}{3}$ or $\dfrac{g}{6}$ N.

The magnitude of F_{max} $= \mu R$

$$= \frac{\mu g\sqrt{2}}{3} - \frac{1}{\sqrt{2}}\,\frac{g\sqrt{2}}{3} = \frac{g}{3}\text{ N.}$$

Hence the applied force of 6 N is greater than the sum of the forces acting down the plane. Motion takes place up the plane and the equation of motion is

$$6 - \tfrac{1}{2}g\sin\theta - F_{max} = \text{mass} \times \text{acceleration}$$

$$\therefore\quad 6 - \frac{g}{6} - \frac{g}{3} = \tfrac{1}{2} \times a$$

$$\therefore\quad a = 2{\cdot}2\text{ m/s}^2$$

Exercise 6B

1. Each of the following diagrams shows a body of mass 10 kg released from rest on a rough inclined plane. R is the normal reaction and F the frictional force exerted on the body by the plane. In each case, find the magnitude of F and state whether the body will remain at rest or will begin to slip down the plane. For (a), (b) and (c), $\mu = \frac{1}{2}$ and for (d), (e) and (f), $\mu = \frac{1}{4}$.

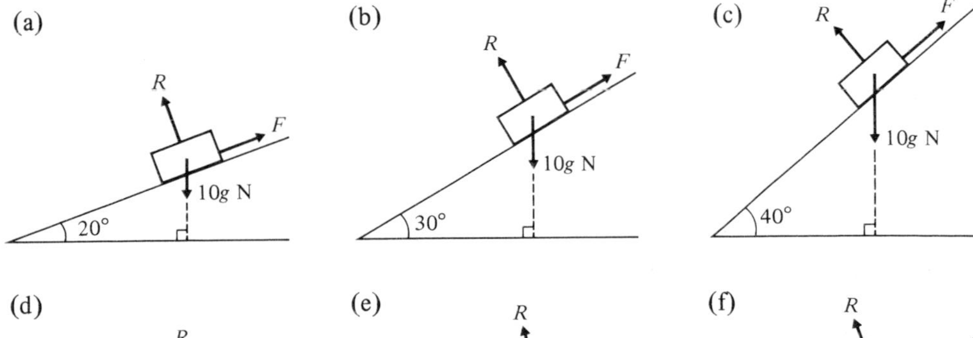

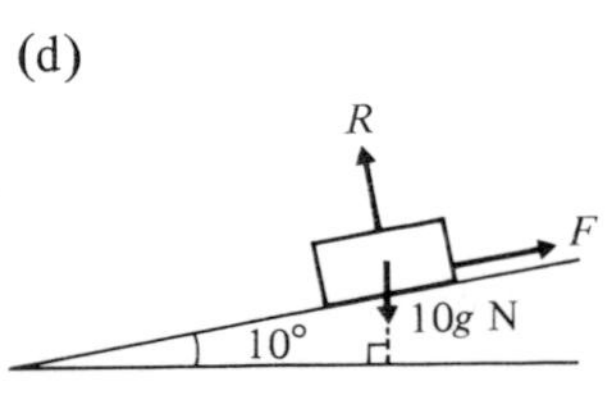

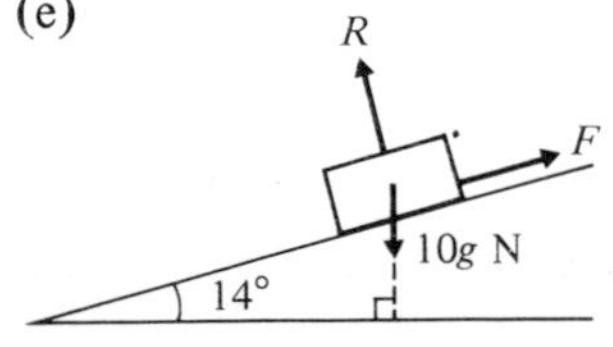

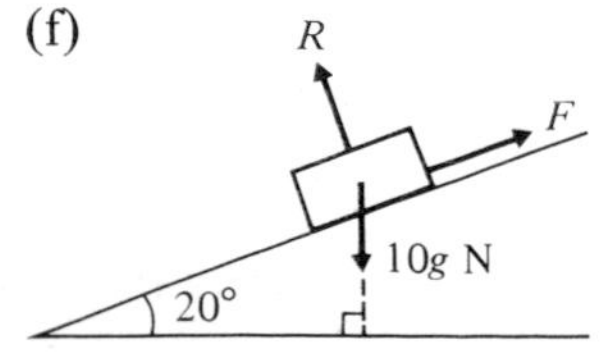

2. Each of the following diagrams shows a body of mass 5 kg on a rough inclined plane, coefficient of friction $\frac{1}{7}$; R is the normal reaction and F the frictional force exerted on the body, by the plane. In each case, find the magnitude of the force X if it just prevents the body from slipping down the plane.

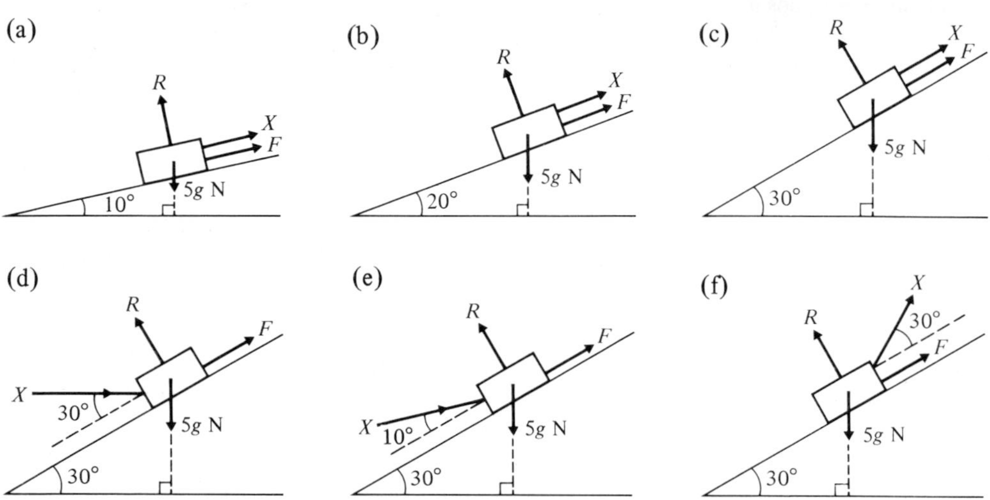

3. Each of the following diagrams shows a body of mass 3 kg on a rough inclined plane, coefficient of friction $\frac{1}{3}$; R is the normal reaction and F the frictional force exerted on the body by the plane. In each case, find the magnitude of the force X if the body is just on the point of moving up the plane.

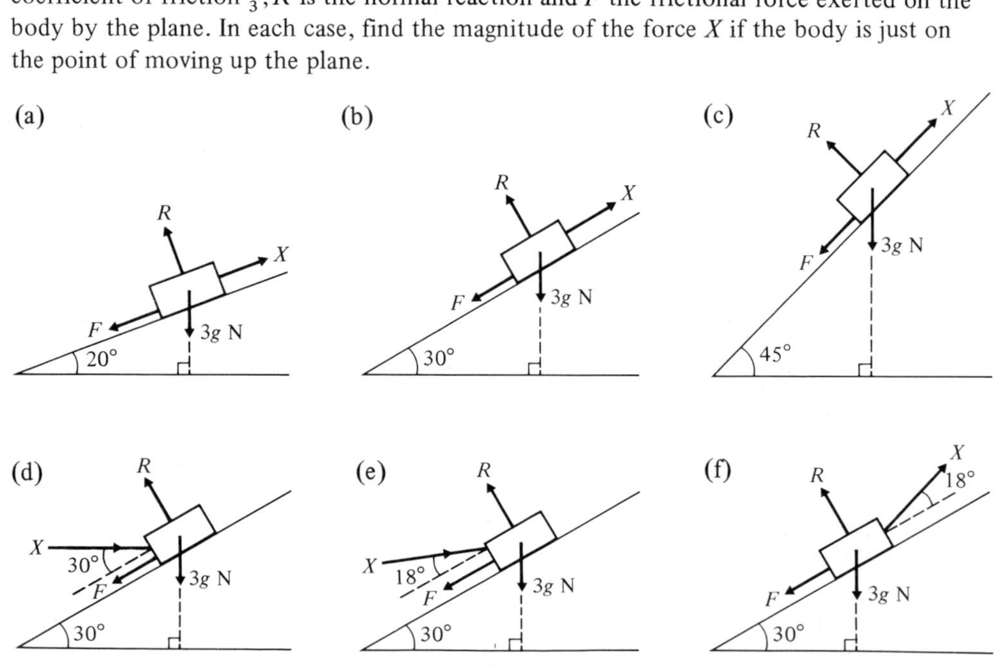

4. In each of the following situations, the forces acting on the body of mass 2 kg cause it to accelerate along the rough inclined plane as indicated. R is the normal reaction and F the frictional force exerted on the body by the plane. Find the value of μ for each situation.

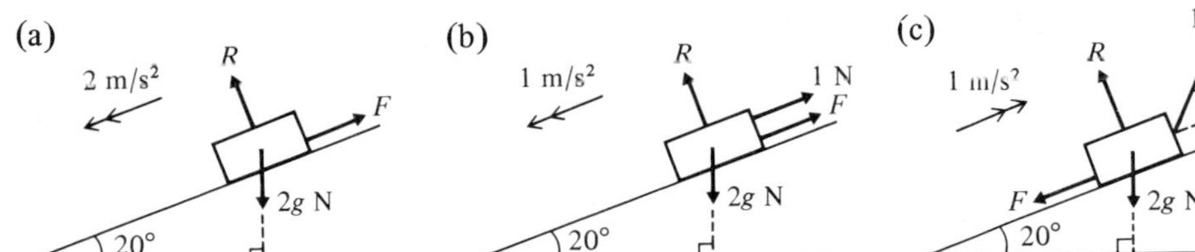

5. A body of mass 500 g is placed on a rough plane which is inclined at 40° to the horizontal. If the coefficient of friction between the body and the plane is 0·6, find the frictional force acting and state whether motion will occur.
6. A body of mass 5 kg lies on a rough plane which is inclined at 35° to the horizontal. When a force of 20 N is applied to the body, parallel to and up the plane, the body is found to be on the point of moving down the plane, i.e. in limiting equilibrium. Find μ, the coefficient of friction between the body and the plane.
7. A body of mass 2 kg lies on a rough plane which is inclined at 30° to the horizontal. When a horizontal force of 20 N is applied to the body in an attempt to push it up the plane, the body is found to be on the point of moving up the plane, i.e. in limiting equilibrium. Find μ, the coefficient of friction between the body and the plane.
8. A body of mass 2 kg lies on a rough plane which is inclined at $\sin^{-1}\frac{5}{13}$ to the horizontal. A force of 20 N is applied to the body, parallel to and up the plane. If the body accelerates up the plane at 1·5 m/s², find μ, the coefficient of friction between the body and the plane.
9. A parcel of mass 1 kg is placed on a rough plane which is inclined at 30° to the horizontal. The coefficient of friction between the parcel and the plane is 0·25. Find the force that must be applied to the parcel in a direction parallel to the plane so that
   (a) the parcel is just prevented from sliding down the plane,
   (b) the parcel is just on the point of moving up the plane,
   (c) the parcel moves up the plane with an acceleration of 1·5 m/s².
10. A box of mass 6 kg is placed on a rough plane which is inclined at 45° to the horizontal. The coefficient of friction between the box and the plane is 0·5. Find the horizontal force that must be applied to the box so that
    (a) the box is just prevented from sliding down the plane,
    (b) the box is just on the point of moving up the plane,
    (c) the box moves up the plane with an acceleration of $2\sqrt{2}$ m/s².
11. A body of mass 3 kg is released from rest on a rough surface which is inclined at $\sin^{-1}\frac{3}{5}$ to the horizontal. If, after $2\frac{1}{2}$ seconds, the body has acquired a velocity of 4·9 m/s down the surface, find the coefficient of friction between the body and the surface.
12. A particle of mass 250 g is released from rest at the top of a rough plane which is inclined at $\sin^{-1}\frac{3}{5}$ to the horizontal. The coefficient of friction between the particle and the plane is $\frac{11}{18}$ and the plane is of length 2·5 m. Find whether the particle will slide down the plane and, if it does, find its speed on reaching the bottom.

13. A body of mass 4 kg lies on a rough plane which is inclined at 16° to the horizontal. A force of 1 N applied parallel to the plane is just sufficient to prevent the body sliding down the plane. Find the coefficient of friction between the body and the plane.
With the body at the top of the plane, the applied force is removed. Find the time taken for the body to reach the bottom of the plane if the length of the plane is 2 m.

14. A horizontal force of 1 N is just sufficient to prevent a brick of mass 600 g sliding down a rough plane which is inclined at $\sin^{-1}\frac{5}{13}$ to the horizontal. Find the coefficient of friction between the brick and the plane.

15. A body of mass 5 kg is initially at rest at the bottom of a rough inclined plane of length 6·3 m. The plane is inclined at 30° to the horizontal and the coefficient of friction between the body and the plane is $\frac{1}{2\sqrt{3}}$. A constant horizontal force of $35\sqrt{3}$ N is applied to the body causing it to accelerate up the plane. Find the time taken for the body to reach the top and its speed on arrival.

16. (a) A mass m lies on a rough plane which is inclined at angle θ to the horizontal. The coefficient of friction between the mass and the plane is μ. Show that slipping will occur if $\tan\theta > \mu$.
(b) Will slipping occur when a body is placed on a rough plane ($\mu = 0{\cdot}5$) which is inclined at 40° to the horizontal?
(c) Will slipping occur when a body is placed on a rough plane ($\mu = 0{\cdot}25$) which is inclined at 10° to the horizontal?
(d) When a body is placed on a rough plane which is inclined at 30° to the horizontal, the body is found to be in limiting equilibrium. Find μ, the coefficient of friction between the body and the plane.

17. A mass of 4 kg lies on a rough plane which is inclined at 30° to the horizontal. A light string has one end attached to this mass, passes up the line of greatest slope, over a smooth pulley fixed at the top of the plane and carries a freely hanging mass of 1 kg at its other end. The tension in the string is just sufficient to prevent the 4 kg mass from sliding down the slope. Find the coefficient of friction between the 4 kg mass and the plane.

18. The diagram shows a mass of 1 kg lying on a rough inclined plane ($\mu = \frac{1}{4}$). From this mass, a light inextensible string passes up the line of greatest slope and over a smooth fixed pulley to a mass of 4 kg hanging freely. The plane makes an angle θ with the horizontal where $\sin\theta = \frac{3}{5}$. Show that the 1 kg mass will slide up the plane and find the velocity with which the 4 kg mass hits the floor.

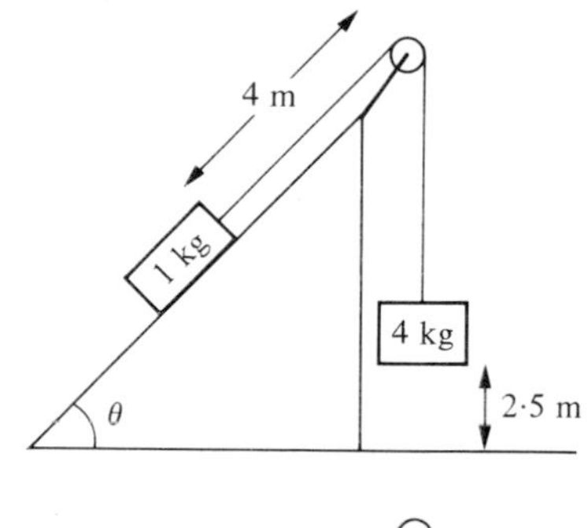

19. The diagram shows a body A of mass 13 kg lying on a rough inclined plane, coefficient of friction μ. From A, a light inextensible string passes up the line of greatest slope and over a smooth fixed pulley to a body B of mass m kg hanging freely. The plane makes an angle θ with the horizontal where $\sin\theta = \frac{5}{13}$. When $m = 1$ kg and the system is released from rest, B has an upward acceleration of a m/s^2; when $m = 11$ kg and the system is released from rest, B has a downward acceleration of a m/s^2. Find a and μ.

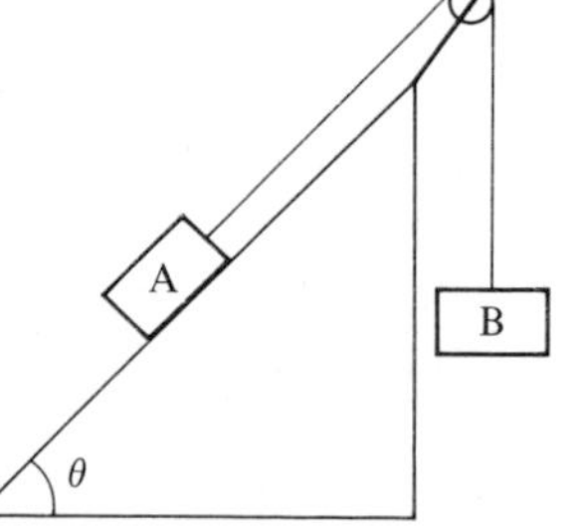

20. Masses of 5 kg and 15 kg are held at rest on inclined surfaces as shown in the diagram. The masses are connected by a light, taut, inextensible string passing over a smooth fixed pulley. The coefficient of friction between each mass and the surface with which it is in contact is 0·25. The inclination of the plane is such that $\sin\theta = \frac{3}{5}$. When the system is released from rest, the 15 kg mass accelerates down the slope. Find the magnitude of this acceleration and the tension in the string.

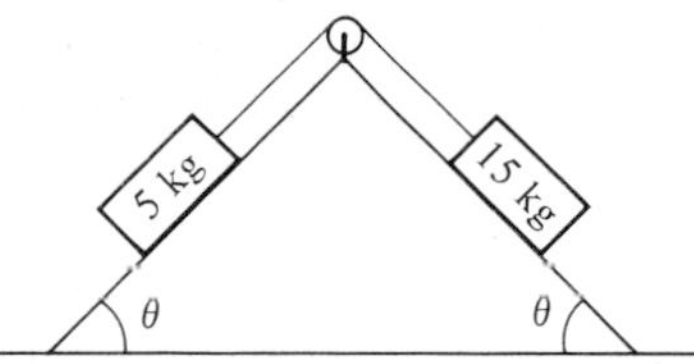

21. A force F acting parallel to and up a rough plane of inclination θ, is just sufficient to prevent a body of mass m from sliding down the plane. A force $4F$ acting parallel to and up the same rough plane causes the mass m to be on the point of moving up the plane. If μ is the coefficient of friction between the mass and the plane, show that $5\mu = 3\tan\theta$.
22. A horizontal force X is just sufficient to prevent a body of mass m from sliding down a rough plane of inclination θ. A horizontal force $4X$ applied to the same mass on the same rough plane, causes the mass to be on the point of moving up the plane. If μ is the coefficient of friction between the mass and the plane show that $5\mu\tan^2\theta - 3(\mu^2 + 1)\tan\theta + 5\mu = 0$.

Angle of friction

When a state of limiting equilibrium exists, the frictional force F has its maximum value $F_{max} = \mu R$, where R is the normal reaction between the surfaces. If this frictional force F_{max} and the normal reaction are compounded into a single force called the resultant reaction, then the angle between the normal reaction and the resultant reaction is called the angle of friction λ.

It can be seen that

$$\tan\lambda = \frac{F_{max}}{R} = \frac{\mu R}{R} = \mu$$

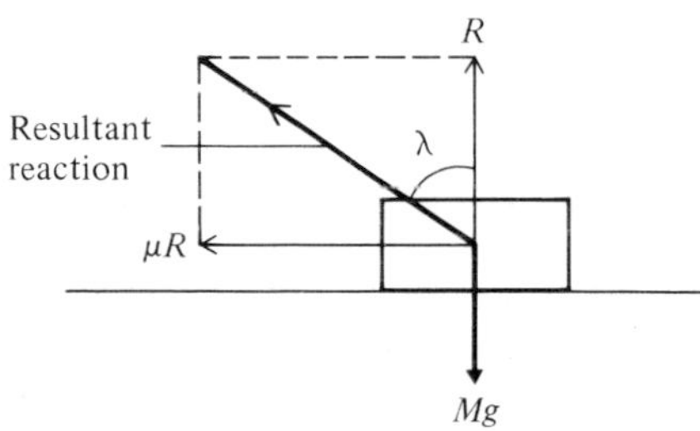

Hence the angle of friction $\lambda = \tan^{-1}\mu$.

In certain problems, use of the resultant reaction and the angle of friction provides a neat and sometimes shorter solution.

Example 7

When a horizontal force of 14·7 N is applied to a body of mass 4 kg which is resting on a rough horizontal plane, the body is found to be in limiting equilibrium. Calculate the resultant reaction acting on the body and the angle of friction.

This question can be solved in several different ways.

Figure 1 shows the forces acting on the body, including the normal reaction R and the frictional force μR, and this is used in Method 1 as explained below.
Figure 2 shows the resultant reaction P and the angle of friction λ, and this is used in Method 2.

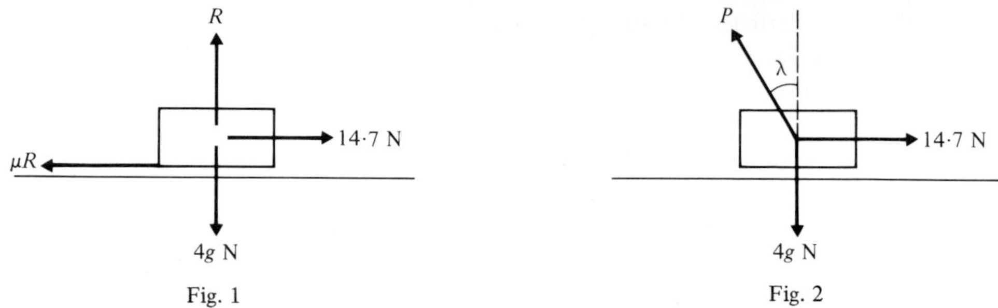

Fig. 1 Fig. 2

Method 1, using Fig. 1

Since the body is in limiting equilibrium:

Resolving vertically $R = 4g$

Resolving horizontally $\mu R = 14{\cdot}7$

$$\therefore \quad \mu \times 4g = 14{\cdot}7$$

$$\therefore \quad \mu = \frac{14{\cdot}7}{4g} = \tfrac{3}{8}$$

If $\mu = \frac{3}{8}$ then $\tan\lambda = \frac{3}{8}$ and $\lambda = 20{\cdot}55°$.

The resultant reaction

$$= \sqrt{(R^2 + \mu^2 R^2)}$$
$$= R\sqrt{(1 + \mu^2)}$$
$$= 4g\sqrt{(1 + \tfrac{9}{64})} = \frac{g}{2}\sqrt{73} \text{ N}$$

The resultant reaction is $\frac{g}{2}\sqrt{73}$ N and the angle of friction is $20{\cdot}55°$.

Method 2, using Fig. 2

Resolving vertically $P\cos\lambda = 4g$ . . . [1]

Resolving horizontally $P\sin\lambda = 14{\cdot}7$ . . . [2]

Dividing equation [2] by equation [1]:

$$\tan\lambda = \frac{14{\cdot}7}{4g} = \tfrac{3}{8}$$

$$\therefore \quad \lambda = 20{\cdot}55°$$

Substituting in equation [1]:

$$P\cos 20{\cdot}55° = 4g$$

$$\therefore \quad P = 41{\cdot}9 \text{ N}$$

The resultant reaction is 41·9 N and the angle of friction is 20·55°.

Since the situation as shown in Figure 2 is one of three forces acting on a body and producing a state of equilibrium, this problem could also be solved by the use of Lami's Theorem.

Method 3, using Fig. 2

Applying Lami's Theorem:

$$\frac{P}{\sin 90} = \frac{14{\cdot}7}{\sin(180-\lambda)} = \frac{4g}{\sin(90+\lambda)}$$

$$\therefore \quad \frac{14{\cdot}7}{\sin\lambda} = \frac{4g}{\cos\lambda}$$

$$\therefore \quad \tan\lambda = \frac{14{\cdot}7}{4g} \qquad \text{and} \quad \lambda = 20{\cdot}55^\circ$$

$$\text{and also} \quad \frac{P}{\sin 90} = \frac{14{\cdot}7}{\sin\lambda} \qquad \therefore \quad P = \frac{14{\cdot}7}{\sin 20{\cdot}55^\circ} = 41{\cdot}9\text{ N}$$

It should also be noted that, if the question had only required λ to be found, then resolving perpendicular to the resultant reaction gives a very neat solution:

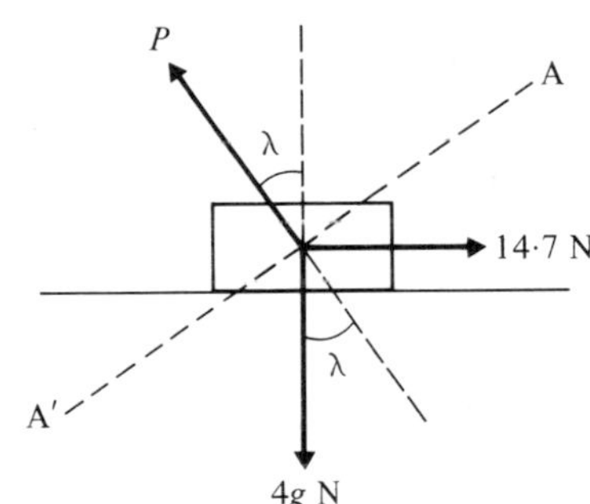

Resolving along AA′, perpendicular to P

$$4g\sin\lambda = 14{\cdot}7\cos\lambda$$

$$\therefore \quad \tan\lambda = \frac{14{\cdot}7}{4g}$$

$$\therefore \quad \lambda = 20{\cdot}55^\circ$$

The reader should be aware of all these possible methods of solution and select the most suitable method for each question.

Example 8

A body of mass 2 kg lies on a rough plane which is inclined at 40° to the horizontal. The angle of friction between the plane and the body is 15°. Find the greatest force which can be applied to the body, parallel to and up the plane, without motion occurring.

Let the applied force be X.
X will have its maximum value when the body is on the point of moving up the plane, i.e. it is in limiting equilibrium.

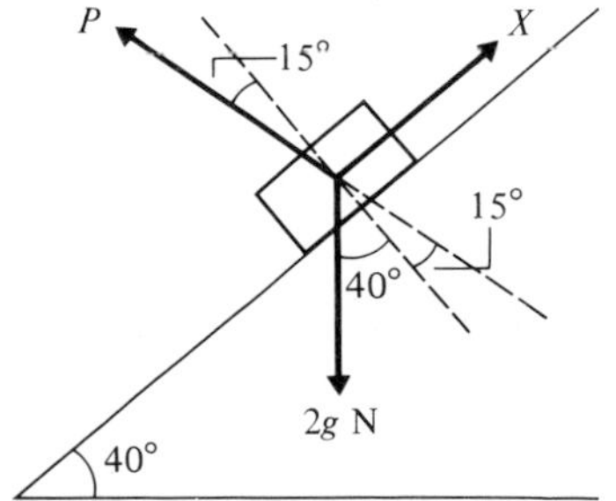

Resolving in a direction perpendicular to the line of the resultant reaction P:

$$X\sin(90-15) = 2g\sin(40+15)$$

$$\therefore \quad X = \frac{2g\sin 55}{\sin 75}$$

$$\therefore \quad X = 16{\cdot}6\text{ N}$$

The greatest force which can be applied to the body without motion occurring is 16·6 N.

It should be noted that this result could also have been obtained by the use of Lami's Theorem.

Example 9

A mass m rests in limiting equilibrium on a rough plane inclined at θ to the horizontal. If the angle of friction is λ, show that if the mass is on the point of moving *down* the plane, then $\theta = \lambda$.

Resolving parallel to the plane

$$mg \sin \theta = P \sin \lambda \qquad \ldots [1]$$

Resolving perpendicular to the plane

$$mg \cos \theta = P \cos \lambda \qquad \ldots [2]$$

Dividing equation [1] by equation [2]

$$\tan \theta = \tan \lambda$$

$$\therefore \quad \theta = \lambda$$

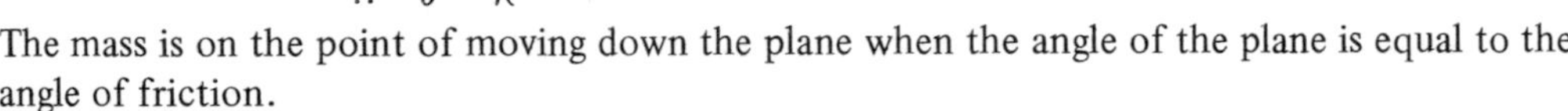

The mass is on the point of moving down the plane when the angle of the plane is equal to the angle of friction.

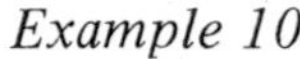

Example 10

If a force Q, inclined at an angle α to the surface of the plane, is applied to the mass in Example 9, find the minimum value of Q and the corresponding angle α when the mass is on the point of moving up the plane.

As the mass is on the point of moving up the plane, the resultant reaction will act as shown in the diagram.

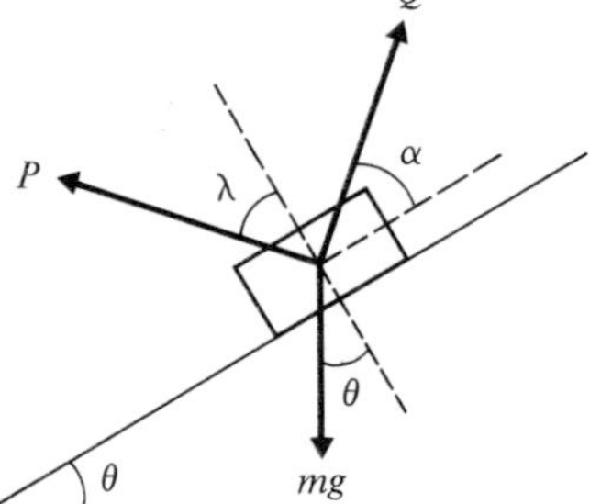

Using Lami's Theorem

$$\frac{P}{\sin(90 + \theta + \alpha)} = \frac{Q}{\sin(180 - (\lambda + \theta))} = \frac{mg}{\sin(90 - \alpha + \lambda)}$$

$$\therefore \quad \frac{Q}{\sin(\lambda + \theta)} = \frac{mg}{\cos(\alpha - \lambda)}$$

$$\text{or} \quad Q = \frac{mg \sin(\lambda + \theta)}{\cos(\alpha - \lambda)}$$

This expression for Q has a minimum value when $\cos(\alpha - \lambda) = 1$, or when $\alpha = \lambda$.
In this case $Q = mg \sin(\lambda + \theta)$ and the force must be applied at an angle λ to the surface of the plane.

The minimum value of Q is $mg \sin(\lambda + \theta)$ when the angle $\alpha = \lambda$.

Exercise 6C

1. The coefficient of friction for two surfaces in contact is 0·2. Find the angle of friction for the two surfaces.
2. The angle of friction for two surfaces in contact is 30°. Find the coefficient of friction for the two surfaces.

3. Each of the following diagrams shows a body on a rough plane. P is the resultant reaction of the plane on the body and λ is the angle of friction for the two surfaces in contact. Parts (a), (b) and (c) each involve a body, of mass 5 kg, on the point of moving along a horizontal surface. Find P and λ.

(a)

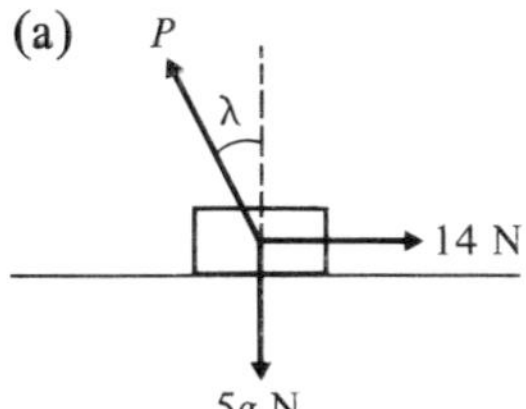

(b)

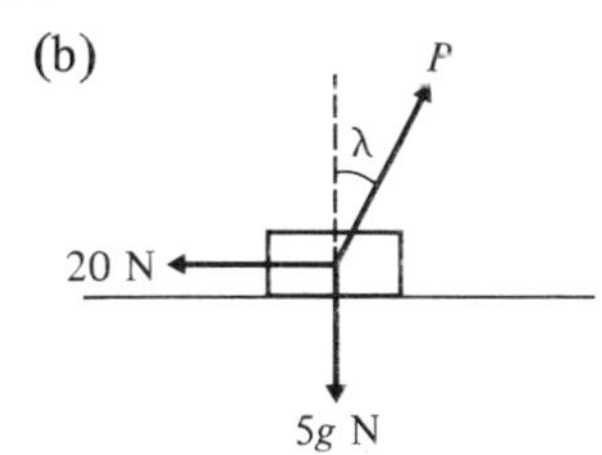

(c)

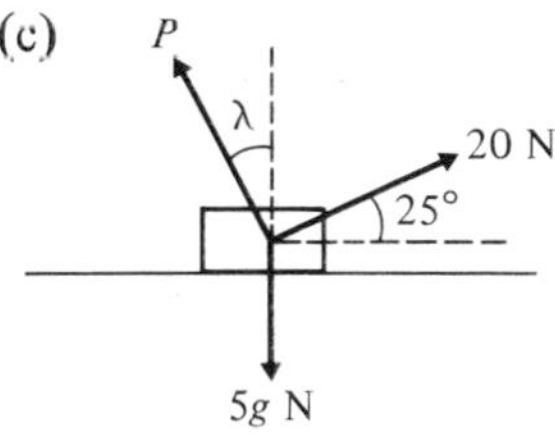

Parts (d), (e) and (f) each involve a body, of mass 2 kg, on the point of moving down a slope. If $\lambda = 25°$, find the magnitude of force X.

(d)

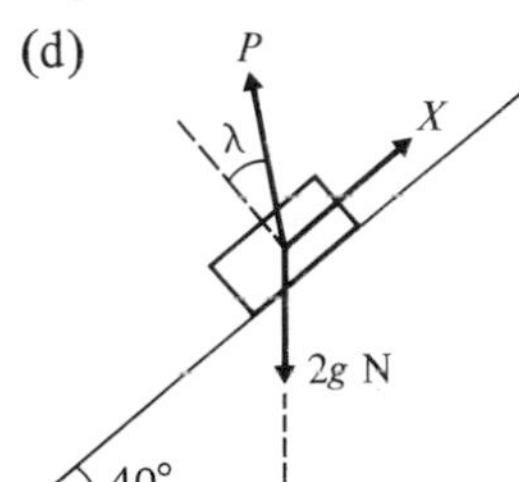

(e)

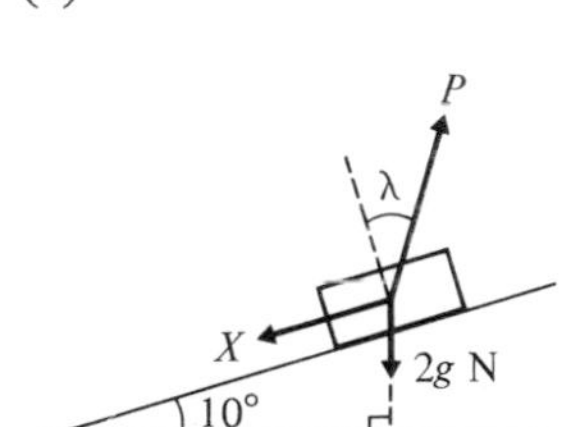

(f)

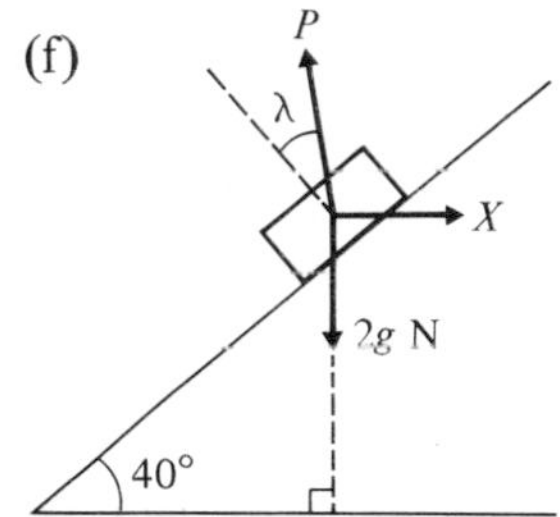

Parts (g), (h) and (i) each involve a body of mass 10 kg on the point of moving up the plane. If $\lambda = 18°$, find the magnitude of force X.

(g)

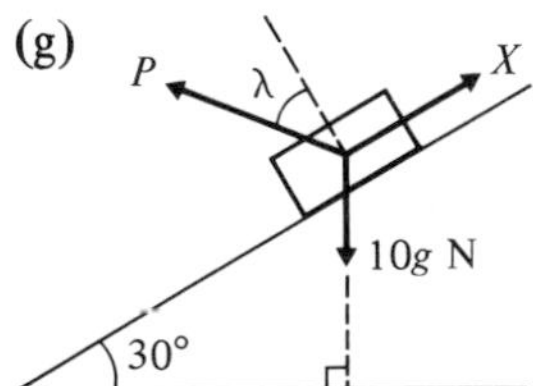

(h)

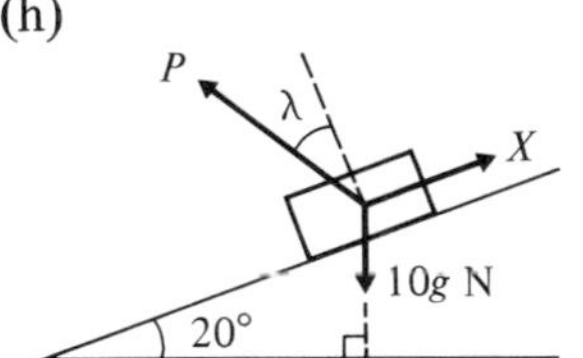

(i)

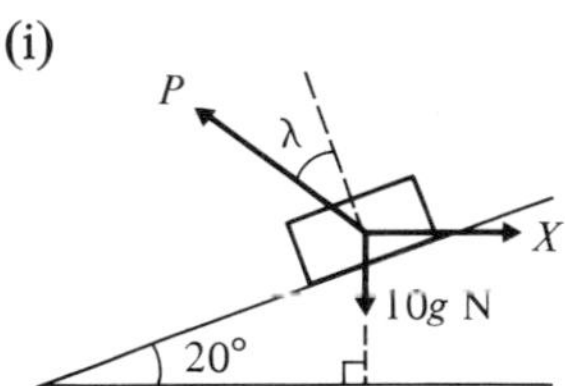

Parts (j), (k) and (l) each involve a body of mass 5 kg accelerating along the plane as indicated. Find P and λ.

(j)

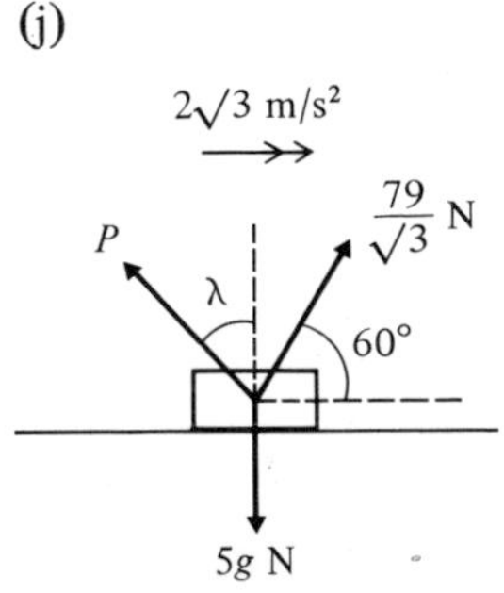

(k)

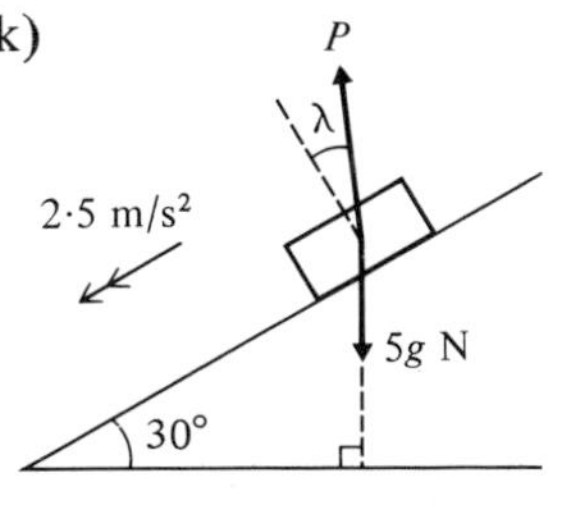

(l)

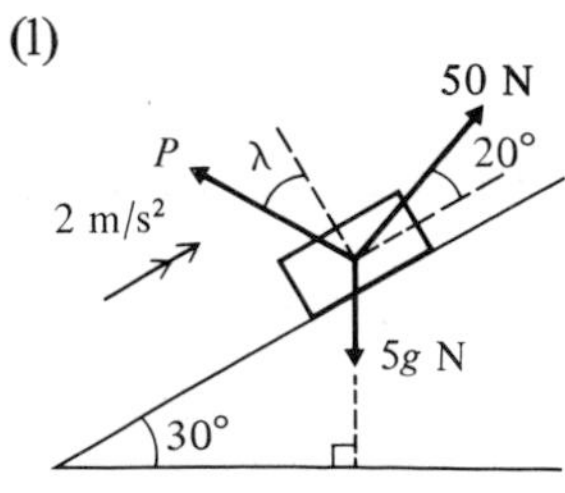

4. When a horizontal force of 4·9 N is applied to a body of mass 2 kg which is resting on a rough horizontal plane, the body is found to be in limiting equilibrium. Calculate the resultant reaction acting on the body and the angle of friction.
5. A boy pulls a sledge of mass 2 kg across a rough horizontal surface by means of a rope inclined at $\sin^{-1}\frac{3}{5}$ above the horizontal. The tension in the rope is 8N and the angle of friction for the two surfaces is 8°. Find the resultant reaction that the surface has on the sledge and the acceleration of the sledge.
6. (a) A body of mass m is placed on a rough plane which is inclined at θ degrees to the horizontal. The angle of friction between the body and the plane is λ. Show that the body will slip down the plane if $\theta > \lambda$.
   (b) State whether slipping will occur when a body is placed on a rough plane which is inclined at 30° to the horizontal, if the angle of friction between the body and the plane is (i) 20°, (ii) 40°.
   (c) When a body is placed on a rough plane which is inclined at 25° to the horizontal, the body is found to be in limiting equilibrium. Find λ, the angle of friction between the body and the plane. What is the coefficient of friction between the body and the plane?
7. A body of mass 5 kg lies on a rough plane which is inclined at 35° to the horizontal. The angle of friction between the plane and the body is 20°. Find the magnitude of the least force that must be applied to the body, in a direction parallel to and up the plane, in order to prevent motion down the plane.
8. A body of mass 4 kg lies on a rough plane which is inclined at 30° to the horizontal. The angle of friction between the plane and the body is 15°. Find the magnitude of the least horizontal force that must be applied to the body to prevent motion down the plane.
9. A body of mass 2 kg lies on a rough plane which is inclined at 40° to the horizontal. The angle of friction between the plane and the body is 15°. Find the greatest horizontal force that can be applied to the body without motion occurring.
10. A body of mass 3 kg lies on a rough plane which is inclined at 20° to the horizontal. A force of 28 N applied to the body, parallel to and up the slope, causes the body to accelerate up the slope at 1·5 m/s². Find λ, the angle of friction between the body and the plane.
    If the applied force were subsequently removed, the body would travel on up the slope, eventually coming to rest. Would it then slip down the slope or would it remain at rest?
11. A light string is attached to a body of mass 5 kg lying on a rough horizontal surface. The angle of friction between the body and the surface is 20° and the string is pulled upwards at an angle θ to the horizontal. If the body is on the point of moving, find the tension in the string when θ is (a) 10°, (b) 20°, (c) 40°.
12. A body of mass m lies on a rough horizontal surface. The angle of friction between the body and the surface is λ. A light string is attached to the body and is pulled upwards, at an angle θ with the horizontal. If the body is on the point of moving, show that the least value for the tension in the string is $mg \sin \lambda$ and that it occurs when $\theta = \lambda$.

Questions **13** to **15** refer to Fig. 1 which shows a mass m lying on a rough plane inclined at an angle θ to the horizontal. The angle of friction between the body and the plane is λ, with $\theta > \lambda$.
Force X is applied to the body and makes an angle ϕ with the plane. The body is on the point of moving down the plane.

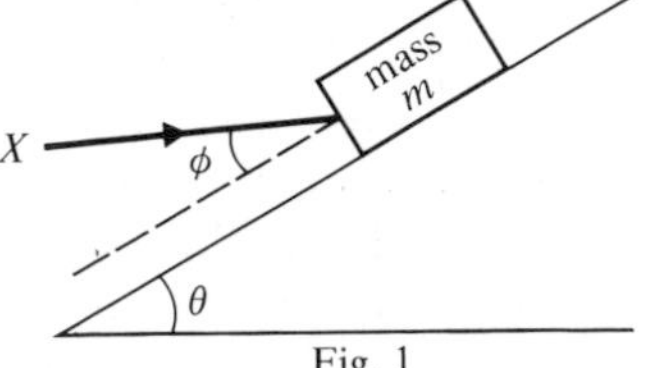

Fig. 1

13. If $\lambda = 18°$, $\theta = 30°$ and $m = 2$ kg, find the magnitude of X when ϕ is (a) $10°$, (b) $13°$, (c) $30°$.
14. If $\lambda = 15°$, $\theta = 40°$ and $m = 5$ kg find the magnitude of X when ϕ is (a) $10°$, (b) $15°$, (c) $30°$.
15. Show that the least force X sufficient to prevent motion down the plane is $mg \sin(\theta - \lambda)$ and that it occurs when $\phi = \lambda$.

Questions **16** to **18** refer to Fig. 2 which shows a mass m lying on a rough plane inclined at an angle θ to the horizontal. The angle of friction between the body and the plane is λ.
Force X is applied to the body and makes an angle ϕ with the plane. The body is on the point of moving up the plane.

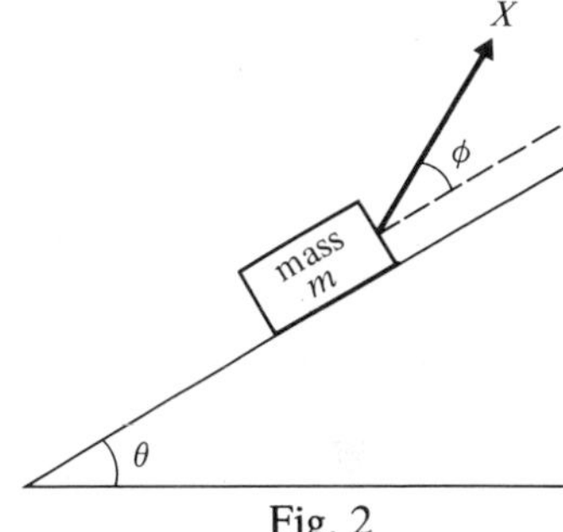

Fig. 2

16. If $\lambda = 30°$, $\theta = 20°$ and $m = 2$ kg, find the magnitude of X when ϕ is (a) $20°$, (b) $30°$, (c) $40°$.
17. If $\lambda = 20°$, $\theta = 10°$ and $m = 5$ kg, find the magnitude of X when ϕ is (a) $20°$, (b) $30°$, (c) $40°$.
18. Show that the least force X sufficient to ensure that the body is on the point of moving up the plane is $mg \sin(\theta + \lambda)$ and that it occurs when $\phi = \lambda$.

Exercise 6D Examination questions

1. A stone is sliding in a straight line across a horizontal ice rink. Given that the initial speed of the stone is 6 m/s and that it slides 24 m before coming to rest, calculate the coefficient of friction between the stone and the ice. (London)

2. A block of mass 5·2 kg is placed on a rough plane inclined at an angle α to the horizontal where $\sin \alpha = 0{\cdot}6$. The coefficient of friction between the block and the plane is 0·4. The block is just prevented from sliding down the plane by a horizontal force P N.
Draw a diagram showing all the forces acting on the block and calculate the value of P. (Take g as 10 m/s^2.) (Cambridge)

3. A rough plane is inclined at an angle α to the horizontal, where $\tan \alpha = \frac{3}{4}$. A particle slides with acceleration 3·5 m/s^2 down a line of greatest slope of this inclined plane. Calculate the coefficient of friction between the particle and the inclined plane. (London)

4. Three particles A, B, C are of masses 4, 4, 2 kg respectively. They lie at rest on a horizontal table in a straight line, with particle B attached to the mid-point of a light inextensible string. The string has particle A attached at one end and particle C at the other, and is taut. A force of 60 N is applied to A in the direction CA produced, and a force of 15 N is applied to C in the opposite direction. Find the acceleration of the particles and the tension in each part of the string (a) if the table is smooth, (b) if the coefficient of friction between each particle and the table is $\frac{1}{4}$. (Take g as 10 m/s^2.) (London)

5. Two points A and B are on a rough horizontal table at a distance $2d$ apart. A particle Q leaves B with initial speed $2u$ and moves along the line AB produced until it comes to rest at the point C. The coefficient of friction between Q and the table is $\frac{1}{3}$.
   (i) Show that $BC = 6u^2/g$ and find the time taken by Q to travel from B to C.
   At the same instant as Q leaves B, a second particle P leaves A with initial speed $5u$ and moves towards B. The coefficient of friction between P and the table is $\frac{1}{3}$ and Q comes to rest at C before P collides with Q.
   (ii) Show that $d > 9u^2/g$.
   Given that $d = 12u^2/g$, calculate
   (iii) the speed of P when it collides with Q,
   (iv) the time taken by P to move from A to C. (A.E.B)

6. A particle of mass m is on a rough horizontal plane. A force with its line of action making an angle θ with the plane is applied to the particle. Show that, if this force is just sufficient to pull the particle along the plane, the magnitude, P of the force, is $[mg \sin \lambda]/\cos(\theta - \lambda)$, where $\tan \lambda$ is the coefficient of friction.
   State the least value of P.
   The particle is now placed on the same plane which is tilted at an angle α to the horizontal. A force of magnitude $mg \sin \lambda$ acting along the line of greatest slope is just sufficient to move the particle up the plane. Show that $\sin(\lambda + \alpha) = \sin \lambda \cos \lambda$. (London)

7. The diagram shows the vertical cross-section ABC of a fixed wedge. AB is rough with coefficient of friction $\frac{1}{6}\sqrt{3}$ and is inclined at $30°$ to the horizontal. BC is smooth and vertical. A particle P, of mass m, moves on AB and a particle Q, of mass M, moves on BC. The two particles are connected by a light smooth inextensible string which is initially taut and which passes over a smooth peg at B. The system is released and it is observed that Q descends with acceleration a. Find a and the tension in the string in terms of m, M and g. Show that this motion is only possible if $M > \frac{3}{4}m$.

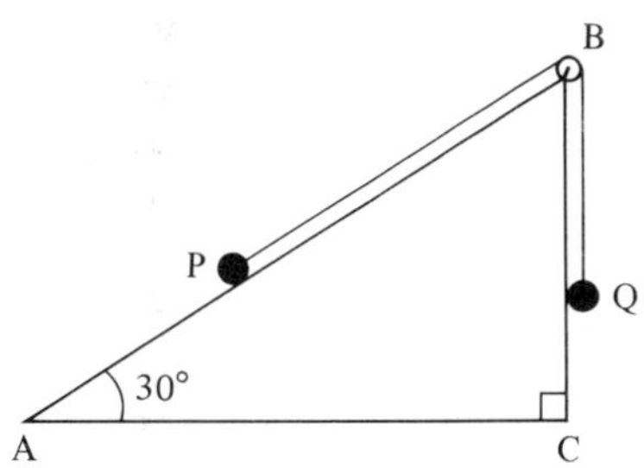

   In a similar situation (but with different values of M and m) it is observed that Q ascends. Find a similar inequality relating M and m for this to be possible.

   In each of the following cases state briefly what happens and find the tension in the string and the normal and frictional components of the reaction between P and AB in terms of M and g: (i) $m = 5M$, (ii) $m = 3M$, (iii) $m = M$. (Cambridge)

7

Moments

Moment of a force

From our everyday experience we know that:

(i) it is easier to undo a tight nut using a long spanner when the force is applied at the end of the spanner, rather than by using a short spanner;

(ii) if a boy sits on one end of a see-saw which is pivoted at its centre, he can be balanced by a heavier boy sitting near to the centre of the see-saw;

(iii) a door is more easily closed by pushing on the edge further from the hinges, rather than by pushing at a point part way across the door, such as Q.

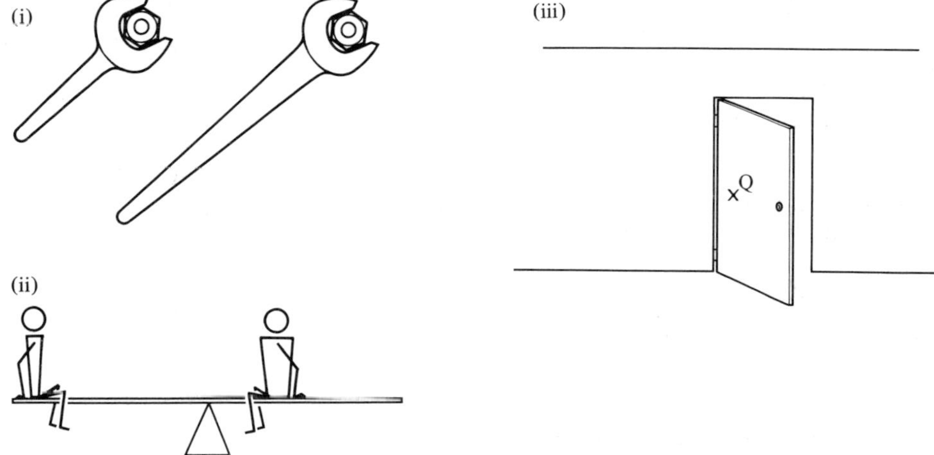

In each of these examples, the application of the force is causing or tending to cause a body to rotate about an axis, i.e. rotational motion. Previously, only motion along a line has been considered, i.e. translational motion.

The diagram shows a rod AB pinned to a horizontal table by a pin through A. When a horizontal force of 5 N is applied at the point B, perpendicular to BA, the rod will turn about the axis through A. The turning effect of the force will be lessened if instead of being applied at B, it is applied at C or at D. This turning effect is known as the moment of the force about the point A.

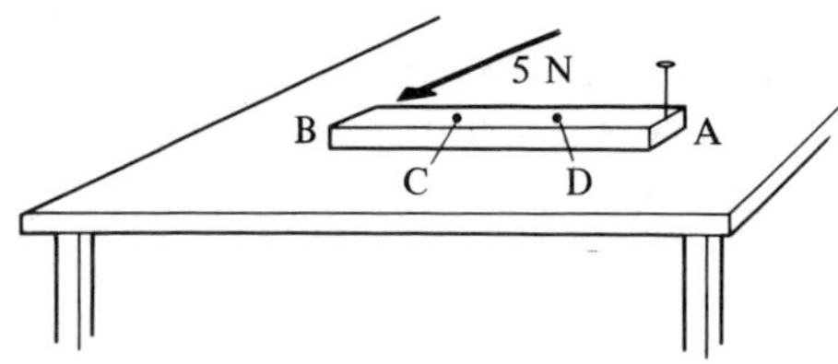

Although the phrase 'the moment of the force about the point A' has just been used, strictly speaking moments are always taken about an axis, and not about a point. In the last case, the turning effect (or moment) of the force about the vertical axis through A has really been considered. However in the questions which follow, all the forces act in one plane, and so for simplicity, the phrase 'moment about a point' will be used.

Definition

The moment of a force about a point is found by multiplying the magnitude of the force by the perpendicular distance from the point to the line of action of the force.

The moment of the force P about the point X is $P \times a$

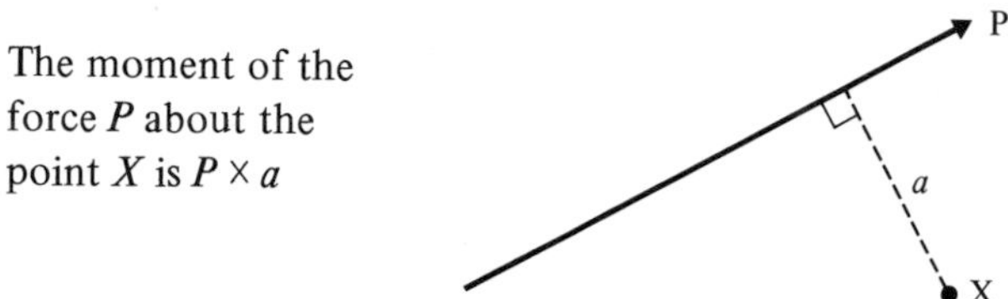

Hence a force will have no moment about a point on its line of action.

If the force is measured in newtons and the distance in metres, the moment of the force is measured in newton metres (N m).

Sense of rotation

A nut is usually rotated in an anticlockwise direction when being undone.
All rotations should have their sense clearly stated: the moment of a force about a point has both magnitude and direction.

Example 1

Consider the rod AB shown in the diagram. Find the moment about A of the force of 5 N when it is applied at each of the points B, C and A.

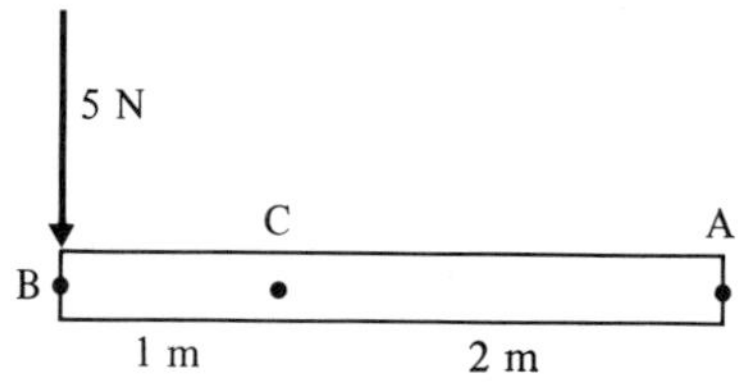

When the force of 5 N is applied at B,

$$\begin{aligned}\text{moment about A} &= \text{force} \times \text{perpendicular distance from A to force}\\ &= 5\text{ N} \times 3\text{ m} = 15\text{ N m anticlockwise}\end{aligned}$$

When the force of 5 N is applied at C,

$$\text{moment about A} = 5\text{ N} \times 2\text{ m} = 10\text{ N m anticlockwise}$$

When a force of 5 N is applied at A,

$$\text{moment about A} = 5\text{ N} \times 0\text{ m} = 0$$

A force through A has no turning effect about the point A and therefore no moment about A: it is like trying to close a door by pressing on the hinges.

It is usual when taking moments about an axis through, say, the point K to write $\overset{\curvearrowright}{\text{K}}$ to stand for 'taking moments about the point K'.

Example 2

In the diagram, find the moment of the force at C about each of the points A, B and C. State the direction of the moment in each case.

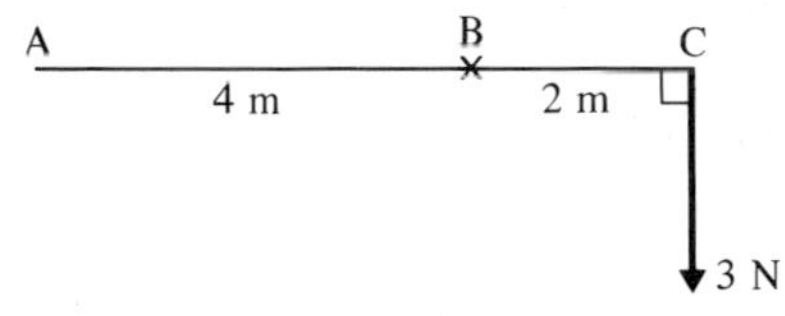

Taking moments

$\overset{\curvearrowright}{A}$ $3 \text{ N} \times 6 \text{ m} = 18 \text{ N m}$ clockwise

$\overset{\curvearrowright}{B}$ $3 \text{ N} \times 2 \text{ m} = 6 \text{ N m}$ clockwise

$\overset{\curvearrowright}{C}$ $3 \text{ N} \times 0 = 0$

In the third case, the force has no turning effect (and therefore no moment) about the point C since the force passes through the point C.

Algebraic sum of moments

If a number of coplanar forces act on a body, their moments about any point may be added provided due regard is given to the sense of each moment.

Example 3

Three forces acting on a body have moments of 15 N m clockwise, 10 N m anticlockwise and 13 N m clockwise, about a point X. Find the sum of these moments in magnitude and direction.

A moment of 10 N m anticlockwise = a moment of −10 N m clockwise

The sum of the moments = (15 N m − 10 N m + 13 N m) clockwise
= 18 N m clockwise

Example 4

A rod AB is free to rotate about an axis through the point O, perpendicular to the plane on which the rod rests. Forces of P and Q newtons act as shown. Find the combined turning effect about the point O of these forces if
(a) $P = 6$ N and $Q = 5$ N, (b) $P = 7$ N and $Q = 3\frac{1}{2}$ N.

Taking moments about O

(a) $\overset{\curvearrowright}{O}$ (5×4) N m clockwise + (6×2) N m anticlockwise
= (20 N m − 12 N m) clockwise
= 8 N m clockwise

(b) $\overset{\curvearrowright}{O}$ $(3\frac{1}{2} \times 4)$ N m clockwise + (7×2) N m anticlockwise
= 14 N m − 14 N m
= 0

Example 5

Find the moment, or the sum of the moments, about the point O of the forces shown in each of the following diagrams.

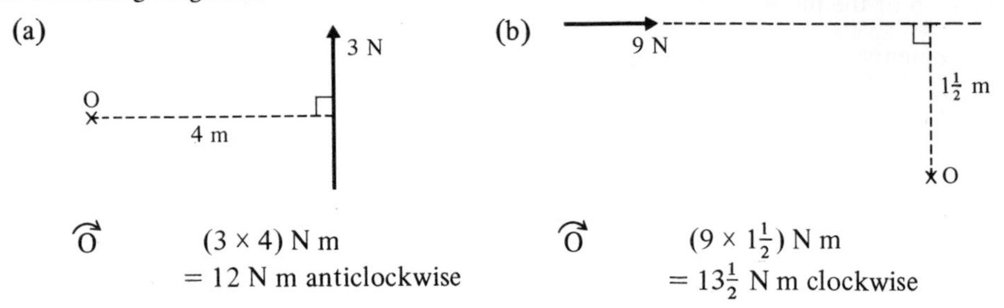

(a) $\overset{\curvearrowleft}{\mathrm{O}}$ (3×4) N m
$= 12$ N m anticlockwise

(b) $\overset{\curvearrowleft}{\mathrm{O}}$ $(9 \times 1\frac{1}{2})$ N m
$= 13\frac{1}{2}$ N m clockwise

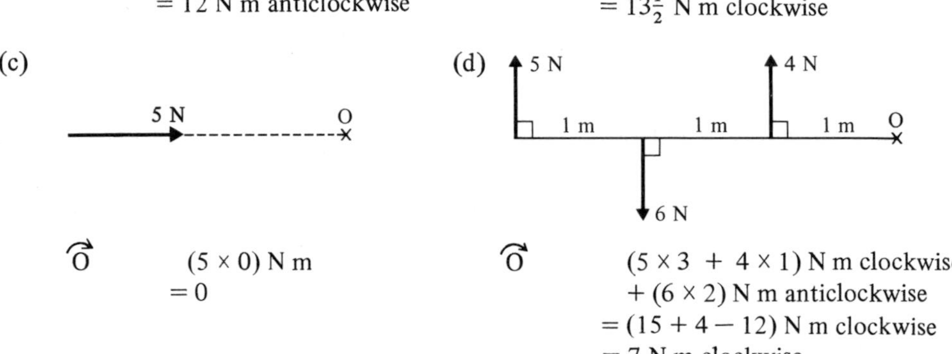

(c) $\overset{\curvearrowleft}{\mathrm{O}}$ (5×0) N m
$= 0$

(d) $\overset{\curvearrowleft}{\mathrm{O}}$ $(5 \times 3 + 4 \times 1)$ N m clockwise
$+ (6 \times 2)$ N m anticlockwise
$= (15 + 4 - 12)$ N m clockwise
$= 7$ N m clockwise

Example 6

Find the moment about the origin of a force of $3\mathbf{j}$ N acting at the point which has position vector $4\mathbf{i}$ m.

Drawing a diagram and taking moments about the point O

$\overset{\curvearrowleft}{\mathrm{O}}$ (4×3) N m
$= 12$ N m anticlockwise

Example 7

Find the moment about the point P with position vector $(-3\mathbf{i} + 4\mathbf{j})$ m of a force $(6\mathbf{i} + 5\mathbf{j})$ N acting at a point Q which has position vector $(2\mathbf{i} + \mathbf{j})$ m.

Again drawing a diagram

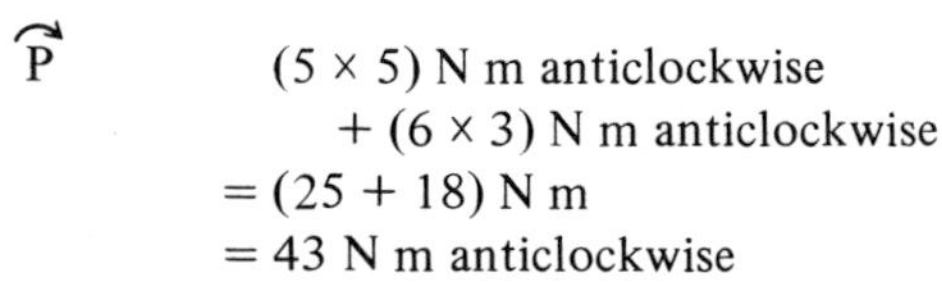

$\overset{\curvearrowleft}{\mathrm{P}}$ (5×5) N m anticlockwise
$+ (6 \times 3)$ N m anticlockwise
$= (25 + 18)$ N m
$= 43$ N m anticlockwise

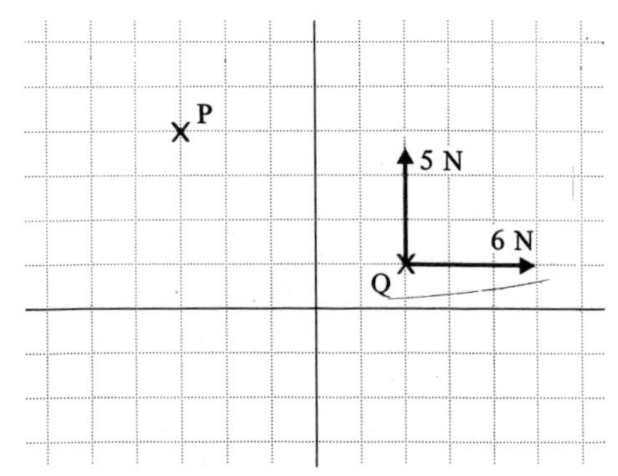

Exercise 7A

1. Three forces acting on a body have moments of 7 N m clockwise, 12 N m anticlockwise and 15 N m clockwise, about a point X. Find the sum of these moments in magnitude and direction.
2. Four forces acting on a body have moments of 8 N m clockwise, 5 N m anticlockwise, 17 N m clockwise and 22 N m anticlockwise, about a point X. Find the sum of these moments in magnitude and direction.

For each of the questions **3** to **23** find the moment (or the sum of the moments) about the point A of the forces shown.

3.

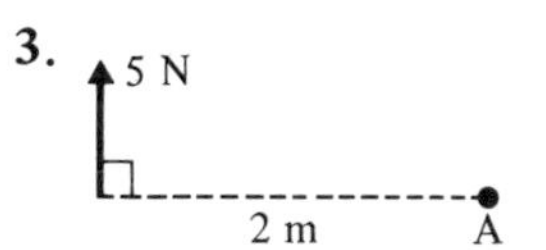

4.

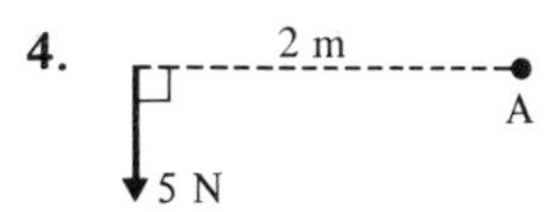

5.

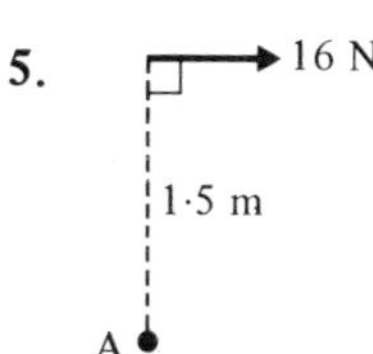

6.

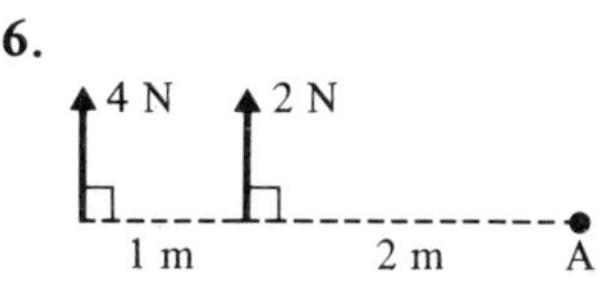

7.

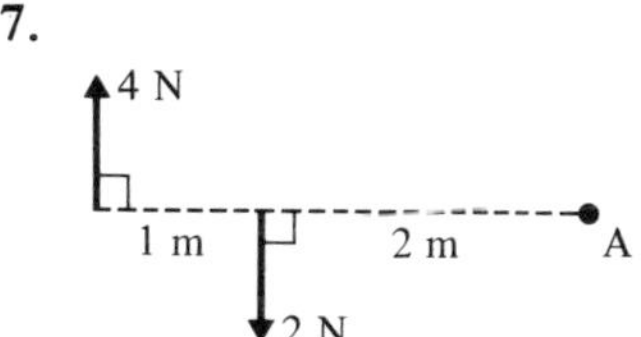

8.

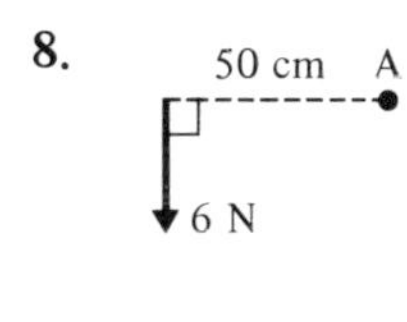

9.

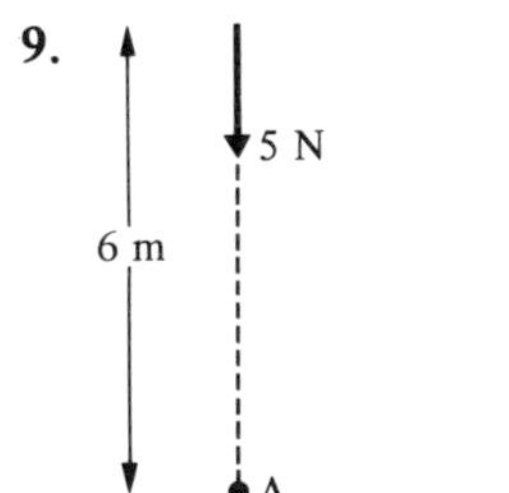

10.

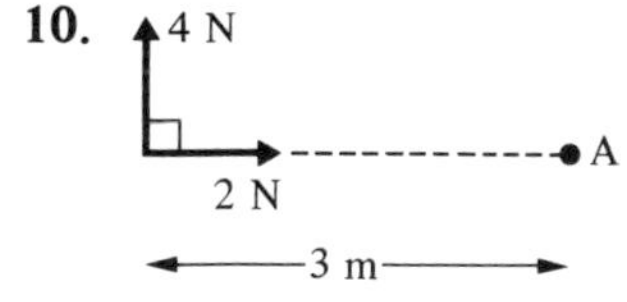

11.

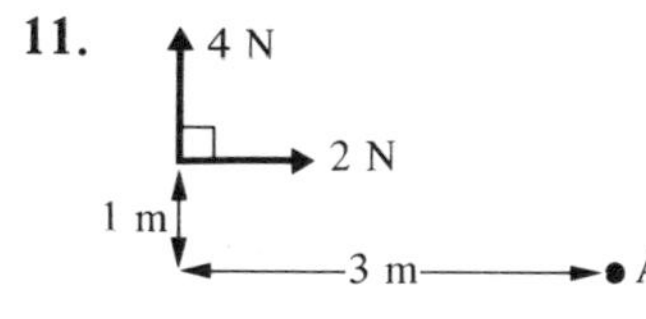

12.

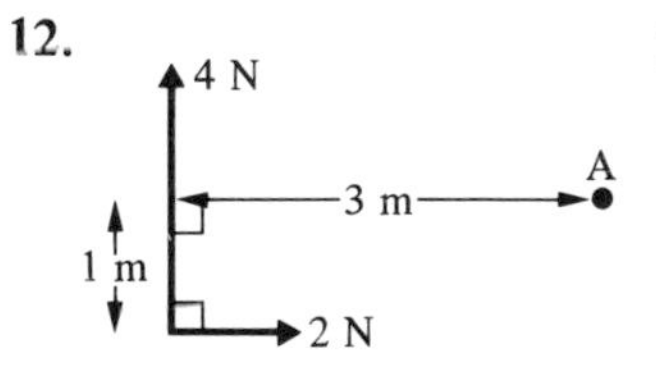

13.

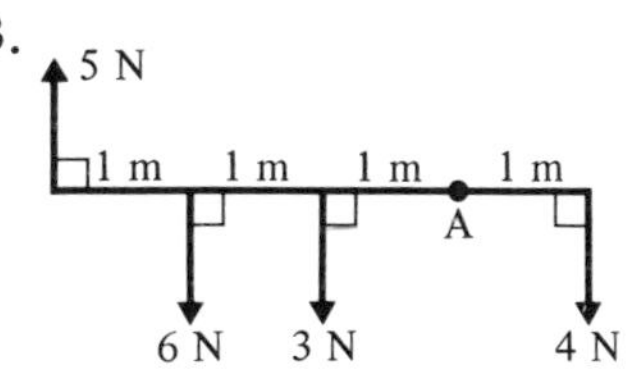

14.

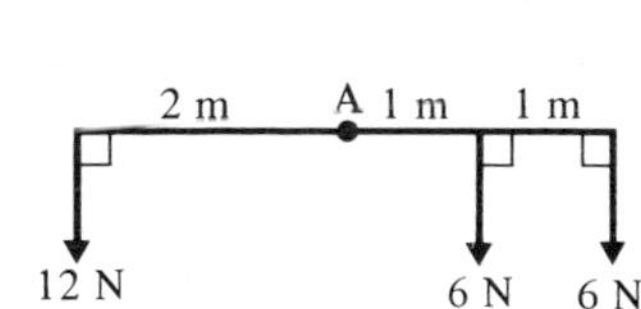

15.

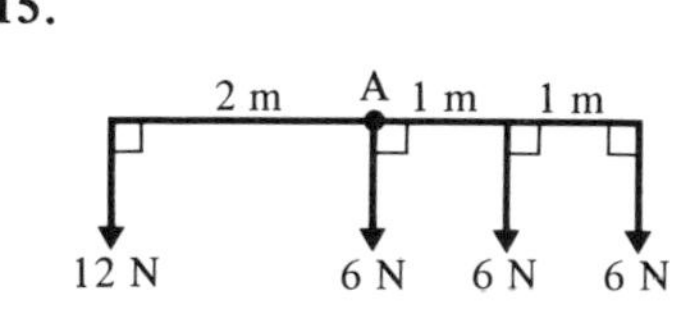

16.

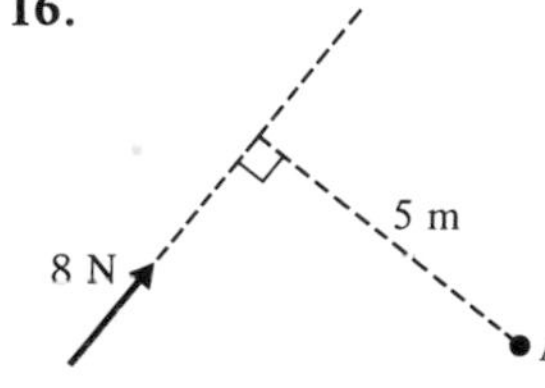

17.

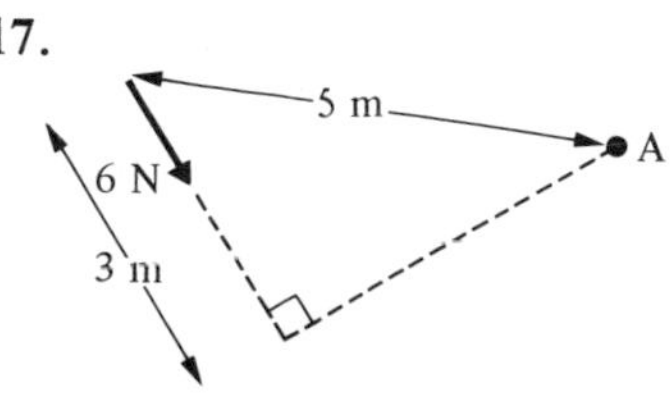

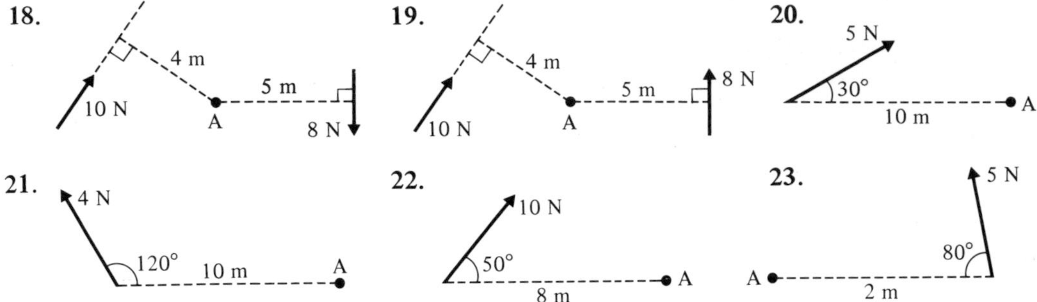

24. Find the moment about the origin of a force of $4\mathbf{j}$ N acting at the point which has position vector $5\mathbf{i}$ m.

25. Find the moment about the origin of a force of $4\mathbf{j}$ N acting at the point which has position vector $-5\mathbf{i}$ m.

26. Find the moment about the origin of a force of $3\mathbf{i}$ N acting at the point which has position vector $(2\mathbf{i} + 3\mathbf{j})$ m.

27. Find the moment about the origin of a force of $(4\mathbf{i} + 2\mathbf{j})$ N acting at the point which has position vector $(3\mathbf{i} + 2\mathbf{j})$ m.

28. A force of $(3\mathbf{i} - 2\mathbf{j})$ N acts at the point which has position vector $(5\mathbf{i} + \mathbf{j})$ m. Find the moment of this force about the point which has position vector $(\mathbf{i} + 2\mathbf{j})$ m.

29. A force of $(2\mathbf{i} + \mathbf{j})$ N acts at the point which has position vector $(2\mathbf{i} + 2\mathbf{j})$ m and a force of $5\mathbf{i}$ N acts at the point which has position vector $(-2\mathbf{i} + \mathbf{j})$ m. Find the sum of the moments of these forces about the origin.

30. A force of $(3\mathbf{i} + 2\mathbf{j})$ N acts at the point which has position vector $(5\mathbf{i} + \mathbf{j})$ m and a force of $(\mathbf{i} + \mathbf{j})$ N acts at the point which has position vector $(2\mathbf{i} + \mathbf{j})$ m. Find the sum of the moments of these forces about the point which has position vector $(\mathbf{i} + 3\mathbf{j})$ m.

31. If a line AB represents the force P, both in magnitude and direction, show that the moment of force P about some point O is represented in magnitude by twice the area of the triangle AOB.

Parallel forces and couples

Example 8

Two forces, each of 5 N, are applied at the ends A and B of a rod, 3 m in length. The forces are parallel but act in opposite directions.
Find the sum of the moments of the forces about each of the points A, B and P shown in the diagram.

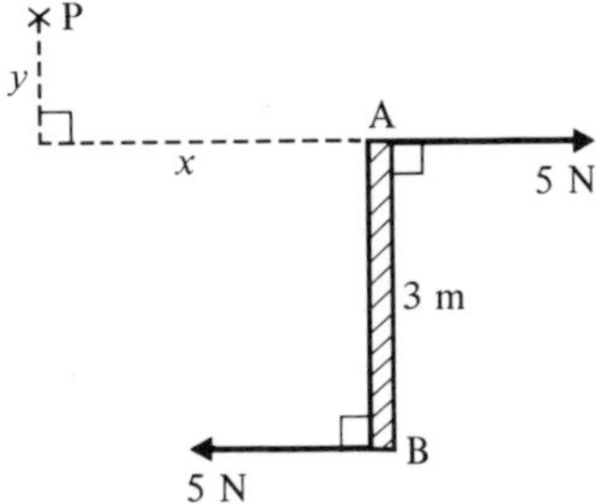

Since the forces are equal in magnitude but act in opposite directions, they have no *translational* effect. However the forces will have a *turning* effect.

Taking moments

A ↻ $(5 \times 3) + (5 \times 0)$ N m
$= 15$ N m clockwise

B ↻ $(5 \times 3) + (5 \times 0)$ N m
$= 15$ N m clockwise

$\overset{\frown}{\text{P}}$ $\quad 5 \times (y + 3)$ N m clockwise + $(5 \times y)$ N m anticlockwise
$= (5y + 15 - 5y)$ N m clockwise
$= 15$ N m clockwise

Note that the moments about each of the points is the same.

Like and unlike forces

Forces which are parallel and act in the *same* direction are said to be *like* forces.

Forces which are parallel and act in *opposite* directions are said to be *unlike* forces.

Definition of a couple

Two unlike forces of equal magnitude, not acting along the same line, are said to form a *couple.* A couple has a turning effect but cannot produce a translatory effect.

Moment of a couple

If the magnitude of each force forming a couple is P newtons and the perpendicular distance between their lines of action is a metres, the magnitude of the moment of the couple is

$$P \times a \text{ N m}$$

The turning effect of a couple is *independent* of the point about which the turning is taking place.

This statement, which was illustrated in Example 8, can be verified more generally by considering the sum of the moments of the forces about each of the points O_1 and O_2 in the diagram below.

$\overset{\frown}{O_1}$ $\quad P \times (a + b)$ clockwise + $(P \times b)$ anticlockwise
$= (Pa + Pb - Pb)$ N m clockwise
$= Pa$ N m clockwise

$\overset{\frown}{O_2}$ $\quad P \times c$ clockwise $+ P \times (a - c)$ clockwise
$= (Pc + Pa - Pc)$ N m clockwise
$= Pa$ N m clockwise

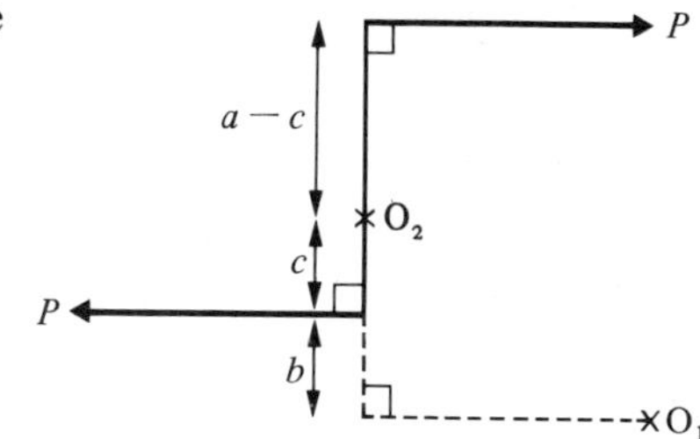

Here it is found that the moment of this couple is the same about all points in the plane of the forces forming the couple.

It is also possible for three (or more) parallel forces to form a couple, as is illustrated in the following example.

Example 9

Show that the system of forces given in the diagram forms a couple and find the moment of this couple.

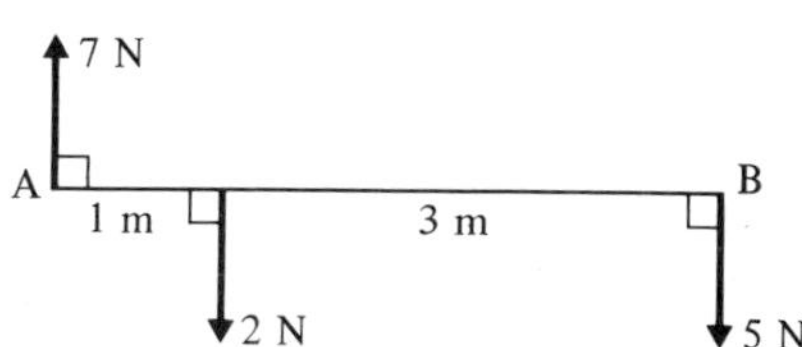

Resolving the forces in a direction parallel to the 7 N force

$$7 \text{ N} - 2 \text{ N} - 5 \text{ N} = 0$$

In this case the forces balance in this direction and they have no components in any other direction; consequently they have no translatory effect.

Hence, either (a) their moments about any point also balance and therefore the forces have no turning effect,

or (b) the forces have a turning effect, that is they form a couple.

Taking moments

$\curvearrowright A$ $\quad (7 \times 0) + (2 \times 1) + (5 \times 4)$ N m clockwise
$= 22$ N m clockwise

Therefore these forces form a couple with a moment of 22 N m in a clockwise sense.

Example 10

Forces of 5 N, 2 N, 5 N and 2 N act along the sides BA, BC, DC and DA respectively, of the square ABCD, in the directions indicated by the order of the letters. The side of the square is 3 m. Show that the forces form a couple and find the moment of this couple by taking moments about (a) the centre of the square and (b) the point A.
Find the moment of the couple which must be applied to the system in order to produce equilibrium.

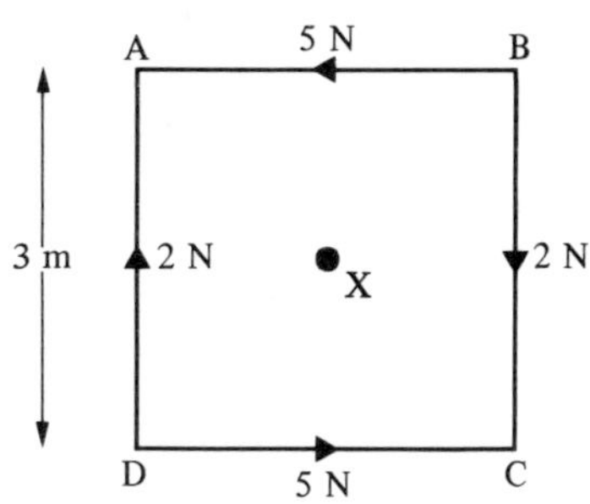

Resolving parallel to AB

$-5\text{ N} + 5\text{ N}$
$= 0$

Resolving parallel to AD

$-2\text{ N} + 2\text{ N}$
$= 0$

Hence the system has no translatory effect, and is therefore either in equilibrium or reduces to a couple.

Taking moments

(a) $\curvearrowright X$ $\quad (2 \times 1\frac{1}{2}) - (5 \times 1\frac{1}{2}) + (2 \times 1\frac{1}{2}) - (5 \times 1\frac{1}{2})$ N m clockwise
$= (3 - 7\frac{1}{2} + 3 - 7\frac{1}{2})$ N m
$= -9$ N m clockwise
$= 9$ N m anticlockwise

(b) $\curvearrowright A$ $\quad (2 \times 3) - (5 \times 3)$ N m clockwise
$= 6 - 15$ N m clockwise
$= 9$ N m anticlockwise

As was to be expected, the moments about the points X and A are the same, since the system is equivalent to a couple. The moment of the couple is 9 N m in an anticlockwise sense.
In order to produce a state of equilibrium, a couple of the same magnitude, but opposite in sense is required.

Couple required = 9 N m clockwise

Exercise 7B

1. Find which of the following systems will reduce to a couple, and in these cases, find the moment of the couple:

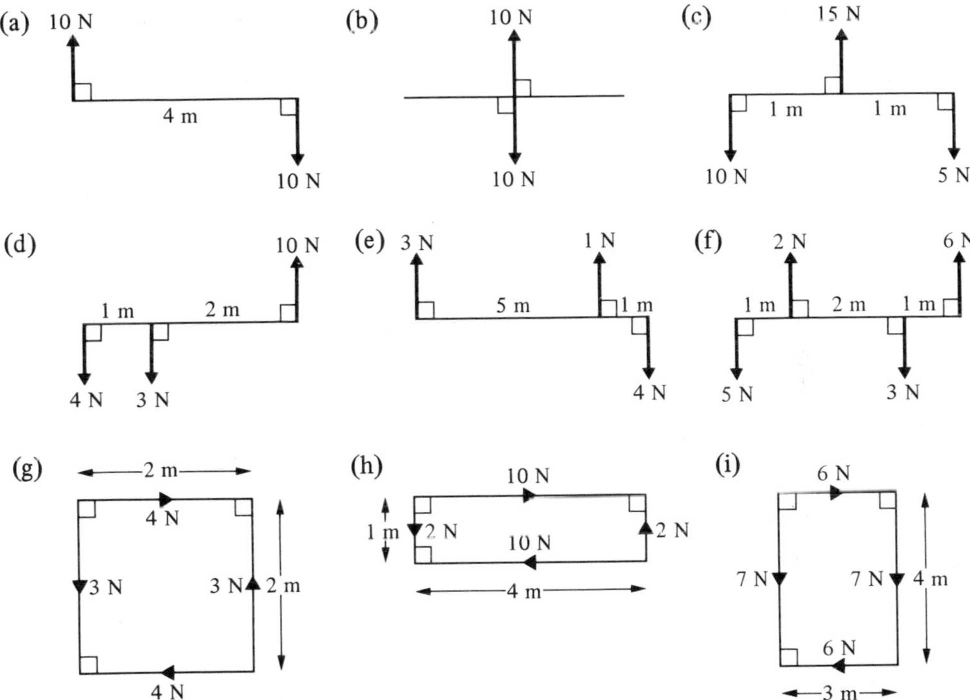

2. Find the moment of the couple applied to a corkscrew by two equal and opposite forces of 25 N acting on it, along lines 7 cm apart.
3. ABCD is a rectangle with AB = 5 m and BC = 2 m. A force of 3 N acts along each of the four sides AB, BC, CD and DA in the direction indicated by the order of the letters. Show that the forces form a couple and find its moment.
4. ABCD is a rectangle with AB = 6 m and BC = 2 m. Forces of 5 N, 5 N, X N and X N act along CB, AD, AB and CD respectively. The directions of the forces are given by the order of the letters. If the system is in equilibrium, find X.
5. ABCD is a square of side 40 cm. Forces of 20 N, 15 N and 20 N act along the sides AB, BC and CD respectively and a force Y acts along DA. The directions of the forces are given by the order of the letters. If the system is equivalent to a couple, find the magnitude of Y and the moment of the couple.
6. ABCD is a square of side 60 cm. Forces of 6 N, 2 N, 6 N and 2 N act along the sides AB, CB, CD and AD respectively, in the directions indicated by the order of the letters. Show that the forces form a couple and find the moment of the couple that must be applied to the system in order to produce equilibrium.
7. ABCD is a rectangle with AB = 8 m and BC = 3 m. Forces, each of 4 N, act along AB and CD in the directions indicated by the order of the letters. Show that the system reduces to a couple and find the moment of the couple. It is now required to reduce the system to equilibrium by applying a force P N along AD and another force P N along CB. Find the value of P.

8. A light rod of length 50 cm lies on a horizontal table. A man holds the ends of the rod. The rod is subjected to a couple, of moment 40 N m, causing it to rotate upon the table. What force perpendicular to the rod must the man exert through each hand in order to prevent the rotation?
9. A force of $(3\mathbf{i} - 5\mathbf{j})$ N acts at the point which has position vector $(6\mathbf{i} + \mathbf{j})$ m and a force of $(-3\mathbf{i} + 5\mathbf{j})$ N acts at the point which has position vector $(4\mathbf{i} + \mathbf{j})$ m. Show that these forces reduce to a couple and find the moment of the couple.
10. A force of $(4\mathbf{i} + 3\mathbf{j})$ N acts at the point which has position vector $(6\mathbf{i} + 3\mathbf{j})$ m and a force of $(-4\mathbf{i} - 3\mathbf{j})$ N acts at the point which has position vector $(3\mathbf{i} - \mathbf{j})$ m. Show that these forces reduce to a couple and find the moment of the couple.
11. Forces of $(\mathbf{i} + \mathbf{j})$ N, $(-4\mathbf{i} + \mathbf{j})$ N and $(3\mathbf{i} - 2\mathbf{j})$ N act at the points having position vectors $(2\mathbf{i} + 2\mathbf{j})$ m, $(-\mathbf{i} + 4\mathbf{j})$ m and $(4\mathbf{i} - 2\mathbf{j})$ m respectively. Show that these forces reduce to a couple and find the moment of the couple.
12. Forces of $(a\mathbf{i} + b\mathbf{j})$ N and $(6\mathbf{i} - 4\mathbf{j})$ N act at the points having position vectors $(-2\mathbf{i} - 2\mathbf{j})$ m and $(3\mathbf{i} - \mathbf{j})$ m respectively. If these forces reduce to a couple, find a and b and the moment of the couple.
13. Forces of $6\mathbf{j}$ N and $-6\mathbf{j}$ N act at the origin and at position vector $2\mathbf{i}$ m respectively. Show that the moment of the forces about any point P (x, y) is independent of x and y.

Replacement of parallel forces by a single force

In the last section, systems of parallel forces which reduced to couples were considered. These systems produced a turning effect but did not have a translatory effect.
Systems of parallel forces which do not reduce to a couple must now be considered.

Any number of parallel forces, which are not equivalent to a couple, may be replaced by a single force. This resultant force will be parallel to the given forces, and it must have the same translational and rotational effects as the given forces.

The same translational effect is ensured if the magnitude of the resultant is obtained by resolving in the direction of the given forces.
To ensure the same rotational effect, use is made of the very important Principle of Moments, which states that:

> When a number of coplanar forces act on a body, the algebraic sum of the moments of these forces, about any point in their plane, is equal to the moment of the resultant of these forces about that point.

When finding the resultant of a system of forces, it is advisable to draw two diagrams, one showing the given forces and a second to show the equivalent resultant force.

Example 11

Two unlike parallel forces of 1 N and 3 N are 4 m apart. Find the magnitude, direction and line of action of the resultant of these forces.

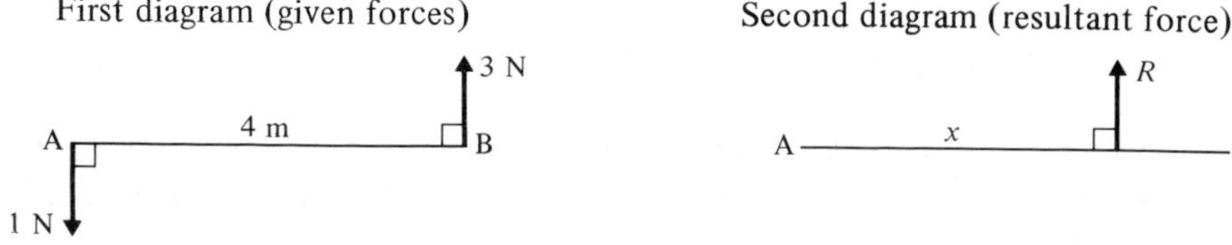

Let A and B be the points in which a line drawn at right angles to both forces meets their lines of action. Let R act at a distance of x metres from A. The resultant R of the given forces will act in the same direction as the larger force.

R must have the same translational effect thus, resolving at right angles to AB

$$R = 3 - 1 \qquad \therefore \quad R = 2 \text{ N}$$

R must have the same rotational effect thus, taking moments

$$\overset{\curvearrowright}{\text{A}} \quad 3 \times 4 = R \times x \qquad \therefore \quad x = 6 \text{ m}$$

The resultant acts at a distance of 6 m from A, is 2 N in magnitude and acts in the same direction as the 3 N force.

Example 12

Two like parallel forces of 2 N and 5 N are 21 m apart. Find the magnitude, direction and line of action of the resultant of these forces.

Draw two diagrams as before. The resultant R will act in the same direction as the given forces.

For R to have the same translational effect

$$R = 2 + 5 \qquad \therefore \quad R = 7 \text{ N in the same direction as the given forces}$$

For R to have the same rotational effect

$$\overset{\curvearrowright}{\text{A}} \quad 5 \times 21 = R \times x \qquad \therefore \quad 105 = 7x \text{ or } x = 15 \text{ m}$$

The magnitude of the resultant is 7 N, acting in the same direction as the given forces, and its line of action is 15 m from the 2 N force.

Position of resultant

It should be carefully noted that the resultant of two like forces will have its line of action between the two forces; if the forces are unlike, the resultant will lie outside the given forces.

Exercise 7C

1. In each of the following cases, find the magnitude of the resultant of the forces shown and the distance of its line of action from A.

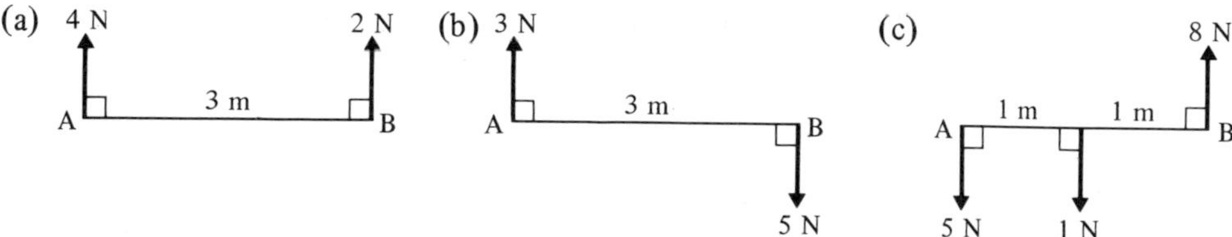

2. The lines of action of two like parallel forces of 3 N and 2 N are 10 m apart. Find the magnitude of the resultant of the forces and the distance between its line of action and that of the 3 N force.
3. The lines of action of two like parallel forces of 8 N and 12 N are 5 m apart. Find the magnitude of the resultant of the forces and the distance between its line of action and that of the 8 N force.

4. The lines of action of two unlike parallel forces of 5 N and 3 N are 4 m apart. Find the magnitude of the resultant of the forces and the distance between its line of action and that of the 3 N force.
5. The lines of action of two unlike parallel forces of 8 N and 12 N are 5 m apart. Find the magnitude of the resultant of the forces and the distance between its line of action and that of the 8 N force.
6. Two like vertical forces of 4 N and Q N act at points A and B respectively where AB is horizontal and of length 6 m. Their resultant is a force of P N and acts at a point X, between A and B, where AX = 2 m. Find P and Q.
7. Two unlike vertical forces of 10 N and Q N act at points A and B respectively where AB is horizontal and of length 3 m. The resultant of the two forces is a force of P N which acts at a point X on the line BA produced such that XA = 4·5 m. Find P and Q.
8. A, B and C are three points on a horizontal line with B between A and C. AB = 1 m and BC = 3 m. Forces of 5 N, 3 N and 2 N act at A, B and C respectively in a direction vertically downwards. Find the magnitude of the single force that could replace these forces and the distance of its line of action from A.
9. A, B and C are three points on a horizontal line with B between A and C; AB = 1 m and AC = 3 m. Forces of 2 N and 0·5 N act vertically downwards at A and C respectively and a force of 4 N acts vertically upwards at B. Find the magnitude of the resultant of these three forces and the distance of its line of action from A.
10. A, B and C are three points on a horizontal line with AC = 4 m and point B situated between A and C. Vertical forces of 4 N, 1 N and 3 N act at A, B and C respectively. The resultant of these forces is a force of 6 N, vertically upwards, acting through the point X on AB such that AX = $1\frac{1}{2}$ m. State the direction of the forces at A, B and C and find the distance AB.
11. Show that the resultant of two like parallel forces, of magnitudes P N and Q N, is a force in the same direction as the two forces and whose line of action divides the distance between the two forces internally, in the ratio $Q : P$.
12. Show that the resultant of two unlike parallel forces, of magnitudes P N and Q N where P is greater than Q, is a force in the direction of the larger of the two forces and whose line of action divides the distance between the two forces externally, in the ratio $Q : P$.

Parallel forces in equilibrium

A system of parallel forces is either equivalent to a couple or it can be replaced by a single force or resultant.

Consider a system of parallel forces in equilibrium. Such forces cannot be equivalent to a couple because, if they were, they would have a turning effect and would not therefore be in equilibrium. The second possibility is that the forces could be replaced by a resultant. However, as the forces are in equilibrium, this resultant force must be zero. Furthermore, if the resultant force is zero, then the moment of this resultant force about any point must also be zero and so, by the Principle of Moments, the algebraic sum of the moments of the forces about any point must be zero.

Thus for parallel forces in equilibrium

(i) the resultant force in any direction is zero,

and (ii) the algebraic sum of the moments of the forces about any point is zero (i.e. we can equate clockwise and anticlockwise moments.)

Example 13

A uniform beam, of length 2 m and mass 4 kg, has a mass of 3 kg attached at one end and a mass of 1 kg attached at the other end. Find the position of the support if the beam rests in a horizontal position.

Suppose the beam is supported at the point C, s metres from the 3 kg mass, and that the support exerts an upward force of P newtons on the beam. The forces due to the two masses are shown on the diagram. Since the beam is uniform, it is assumed that its weight will act at its centre.

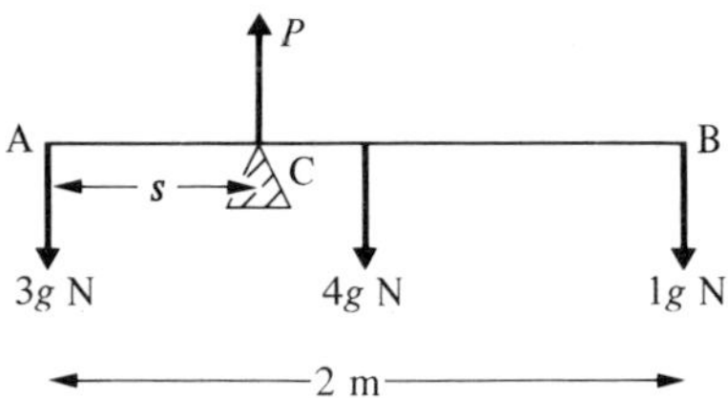

Method 1

Equating the clockwise and anticlockwise moments

$$\overset{\curvearrowright}{\text{C}} \quad 3g \times s = 4g \times (1 - s) + 1g \times (2 - s)$$
$$\therefore \quad 8gs = 4g + 2g$$
$$\therefore \quad s = \tfrac{3}{4}\text{ m}$$

Method 2

Equating the forces in a vertical direction

$$3g + 4g + 1g = P$$
$$\therefore \quad P = 8g$$
$$\overset{\curvearrowright}{\text{A}} \quad 4g \times 1 + 1g \times 2 = P \times s$$
$$\therefore \quad 6g = 8gs$$
$$\therefore \quad s = \tfrac{3}{4}\text{ m}$$

Thus the beam must be supported at a point $\frac{3}{4}$ m from the 3 kg mass.

The first method has the advantage that the unknown force at the support is not needed, since moments are taken about this point.

Example 14

A light horizontal beam of length 2 m rests with its ends A and B on smooth supports. The beam carries masses of 5 kg and 2 kg at distances of 60 cm and 150 cm respectively, from A. Find the reaction at each support.

Since the beam is light, its mass can be ignored. Suppose the reactions are P and Q newtons.

Equating the clockwise and anticlockwise moments

$$\overset{\curvearrowright}{\text{A}} \quad 5g \times 0{\cdot}6 + 2g \times 1{\cdot}5 = Q \times 2$$
$$\therefore \quad Q = 3g$$

The force P does not appear in this equation since it has no moment about the point A.

$$\overset{\curvearrowright}{\text{B}} \quad 2g \times 0{\cdot}5 + 5g \times 1{\cdot}4 = P \times 2$$
$$\therefore \quad P = 4g$$

The reactions at the supports are $4g$ and $3g$ newtons.

If the forces acting on the beam in a vertical direction are considered:

$$P + Q = 5g + 2g$$

and this equation can be used, either to check the values of P and Q, or in place of one of the the moment equations.

Exercise 7D

Questions **1** to **8** involve light horizontal rods in equilibrium. Each diagram shows the forces acting on the rods. Find the magnitudes of the forces P and Q and the distance x, as applicable.

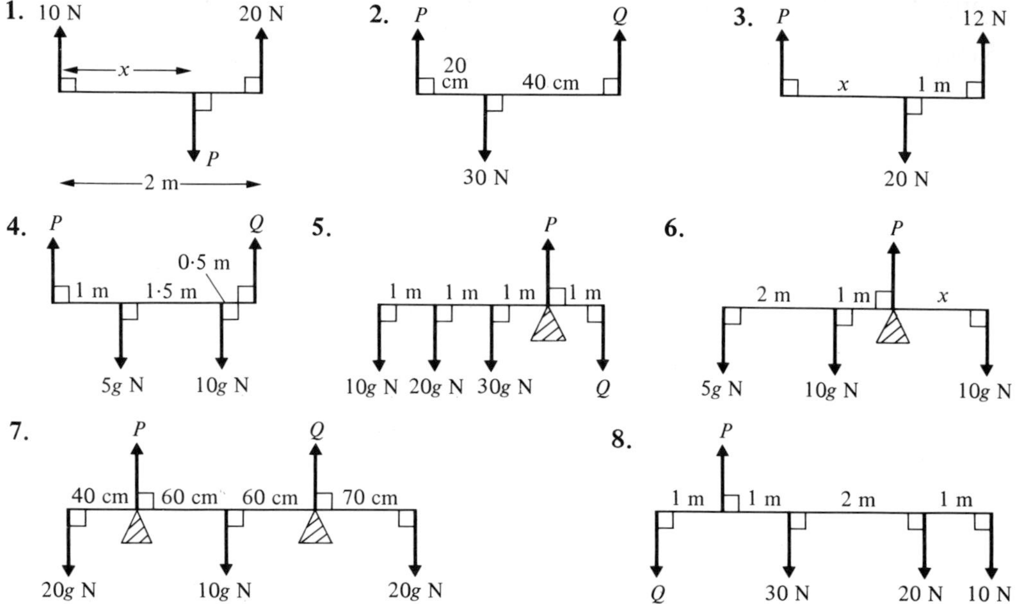

Some of the following questions involve *uniform* beams. This means that the weight of the beam can be taken as acting at the centre point of the beam.

9. A uniform beam of length 4 m and mass 50 kg rests horizontally, supported at each end. A mass of 20 kg is placed on the beam, 1 m from one end. Find the reactions at the supports.
10. A uniform beam of length 6 m and mass 8 kg has a mass of 10 kg attached at one end and a mass of 3 kg attached at the other end. Find the position of the support if the beam rests in a horizontal position.
11. A play ground see-saw consists of a uniform beam of length 4 m supported at its mid-point. If a girl of mass 25 kg sits at one end of the see-saw, find where her brother of mass 40 kg must sit if the see-saw is to balance horizontally.
12. A broom consists of a uniform broomstick of length 120 cm and mass 4 kg with a broom head of mass 6 kg attached at one end. Find where a support should be placed so that the broom will balance horizontally.
13. A non-uniform beam AB is of length 4 m and its weight of 5 N can be considered to act at a point 1·8 m from the end A. The rod rests horizontally on smooth supports at A and B. Find the reactions at the supports.
14. A uniform beam AB of mass 10 kg and length 4 m rests horizontally on two supports, one at A and the other 1 m from B. Where must a boy of mass 50 kg stand on the beam if he wishes to make the reactions at the supports equal?
15. A non-uniform rod AB of length 4 m is supported horizontally on two supports, one at A and the other at B. The reactions at these supports are $5g$ N and $3g$ N respectively. If instead the rod were to rest horizontally on one support find how far from end A this support would have to be placed.

16. A pole vaulter uses a uniform pole of length 4 m and mass 5 kg. He holds the pole horizontally by placing one hand at one end of the pole and the other hand at a position on the pole 80 cm away. Find the vertical forces exerted by his hands.

17. Three uniform rods of mass 2, 4 and 8 kg and each of length 20 cm, are joined together in the order mentioned to form one long rigid rod of length 60 cm. This rod is then suspended horizontally by a vertical string attached to the rod at a point x cm from its mid-point. Find the value of x and the tension in the string.

18. The diagram shows a spade which consists of a handle, a uniform shaft and a uniform rectangular blade. The handle is of mass 0·5 kg, the shaft of mass 2 kg and the blade of mass 2 kg.

(a) If the spade is to rest horizontally on one support, where should this support be placed?

(b) If a man carries the spade horizontally with one hand on the handle and the other at a distance of 72 cm from the handle, find the vertical forces exerted by his hands when a brick of mass 6 kg is placed at the centre of the blade.

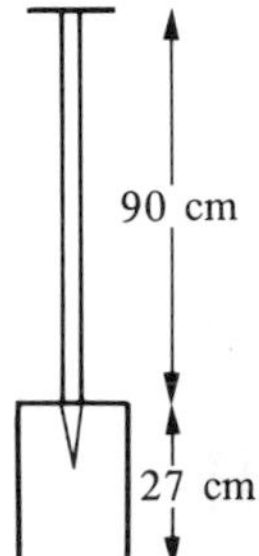

Non-parallel forces in equilibrium

The results stated for parallel forces in equilibrium also apply to non-parallel forces which are in equilibrium.

In particular, for any system of forces in equilibrium the algebraic sum of the moments of the forces about any point must be zero.

Problems involving forces which are not parallel can be solved by using this principle.

Example 15

A pendulum AM consists of a string, of length 2 m, and a bob of mass 5 kg. The pendulum is suspended from A and held in equilibrium by a horizontal force F applied at M so that the string makes an angle of 30° with the vertical. Find the force F.

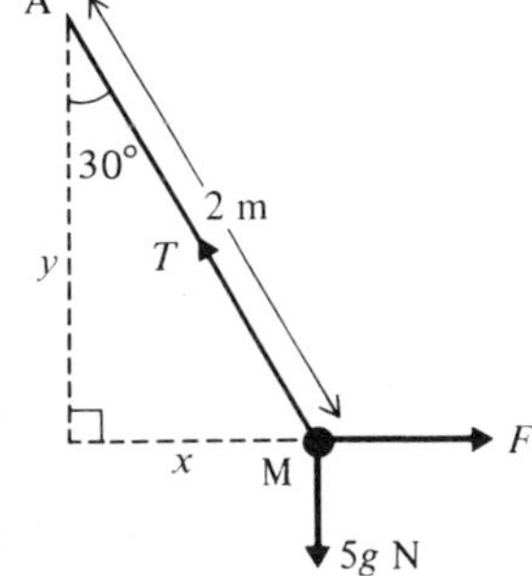

Suppose the horizontal and vertical distances of M from A are x and y respectively. Taking moments about the point A for all the forces acting on the bob

$$\text{A}\circlearrowright \quad 5g \times x = F \times y$$

$$\therefore \quad 5g \times 2\cos 60 = F \times 2\cos 30$$

$$\therefore \quad F = \frac{5g\sqrt{3}}{3}\ \text{N}$$

$$= 28{\cdot}3\ \text{N}$$

Note that the tension T in the string does not appear in the moment equation since its line of action passes through the point A about which moments are taken. It therefore has no moment about this point.

Moment of a force as sum of moments

When finding the moment of a force about a point, it is sometimes useful first to resolve the force into its components.

Suppose the force P acts at the point N at a distance s from the point O and in a direction making an angle θ with ON produced.

Resolving the force P

component in direction ON is $P\cos\theta$
component at right angles to ON is $P\sin\theta$

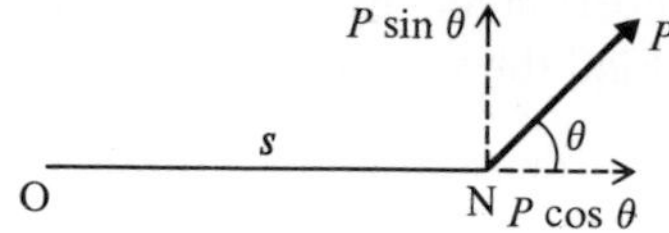

Moment of P about O = sum of moments of these components about O

$$= P\cos\theta \times 0 + P\sin\theta \times s$$
$$= Ps\sin\theta$$

Example 16

A uniform beam AB of mass 100 kg and length $2l$ m has its lower end A resting on rough horizontal ground. It is held in equilibrium at an angle of 20° with the horizontal by a rope attached to the end B which makes an angle of 40° with BA. Find the tension in the rope.

Let T be the tension in the rope. There are two forces acting on the beam, other than the action between the foot of the beam and the ground.
If moments are taken about the point A, the force at this point will have no moment and will not therefore appear in the equation.

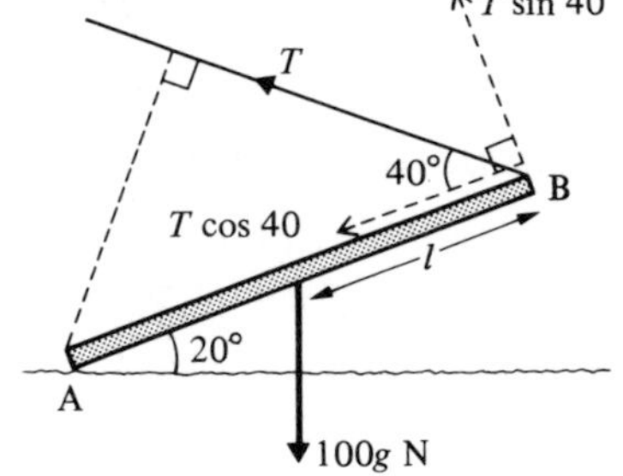

Resolving T into its components $T\cos 40$ and $T\sin 40$ along and at right angles to the beam respectively:

A↷ $\quad 100g \times l\cos 20 = T\sin 40 \times 2l + T\cos 40 \times 0$

$$\therefore \quad T = \frac{100g\cos 20}{2\sin 40} = 716\text{ N}$$

The tension in the rope is 716 N.

It should be noted that, for these bodies in equilibrium, the algebraic sum of the forces resolved in any direction must be zero. However, in examples 15 and 16 above, and in Exercise 7 E, this fact need not be used as the required answers can be obtained by taking moments about a suitably chosen point.

Exercise 7E

1. Each of the following diagrams shows a uniform beam, of mass 5 kg and length 4 m, freely hinged at a point A and resting horizontally in equilibrium. Find the magnitude of the force T in each case.

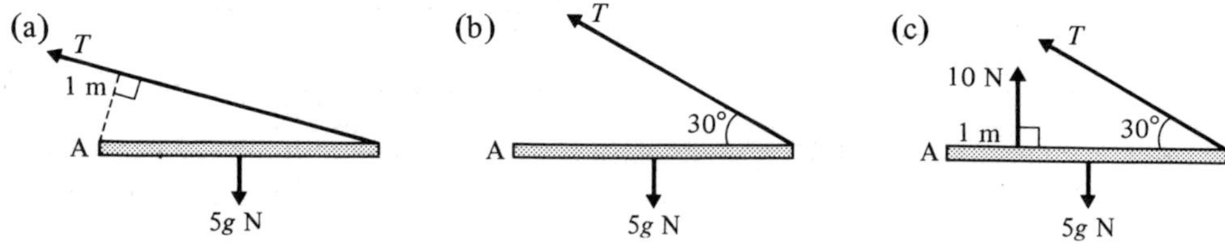

2. Each of the following diagrams shows a pendulum consisting of a light string with one end freely pivoted at A and the other end carrying a bob of mass 1 kg. The pendulum is held in equilibrium at an angle to the vertical by a force F. Find the magnitude of F in each case.

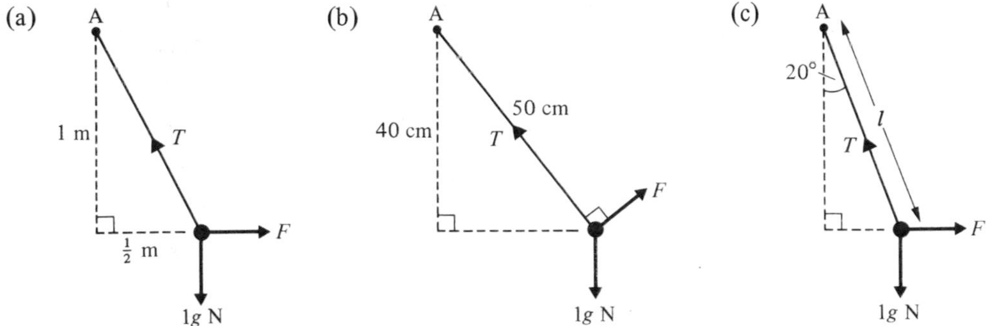

3. Each of the following diagrams shows a uniform beam of mass 5 kg and length 4 m with one end freely hinged at a point A. The beam rests in equilibrium at an angle to the horizontal. Find the magnitude of the force T in each case.

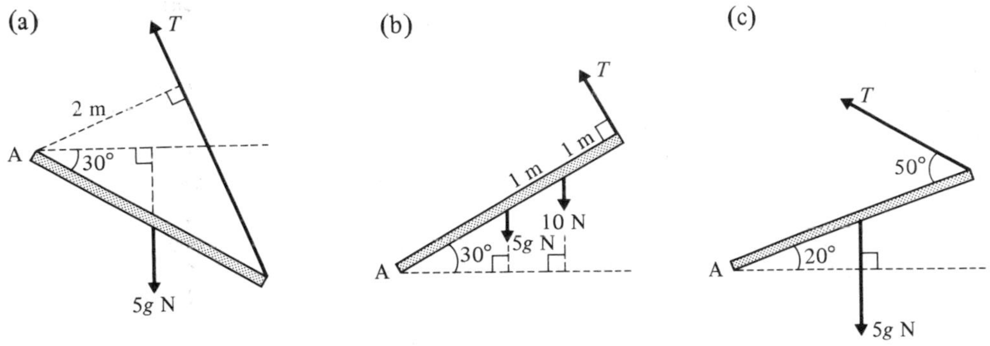

4. A pendulum consists of a light string AB of length 60 cm with end A fixed and a bob of mass 3 kg attached to B. Find the horizontal force that must be applied to the bob to keep the string at an angle of 25° to the downward vertical.
5. A light string AB of length 80 cm has end A fixed and a particle of mass 4 kg attached to B. Find the magnitude of the least force that must be applied to the particle so that it is held at a distance of 40 cm from the vertical through A with the string taut.
6. A uniform rod AB of mass 10 kg and length 2 m has its lower end A freely hinged at a fixed point and a particle of mass 4 kg attached to B. A horizontal string is attached to a point X on the rod where AX = 1·5 m. If the system rests in equilibrium, with the beam making an angle of 45° with the vertical, find the tension in the string.
7. A uniform horizontal shelf of mass 5 kg is freely hinged to a vertical wall and is supported by a chain CD as shown in the diagram:

The tension in the chain is 98 N, AD = 15 cm and angle CDA = 50°. Find the length AB.

8. A uniform rod AB of mass m hangs vertically with end A freely hinged to a fixed point. The rod is pulled aside by a horizontal force F, applied at B, until it makes an angle of 30° with the downward vertical. Show that $F = \frac{mg}{2\sqrt{3}}$.

9. The diagram shows a uniform rectangular paving slab ABCD of mass 15 kg held in equilibrium by a horizontal force F applied at the corner C. Corner A rests on the ground; AB = 30 cm and BC = 40 cm. AD makes an angle of 30° with the horizontal. Find the magnitude of F.

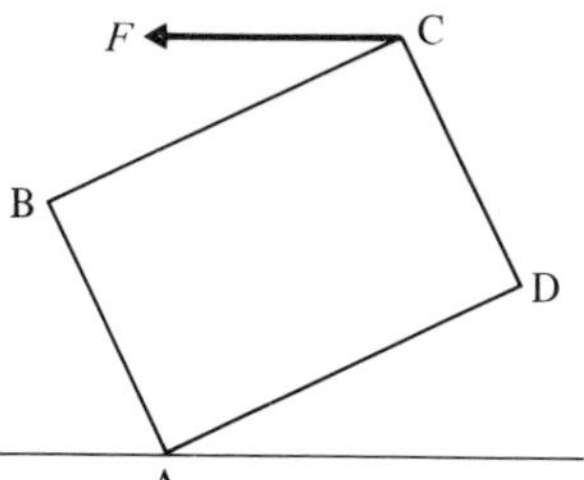

10. A non-uniform beam AB is of mass 100 kg and length 12 m. X and Y are points on the beam such that AX = 4 m and AY = 7 m. The weight of the beam can be considered to act at point X. The beam has its lower end A resting on rough horizontal ground and is kept in equilibrium, at an angle of 26° to the horizontal, by a rope attached to point Y and making an angle of 55° with YA. Find the tension in the rope.

11. A uniform beam AB of mass m and length $2l$ has its lower end A resting on rough horizontal ground and is kept in equilibrium, at an angle of 45° to the horizontal, by a rope attached to end B. If the rope makes an angle of 60° with BA and T is the tension in the rope, show that $T = \frac{mg}{\sqrt{6}}$.

Equivalent systems of forces

It has already been seen that a system of parallel forces is either equivalent to a couple or can be replaced by a single force or resultant. In fact *any* system of coplanar forces is either equivalent to a couple or can be replaced by a single force or resultant. In the latter case the magnitude, direction and line of action of the resultant can be found by resolving and taking moments.

Example 17

The forces 4, 2, 3 and 5 N act along the sides AB, BC, CD and DA respectively of the rectangle ABCD in which AB = 6 cm and BC = 3 cm. The forces act in the directions indicated by the order of the letters. Forces X and Y, acting as shown in the second diagram, are equivalent to the given system of forces. Find the magnitude of X and Y and also the distance x. Hence find the magnitude and the direction of the single force which is equivalent to the given system.

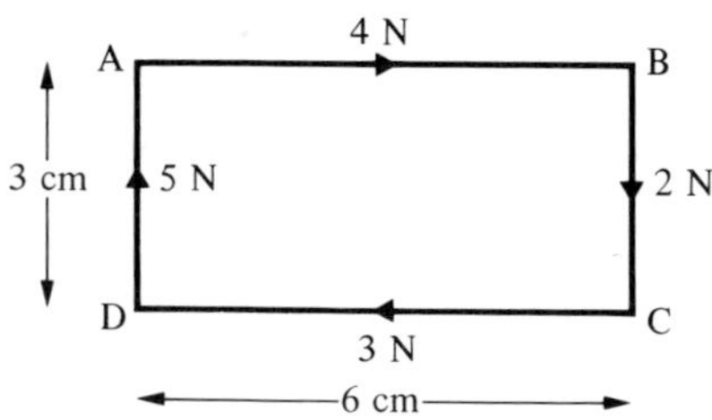

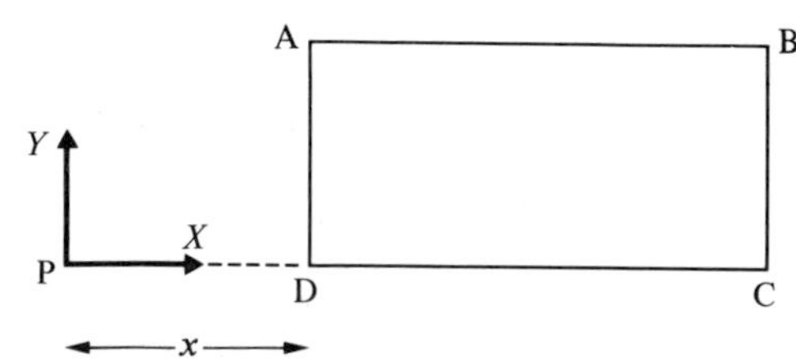

Resolving both systems parallel to AB $\qquad 4 - 3 = X$

$\therefore \quad X = 1$ N

Resolving both systems parallel to DA $\qquad 5-2=Y$

$\therefore \quad Y = 3\text{ N}$

Equating the moments of the two systems $\overset{\curvearrowright}{\text{D}}$ $\qquad 2(6)+4(3) = Y\times x$

$\therefore \quad 24 = 3x$

$\therefore \quad x = 8\text{ cm}$

$X = 1$ N, $Y = 3$ N and the distance $x = 8$ cm.

The single force R equivalent to the system is given by

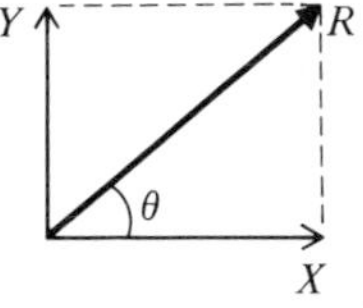

$$R^2 = X^2 + Y^2$$
$$\therefore \quad R^2 = 1^2 + 3^2$$
$$\therefore \quad R = 3{\cdot}16\text{ N}$$

and $\tan\theta = \frac{3}{1} \quad \therefore \theta = 71{\cdot}57^\circ$

The single force is 3·16 N acting at an angle of $71{\cdot}57^\circ$ to the line DC.

Use of components when taking moments

The following example illustrates the use of the components of a force when the moment of a force about a point is required.

Example 18

The square ABCD has sides of length a. Find the moment of the force $3\sqrt{2}$ N, acting along DB, about the point O as shown in the diagram.

(i) The moment, by definition, is:

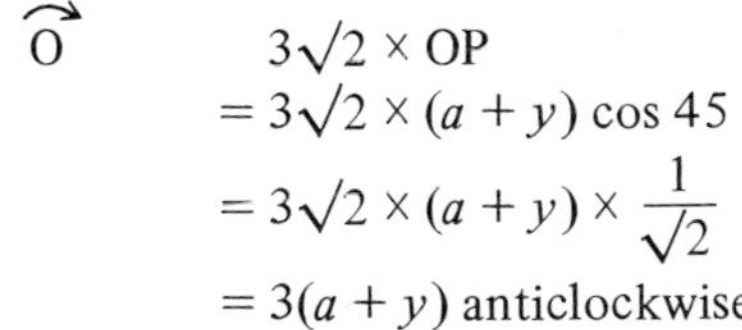

$$\overset{\curvearrowleft}{\text{O}} \qquad 3\sqrt{2}\times \text{OP}$$
$$= 3\sqrt{2}\times(a+y)\cos 45$$
$$= 3\sqrt{2}\times(a+y)\times\frac{1}{\sqrt{2}}$$
$$= 3(a+y)\text{ anticlockwise}$$

(ii) The force acting along DB can be resolved into two components acting at D and the algebraic sum of the moments of these components about O is:

$$\overset{\curvearrowleft}{\text{O}} \qquad 3\sqrt{2}\sin 45\times y \;+\; 3\sqrt{2}\cos 45\times a$$
$$= 3\sqrt{2}\times\frac{1}{\sqrt{2}}\times y \;+\; 3\sqrt{2}\times\frac{1}{\sqrt{2}}\times a$$
$$= 3(a+y)\text{ anticlockwise}$$

(iii) Alternatively the force acting along DB can be resolved into two components at B

$$\overset{\curvearrowleft}{\text{O}} \qquad 3\sqrt{2}\sin 45\times(a+y) \;+\; 3\sqrt{2}\cos 45\times 0$$
$$= 3\sqrt{2}\,\frac{1}{\sqrt{2}}\times(a+y)$$
$$= 3(a+y)\text{ anticlockwise}$$

It is seen that the same result is obtained each time, but care is necessary in distinguishing between case (ii) and case (iii).

Example 19

The forces 2, 4, 2, 5 and $3\sqrt{2}$ N act along the sides AB, BC, CD, DA and DB respectively of the square ABCD in the directions indicated by the order of the letters. Find the magnitude and direction of the force R which is equivalent to this system of forces. If the line of action of the force R cuts BA produced at a point P, x cm from A, and the length of each side of the square is 10 cm, find x.

Draw the two diagrams showing the given system of forces and the required force R.

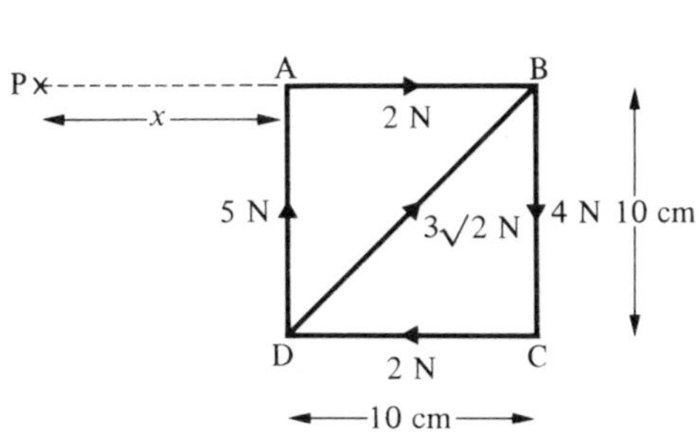

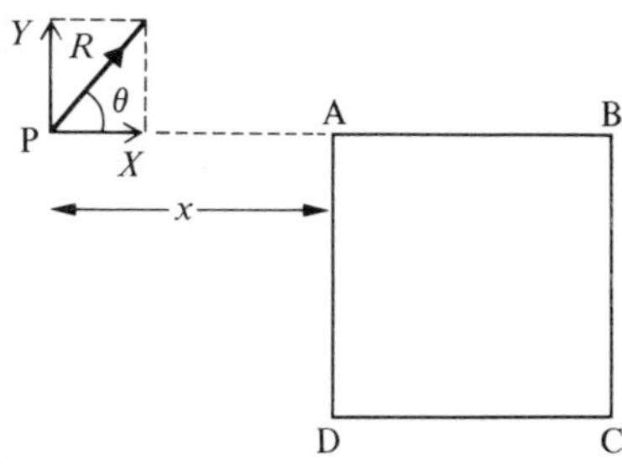

Resolving both systems parallel to AB $\quad 2 + 3\sqrt{2}\cos 45 - 2 = X$

$$\therefore \quad 3\text{ N} = X \qquad \ldots [1]$$

Resolving both systems parallel to DA $\quad 5 + 3\sqrt{2}\sin 45 - 4 = Y$

$$\therefore \quad 4\text{ N} = Y \qquad \ldots [2]$$

Squaring and adding equations [1] and [2]

$$R^2 = X^2 + Y^2$$
$$= 9 + 16$$
$$= 25$$
$$\therefore \quad R = 5\text{ N}$$

also $\tan\theta = \dfrac{Y}{X} = \frac{4}{3}$

$$\therefore \quad \theta = 53{\cdot}13^\circ$$

Equating the moments of the two systems about the point B

$$\overset{\curvearrowright}{\text{B}} \quad (5 \times 10) + (2 \times 10) = Y(x + 10)$$

substituting $Y = 4$, $\quad \therefore \quad x = 7\frac{1}{2}$ cm

The equivalent force is 5 N acting at an angle of $53{\cdot}13^\circ$ to the libe AB and cutting BA produced $7\frac{1}{2}$ cm from A.

Note that alternatively moments can be taken about the point P. In this case the moment of the system consisting of the resultant R will be zero.

Hence, equating moments of the two systems about the point P,

$$\overset{\curvearrowright}{\text{P}} \quad (2 \times 10) + 4(10 + x) - 5x - 3\sqrt{2}\sin 45(x + 10) = R \times 0$$

$$\therefore \quad 20 + 40 + 4x - 5x - 3x - 30 = 0$$

and again $x = 7\frac{1}{2}$ cm is obtained.

Exercise 7F

1. For each of the following, find the magnitude and direction of the single force equivalent to the system of forces shown. Find also where the line of action of this single force crosses AB.

(a)

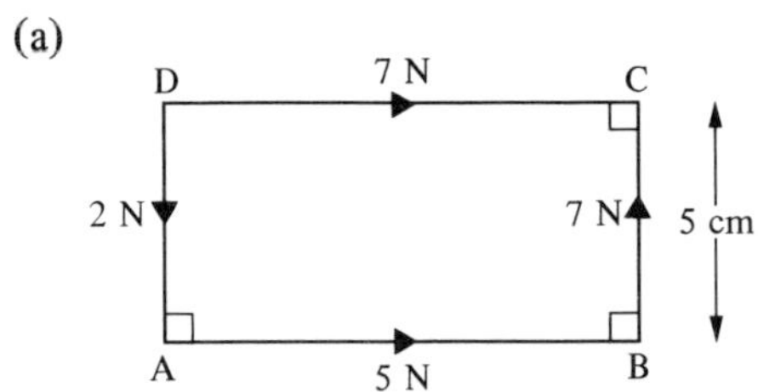

(b)

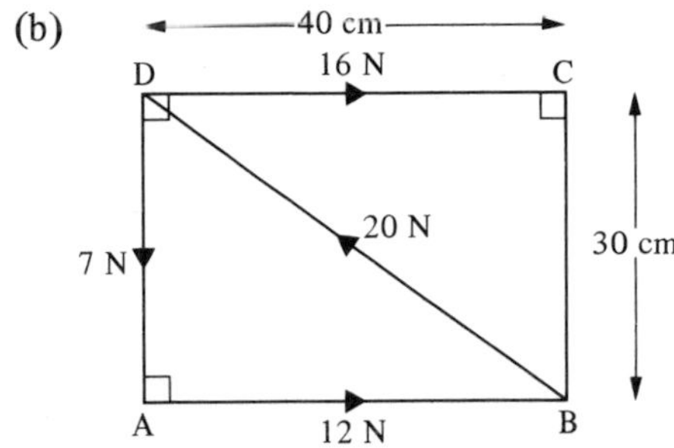

(c)

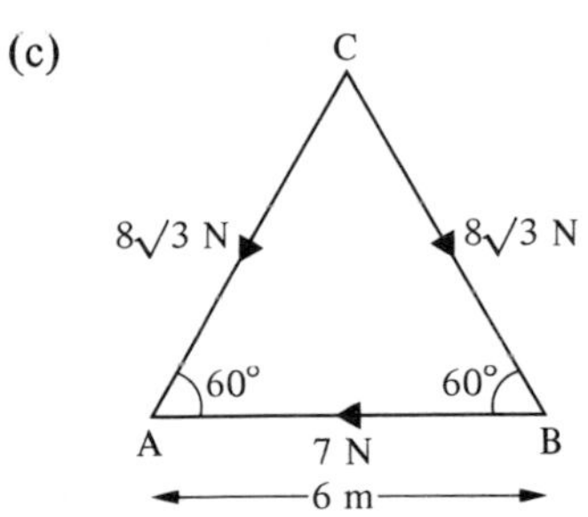

(d)

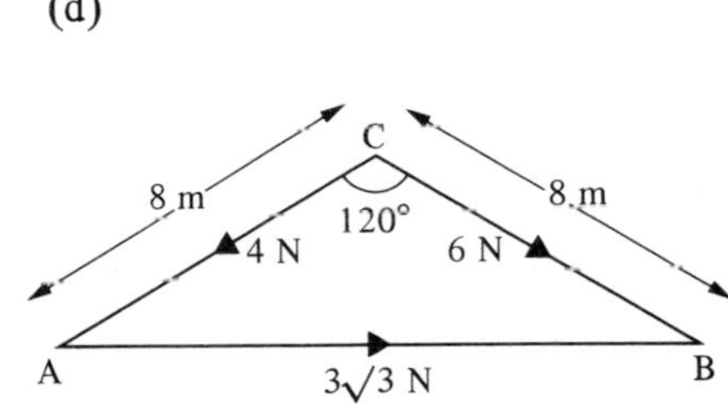

2. Show that each of the following systems of forces are equivalent to a couple and in each case find the moment of the couple.

(a)

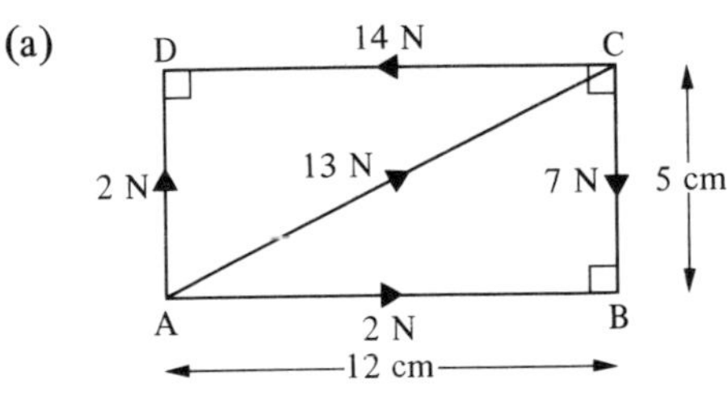

(b)

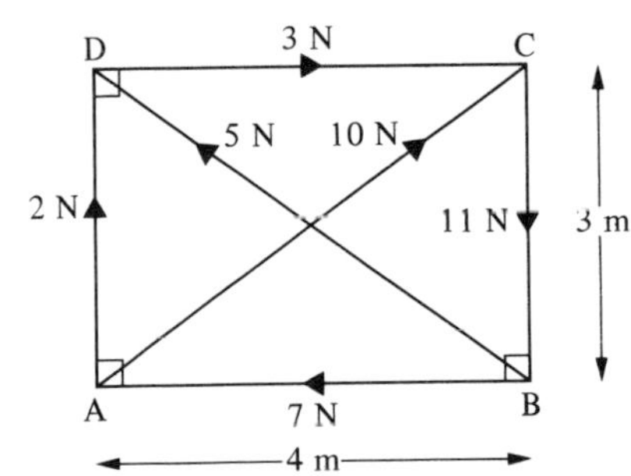

(c)

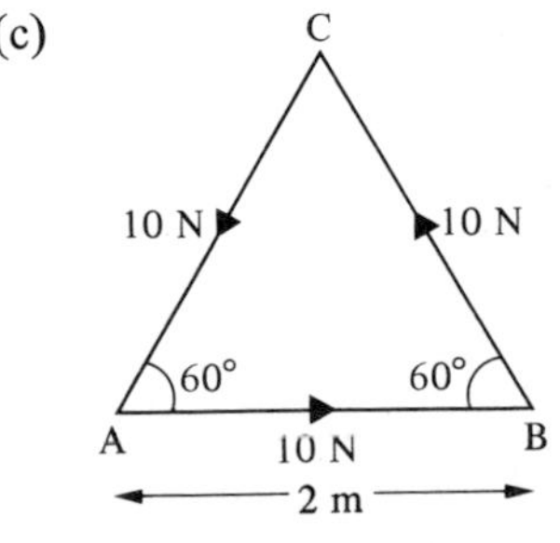

(d)

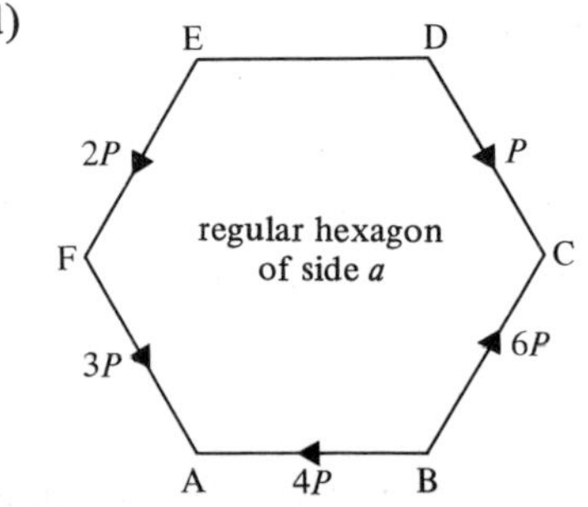

In the following questions involving geometrical figures, the forces act along the given lines in the directions indicated by the order of the letters.

3. ABCD is a rectangle with AB = 1·5 m and AD = 1 m. Forces of 2 N, 1 N, 1 N and 3 N act along AB, BC, DC and AD respectively. Calculate the magnitude and direction of the single force that could replace this system of forces and find where its line of action cuts AB.
4. ABCD is a square of side 5 m. Forces of 4 N, 6 N, 8 N and 10 N act along BD, DC, CA and CB respectively. When a force P acts along AD and a force Q acts along AB, the whole system is equivalent to a couple. Find the magnitudes of P and Q and the moment of the couple.
5. ABCD is a square of side a. Forces of 1 N, 4 N, 3 N and 6 N act along AB, CB, DC and AD respectively. Calculate the magnitude and direction of the single force that could replace this system of forces and find where its line of action cuts AB.
6. ABCD is a rectangle with AB = 3 m and $C\hat{A}B = 30°$. Forces of 10 N, 20 N and 20 N act along AC, AD and DB respectively. Calculate the magnitude and direction of the single force that could replace this system of forces and find where its line of action cuts AB.
7. ABC is an equilateral triangle of side a. Forces of 10 N, 6 N and 10 N act along AB, CB and AC respectively. Find the magnitude and direction of the single force equivalent to this system and find where its line of action cuts AB.
8. Point O is the origin and points A, B and C have position vectors $3\mathbf{i}$, $3\mathbf{i} + 2\mathbf{j}$ and $2\mathbf{j}$ respectively. A force of $(\mathbf{i} + 4\mathbf{j})$ N acts at A, $5\mathbf{i}$ N at B and $(-2\mathbf{i} + 2\mathbf{j})$ N at C. Find the single force that could replace this system and find the position vector of the point where its line of action cuts OA.
9. Point O is the origin and points A and B have position vectors $4\mathbf{i}$ and $4\mathbf{j}$ respectively. A force of $(5\mathbf{i} + 7\mathbf{j})$ N acts at A, a force of $(-6\mathbf{i} + 3\mathbf{j})$ N acts at B and a force of $(4\mathbf{i} - 6\mathbf{j})$ N acts at O. Find the single force equivalent to this system and find the position vector of the point where its line of action cuts OA.
10. ABC is an isosceles triangle, right-angled at A with AB = 1 m. Forces of 8 N, 4 N and 6 N act along BA, BC and CA respectively. Find the single force that could replace this system and find where its line of action cuts AB.
11. ABCDEF is a regular hexagon of side 2 m. Forces of 2 N, 3 N, 4 N and 5 N act along AC, AE, AF and ED respectively. Find the single force equivalent to this system and find where its line of action cuts AB.
12. ABCDE is a regular pentagon of side 2 m. Forces of 5 N act along AB, BC and AD. Find the single force that could replace these three forces and find where its line of action cuts AB.
13. ABCD is a rectangle with AB = 10 cm and $C\hat{A}B = 20°$. Forces of 5 N act along BA, CD and AD and forces of 10 N and 20 N act along DB and CA respectively. Find the single force equivalent to this system and find where its line of action cuts AB.
    If the 5 N force along AD were along BC instead, how would this affect your answers?
14. ABCD is a rectangle with AB = 70 cm and AD = 20 cm. Forces of 5 N, 2 N, 3 N and 6 N act along AB, BC, CD and AD respectively. Find the single force that could replace this system and find where its line of action cuts AB.
    The force along AB is now replaced by a force through A which reduces the system to a couple. Find the magnitude and direction of this force and the moment of the couple.
15. ABCD is a rectangle with AB = 4 m and BC = 3 m. Forces of 4 N, 5 N and 10 N act along CB, DC and DB respectively. Find the single force equivalent to this system and find where its line of action cuts AB.
    A couple of moment 15 N m, in the sense ABCD, is now introduced to the system. Find

the magnitude and direction of the single force that will replace this new system and show that its line of action passes through B.

16. Point O is the origin and points A, B, C and D have position vectors $(3\mathbf{i}+\mathbf{j})$ m, $(\mathbf{i}+3\mathbf{j})$ m, $(-2\mathbf{i}+\mathbf{j})$ m and $(-2\mathbf{i}-2\mathbf{j})$ m respectively. Forces of $(3\mathbf{i}+3\mathbf{j})$ N, $(4\mathbf{i}-5\mathbf{j})$ N, $(-5\mathbf{i}+2\mathbf{j})$ N and $(2\mathbf{i}+3\mathbf{j})$ N act at points A, B, C and D respectively. Find the single force that could replace this system and find where its line of action cuts the horizontal axis through O. A couple of moment a N m anticlockwise and a force $(b\mathbf{i}+c\mathbf{j})$ N acting through the point which has position vector $(2\mathbf{i}+\mathbf{j})$ m are now added to the system. If these reduce the system to equilibrium find a, b and c.

Exercise 7G Examination questions

1. In this question the unit of distance is the metre and the unit of force is the newton.

   The force $\mathbf{P}=2\mathbf{i}$ acts through the point with position vector $5\mathbf{j}$. The force $\mathbf{Q}=6\mathbf{i}$ acts through the point with position vector $\mathbf{j}$. The resultant of $\mathbf{P}$ and $\mathbf{Q}$ acts through the point with position vector $n\mathbf{j}$. Find the value of n and the magnitude of the resultant of $\mathbf{P}$ and $\mathbf{Q}$. (London)

2. Three forces are represented by the vectors $-2\mathbf{i}-3\mathbf{j}$, $3\mathbf{i}+4\mathbf{j}$ and $-\mathbf{i}-\mathbf{j}$. The forces act at the points (2, 0), (0, 3) and (1, 1) respectively. Show that the three forces combine to form a couple, and find the magnitude of the couple. (A.E.B)

3. A straight uniform rigid rod AB is of length 8 m and mass 10 kg. The rod is supported at the point X, where AX = 5 m, and, when downward vertical forces of magnitudes P and $4P$ newtons are applied at A and B respectively, the rod rests in equilibrium with AB horizontal. Calculate (a) the value of P, (b) the force, in N, exerted on the support at X. (London)

4. A thin non-uniform beam AB, of length 6 m and mass 50 kg, is in equilibrium resting horizontally on two smooth supports which are respectively 2 m and 3·5 m from A. The thrusts on the two supports are equal. Find the position of the centre of gravity of the beam. The original supports are removed and a load of 10 kg is attached to the beam at B. The loaded beam rests horizontally on two new smooth supports at A and C, where C is a point on the beam 1 m from B. Calculate the thrusts on each of the new supports. (A.E.B)

5. (a) Forces $\overrightarrow{AB}$, $2\overrightarrow{BC}$, $\overrightarrow{CA}$ act along the sides of a triangle ABC. Prove that they are equivalent to a force $\overrightarrow{BC}$ acting through A, together with a couple represented in magnitude by 4Δ, where Δ is the area of ΔABC.

   (b) Forces act along the sides of a square, side a, as shown, together with a clockwise couple $2Pa$. Calculate the single resultant force and the distance from A where its line of action meets AB.

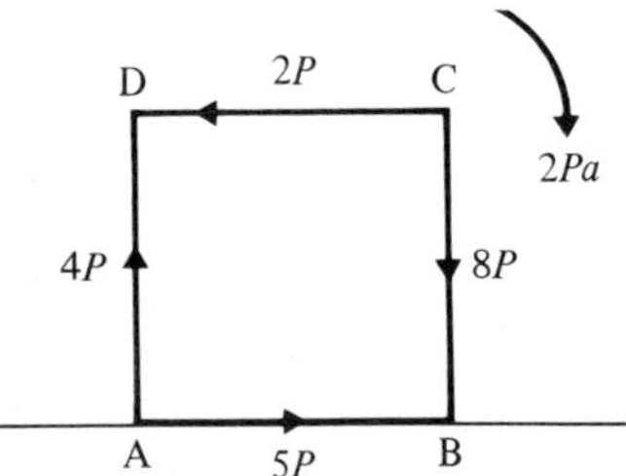

   Forces R, S, are added to the system along $\overrightarrow{BC}$ and $\overrightarrow{CD}$ respectively. Calculate the values of R and S if the single resultant force for the new system acts along BD. (Cambridge)

6. OABC is a square of side 1 m. Forces of magnitude 2, 3, 4 and $5\sqrt{2}$ newtons act along OA, AB, CB and AC respectively in directions indicated by the order of the letters and a couple of moment 7 N m acts in the plane of the square in the sense OCBA. Find the magnitude and direction of the resultant of the system and the equation of its line of action referred to OA and OC as axes. What is the magnitude and direction of the least force introduced at A if the resultant of the original system and this new force is to pass through O? (S.U.J.B)

7. An equilateral triangle ABC has sides of length $2a$, and D, E, F are the midpoints of BC, CA, AB respectively. Forces of magnitude P, P, $2P$, $2P$ act along AB, FC, AC, BE respectively, the direction of each force being indicated by the order of the letters. Show that the resultant of this system of forces acts at an angle $\tan^{-1}(7 + 4\sqrt{3})$ to AB, and calculate its magnitude.
The line of action of the resultant cuts AB at H. Show that AH is approximately $0{\cdot}8a$.
The system is equivalent to three forces acting along AC, BC, DA. Calculate the magnitudes of each of these forces. (Cambridge)

8. The points A, B and C have coordinates $(9a, 0)$, $(0, -4a)$ and $(6a, 4a)$ respectively referred to the coordinate axes Ox and Oy.
   (a) Forces $12P$, $15P$, $5P$ and P act along $\overrightarrow{OA}$, $\overrightarrow{BC}$, $\overrightarrow{CA}$ and $\overrightarrow{OB}$ respectively. Calculate the magnitude of the resultant of these forces and the equation of the line of action of this resultant.
   (b) Forces $15P$, S and T act along $\overrightarrow{BC}$, $\overrightarrow{OA}$ and $\overrightarrow{CA}$ respectively. Given that these forces reduce to a couple, calculate (i) the values of S and T, in terms of P, (ii) the magnitude of the couple, in terms of a and P, (iii) the sense of the couple. (A.E.B)

9. The centre of a regular hexagon ABCDEF of side a is O. Forces of magnitude P, $2P$, $3P$, $4P$, mP and nP act along $\overrightarrow{AB}$, $\overrightarrow{BC}$, $\overrightarrow{CD}$, $\overrightarrow{DE}$, $\overrightarrow{EF}$ and $\overrightarrow{FA}$ respectively. Given that the resultant of these six forces is of magnitude $3P$ acting in a direction parallel to $\overrightarrow{EF}$,
   (i) determine the values of m and n,
   (ii) show that the sum of the moments of the forces about O is $9Pa\sqrt{3}$.
   The mid-point of EF is M.
   (iii) Find the equation of the line of action of the resultant referred to OM as x-axis and OA as y-axis.

   The forces mP and nP acting along $\overrightarrow{EF}$ and $\overrightarrow{FA}$ are removed from the system. The remaining four forces and an additional force Q, which acts through O, reduce to a couple. Calculate the magnitude of Q and the moment of the couple. (A.E.B)

8
Centre of gravity

Attraction of the Earth

Every particle is attracted towards the centre of the Earth, and this force of attraction is the force previously referred to as the weight of the particle.
As was explained in Chapter 3, a particle of mass m kg has a vertical force of mg newtons acting upon it, i.e. its weight is mg newtons.

For a number of particles m_1, m_2 and m_3, these forces may be considered to be parallel, all being directed towards the centre of the earth.
If the relative positions of these particles are fixed and known, then the resultant of these parallel forces can be determined.

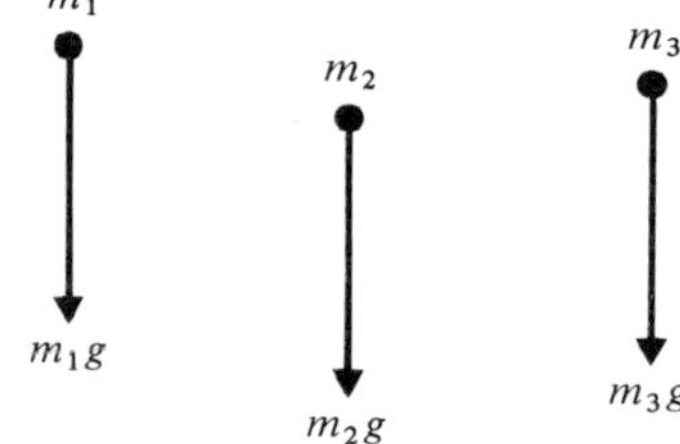

Centre of gravity of a system of particles

The centre of gravity of a number of particles is the point through which the line of action of the resultant of these parallel forces always passes, i.e. it is the point through which the resultant weight of the system acts.
In particular, if two equal particles are at the points A and B, then the centre of gravity of the system made up of these two particles will be the mid-point of the line AB.
Whatever the positions of the two particles, the resultant of the two forces acting on the particles is known to pass through the point G, where AG = GB.
Hence G is the centre of gravity.

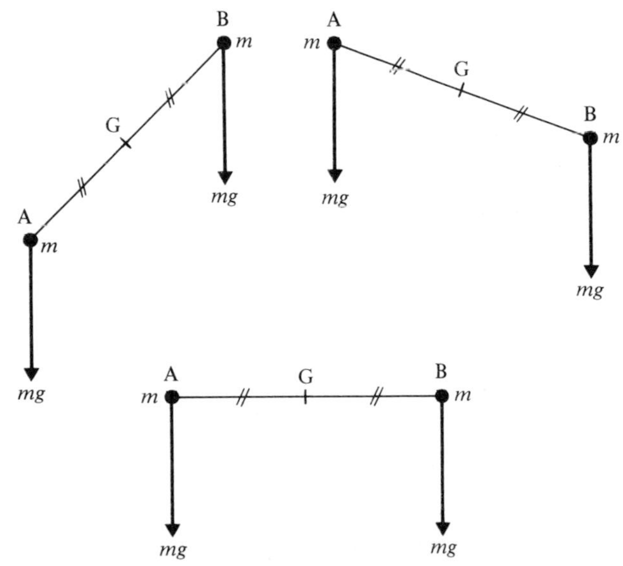

Example 1

Find the position of the centre of gravity of three particles of mass 4 kg, 10 kg and 6 kg, which lie on the x-axis at the points (3, 0), (4, 0) and (7, 0) respectively.

Draw two diagrams, the first showing the forces due to the masses of the particles and the second showing the resultant force Mg due to the total mass M.

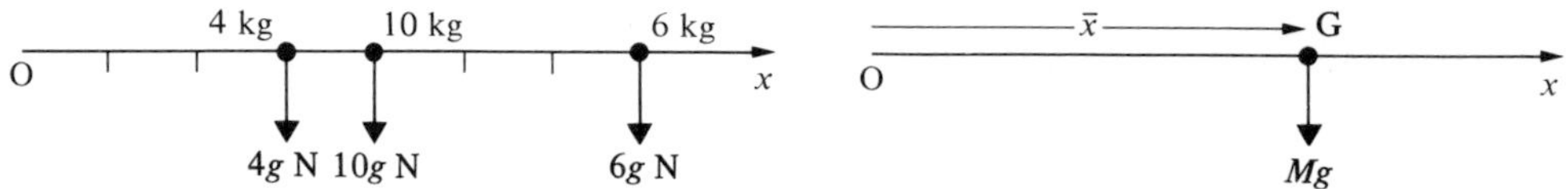

Suppose that the resultant weight Mg acts through a point G on the x-axis, at a distance $\bar{x}$ from the origin.

Resolving vertically $\quad 4g + 10g + 6g = Mg$
$\therefore \quad M = 20$ kg

Equating the moments of the two systems about the point O

$$(4g \times 3) + (10g \times 4) + (6g \times 7) = Mg \times \bar{x}$$

substituting for M gives $\quad \bar{x} = 4{\cdot}7$

The centre of gravity is at a point on the x-axis, 4·7 units from the origin.

Note that, in this example, all the particles lie on one line, the x-axis. It is therefore apparent that the centre of gravity is also on this line.

When a system of particles is such that the particles are not at collinear points, then the position of the centre of gravity relative to two axes must be considered.

Example 2

Find the coordinates of the centre of gravity of the given system of particles, Fig. 1.

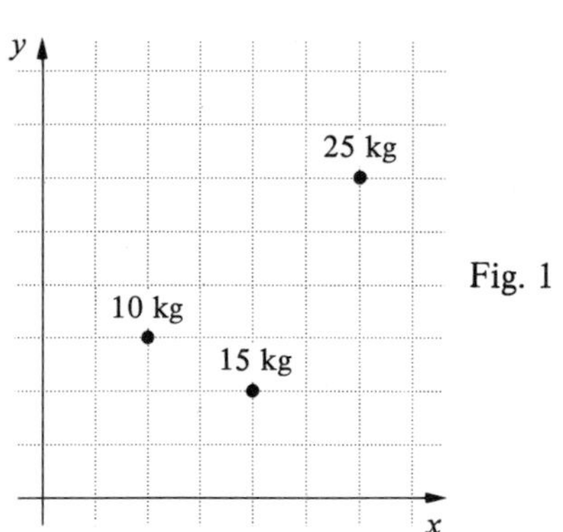

Fig. 1

Draw a second diagram showing the total mass of the particles at the point $(\bar{x}, \bar{y})$, Fig. 2.

Resolving vertically $\quad 10g + 15g + 25g = Mg$
$\therefore \quad M = 50$ kg

Equating the moments of the two systems about the y-axis

$$(10g \times 2) + (15g \times 4) + (25g \times 6) = Mg \times \bar{x}$$

substituting for M gives $\quad \bar{x} = 4{\cdot}6$

Equating the moments of the two systems about the x-axis

$$(10g \times 3) + (15g \times 2) + (25g \times 6) = Mg \times \bar{y}$$

substituting for M gives $\quad \bar{y} = 4{\cdot}2$

The centre of gravity is at the point with coordinates (4·6, 4·2).

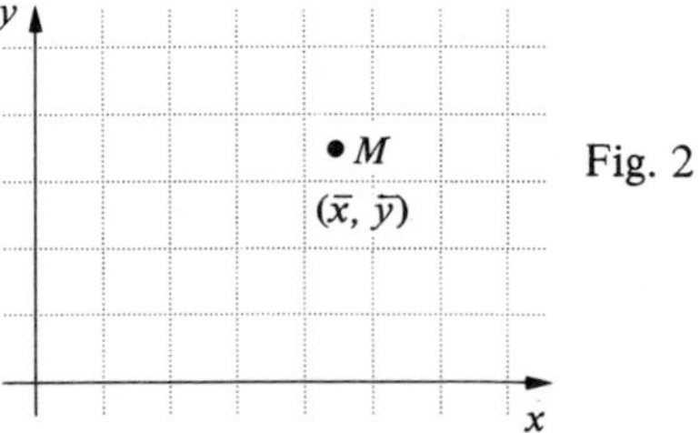

Fig. 2

Example 3

The rectangle ABCD has AB = 20 cm and AD = 30 cm. Particles of mass 4 kg, 4 kg, 5 kg and 2 kg are placed at the points A, B, C and D respectively. Find the position of the centre of gravity of the system of particles.

Draw two diagrams.

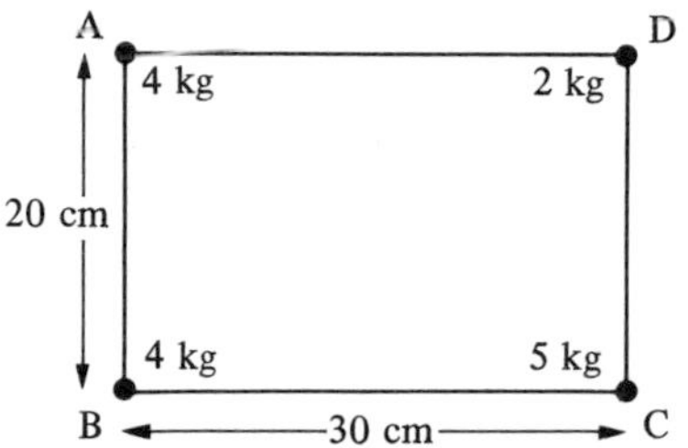

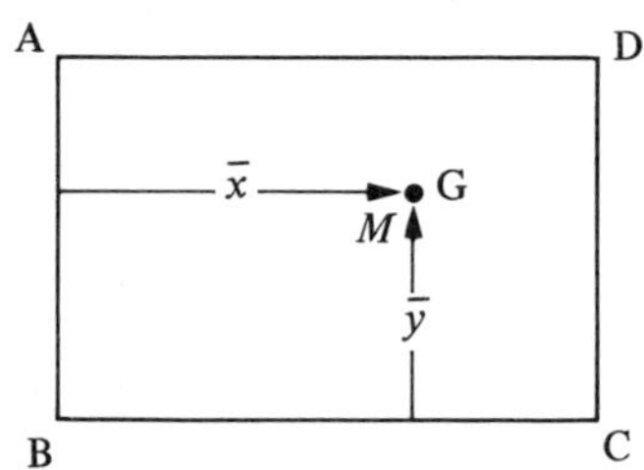

Suppose that the centre of gravity of the particles is at the point G, where $\bar{x}$ and $\bar{y}$ are the distances of G from the sides AB and BC respectively.
Hence the resultant weight of the particles acts through the point G, perpendicular to the plane of the rectangle ABCD.

Resolving perpendicular to the plane ABCD

$$4g + 4g + 5g + 2g = Mg$$
$$\therefore \quad M = 15 \text{ kg}$$

Equating moments about the axis AB

$$(4g \times 0) + (4g \times 0) + (5g \times 30) + (2g \times 30) = Mg \times \bar{x}$$
substituting for M gives $\quad \bar{x} = 14$ cm

Equating moments about the axis BC

$$(4g \times 0) + (4g \times 20) + (5g \times 0) + (2g \times 20) = Mg \times \bar{y}$$
substituting for M gives $\quad \bar{y} = 8$ cm

The centre of gravity is at a point 14 cm from AB and 8 cm from BC, and lies within the rectangle.

Example 4

Find the position vector of the centre of gravity of particles of mass 0·5 kg, 1·5 kg and 2 kg which are at the points with position vectors (6**i** − 3**j**), (2**i** + 5**j**) and (3**i** + 2**j**) respectively.

Draw two diagrams

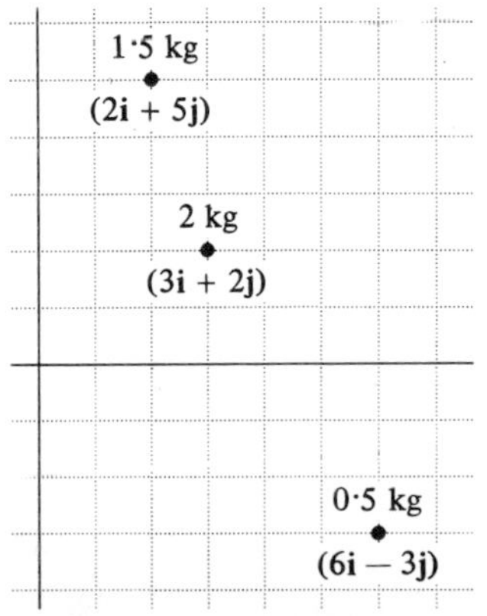

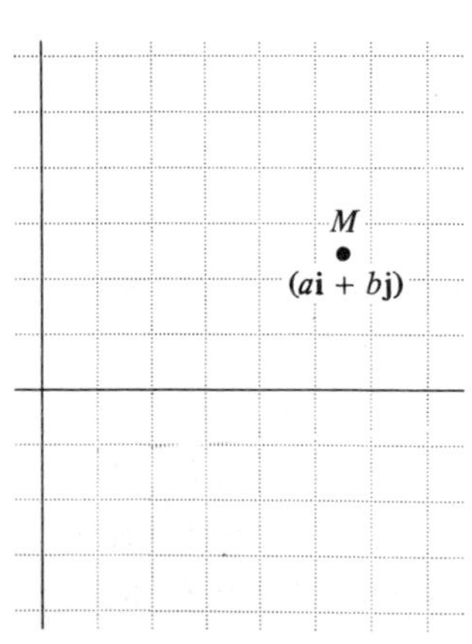

Suppose that the resultant weight acts at the point with position vector $(a\mathbf{i} + b\mathbf{j})$.

Resolving perpendicular to the plane of the axes

$$0{\cdot}5g + 1{\cdot}5g + 2g = Mg \qquad \therefore \quad M = 4 \text{ kg}$$

Equating moments about each axis in turn

$$(1{\cdot}5g \times 2) + (2g \times 3) + (0{\cdot}5g \times 6) = Mg \times a$$

substituting for M $\quad \therefore \quad a = 3$

and $\quad (1{\cdot}5g \times 5) + (2g \times 2) - (0{\cdot}5g \times 3) = Mg \times b \qquad \therefore \quad b = 2{\cdot}5$

The centre of gravity is at the point with position vector $(3\mathbf{i} + 2{\cdot}5\mathbf{j})$.

The effect of the negative sign in the position vector of one of the particles should be carefully noted.

General result

Although each example should, at this stage, be considered from first principles, it is possible to state a general result for the position of the centre of gravity of a system of particles m_1, m_2, m_3, . . . at the points with coordinates (x_1y_1), (x_2y_2), (x_3y_3), . . .

The centre of gravity is at the point $(\bar{x}, \bar{y})$ given by

$$m_1x_1 + m_2x_2 + m_3x_3 + \ldots = (m_1 + m_2 + m_3 + \ldots)\bar{x}$$

$$\text{or} \qquad \bar{x} = \frac{\Sigma(m_ix_i)}{\Sigma m_i}$$

and similarly for $\bar{y}$.

Exercise 8A

1. Find the coordinates of the centre of gravity of each of the following systems of particles. The grid squares on each graph are unit squares.

(a)

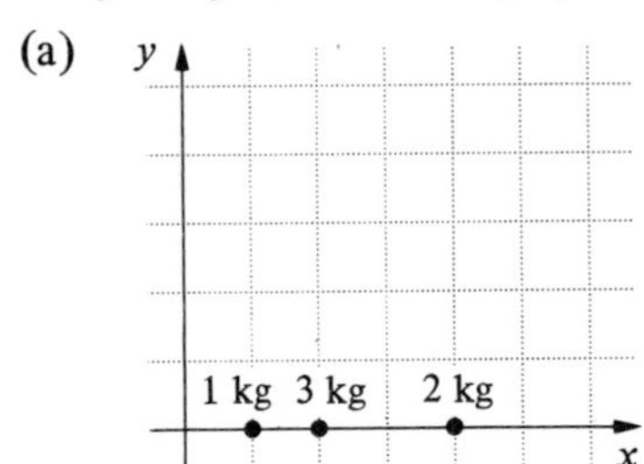

(b)

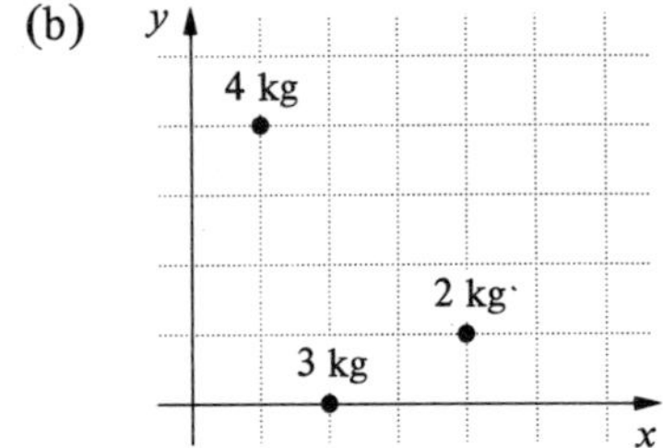

(c)

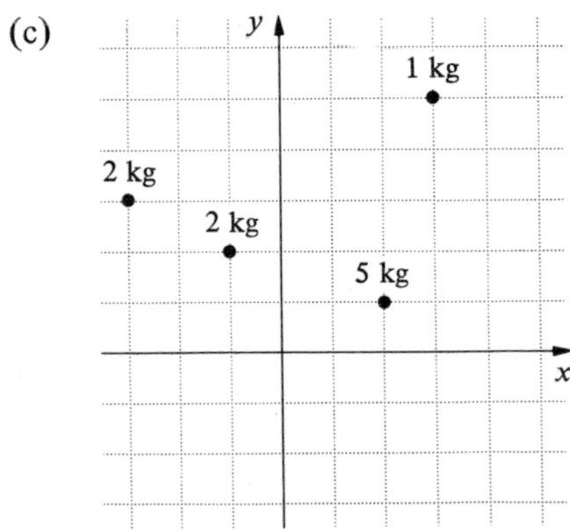

(d)

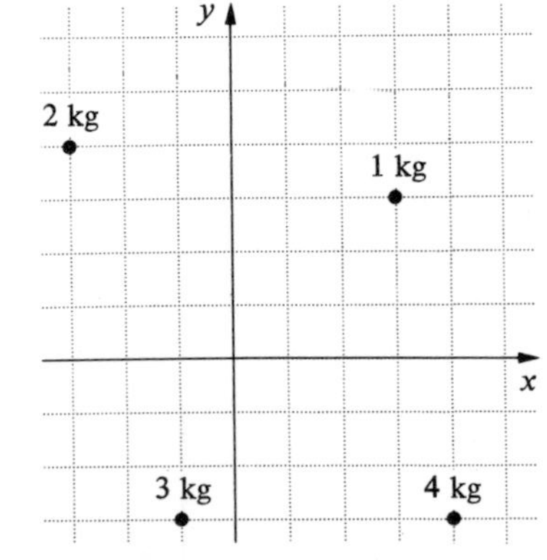

2. Find the position of the centre of gravity of three particles of mass 1 kg, 5 kg and 2 kg which lie on the y-axis at the points (0, 2), (0, 4) and (0, 5) respectively.
3. Find the coordinates of the centre of gravity of four particles of mass 5 kg, 2 kg, 2 kg and 3 kg situated at (3, 1), (4, 3), (5, 2) and (−3, 1) respectively.
4. Find the coordinates of the centre of gravity of four particles of mass 60 g, 30 g, 70 g and 40 g situated at (4, 3), (6, 5), (−6, 5) and (−5, −2) respectively.
5. Three particles of mass 2 kg, 1 kg and 3 kg are situated at (4, 3), (1, 0) and (a, b) respectively. If the centre of gravity of the system lies at (0, 2), find the values of a and b.
6. The rectangle ABCD has AB = 4 cm and AD = 2 cm. Particles of mass 3 kg, 5 kg, 1 kg and 7 kg are placed at the points A, B, C and D respectively. Find the distance of the centre of gravity of the system from each of the lines AB and AD.
7. The rectangle EFGH has EF = 3 m and EH = 2 m. Particles of mass 2 g, 3 g, 6 g and 1 g are placed at the mid-points of the sides EF, FG, GH and EH respectively. Find the distance of the centre of gravity of the system from each of the lines EF and EH.
8. Find the position vector of the centre of gravity of particles of mass 2 kg, 1 kg, 3 kg and 2 kg which are at the points with position vectors $(6\mathbf{i} + 6\mathbf{j})$, $(3\mathbf{i} + 5\mathbf{j})$, $(7\mathbf{i} + 3\mathbf{j})$ and $(2\mathbf{i} - \mathbf{j})$ respectively.
9. Find the position vector of the centre of gravity of particles of mass 50 g, 60 g, 20 g and 20 g which are at the points with position vectors $(5\mathbf{i} - 7\mathbf{j})$, $(-3\mathbf{i} + 2\mathbf{j})$, $(3\mathbf{i} - 5\mathbf{j})$ and $(\mathbf{i} - 6\mathbf{j})$ respectively.
10. Particles of mass 1 kg, 2 kg, 3 kg and 4 kg lie at the points with position vectors $6\mathbf{i}$, $(\mathbf{i} - 5\mathbf{j})$, $(3\mathbf{i} + 2\mathbf{j})$ and $(a\mathbf{i} + b\mathbf{j})$ respectively. If the centre of gravity of this system lies at the point with position vector $(2\frac{1}{2}\mathbf{i} - 2\mathbf{j})$, find the value of a and b.
11. Particles of mass $2m$, m, $5m$ and $2m$ are situated at (4, −5), (1, 2), (3, −6) and (0, 3) respectively. Find the coordinates of the centre of gravity of the system.
12. Three particles of mass 1 kg, 2 kg and m kg are situated at (5, 2), (1, 5) and (1, −2) respectively. If the centre of gravity of the system lies at $(2, \bar{y})$, find the values of m and $\bar{y}$.
13. Particles of mass 2 kg, 1 kg and 3 kg lie on the y-axis at the points (0, 7), (0, 4) and (0, −2) respectively. Where must a 6 kg mass be placed to ensure that the centre of gravity of the entire system lies at the origin?
14. Particles of mass $5m$, $4m$ and $3m$ are placed at the points (−5, 0), $(4, \frac{1}{2})$ and (−4, −3) respectively. Where must a particle of mass $7m$ be placed to ensure that the centre of gravity of the entire system lies at the origin?
15. PQRS is a rectangle with PQ = 8 cm and PS = 6 cm. Particles of mass 2 g, 2 g and 3 g are placed at points P, Q and R respectively. Find the mass that must be placed at S for the centre of gravity of the entire system to lie 3 cm from the line PQ.
With this mass in place, find the distance of the centre of gravity of the system from the line PS.

Centre of gravity of a rigid body

A rigid body is made up of a large number of particles. The position of its centre of gravity may be found by considering the constituent particles. If the rigid body has an axis of symmetry, then the centre of gravity will lie on that axis, since there will be an equal amount of matter, i.e. number of particles, in similar positions on either side of the line of symmetry.
The position of the centre of gravity of some rigid bodies can be determined by considering their symmetry.

Uniform

A *uniform* body is one in which equal volumes have the same masses. It is of uniform or constant density throughout. A uniform rod is therefore such that its length is proportional to its mass.

Lamina

A *lamina* is a flat body, the thickness of which is negligible compared with its other two dimensions, its length and breadth. Thus a piece of card, a sheet of paper, or a thin metal sheet may be taken as examples of laminae.
A uniform lamina is therefore one in which equal areas of the lamina have equal masses, i.e. the whole of the lamina is of the same material.

Uniform rod

The centre of gravity lies at its middle point, G mid-way between its ends A and B.

Uniform rectangular lamina

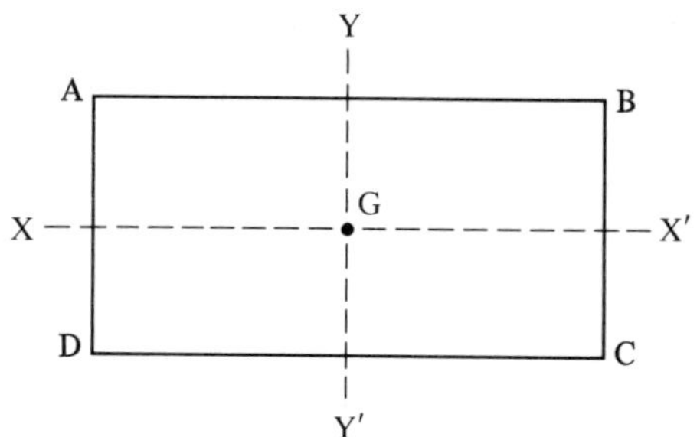

The line XX′ which bisects AD and BC, is an axis of symmetry, as also is the line YY′ which bisects the lines AB and DC.
The centre of gravity must lie on each of these lines of symmetry, and is therefore at their intersection G.

Uniform circular lamina

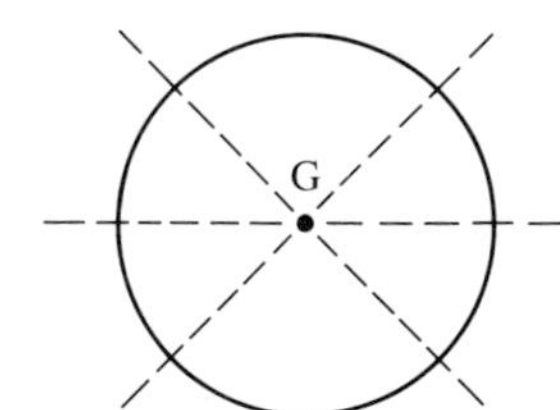

Since all diameters of a circle are lines of symmetry, the centre of gravity lies at the centre of the circular lamina.

The position of the centre of gravity of a solid can also be determined by consideration of its symmetry.

Uniform sphere

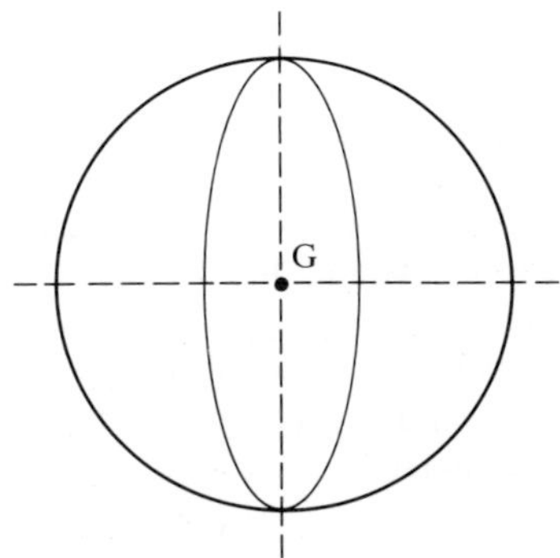

Since this solid has an infinite number of planes of symmetry, all of which contain the centre of the sphere, this is therefore its centre of gravity.

Uniform right circular cylinder

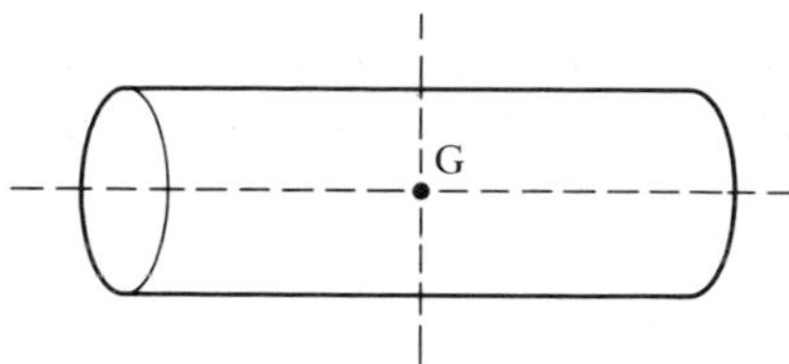

The cross section is a circle and this means that the cylinder has an infinite number of planes of symmetry, all of which contain the axis of the cylinder. Hence the centre of gravity must lie on this axis. The cylinder also has a plane of symmetry which bisects the axis of the cylinder and is parallel to the plane ends of the cylinder. This plane of symmetry intersects the other planes of symmetry, on the axis of the cylinder, at the point G mid-way between the ends of the cylinder. Point G is therefore the centre of gravity of the cylinder.

Uniform triangular lamina

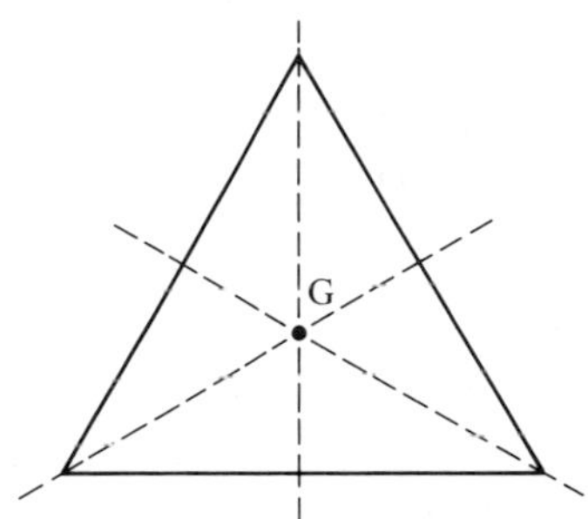

This does not, in general, have an axis of symmetry, although an equilateral triangular lamina will indeed have three. These three axes are the medians of the triangle, and their point of intersection is the centre of gravity of the triangular lamina; it lies on the median and two-thirds of the distance from the vertex to the other side of the triangle.

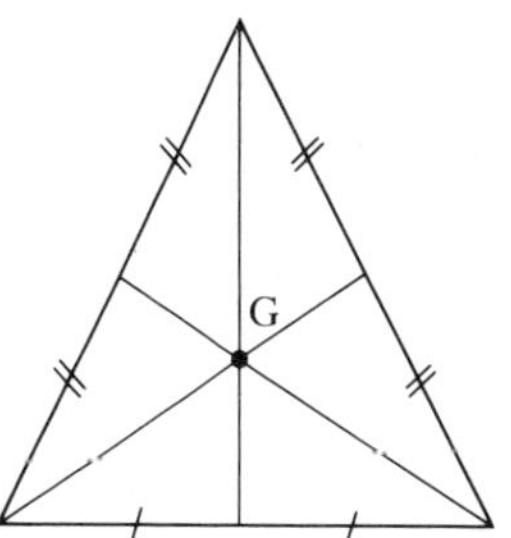

In the case of an isosceles triangular lamina, there is one axis of symmetry and the centre of gravity lies on it. The centre of gravity will again lie at the point of intersection of the medians of the triangle, and this result is true for all triangles as the following reasoning shows.

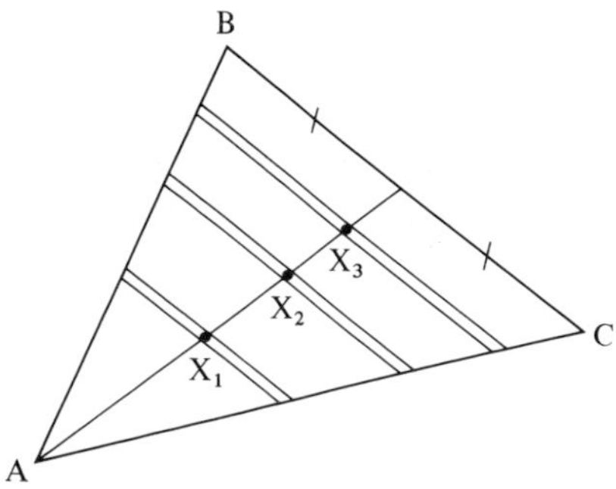

Suppose the triangle ABC is divided into a large number of thin parallel strips, all parallel to the side BC of the triangle. The centre of gravity of each of these strips will be at its centre point $X_1, X_2, \ldots$ But these points X all lie on the median drawn from the point A. Hence the centre of gravity of the triangle must also lie on this median.
In a similar way the centre of gravity can be shown to lie on each of the other two medians.
Therefore the centre of gravity is at the intersection of the medians, i.e. two-thirds of the distance from the vertex to the mid-point of the opposite side.

Example 5

Write down the coordinates of the centre of gravity of each of the following triangular laminae.

(a)

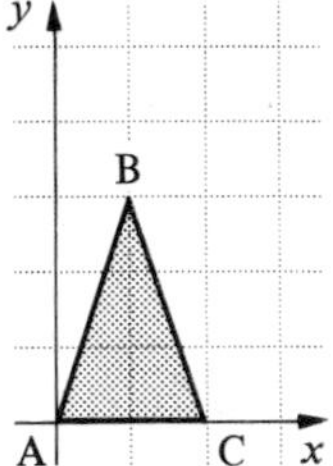

(b)

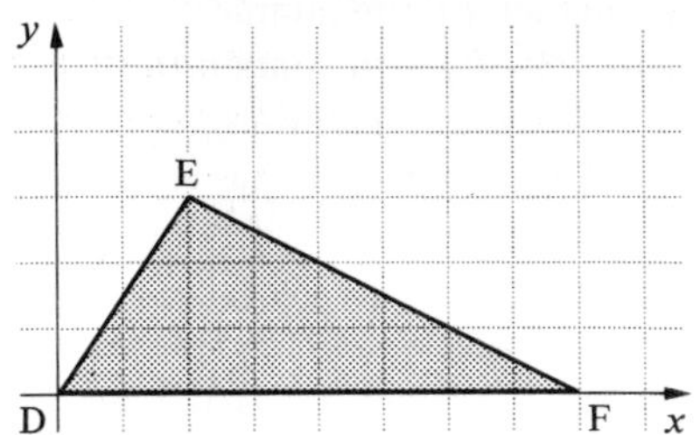

(c)

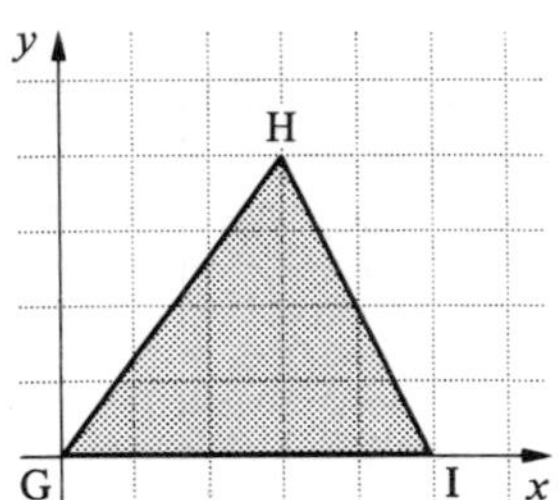

(d)

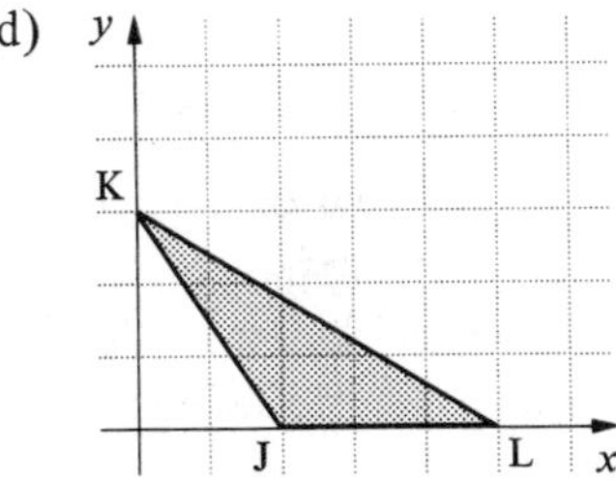

(a) The centre of gravity will lie at the intersection of the medians, i.e. at a point two-thirds of the distance from a vertex to the mid-point of the opposite side.
By considering the median through B, the centre of gravity is seen to be at the point (1, 1).

$\therefore$ centre of gravity is at (1, 1)

(b) The mid-point of EF is at $(5, 1\frac{1}{2})$. Call this point M.
The point that is two-thirds of the way from D (0, 0) to M $(5, 1\frac{1}{2})$ is $(3\frac{1}{3}, 1)$.

$\therefore$ centre of gravity is at the point $(3\frac{1}{3}, 1)$

(c) The mid-point of HI is at (4, 2).
The point that is two-thirds of the way from (0, 0) to (4, 2) is $(2\frac{2}{3}, 1\frac{1}{3})$.

$\therefore$ centre of gravity is at the point $(2\frac{2}{3}, 1\frac{1}{3})$

(d) The mid-point of JL is at $(3\frac{1}{2}, 0)$. The centre of gravity is two-thirds of the way from K (0, 3) to $(3\frac{1}{2}, 0)$.
x-coordinate of centre of gravity is $\frac{2}{3} \times 3\frac{1}{2} = 2\frac{1}{3}$
y-coordinate of centre of gravity is $3 - (\frac{2}{3} \times 3) = 1$

$\therefore$ centre of gravity is at the point $(2\frac{1}{3}, 1)$

It can also be shown that if three equal particles are placed at the vertices A, B, C of a triangle, then the centre of gravity of these particles is at the same point as the centre of gravity of the uniform triangular lamina ABC.

Example 6

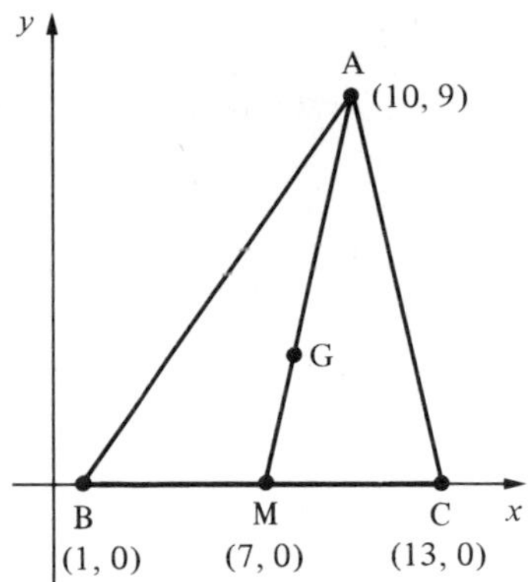

The triangle ABC has its vertices at the points with coordinates (10, 9), (1, 0) and (13, 0). Particles, each of mass m are placed at the points A, B and C. Suppose M is the mid-point of BC.

The centre of gravity of the lamina lies on the median AM. If G is such that $AG = \frac{2}{3}AM$, then G is the centre of gravity.

The centre of gravity of the particles is found by finding the position of the resultant of the forces due to their masses.

The resultant of mg at (1, 0) and mg at (13, 0) will be a force of $2mg$ at the mid-point, M, of BC, i.e. at the point (7, 0).

The resultant of $2mg$ at M and mg at A will be at a point P such that

$$2mg \times MP = mg \times AP$$
$$\text{or} \quad AP = 2MP \quad \text{i.e. } AP = \tfrac{2}{3}AM$$

Hence the point P and the point G are the same point, and the centre of gravity of the lamina and the three particles are at the same point.

The coordinates of the centre of gravity can be determined as follows.

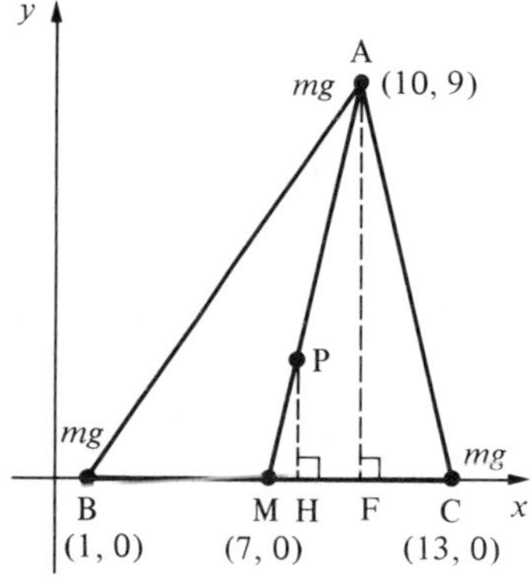

Through the point P, draw lines PH and AF parallel to the y-axis

$$\text{As} \quad MP = \tfrac{1}{3}(MA)$$
$$\therefore \quad HP = \tfrac{1}{3}(FA)$$
$$= \tfrac{1}{3}(9) = 3$$
$$MH = \tfrac{1}{3}(MF)$$
$$= \tfrac{1}{3}(10 - 7) = 1$$

The coordinates of the centre of gravity are (8, 3).

Exercise 8B

Write down the coordinates of the centre of gravity of each of the following uniform laminae. Each grid consists of unit squares.

1. 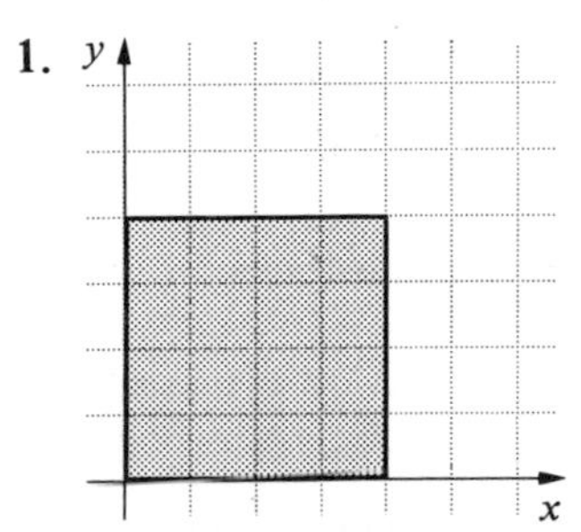

2. 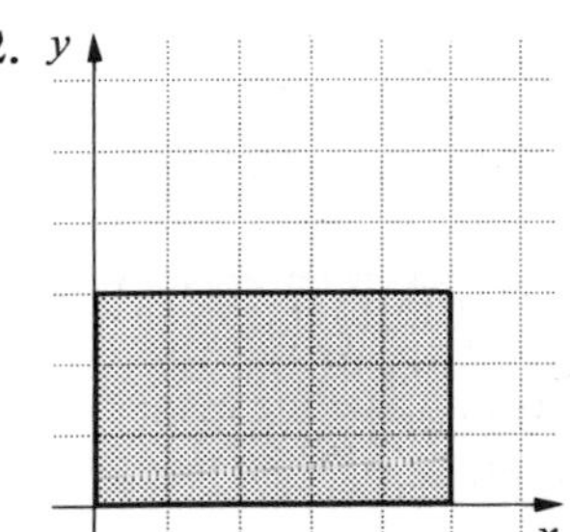

3. 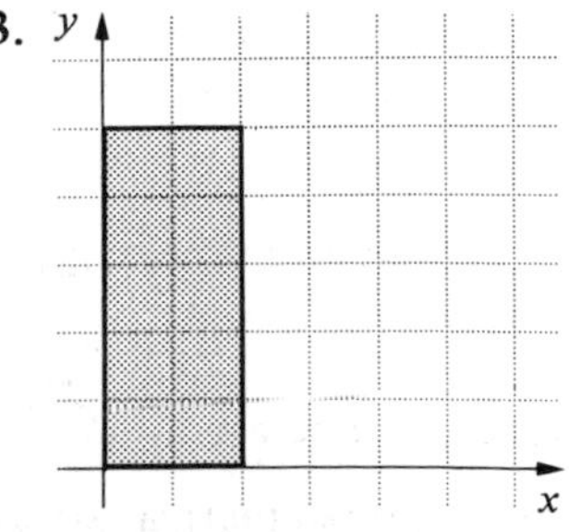

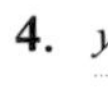

4.

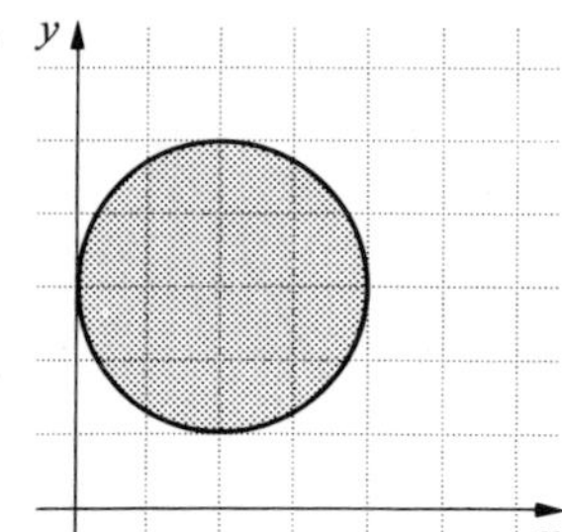

5.

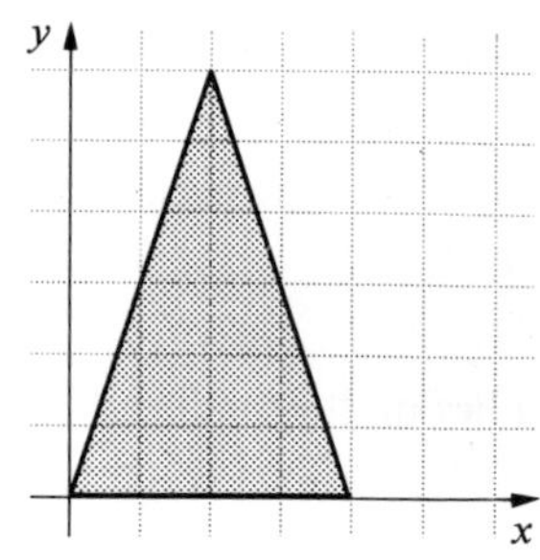

6.

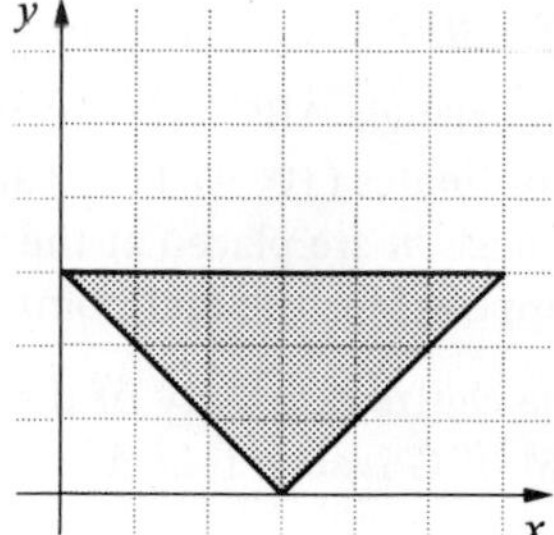

7.

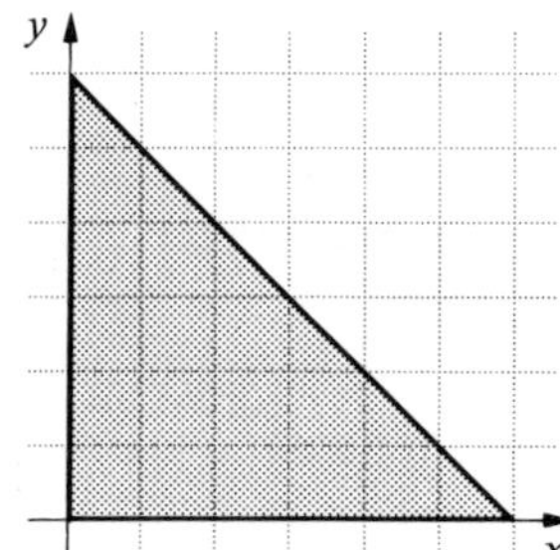

8.

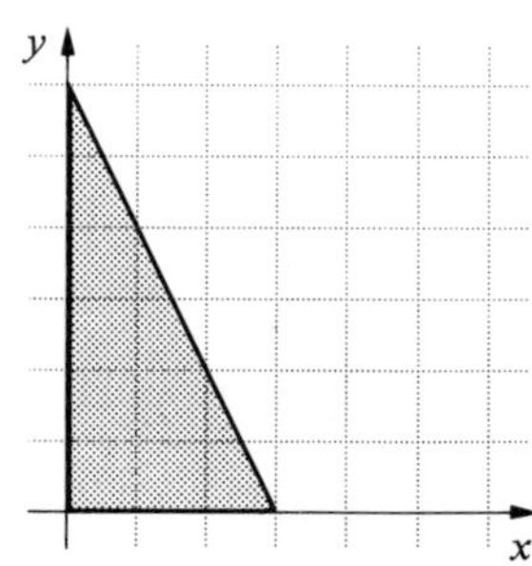

9.

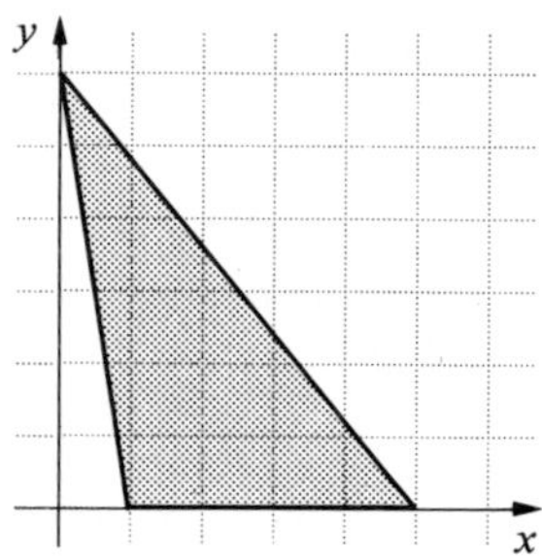

10.

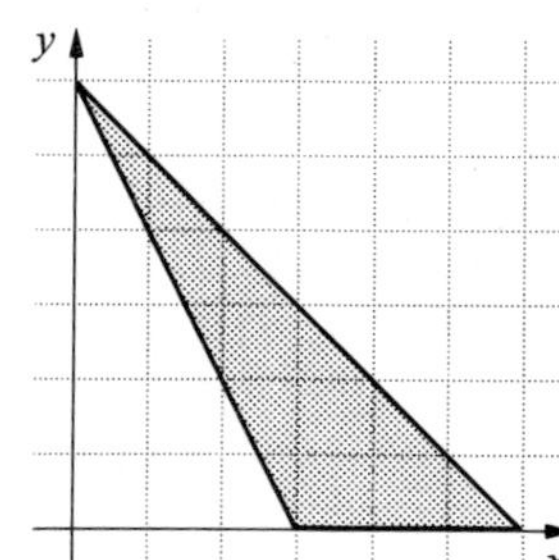

11.

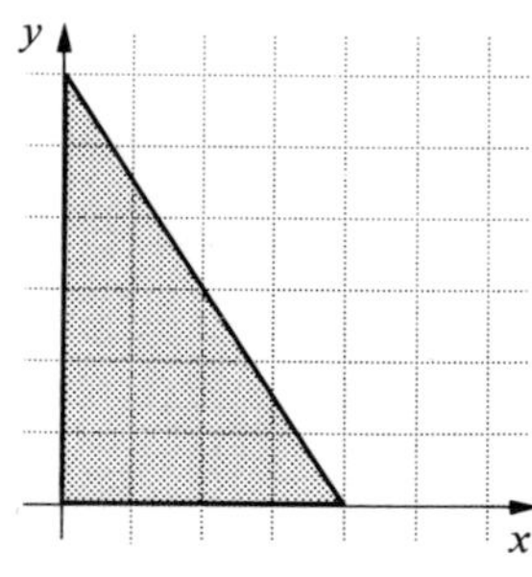

12.

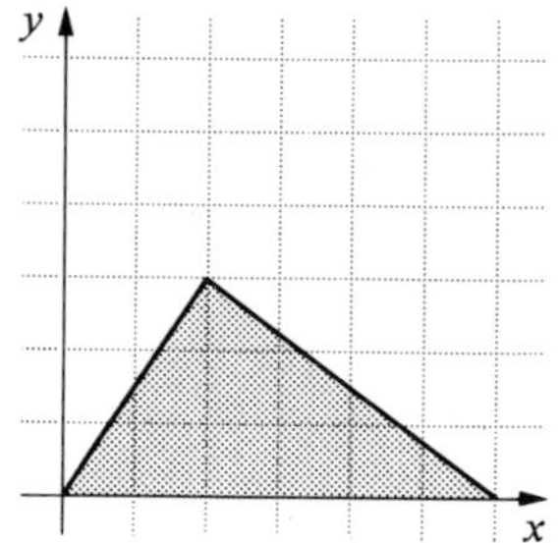

Composite laminae

A lamina may consist of two or more regular laminae joined together.

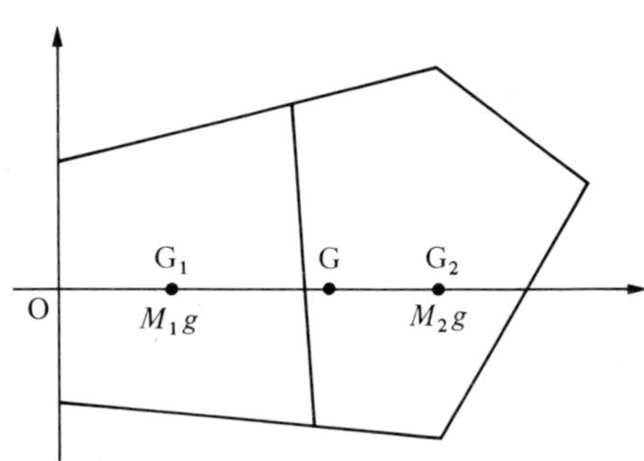

The forces due to the masses of the two laminae will act at their centres of gravity G_1 and G_2. The resultant of these two forces will act through a point, G, on the line G_1G_2.

Suppose the masses of the laminae are M_1 and M_2; take moments about the point O:

$$M_1g \times (G_1O) + M_2g \times (G_2O) = (M_1 + M_2)g \times GO$$

The position of the centre of gravity of the composite body can then be determined from this equation.
It is advisable to take moments about an axis forming a boundary of the body so that all the moments are acting in the same sense, rather than using an axis which intersects the body.

Hence it is seen that the centre of gravity of a composite lamina can be found in an exactly similar way to the centre of gravity of a system of particles. It is only necessary to know the masses, and the positions of the centre of gravity, of each of the constituent parts.

Example 7

Two uniform rectangular laminae, of mass per unit area m, are joined together as is shown in the diagram. Find the distance of the centre of gravity of the complete lamina from the edge AB.

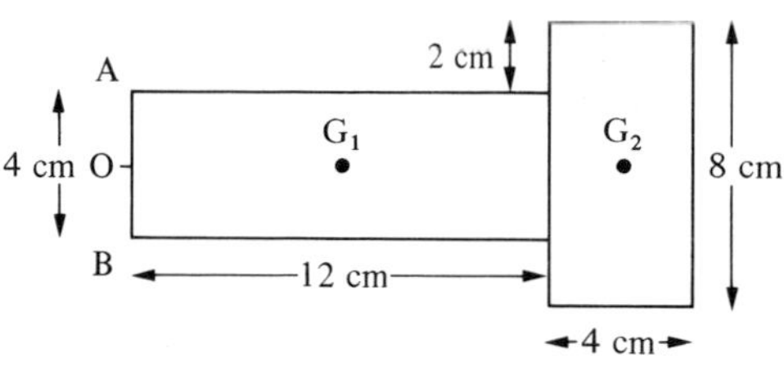

The centres of gravity of the rectangles are at G_1 and G_2 and lie on the line of symmetry of the lamina. Suppose O is the mid-point of AB.

Draw two diagrams

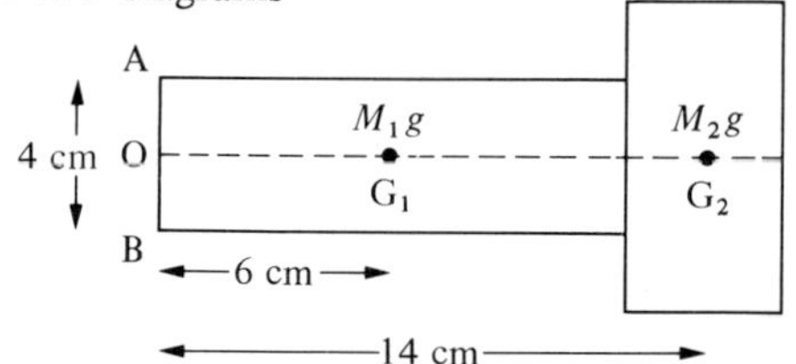

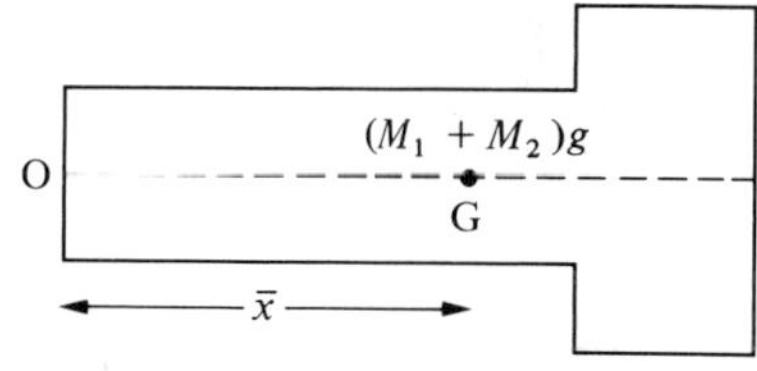

In this example the composite lamina has a line of symmetry and the point G will lie on it. Hence G lies on the line G_1G_2 at a distance $\bar{x}$ from O. Taking moments about O

$$\overset{\curvearrowright}{O} \qquad M_1g \times 6 + M_2g \times 14 = (M_1 + M_2)g\,\bar{x}$$

$$\text{but } M_1 = 48m \qquad \text{and } M_2 = 32m$$

$$\therefore \quad 48mg \times 6 + 32mg \times 14 = 80mg\,\bar{x}$$

$$\therefore \quad 288 + 448 = 80\,\bar{x}$$

$$\therefore \quad \bar{x} = 9{\cdot}2 \text{ cm}$$

The centre of gravity of the complete lamina is 9·2 cm from O, the mid-point of AB.

In the case of a less regular composite lamina, which has not an axis of symmetry, it is necessary to take moments about two axes so as to find both coordinates of the centre of gravity.

Example 8

Three uniform laminae, all made of the same material, are joined together as shown in the diagram. ABCD is a rectangle and the triangles are isosceles. If the laminae have the dimensions shown, find the position of the centre of gravity of the composite lamina.

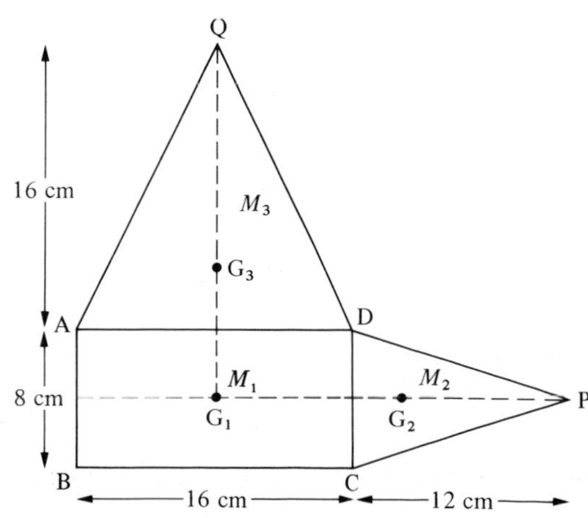

The masses and the positions of the centre of gravity of each of the constituent laminae are known. Suppose the mass per unit area of each lamina is m.

Draw two diagrams.

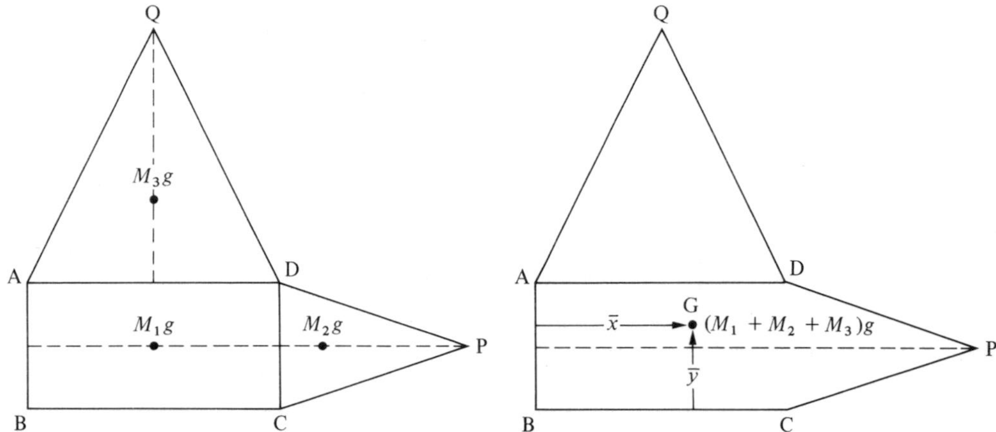

Taking moments about the axis AB and using

$$M_1 = 16 \times 8m, \quad M_2 = \tfrac{1}{2} \times 8 \times 12m \quad \text{and } M_3 = \tfrac{1}{2} \times 16 \times 16m$$

$$(16 \times 8mg) \times (8) + (\tfrac{1}{2} \times 8 \times 12mg) \times (16 + 4) + (\tfrac{1}{2} \times 16 \times 16mg) \times (8) = (304mg)\bar{x}$$

$$\therefore \quad \bar{x} = 9{\cdot}89 \text{ cm}$$

Taking moments about the axis BC

$$(16 \times 8mg) \times (4) + (\tfrac{1}{2} \times 8 \times 12mg) \times (4) + (\tfrac{1}{2} \times 16 \times 16mg) \times (8 + 5\tfrac{1}{3}) = (304mg)\bar{y}$$

$$\therefore \quad \bar{y} = 7{\cdot}93 \text{ cm}$$

The centre of gravity is 9·89 cm from AB and 7·93 cm from BC.

It should be noted that in the case of a uniform lamina, the constants m and g will cancel from each term of the moments equation; nevertheless it is essential that they should be written in the original equation.

The method used to find the centre of gravity of a composite lamina may also be used to determine the centre of gravity of a composite three-dimensional body, provided that the centres of gravity and the masses of the various parts are known.

Example 9

A solid right circular cylinder and a solid right circular cone are joined together at their plane faces as shown in the diagram. The solids are made of the same uniform material. Given that the centre of gravity of a cone is at a distance $\dfrac{h}{4}$ from its plane base where h is the height of the cone, find the position of the centre of gravity of the composite body.

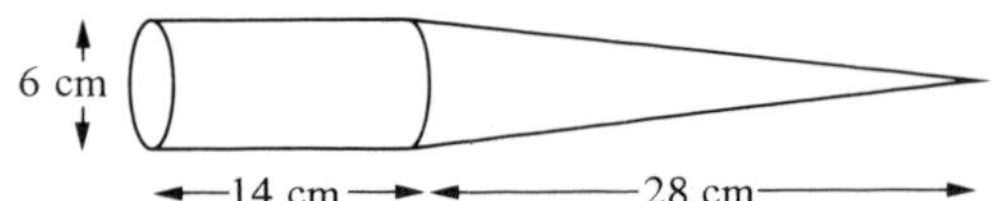

Draw two diagrams.

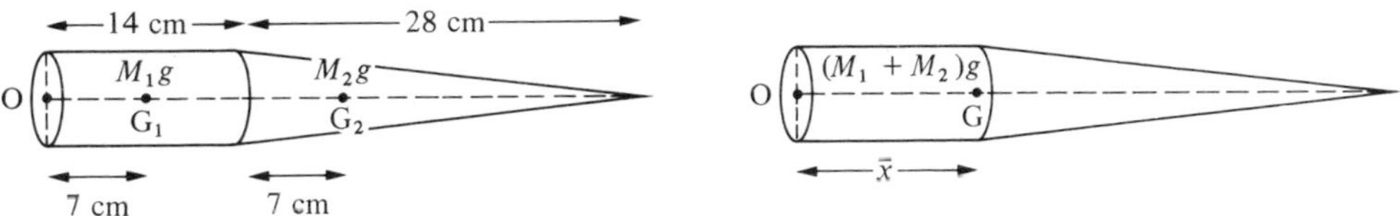

Let the mass per unit volume, i.e. the density, of the cylinder and the cone be σ;

then $\quad M_1g$ (cylinder) $= \pi \times 3^2 \times 14\sigma g \qquad$ and $\quad M_2g$ (cone) $= \frac{1}{3}\pi \times 3^2 \times 28\sigma g$

$$= 126\pi\sigma g \qquad\qquad = 84\pi\sigma g$$

The centre of gravity of the body will lie on the axis of symmetry through the points O, G_1 and G_2.

Equating moments about the vertical axis through O:

$$126\pi\sigma g \times 7 + 84\pi\sigma g \times (14 + 7) = \bar{x}(126\pi\sigma g + 84\pi\sigma g)$$

$$\therefore \quad \bar{x} = 12{\cdot}6 \text{ cm}$$

The centre of gravity is on the axis of symmetry and 12·6 cm from the base of the cylinder.

It is sometimes required to find the position of the centre of gravity of a lamina from which a portion has been removed, for example a perforated disc or a rectangle from which a triangle has been removed.
The method of the previous examples can still be used. The portion which is removed and the part remaining can be considered as the constituent parts, which together make up the whole lamina.
The essential difference to the calculation is that the unknown centre of gravity will now be that of one of the parts of the whole lamina, rather than that of the composite lamina itself.

Example 10

A uniform circular disc, of mass per unit area m and radius 8 cm, has a circular hole of radius 2 cm made in it. The centre of this hole lies 4 cm from the centre of the disc.

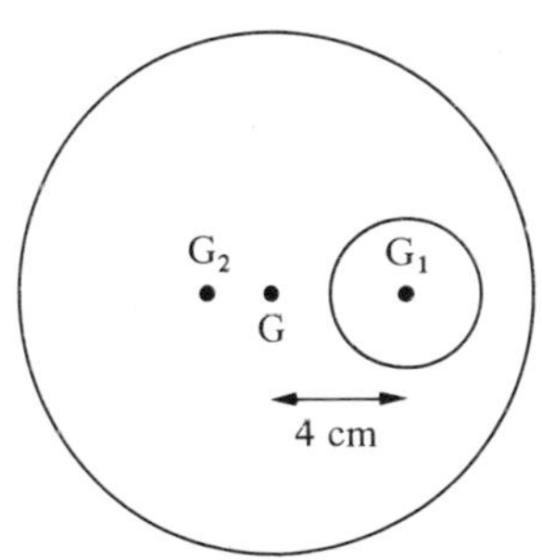

Let G be the centre of gravity of the whole disc, and G_1 and G_2 be the centres of gravity of the disc which is removed and the perforated disc respectively.
G_2 will lie on the diameter passing through G_1.
M_1 and M_2 are the masses of the part removed and of the perforated disc respectively.

Draw two diagrams as shown on the next page

Draw two diagrams.

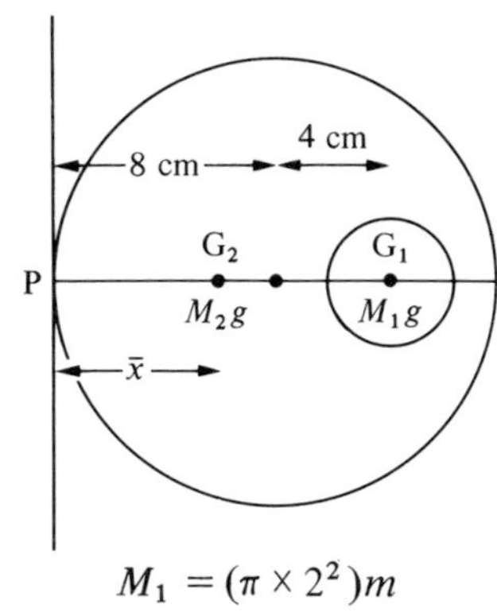

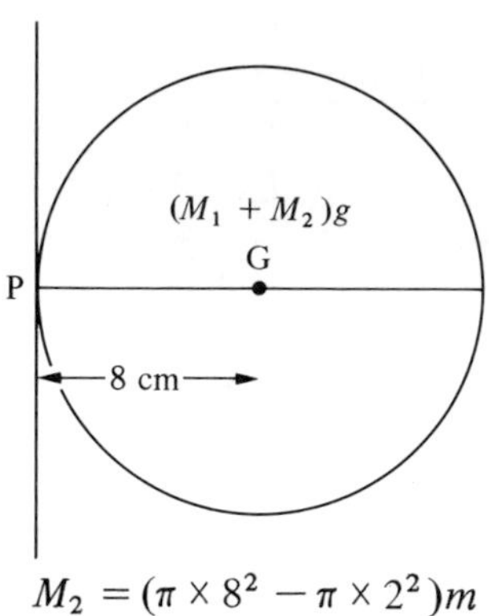

$M_1 = (\pi \times 2^2)m$ $\qquad$ $M_2 = (\pi \times 8^2 - \pi \times 2^2)m$

Taking moments about the axis through P:

$$(\pi \times 2^2 mg) \times (8+4) + (\pi \times 8^2 - \pi \times 2^2)mg \times \bar{x} = (\pi \times 8^2 mg) \times 8$$

$$\therefore \quad \bar{x} = 7{\cdot}73 \text{ cm}$$

The centre of gravity is 0·27 cm from the centre of the original disc and lies on the diameter passing through the centre of the disc and the centre of the hole.

Standard results

The following standard results will be required for some of the questions of Exercise 8C.

Uniform semicircular lamina of radius r: Centre of gravity lies on the axis of symmetry at a distance $\frac{4r}{3\pi}$ from the straight edge.

Uniform solid right circular cone of height h: Centre of gravity lies on the axis of symmetry at a distance $\frac{h}{4}$ from the plane base.

Uniform solid hemisphere of radius r: Centre of gravity lies on the axis of symmetry at a distance $\frac{3r}{8}$ from the plane surface.

Exercise 8C

Questions **1** to **13** show uniform laminae. Find the coordinates of the centre of gravity of each one. Each grid consists of unit squares.

1. 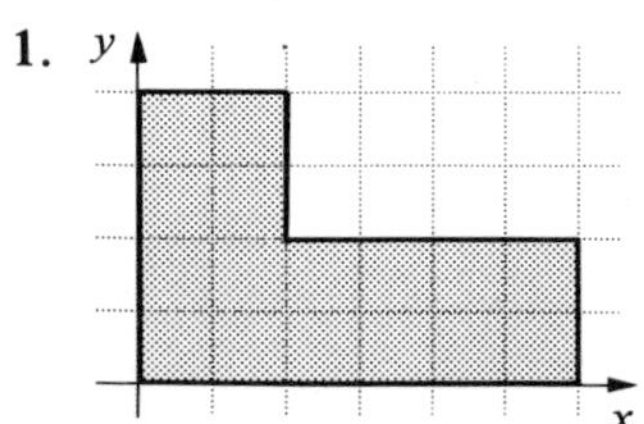

2. 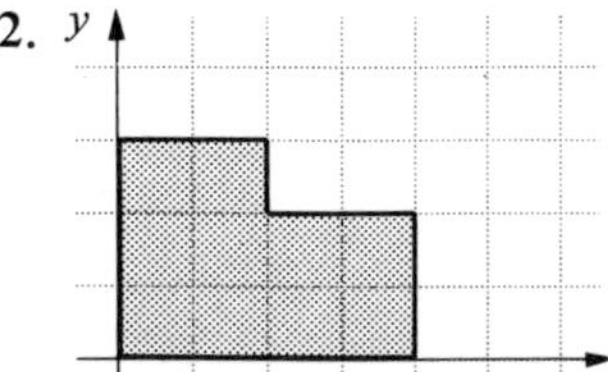

3. 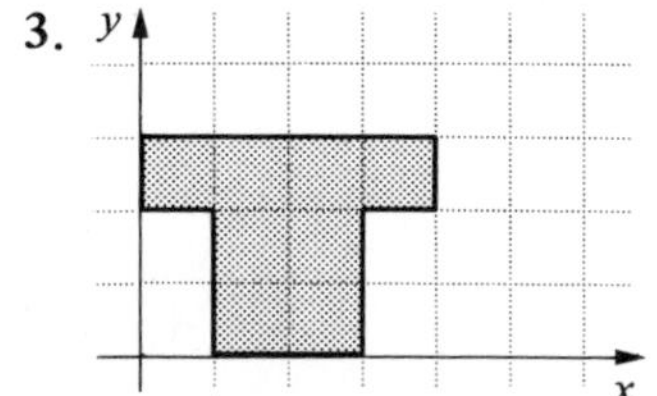

4.

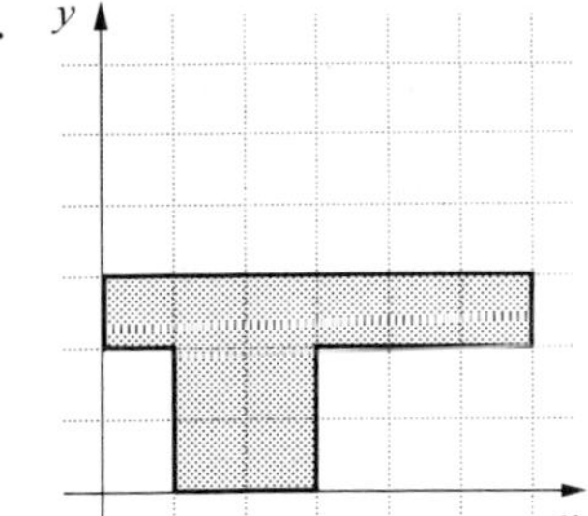

5.

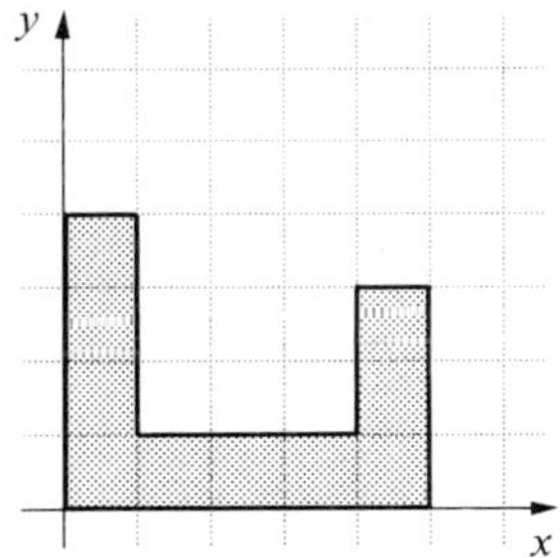

6.

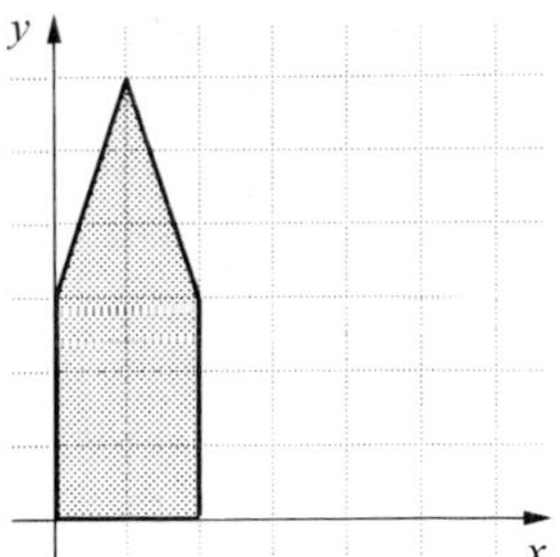

7.

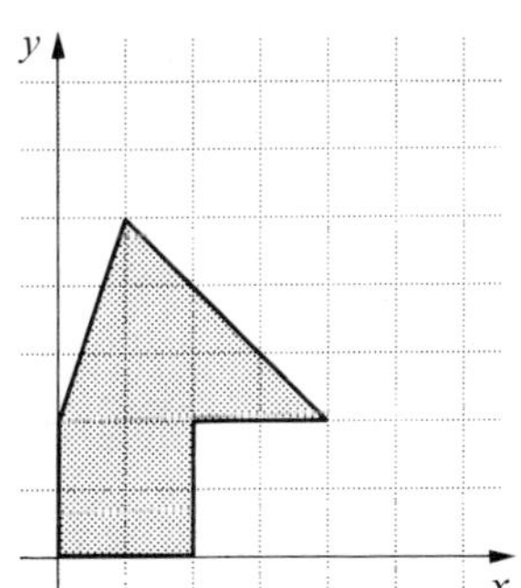

8.

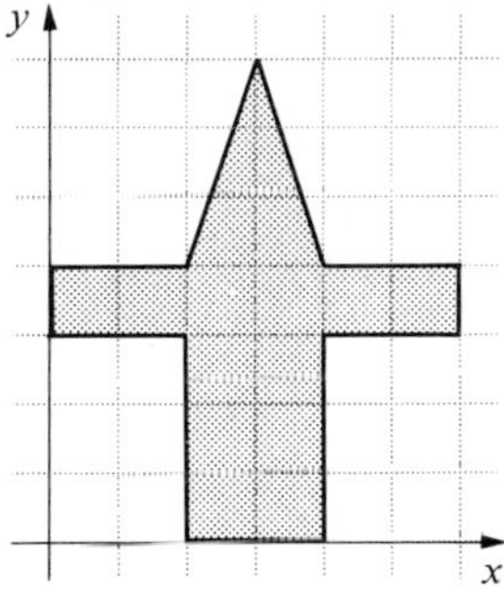

9.

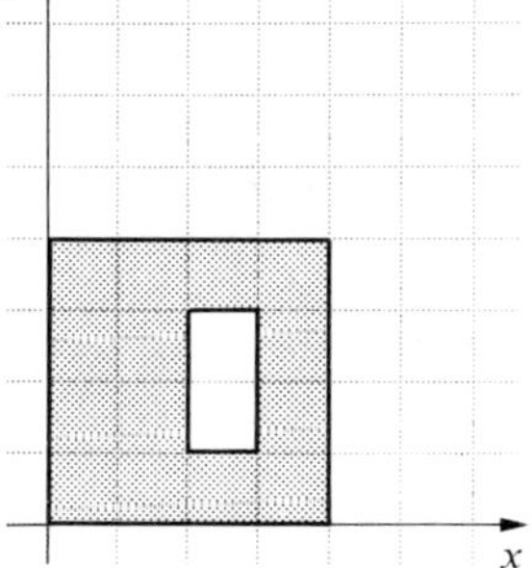

10.

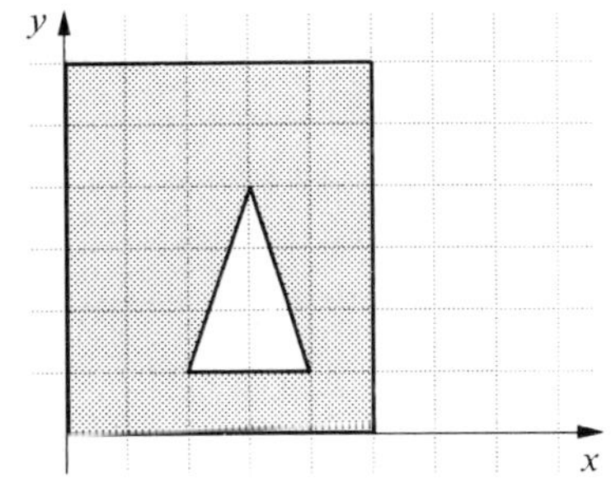

11.

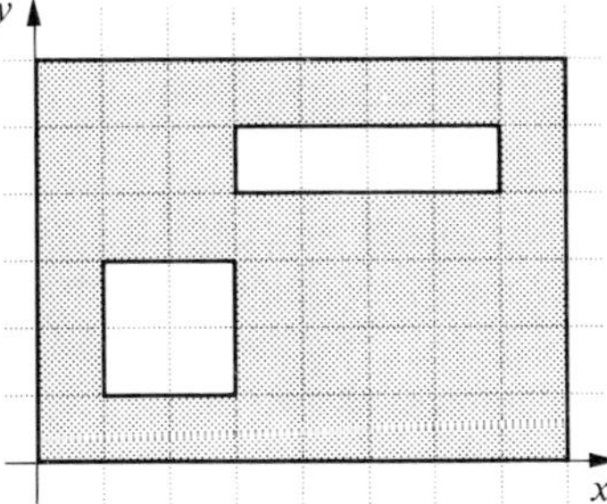

12.

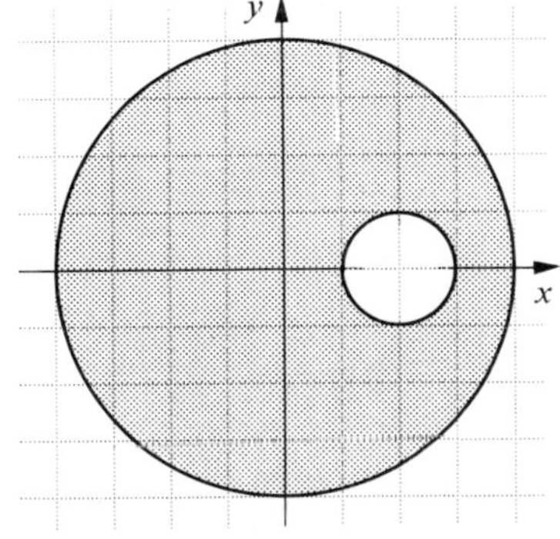

13.

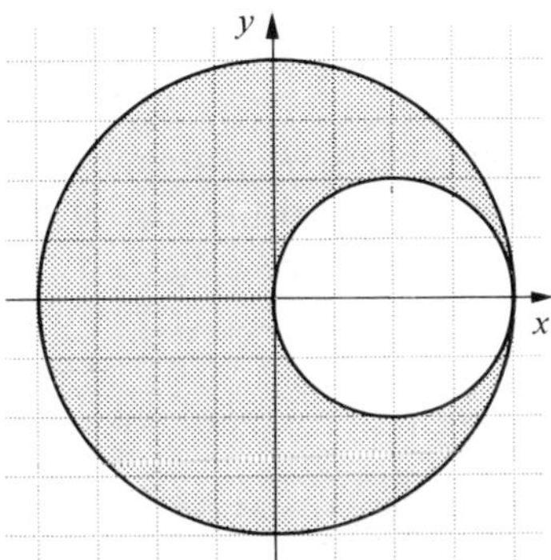

14. A uniform straight wire ABC is such that AB = 4 metres, AC = 6 metres and B is between A and C. If the wire is now bent at B until $A\hat{B}C = 90°$, find the distance that the centre of gravity of this bent wire is from (a) AB, (b) BC.

15. Four uniform rods AB, BC, CD and DA are each 4 metres in length and have masses of 2 kg, 3 kg, 1 kg and 4 kg respectively. If they are joined together to form a square framework ABCD, find the position of its centre of gravity.

16. Two rods AB and BC are joined together at B such that $A\hat{B}C = 60°$. AB is uniform, of length 6 m and mass 4 kg. BC is uniform, of length 4 m and mass 4 kg. Find how far the centre of gravity of ABC is from the point B.

17. Two uniform square laminae, each of side 3 metres, are joined together to form a rectangular lamina, 6 metres by 3 metres. The squares are not made of the same material and the mass per unit area of one of them is twice that of the other. Find the distance of the centre of gravity of the composite body from the common edge of the squares.

18. The diagram shows two uniform squares, ABFG and BCDE, joined together. The mass per unit area of BCDE is twice that of ABFG. Find the distance of the centre of gravity of the composite body from AB and AG.

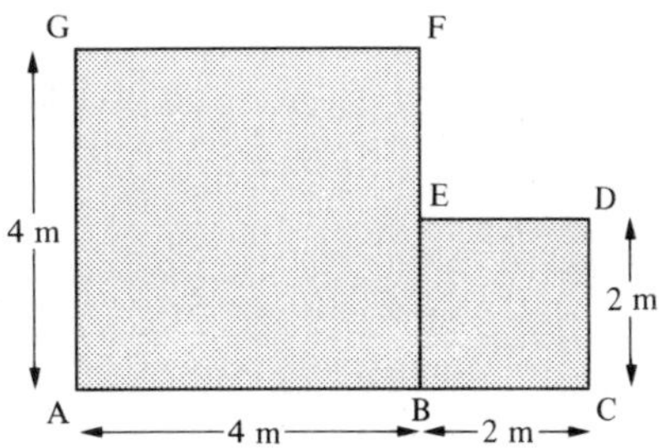

19. A circular lamina, made of uniform material, has its centre at the origin and a radius of 6 units. Two smaller circles are cut from this circle, one of radius 1 unit and centre (−1, −3) and the other of radius 3 units and centre (1, 2). Find the coordinates of the centre of gravity of the remaining shape.

20. Two solid cubes, one of side 4 cm and the other of side 2 cm, are made of the same uniform material. The smaller cube is glued centrally to one of the faces of the larger cube, as shown in the diagram. Find how far the centre of gravity of the composite body is from the common surface of the cubes.

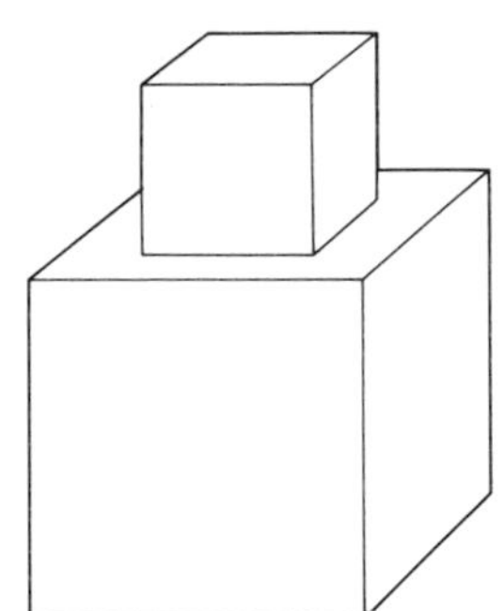

21. A solid cube of side 4 cm is made from uniform material. From this cube a smaller cube of side 2 cm is removed, as shown in the diagram. Find the position of the centre of gravity of the remaining body.

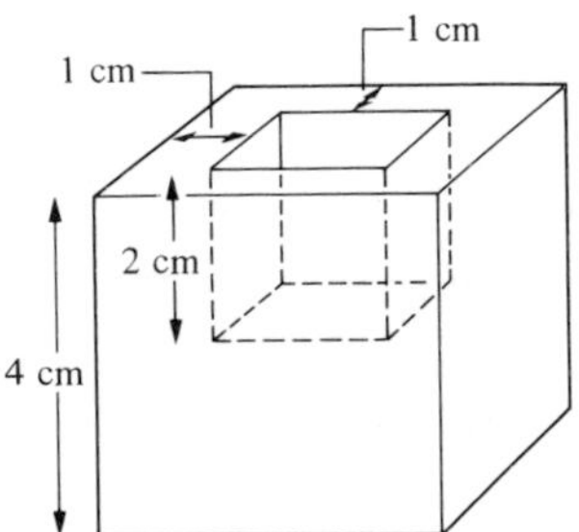

22. The diagram shows a uniform semicircular lamina of radius 6 m, with a semicircular portion of radius 3 m missing. Find the distance of the centre of gravity of the remaining shape from the point A shown in the diagram.

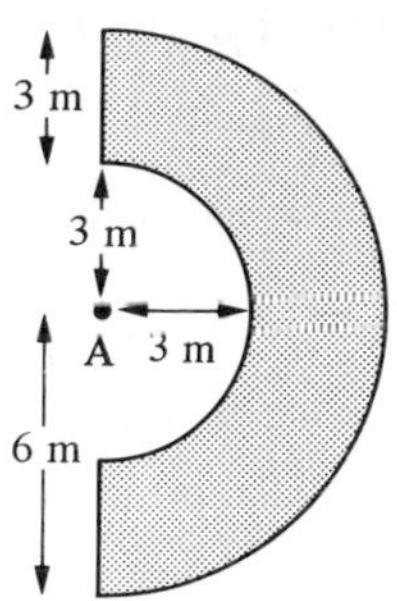

23. A uniform semicircular lamina of radius 6 cm is joined to another uniform semicircular lamina of radius 3 cm. The centres of the straight edges of each lamina coincide, but the laminae do not overlap. If the two laminae are made of the same material, find the position of the centre of gravity of the composite lamina so formed.
24. Repeat question **23** but now with the smaller semicircular lamina having a mass per unit area equal to twice that of the larger one.
25. A solid right circular cylinder has a base radius of 3 cm and a height of 6 cm. The cylinder's circular top forms the base of a solid right circular cone of base radius 3 cm and perpendicular height 4 cm. The cylinder and the cone are made from the same uniform material. Find the position of the centre of gravity of the composite body.

26. A conical hole is made in one end of a right circular cylinder (see diagram). The axis of symmetry of the cone is the same as that of the cylinder. The cylinder is of radius 2 cm and length 6 cm. The conical hole penetrates 4 cm into the cylinder and the circular hole at the end of the cylinder is of radius $1\frac{1}{2}$ cm. Find the position of the centre of gravity of the remaining body.

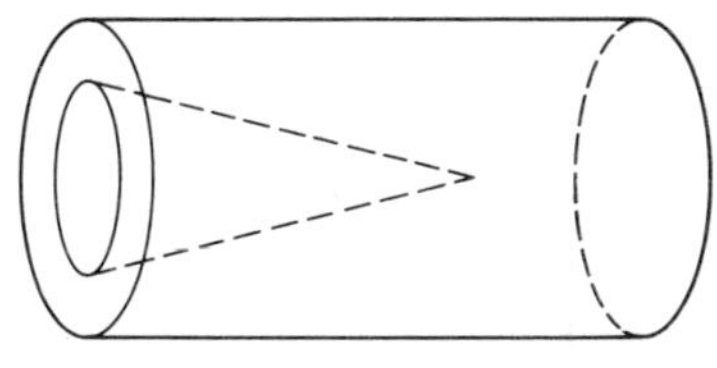

27. A body consists of a solid hemisphere of radius 4 cm joined to a solid right circular cone of base radius 4 cm and perpendicular height 12 cm. The plane surfaces of the cone and hemisphere coincide and both solids are made of the same uniform material. Find the position of the centre of gravity of the body.
28. A body consists of a solid hemisphere of radius r joined to a solid right circular cone of base radius r and perpendicular height h. The plane surfaces of the cone and hemisphere coincide and both solids are made of the same uniform material. Show that the centre of gravity of the body lies on the axis of symmetry at a distance $\dfrac{3r^2 - h^2}{4(h + 2r)}$ from the base of the cone.

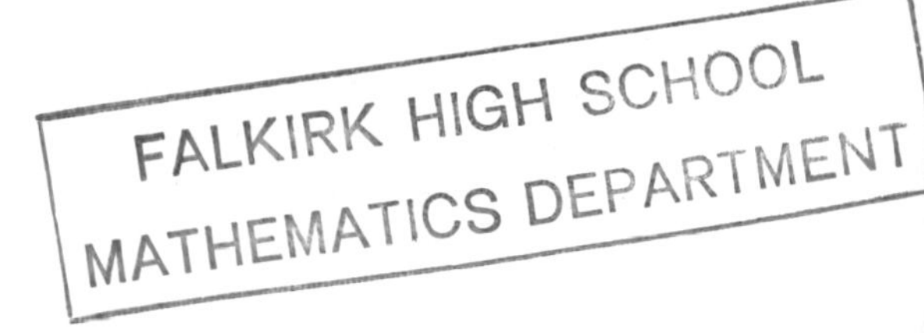
FALKIRK HIGH SCHOOL
MATHEMATICS DEPARTMENT

Equilibrium of a suspended lamina

Consider a rectangular lamina, freely suspended by a string attached at the vertex A.

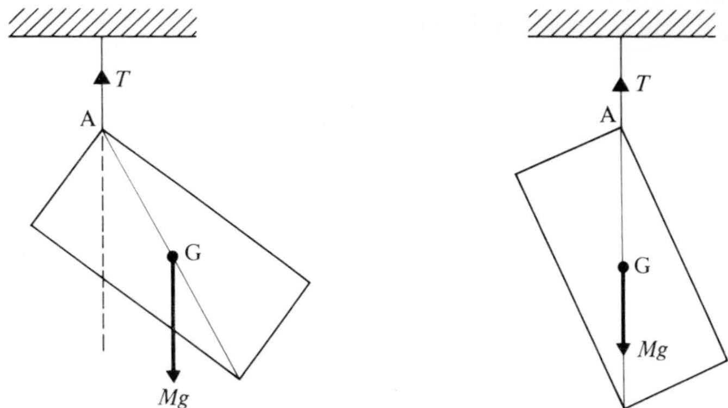

As there is a resultant moment about the point A, the position shown in the first diagram is not a possible position of equilibrium. Since there are only two forces acting on the lamina and their lines of action are parallel, they can only produce equilibrium if they act along the same line. Hence, equilibrium is only possible if the rectangle hangs so that its centre of gravity G lies vertically below the point A, as shown in the second diagram.

Example 11

A rectangular lamina ABCD is freely suspended by one vertex C. If CD = 15 cm, BC = 5 cm and the lamina is uniform, find the angle θ between the side CD and the vertical in the position of equilibrium.

The centre of gravity G of the rectangle ABCD lies on the vertical through the point C, and G is also the mid-point of the diagonal CA.

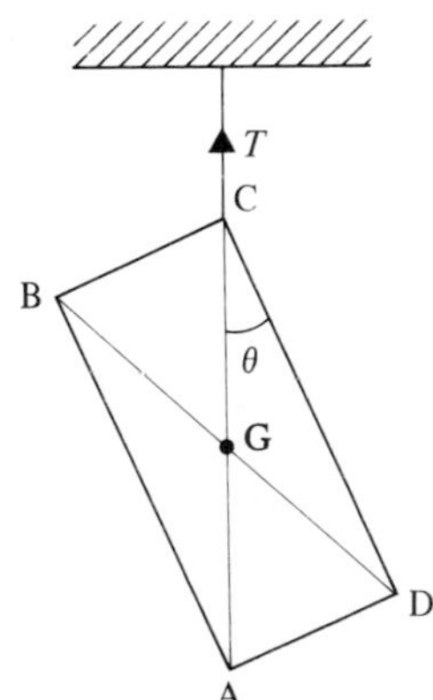

From the triangle ADC,

$$\tan\theta = \frac{AD}{DC}$$

$$= \frac{5}{15}$$

$$\therefore \quad \theta = 18{\cdot}43°$$

The angle between the side CD and the vertical is 18·43° .

Exercise 8D

If the following uniform laminae are freely suspended from point A, find the angle that the line AB makes with the vertical in each case. Each grid consists of unit squares.

1.

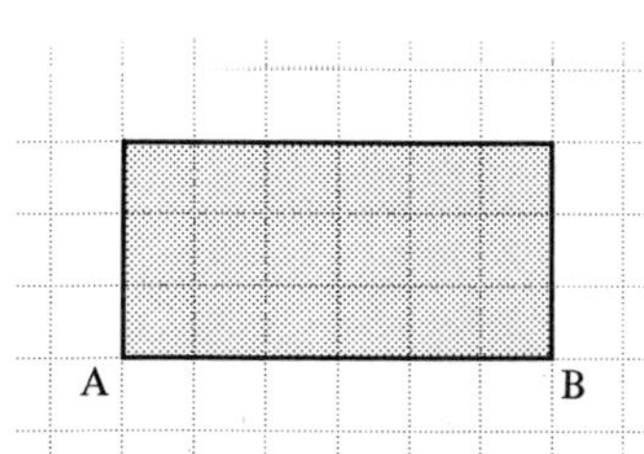

2.

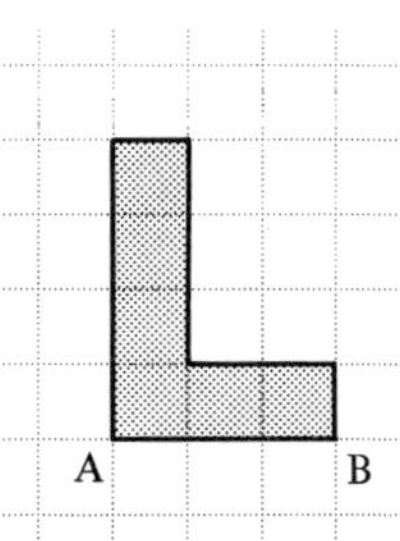

3.

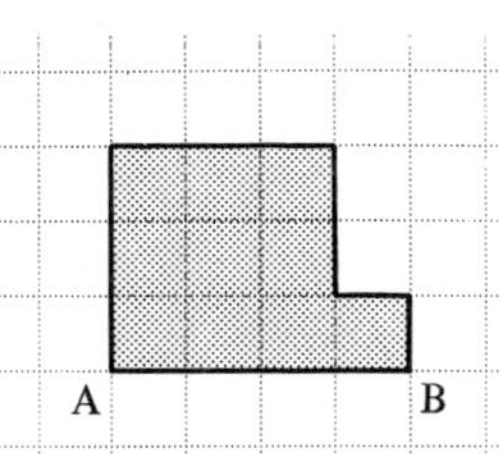

4.

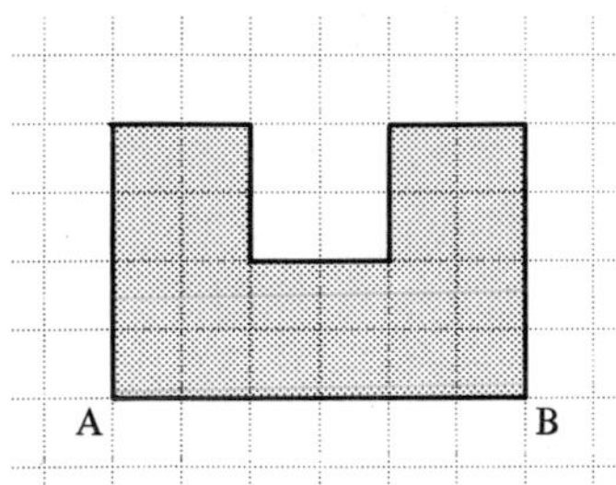

5.

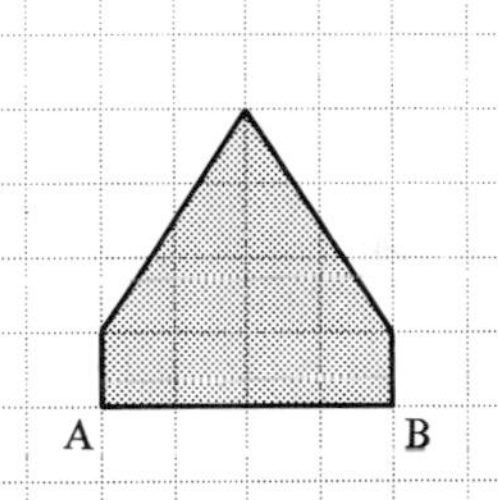

6.

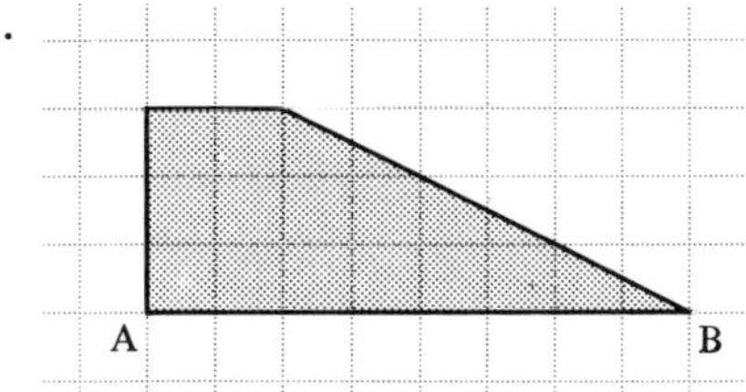

7.

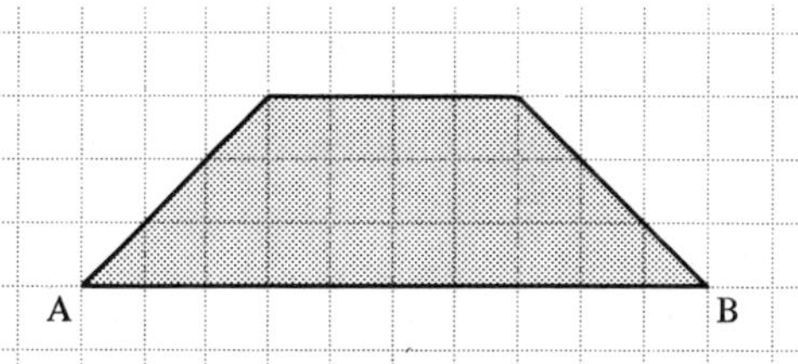

8.

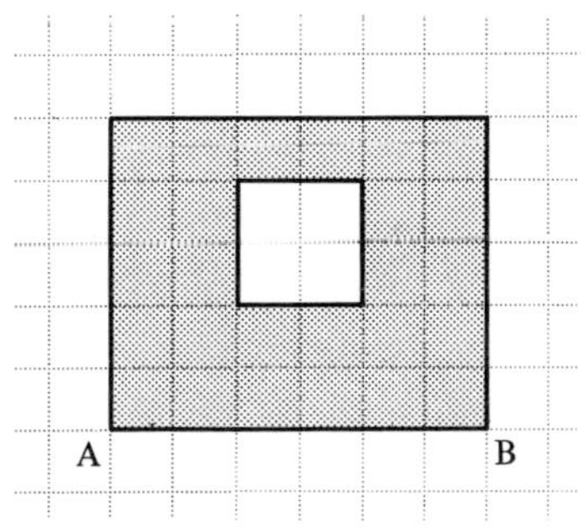

9.

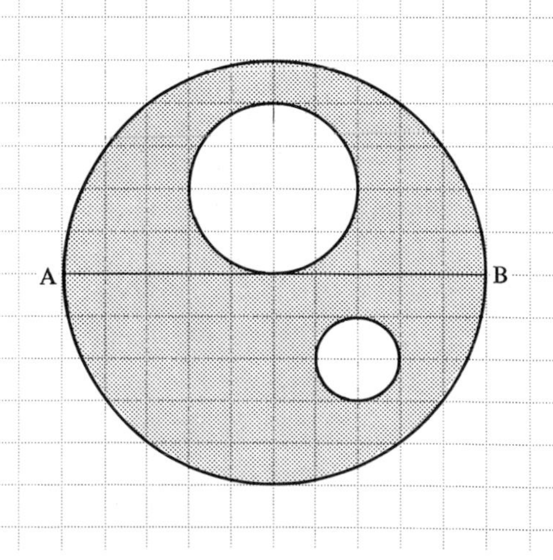

Use of calculus

In some cases it is not possible to treat a lamina as a system involving a finite number of parts, each of which has a known centre of gravity.
By the methods of the calculus, a lamina is divided up into an *infinite* number of elements, all with known centres of gravity.
As was explained previously, a general result may be stated in the form:

$$\bar{x}\sum m_i g = \sum m_i g\, x_i$$

Using the calculus notation, this may be written

$$\bar{x}\int m_i g = \int m_i g\, x_i$$

In words, this may be interpreted as stating that the moment of the whole is equal to the algebraic sum of the moments of the parts about a given axis.

Example 12

A uniform semicircular lamina has a radius of 6 cm and mass per unit area m. Find the position of the centre of gravity.

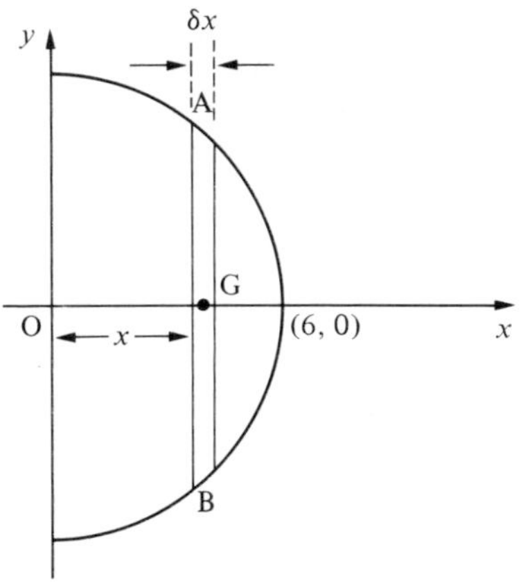

Consider an elementary strip AB of thickness δx parallel to the y-axis and a distance x from it.
The centre of gravity of this strip will be on the x-axis, as will the centres of gravity of all similar parallel strips.
Hence the centre of gravity of the lamina will also lie on the x-axis.

$$AG^2 + x^2 = 6^2 \quad \text{or} \quad AG = \sqrt{(6^2 - x^2)}$$

area of strip $= 2\sqrt{(36 - x^2)} \times \delta x$ and mass $= 2\sqrt{(36 - x^2)} \times \delta x \times m$

area of whole lamina $= \frac{1}{2}\pi 6^2 = 18\pi$ and mass $= 18\pi m$

Equating the moments of the whole lamina and the sum of the moments of the strips:

$$18\pi m\bar{x}g = \sum_{x=0}^{6} 2\sqrt{(36 - x^2)} \times m\delta x \times xg$$

Using calculus

$$18\pi m\bar{x}g = \int_0^6 2mx(36 - x^2)^{\frac{1}{2}} g\,dx$$

$$= 2mg\left[-\tfrac{2}{6}(36 - x^2)^{\frac{3}{2}}\right]_0^6$$

$$= 2mg \times 72$$

$$\therefore \quad \bar{x} = \frac{8}{\pi}\text{ cm}$$

The centre of gravity is at a distance of $\frac{8}{\pi}$ cm from the point O, on the axis of symmetry of the lamina.

In cases where the lamina has not an axis of symmetry, two coordinates will have to be found. Also, in many cases an integral giving the mass of the whole lamina will have to be evaluated rather than as in the above example where the area of the lamina (and hence its mass) was known.

Example 13

Find the centre of gravity of the uniform lamina bounded by the curve $y^2 = 9x$, the x-axis and the ordinates $x = 1$ and $x = 4$, and lying in the first quadrant.

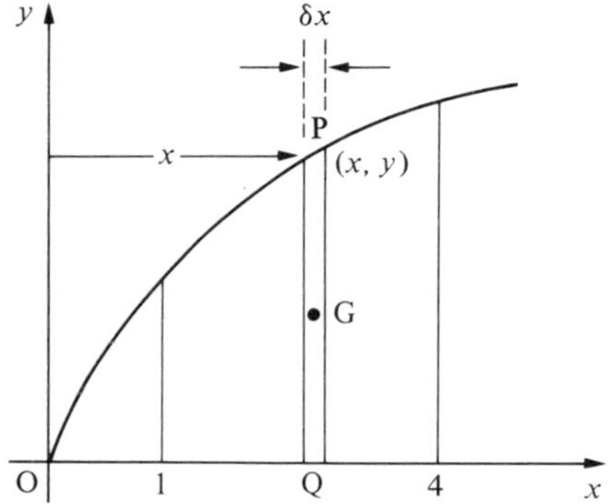

The lamina is divided into elementary strips like PQ parallel to the y-axis, and at a distance x from it.

Let m be the mass per unit area of the lamina, then area of strip $= y\delta x$ and mass of strip $= ym\delta x$. The centre of gravity of this strip will be at G, at a distance x from the y-axis.

Equating the moment of the whole lamina and the sum of the moments of the strips, about the y-axis:

$$\bar{x}\int_1^4 ymg\,dx = \int_1^4 xymg\,dx$$

substituting for y as $3x^{\frac{1}{2}}$ from the equation of the curve

$$\bar{x}\int_1^4 3\,x^{\frac{1}{2}}\,mg\,dx = \int_1^4 x\,3x^{\frac{1}{2}}\,mg\,dx$$

hence

$$\bar{x}\int_1^4 3\,x^{\frac{1}{2}}dx = \int_1^4 3\,x^{\frac{3}{2}}\,dx$$

$$\therefore\quad \bar{x} = 2{\cdot}66$$

In a similar way, the y-coordinate of the centre of gravity is found by taking moments about the x axis. The distance of the centre of gravity of each strip from the x-axis will be $\frac{y}{2}$.

Equating moments about the x-axis:

$$\bar{y}\int_1^4 y\,mg\,dx = \int_1^4 \frac{y}{2} \times y\,mg\,dx$$

substituting for y as $3x^{\frac{1}{2}}$, as before gives

$$\bar{y} = 2{\cdot}41$$

The coordinates of the centre of gravity are (2·66, 2·41).

Solids of revolution

When a plane area is revolved about, say, the x-axis, a solid body is formed. Due to the way in which it is formed, it is sometimes referred to as a solid of revolution. Such bodies will necessarily have an axis of symmetry (the x-axis), and consequently the centre of gravity of the body must lie on the x-axis. There is therefore only one coordinate to determine.

It will usually be possible to divide up the solid into an infinite number of elementary discs, each with its centre of gravity on the x-axis. Then, by taking moments about the y-axis, the x-coordinate of the centre of gravity can be readily determined.

Example 14

A uniform solid cone of height 12 cm and base radius 3 cm is formed by rotating a line about the x-axis, as shown in the diagram. Find the distance of the centre of gravity of the cone from the origin.

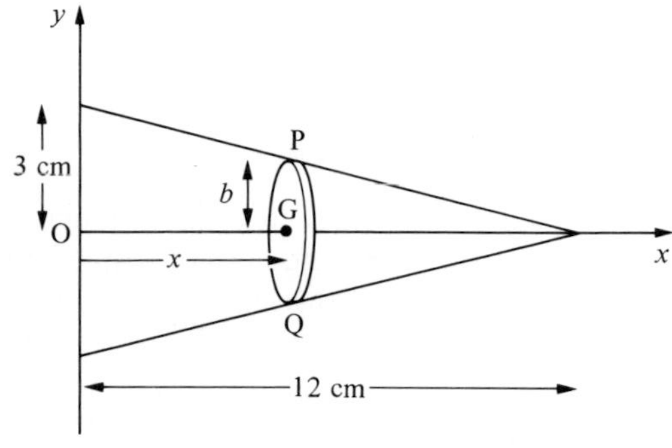

Suppose the density of the cone is σ.
Consider an elementary disc PQ, thickness δx and radius b, with its plane at right angles to the x-axis.
If the distance of PQ from the y-axis is x, then by geometry:

$$\frac{b}{12-x} = \frac{3}{12}$$

$$\text{or} \quad b = \frac{12-x}{4}$$

$$\text{mass of elementary disc PQ} = \pi b^2 \sigma\, \delta x$$

$$\text{moment of PQ about } y\text{-axis} = \pi b^2 \sigma\, g \delta x \times x$$

$$\text{mass of whole cone} = \pi \frac{3^2 \times 12}{3} = 36\pi$$

$$\text{moment of whole cone about } y\text{-axis} = 36\pi\sigma\, g\, \bar{x}$$

Equating the moment of the whole solid with the sum of the moments of the discs about the y-axis:

$$36\pi\sigma\, g\, \bar{x} = \int_0^{12} \frac{\pi}{4^2}(12-x)^2\, \sigma g\, x\, dx$$

$$\therefore \quad 36\,\bar{x} = \tfrac{1}{16}\int_0^{12} (144x - 24x^2 + x^3)\, dx$$

$$\therefore \quad 36\,\bar{x} = 108$$

$$\therefore \quad \bar{x} = 3 \text{ cm}$$

The centre of gravity of the whole cone is 3 cm from the origin.

In the above example, the solid was a cone for which the volume was known. In many cases, the solid of revolution will be such that calculus is needed to find its volume.

Example 15

An area is enclosed by the curve $y^2 = 5x$, the x-axis, the lines $x = 1, x = 3$, and it lies in the first quadrant. The area is rotated about the x-axis through one revolution. Find the coordinates of the centre of gravity of the uniform solid so formed.

The centre of gravity will lie on the axis of symmetry, the x-axis.

Suppose the density of the solid is σ.

Consider an elementary disc PQ, thickness δx and radius y, with its plane at right angles to the x-axis.

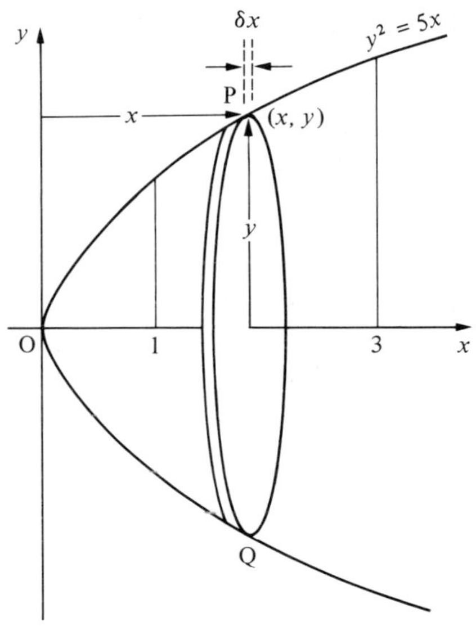

$$\text{mass of elementary disc PQ} = \pi y^2 \sigma\, \delta x$$
$$\text{moment of PQ about } y\text{-axis} = \pi y^2 \sigma\, g \delta x \times x$$

Equating the moment of the whole solid with the sum of the moments of the discs about the y-axis:

$$\bar{x}\int_1^3 \pi y^2 \sigma\, g\, dx = \int_1^3 \pi y^2 \sigma\, g x\, dx$$

substituting for y^2 as $5x$ from the equation of the curve:

$$\bar{x}\int_1^3 \pi 5x \sigma\, g\, dx = \int_1^3 \pi 5x^2 \sigma\, g\, dx$$

$$\therefore \quad 20\,\bar{x} = \tfrac{130}{3}$$

$$\therefore \quad \bar{x} = \tfrac{13}{6}$$

The centre of gravity of the solid of revolution has coordinates $(2\frac{1}{6}, 0)$.

It should be noted that the answers to example 12 and example 14, agree with the standard results stated prior to Exercise 8C. The proof of these and other standard results are covered in questions **1** to **6** of Exercise 8E.

Exercise 8E

1. Uniform semicircular lamina.
   The diagram shows a uniform semicircular lamina of radius r; AB is an elementary strip of the lamina and lies parallel to the y-axis and at a distance x from it. The strip is of thickness δx.
   By considering all such strips, show that the centre of gravity of the lamina lies on its axis of symmetry at a distance of $\dfrac{4r}{3\pi}$ from O.

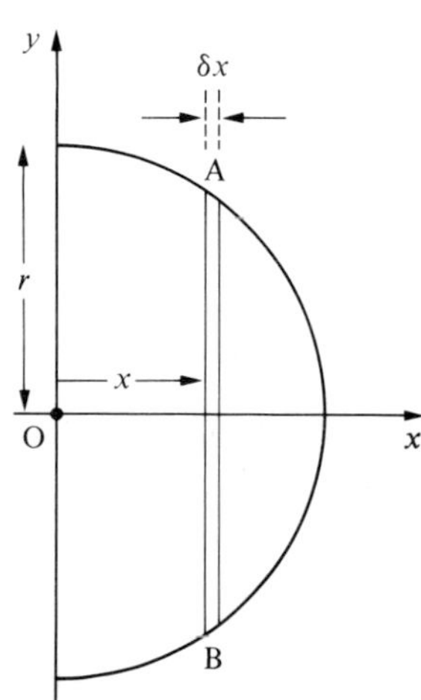

2. Uniform solid right circular cone.
   The diagram shows a uniform solid right circular cone of height h and base radius r; AB is an elementary disc of thickness δx which has its plane parallel to the base of the cone and its centre at a distance x from the y-axis.
   By considering all such discs, show that the centre of gravity of the cone lies on its axis of symmetry at a distance of $\frac{3h}{4}$ from O.

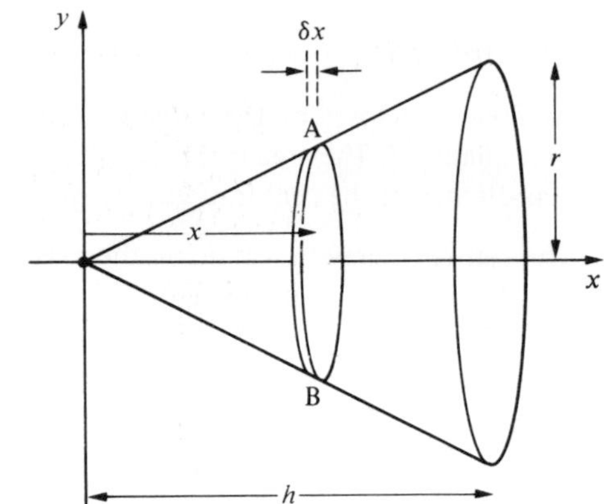

3. Uniform solid hemisphere.
   The diagram shows a uniform solid hemisphere of radius r. AB is an elementary disc of the hemisphere which has its plane parallel to the plane surface of the hemisphere, lies at a distance x from this plane surface and is of thickness δx.
   By considering all such discs, show that the centre of gravity of the hemisphere lies on its axis of symmetry at a distance $\frac{3r}{8}$ from O.

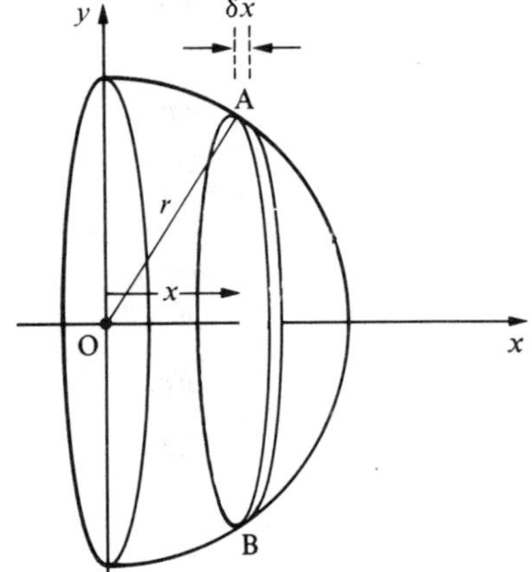

Note: Questions **4, 5** and **6** require an understanding of radians and an ability to integrate trigonometrical functions. The reader may wish to omit these questions at this stage if such pure mathematics topics have not yet been covered.

4. Uniform wire in the shape of a circular arc.
   Fig. (i) shows a uniform wire in the shape of an arc of a circle centre O, radius r.
   The arc subtends an angle of 2α at O.
   Fig. (ii) shows an element of this wire.
   The element subtends an angle of $\delta\theta$ at O.
   All angles are in radians.
   By allowing θ to range from $+\alpha$ to $-\alpha$, all such elements of the wire can be considered.

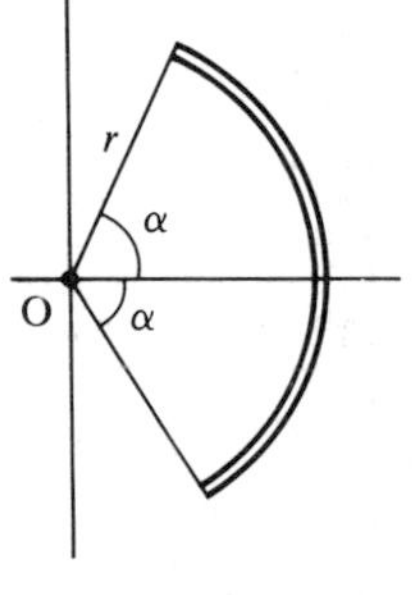

Fig. (i)

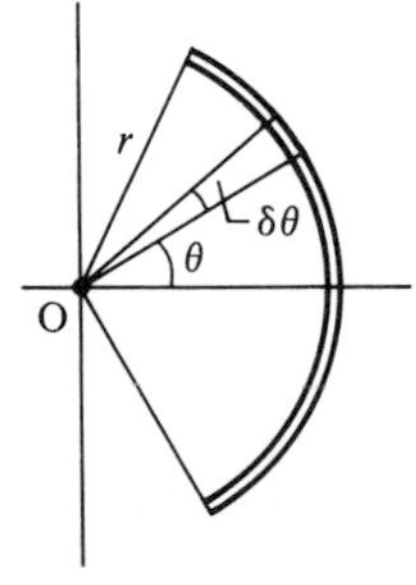

Fig. (ii)

   (a) Show that the centre of gravity of the wire lies on its axis of symmetry at a distance of $\frac{r \sin \alpha}{\alpha}$ from O.
   (b) Show that if the wire were in the form of a semicircular arc of radius r, centre at O, the centre of gravity would lie on the axis of symmetry at a distance $\frac{2r}{\pi}$ from O.

5. Uniform lamina in the shape of a sector of a circle.
Fig. (iii) shows a uniform lamina in the shape of a sector of a circle, centre O and radius r. The sector subtends an angle of 2α at O.
Fig. (iv) shows an element OAB which subtends an angle $\delta\theta$ at O.
All angles are in radians.
By considering this element as a triangle and by allowing θ to range from $+\alpha$ to $-\alpha$, show that the centre of gravity of the sector lies on its axis of symmetry at a point which is $\dfrac{2r \sin \alpha}{3\alpha}$ from O.
Show that this result is compatible with that of question **1** in this exercise.

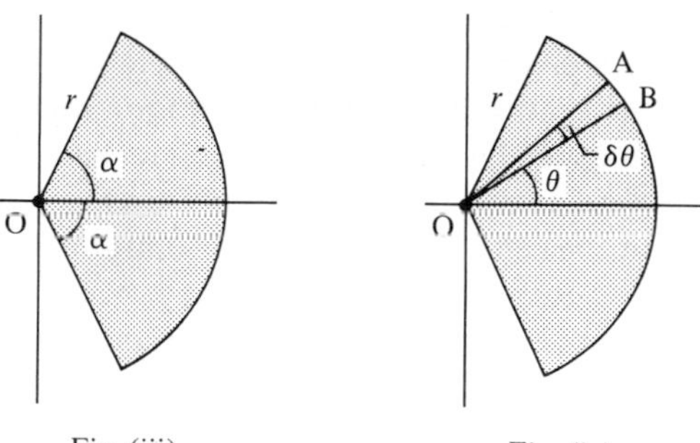

Fig. (iii) Fig. (iv)

6. Uniform hemispherical shell.
The diagram shows a uniform hemispherical shell of radius r. The shaded portion is a small circular element of the shell. This element is parallel to the plane face of the hemisphere and at a distance $r \cos \theta$ from it. The element may be considered to approximate to a circular ring of radius $r \sin \theta$ and thickness $r\delta\theta$. By allowing θ to range from 0 to $\dfrac{\pi}{2}$, these elements would together form the hemispherical shell.
Show that the centre of gravity of the shell lies on the x-axis at a distance $\dfrac{r}{2}$ from O.

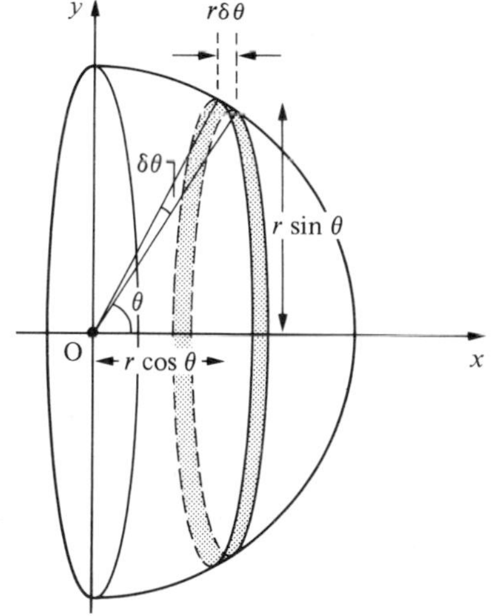

7. Find the coordinates of the centre of gravity of the uniform lamina enclosed by the curve $y = x^2$, the x-axis and the line $x = 2$.
8. Find the coordinates of the centre of gravity of the uniform lamina enclosed between the curve $y = 2x - x^2$ and the x-axis.
9. Find the coordinates of the centre of gravity of the uniform lamina enclosed by the curve $y = x^2 + 2$, the x-axis and the lines $x = 1$ and $x = 2$.
10. Find the coordinates of the centre of gravity of the uniform lamina lying in the first quadrant and enclosed by the curve $y^2 = 8x$, the x-axis and the lines $x = 2$ and $x = 8$.
11. Find the coordinates of the centre of gravity of the uniform lamina enclosed between the line $y = 3x$ and the curve $y = x^2$.
12. Find the coordinates of the centre of gravity of the uniform lamina which lies in the first quadrant and is enclosed by the curves $y = 3x^2$, $y = 4 - x^2$ and the y-axis.
13. The area enclosed by the curve $y^2 = x$, the x-axis, the line $x = 4$ and lying in the first quadrant, is rotated about the x-axis through one revolution. Find the coordinates of the centre of gravity of the uniform solid so formed.

14. The area enclosed by the curve $y^2 = x$, the x-axis, the lines $x = 2, x = 4$ and lying in the first quadrant, is rotated about the x-axis through one revolution. Find the coordinates of the centre of gravity of the uniform solid so formed.
15. The area lying in the first quadrant and enclosed by the curve $y = x^2$ and the lines $y = 0$, $x = 2$ and $x = 4$, is rotated about the x-axis through one revolution. Find the coordinates of the centre of gravity of the uniform solid so formed.
16. The area enclosed by the curve $y = x^2 + 3$, the x-axis, the y-axis and the line $x = 2$ is rotated about the x-axis through one revolution. Find the coordinates of the centre of gravity of the uniform solid so formed.
17. Find the coordinates of the centre of gravity of the uniform lamina enclosed between the curve $y = x^3$, the x-axis and the line $x = 3$.
If this lamina is rotated about the x-axis through one revolution, find the coordinates of the centre of gravity of the uniform solid so formed.

Exercise 8F Examination questions

1. A thin uniform rectangular plate, ABCD, has a mass of 40 g. The side AB is 12 cm long and BC is 20 cm. Particles of mass 10, 5, 15 and 10 g are attached to the plate at the points A, B, C and D respectively. Calculate the distances of the centre of mass of the loaded plate from the sides AB and BC. (A.E.B)

2. A uniform square sheet of cardboard ABCD of side 6 cm has cut from it an isosceles triangle BEC, in which BE = CE, in such a way that the centre of gravity of the remainder is at E. Show that the distance of E from AD is $3(\sqrt{3} - 1)$ cm.
The portion ABECD is freely suspended from B and hangs in equilibrium. Calculate, correct to the nearest degree, the angle which the side BA makes with the vertical. (J.M.B)

3. A uniform rectangular lamina ABCD has AB $= a$ and BC $= 3a$. The point P lies in the edge AD so that DP $= 3s$. The triangle PCD is cut away leaving the trapezium ABCP. Show that the centre of gravity G of this trapezium is at a distance $\dfrac{s^2 - 3as + 3a^2}{2a - s}$ from AB and calculate the distance of G from BC.
The trapezium is suspended from the point P and hangs freely in equilibrium with the edge BC horizontal. Prove that $2s = a(3 - \sqrt{3})$. (A.E.B)

4. Prove that the centre of gravity of a uniform solid hemisphere of radius r is a distance $3r/8$ from the centre.
A child's toy is made up from a uniform and solid right circular cone and hemisphere. The radius of the cone is r, and its height $3r$. The radius of the hemisphere is r. The base of the cone and hemisphere are sealed together. The material from which the hemisphere is made is 3 times as heavy per unit volume as the cone material. Find the distance of the centre of gravity of the toy from the vertex of the cone. (The centre of gravity of a cone is $\frac{1}{4}$ of the way up the central axis from the base.) (S.U.J.B)

5. A uniform semicircular lamina is of radius $3a$. The centre of its bounding diameter BC is O and its centre of gravity is at the point G. Show by integration that G is at a distance $4a/\pi$ from BC.
The uniform plane lamina L shown in Fig. 1 consists of the semicircular lamina and the

isosceles triangular lamina ABC in which OA $= x$ and AB $=$ AC. Given that the centre of gravity of L is at O, show that $x = 3a\sqrt{2}$.

The weight of L is W and a particle of weight W is attached at C. The lamina is smoothly hinged at A to a fixed point and can rotate freely in its own plane which is vertical. Calculate, in terms of a and W, the magnitude of the couple which would be required to keep L in equilibrium with AB horizontal. (A.E.B)

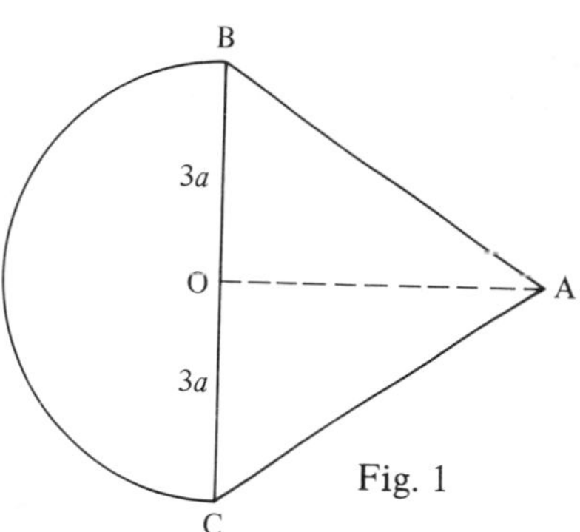

Fig. 1

6. The region defined by the inequalities $0 \leqslant x \leqslant a$, $0 \leqslant y^2 \leqslant 4ax$, $y \geqslant 0$ is rotated through 2π radians about the x-axis to give the solid of revolution V_1. Find, in terms of a and π, (a) the volume of V_1, (b) the position of the centroid of V_1.
The same region is rotated through 2π radians about the y-axis to give the solid of revolution V_2. Find, in terms of a and π, the volume of V_2. (London)

7. Prove by integration that the centre of mass of a uniform solid right circular cone of height h and base radius r is at a distance $\frac{3}{4}h$ from the vertex.
Such a cone is joined to a uniform solid right circular cylinder, of the same material, with base radius r and height l, so that the plane base of the cone coincides with a plane face of the cylinder. Find the centre of mass of the solid thus formed. Show that, if $6l^2 \geqslant h^2$, this solid can rest in equilibrium on a horizontal plane, with the curved surface of the cylinder touching the plane.
Given that $l = h$, show that the solid can rest in equilibrium with its conical surface touching the plane provided that $r \geqslant \frac{1}{4}h\sqrt{5}$. (J.M.B)
[Hint: toppling will occur if the line of action of the weight of the body falls outside the base on which the body stands as in such cases the reaction at the base is incapable of balancing the resulting moments.]

9

General equilibrium of a rigid body

In Chapter 5, forces acting on a particle were considered, and by the definition of a particle this ensured that the forces were concurrent. Thus for equilibrium it was only necessary to show that there was no resultant force acting in any direction.
A rigid body, on the other hand, has size; the forces acting on the body may not be concurrent and so rotation could occur. Thus, for equilibrium we must ensure that there is no resultant force acting *and* that the forces have no turning effect.

Three forces

If a rigid body is in equilibrium under the action of only three forces, these forces must be either concurrent or parallel.

Suppose the forces are P, Q and S and that the lines of action of any two of the three forces, for example P and Q, intersect at the point A. The forces P and Q will have no turning effect about the point A; but if the line of action of the third force S does not pass through A, then the force S will have a turning effect about A and the forces cannot be in equilibrium. Hence, if the forces are in equilibrium, the line of action of S must pass through A: the forces are then concurrent.

Again, suppose that any two of the three forces (say P and Q) have parallel lines of action, then the resultant of these two forces will be parallel to P and Q. Equilibrium is then only possible if the third force S is equal and opposite to the resultant of P and Q, and hence S must be parallel to the other two forces.
It is sometimes possible to use these facts in determining the direction of an unknown third force which is maintaining equilibrium.

When three forces are maintaining equilibrium, it is possible to solve the problem either by the use of the triangle of forces or of Lami's Theorem.

The next section considers the method of resolving in two directions and taking moments as a general method of solution for a rigid body in equilibrium under the action of any number of forces. This general method can also be used for three forces in equilibrium and is often simpler to use than the specific three-force properties mentioned above.

General method

In all problems concerning the equilibrium of a rigid body, the following procedure should be adopted:

(i) interpret the information given and draw a diagram,
(ii) show on the diagram all the forces acting on the body, indicating clearly the directions of these forces,

(iii) equate the clockwise and anticlockwise moments of the forces acting on the body, about any convenient point,

(iv) in each of two perpendicular directions, equate the resolved parts (or components) of the forces acting in one direction to those resolved parts acting in the opposite direction.

Note Careful choice of the point about which moments are taken may well simplify the solution of a particular problem. Usually the best directions in which to resolve the forces are (a) horizontally and vertically, or (b) parallel and at right angles to the surface of an inclined plane. However, there are examples which are more quickly solved by choosing other directions.

Example 1

A uniform rod AB of mass 4 kg and length 80 cm is freely hinged to a vertical wall.
A force P, as shown in the diagram, is applied at the point B and keeps the rod horizontal and in equilibrium. The forces X and Y are the horizontal and vertical components of the reaction at the hinge. Find the magnitudes of the forces X, Y and P.

There are four forces acting on the rod.

Taking moments about A:

$\overset{\curvearrowright}{A}$ $\quad P \times 80 = 4g \times 40$

$\therefore \quad P = 2g$

or $\quad P = 19{\cdot}6$ N

Resolving vertically:

$P + Y = 4g$

$\therefore \quad Y = 4g - 2g$

$Y = 19{\cdot}6$ N

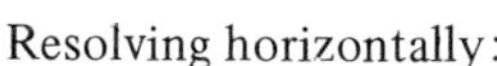

Resolving horizontally:

$X = 0$

The force $X = 0$, $Y = 19{\cdot}6$ N and $P = 19{\cdot}6$ N.

Reaction at hinge

Instead of considering the horizontal and vertical components of the reaction at the hinge A, the force on the rod due to the hinge may be represented by a single force R acting at an angle θ to the vertical. The following example illustrates this method.

Example 2

A uniform rod AB of mass 6 kg and length 4 m is freely hinged at A to a vertical wall.
The force P applied at B as shown in the diagram, keeps the rod horizontal and in equilibrium; R is the force of reaction at the hinge and θ is the angle that the line of action of this force makes with the vertical. Find the magnitude of the forces P and R and the angle θ.

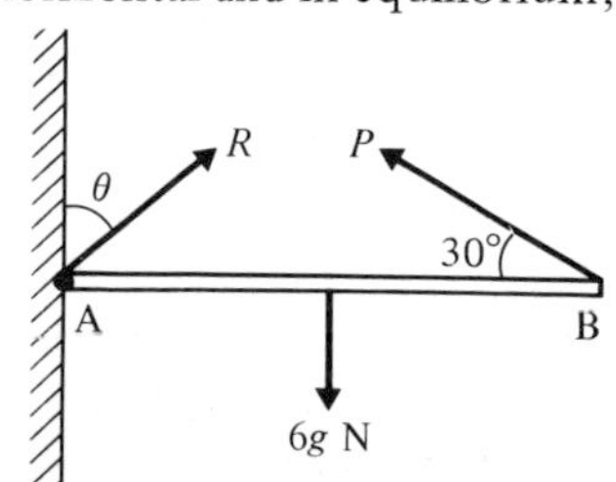

Taking moments about A:

$\overset{\curvearrowright}{A}$ $\quad 6g \times 2 = P \times 4 \sin 30$

$\therefore \quad P = 6g$

Resolving vertically:

$$R \cos \theta + P \cos 60 = 6g$$

substituting for P

$$R \cos \theta = 3g \qquad \ldots [1]$$

Resolving horizontally:

$$R \sin \theta = P \cos 30$$

substituting for P

$$R \sin \theta = 6g \cos 30 \qquad \ldots [2]$$

Dividing equation [2] by equation [1]

$$\frac{R \sin \theta}{R \cos \theta} = \frac{6g \cos 30}{3g}$$

$$\therefore \quad \tan \theta = \sqrt{3} \quad \text{or} \quad \theta = 60^\circ$$

substituting in equation [1]

$$R \cos 60 = 3g$$

$$\therefore \quad R = 6g$$

The force $P = 6g$ N, $R = 6g$ N and the angle $\theta = 60^\circ$.

Alternative method

Since there are only three forces acting on the rod in example 2, the force R must pass through the point of intersection O of the forces P and $6g$, as shown in the diagram below.

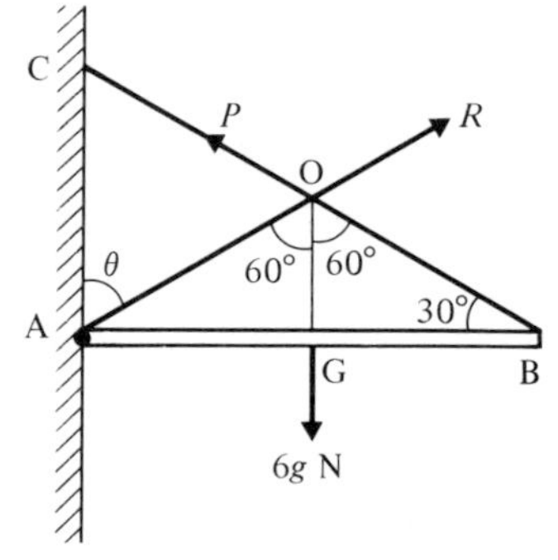

G is the mid-point of AB, and OG is perpendicular to AB; hence the triangle AOB is isosceles and AO = OB.

$$\therefore \quad \text{angle OAB} = 30^\circ$$

$$\text{or} \quad \theta = 60^\circ$$

The angles between the forces are each 120°.

Applying Lami's Theorem at the point O:

$$\frac{6g}{\sin 120} = \frac{P}{\sin 120} = \frac{R}{\sin 120}$$

$$\therefore \quad P = 6g \text{ and } R = 6g \text{ as before}$$

Alternatively from the same diagram, if BO is produced to meet the wall at C, then triangle AOC may be used as a triangle of forces in order to solve the problem.

Example 3

A non-uniform rod of mass 3 kg and length 40 cm rests horizontally in equilibrium, supported by two strings attached at the ends A and B of the rod. The strings make angles of 45° and 60° with the horizontal, as shown in the diagram. Find the tension in each of the strings and the position of the centre of gravity of the rod.

Since the rod is not uniform, the force of $3g$ N is shown acting at a distance s cm from the end A.

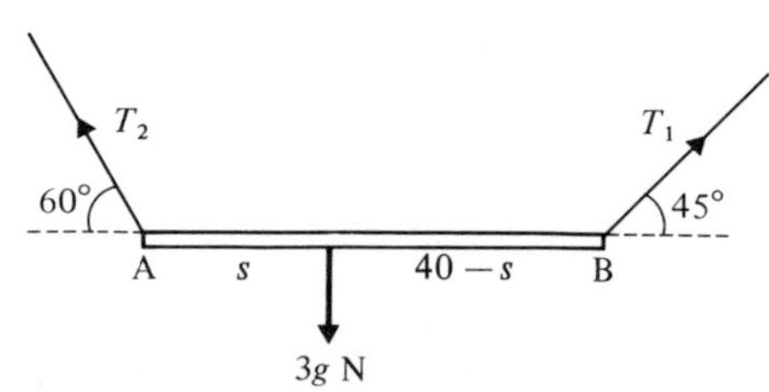

Resolving horizontally:

$$T_2 \cos 60 = T_1 \cos 45$$

$$\therefore \quad T_2 = T_1 \sqrt{2} \qquad \ldots [1]$$

Resolving vertically:

$$T_2 \sin 60 + T_1 \sin 45 = 3g \qquad \ldots [2]$$

Taking moments about A:

$$\overset{\curvearrowright}{A} \qquad 3g \times s = T_1 \sin 45 \times 40 \qquad \ldots [3]$$

From equation [1] and [2] $T_2 = \dfrac{6g}{\sqrt{3}+1}$ and $T_1 = \dfrac{3\sqrt{2}g}{\sqrt{3}+1}$

substituting in equation [3] $\quad s = \dfrac{40}{\sqrt{3}+1}$

The tensions in the strings are 21·5 N and 15·2 N and the centre of gravity is 14·6 cm from end A.

Limiting equilibrium

When a rigid body is in equilibrium under the action of any number of forces, three equations may be obtained by resolving in two directions and by taking moments about a point.
If there is a frictional force acting, it is necessary to be clear whether the body is in limiting equilibrium. It should be remembered that only in the case of limiting equilibrium, when motion is on the point of taking place, does the frictional force F have its maximum value μR.

Example 4

The diagram below shows a uniform rod AB of mass 4 kg with its lower end A resting on a rough horizontal floor, coefficient of friction μ. A string attached to the end B keeps the rod in equilibrium. T is the tension in the string, F is the frictional force at A, and R is the normal reaction at A. Find the magnitudes of the forces T, F and R, and also the least possibl^ value of μ for equilibrium to be possible.

Let the rod be of length $2l$; if motion were to take place, the end A of the rod would tend to move to the left so the frictional force F acts in the opposite direction.

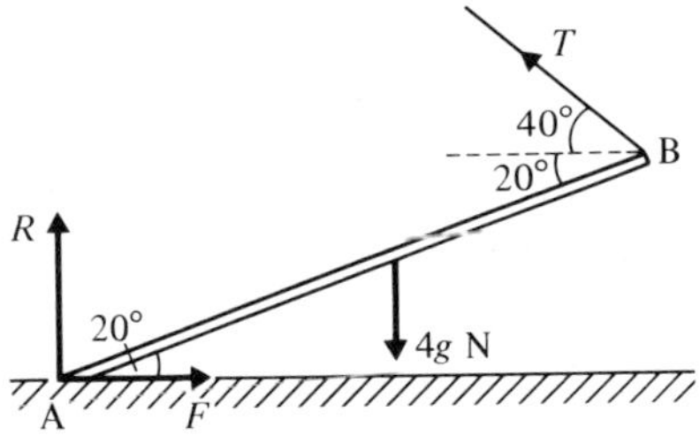

Taking moments about A:

$$\overset{\curvearrowright}{A} \qquad 4g \times l \cos 20 = T \times 2l \sin 60$$

$$\therefore \quad T = \frac{2g \cos 20}{\sin 60}$$

$$= 21{\cdot}3 \text{ N}$$

Resolving horizontally: $\quad T \cos 40 = F$

$$\therefore \quad F = 16{\cdot}3 \text{ N}$$

Resolving vertically: $\quad R + T \sin 40 = 4g$

substituting for T gives: $\quad R = 25{\cdot}5$ N

Since $\mu = \dfrac{F_{max}}{R}$ the least value of μ necessary is $\dfrac{16{\cdot}3}{25{\cdot}5} = 0{\cdot}64$ and the rod would then be in limiting equilibrium.

The force $T = 21{\cdot}3$ N, $F = 16{\cdot}3$ N and $R = 25{\cdot}5$ N and the least possible value of μ is 0·64.

Ladder problems

The situation of a ladder resting against a wall, with the foot of the ladder on the ground, gives rise to a variety of problems. The wall may be rough or smooth, as also may the ground. The ground may, or may not, be horizontal.
It should be remembered that where the ladder rests against a smooth surface, there will only be a normal reaction, R, at that point.
When the surfaces in contact are rough, there is also a frictional force, F, which acts parallel to the surfaces in contact, and in a direction opposite to that in which the ladder would move.

Example 5

A uniform ladder AB, of mass 10 kg and length 4 m, rests with its upper end A against a smooth vertical wall and end B on smooth horizontal ground. A light horizontal string, which has one end attached to B and the other end attached to the wall, keeps the ladder in equilibrium inclined at 40° to the horizontal. The vertical plane containing the ladder and the string is at right angles to the wall. Find the tension T in the string and the normal reactions at the points A and B.

Suppose R and S are the normal reaction at the ground and the wall respectively.

The diagram shows the forces acting on the ladder.

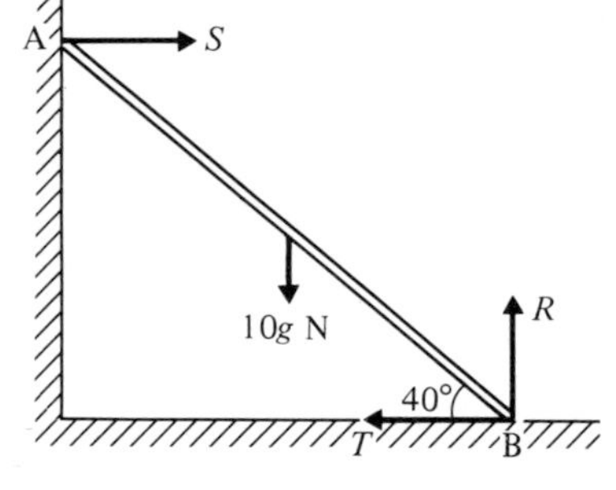

Resolving vertically: $R = 10g$
$= 98$ N

Resolving horizontally: $T = S$

Taking moments about B:

$\overset{\curvearrowright}{\text{B}}$ $S \times 4 \sin 40 = 10g \times 2 \cos 40$

$\therefore \quad S = \dfrac{5g \cos 40}{\sin 40}$

$\therefore \quad S = 58{\cdot}4$ N

and it follows $T = 58{\cdot}4$ N

The tension in the string is 58·4 N and the normal reactions at the top and foot of the ladder are 58·4 N and 98 N respectively.

It should be noted that whether equilibrium is possible or not will depend upon whether the string can take a tension of 58·4 N without breaking.

Rough contact at foot of ladder

If the ladder rests on ground which is rough, then there will be a frictional force F acting on the ladder at this point. The effect of this force is similar to that of the tension in the string in Example 5. The maximum value of this frictional force depends upon the roughness of the contact between the ladder and the ground.

Example 6

The diagram shows a ladder AB of mass 8 kg and length 6 m resting in equilibrium at an angle of 50° to the horizontal with its upper end A against a smooth vertical wall and its lower end B on rough horizontal ground, coefficient of friction μ. Find the forces S, F and R and the least possible value of μ if the centre of gravity G of the ladder is 2 m from B.

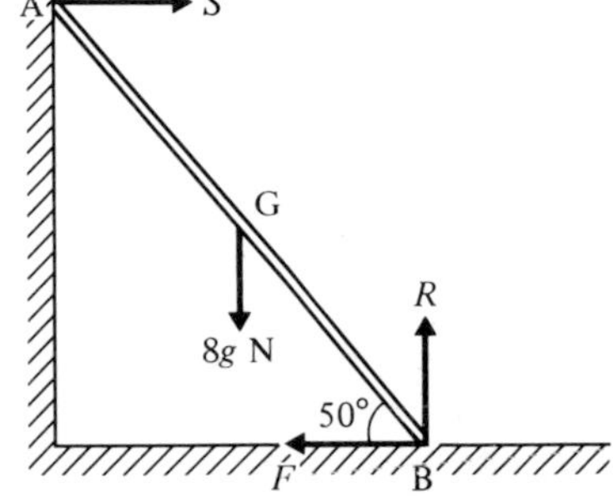

Taking moments about B:

$$\overset{\frown}{B} \qquad S \times 6 \sin 50 = 8g \times 2 \cos 50$$

$$\therefore \quad S = \frac{8g \cos 50}{3 \sin 50}$$

$$\therefore \quad S = 21{\cdot}9 \text{ N}$$

Resolving horizontally: $F = S$

$$\therefore \quad F = 21{\cdot}9 \text{ N}$$

Resolving vertically: $R = 8g$

$$\therefore \quad R = 78{\cdot}4 \text{ N}$$

since $\mu = \dfrac{F_{max}}{R}$ and $\dfrac{F}{R} = \dfrac{21{\cdot}9}{78{\cdot}4} = 0{\cdot}28$ $\quad \mu$ must be at least 0·28

The force $S = 21{\cdot}9$ N, $F = 21{\cdot}9$ N, $R = 78{\cdot}4$ N and μ must be at least 0·28.

Climbing a ladder

Whether or not it is safe to ascend to the top of a ladder will depend upon the magnitude of the frictional force which acts on the foot of the ladder. This will depend upon the roughness of the ground on which the ladder rests.

If the ladder is found to be in limiting equilibrium when a person is part way up a ladder, then any further ascent will cause the ladder to slip.

To determine how far a ladder may be ascended, consider the situation when the climber is at a distance s up the ladder and the ladder is in limiting equilibrium. The following example illustrates the method.

Example 7

A uniform ladder of mass 30 kg and length 5 m rests against a smooth vertical wall with its lower end on rough ground, coefficient of friction $\frac{2}{5}$. The ladder is inclined at 60° to the horizontal. Find how far a man of mass 80 kg can ascend the ladder without it slipping.

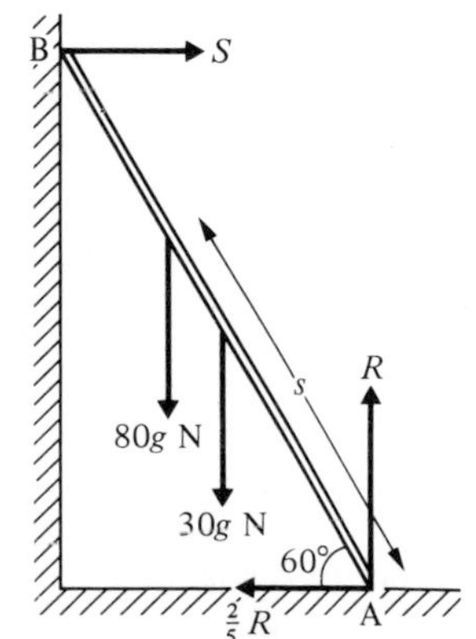

Assume the man can ascend a distance s m from the foot of the ladder, which is then in limiting equilibrium. The maximum frictional force μR will then act at the foot of the ladder.

The forces acting on the ladder are then as shown.

Resolving vertically:

$$R = 30g + 80g = 110g \qquad \ldots [1]$$

Resolving horizontally:

$$S = \tfrac{2}{5}R \qquad \ldots [2]$$

Taking moments about A:.

$\overset{\curvearrowright}{A}$ $30g \times \frac{5}{2} \cos 60 + 80g \times s \cos 60 = S \times 5 \sin 60$ . . . [3]

From equations [1] and [2] $S = \frac{2}{5}(110g) = 44g$

substituting in equation [3]

$$\frac{75g}{2} + 40g\,s = 44g \times 5 \sin 60$$

$$\text{or} \quad s = 3{\cdot}83 \text{ m}$$

The man can climb 3·83 m up the ladder, at which point the ladder will be on the point of slipping.

Example 8

A uniform ladder rests in limiting equilibrium with its top end against a rough vertical wall and its lower end on a rough horizontal floor. If the coefficients of friction at the top and foot of the ladder are $\frac{2}{3}$ and $\frac{1}{4}$ respectively, find the angle which the ladder makes with the floor.

Let the ladder be of length $2l$ and weight W.
The forces acting on the ladder will be as shown in the diagram.
Since the ladder is in limiting equilibrium, both ends of the ladder will be on the point of moving, so the maximum frictional forces will act at both ends.

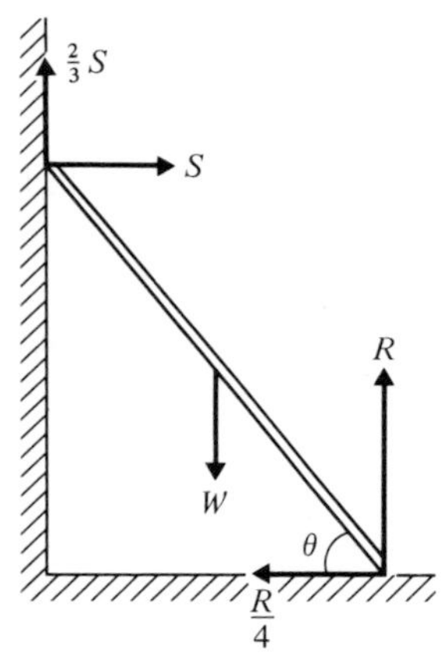

Resolving vertically:

$$\tfrac{2}{3}S + R = W \qquad \ldots [1]$$

Resolving horizontally:

$$S = \tfrac{1}{4}R \qquad \ldots [2]$$

Taking moments about the foot of the ladder:

$$W \times l \cos\theta = S \times 2l \sin\theta + \tfrac{2}{3}S \times 2l \cos\theta \qquad \ldots [3]$$

Eliminating R from equations [1] and [2]:

$$\tfrac{2}{3}S + 4S = W$$
$$\text{or} \quad S = \tfrac{3}{14}W$$

Substituting for S in equation [3]:

$$Wl \cos\theta = \tfrac{3}{14}W \times 2l \sin\theta + \tfrac{2}{3} \times \tfrac{3}{14}W \times 2l \cos\theta$$
$$\therefore \quad \cos\theta = \tfrac{3}{7}\sin\theta + \tfrac{2}{7}\cos\theta$$
$$\therefore \quad \tan\theta = \tfrac{5}{3} \quad \text{or} \quad \theta = 59{\cdot}04^\circ$$

The angle the ladder makes with the floor is 59·04°.

Note that, even though the ladder is in limiting equilibrium when the angle of inclination to the horizontal is 59·04°, it is possible for a person to ascend part way up the ladder without it slipping.
Suppose that a man of weight $3W$ ascends the ladder of Example 8 to a point at a distance s from the foot of the ladder, and that the ladder is then on the point of slipping, i.e. it is in limiting equilibrium.

The three equations then become:

$$\tfrac{2}{3}S + R = W + 3W$$
$$S = \tfrac{1}{4}R$$

and $\quad W \times l \cos\theta + 3W \times s \cos\theta = S \times 2l \sin\theta + \tfrac{2}{3}S \times 2l \cos\theta$

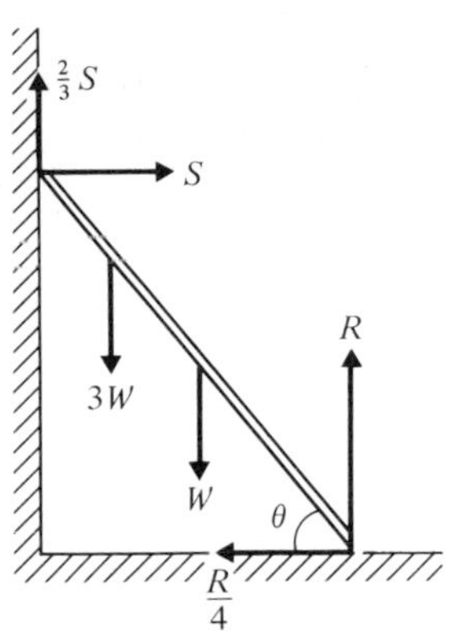

Eliminating R as before:

$$\tfrac{2}{3}S + 4S = 4W$$
$$\therefore \quad S = \tfrac{6}{7}W$$

Then $\quad Wl\cos\theta + 3Ws\cos\theta = \tfrac{6}{7}W \times 2l\sin\theta + \tfrac{4}{7}W \times 2l\cos\theta$

$$\therefore \quad 3s = (\tfrac{12}{7}\tan\theta + \tfrac{1}{7})l$$

substituting for $\tan\theta = \tfrac{5}{3}$

$$\therefore \quad s = l$$

The man can ascend half-way up the ladder.

Practical explanation

The point to notice in this example is that as soon as the man steps on to the foot of the ladder, the normal reaction R at that point is increased. Therefore the maximum value of the frictional force ($F_{max} = \mu R$) is increased. This means that more friction is available at the foot of the ladder and it is no longer in a state of limiting equilibrium. As the man ascends the ladder, his weight has an increasing moment, anticlockwise in the diagram, about the foot of the ladder, and this together with the moment due to the weight of the ladder will eventually balance the maximum clockwise moment of S and $\tfrac{2}{3}S$ about the point A.

This explains why it is safer to ascend a ladder when another person is standing on the foot of the ladder, or a mass is placed on the bottom rung.

Exercise 9A

For all those questions in this exercise which involve a rigid body in contact with a vertical wall, take the vertical plane through the rigid body as being perpendicular to the wall.

1. Each of the following diagrams shows a uniform rod AB of mass 10 kg and length 4 m freely hinged at A to a vertical wall. An applied force P keeps the rod in equilibrium. Forces X and Y are the horizontal and vertical components of the reaction at the hinge. By resolving vertically and horizontally and taking moments, find the magnitudes of the forces X, Y and P.

(a)

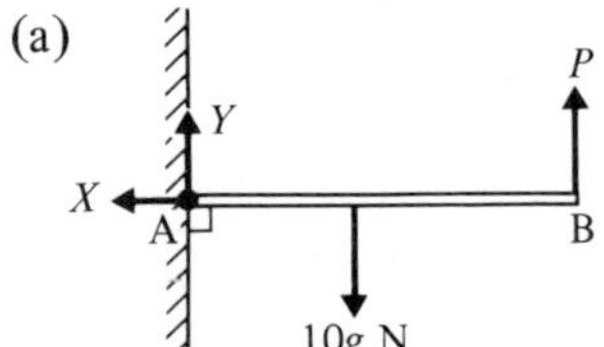

(b)

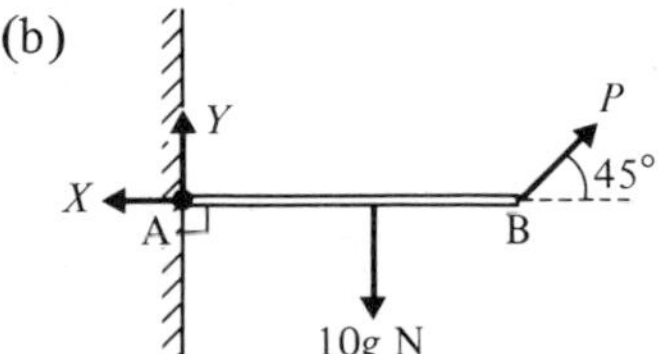

(c)

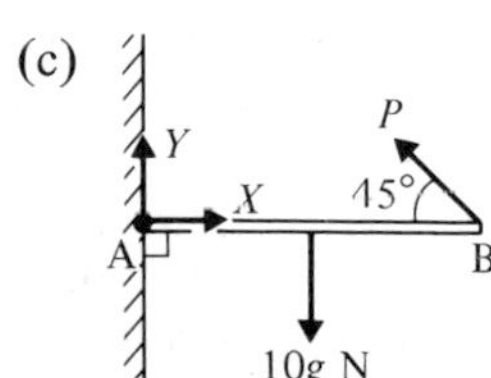

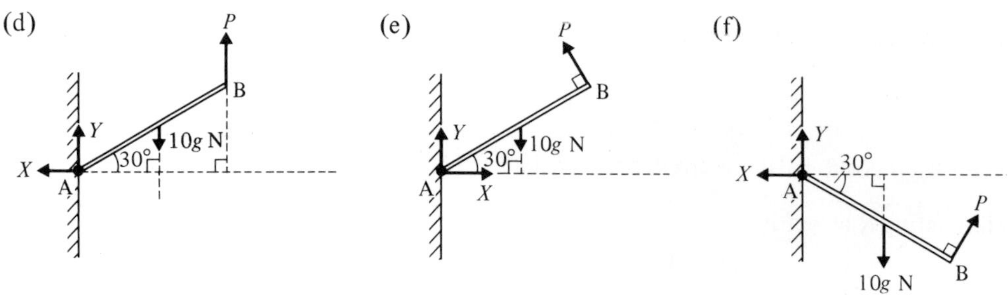

2. Each of the following diagrams shows a uniform rod AB of mass 5 kg and length 6 m freely hinged at A to a vertical wall. An applied force P keeps the rod in equilibrium. R is the force of reaction at the hinge and θ is the angle the line of action this force makes with the wall. For each case, find the magnitudes of the forces P and R and the size of the angle θ.

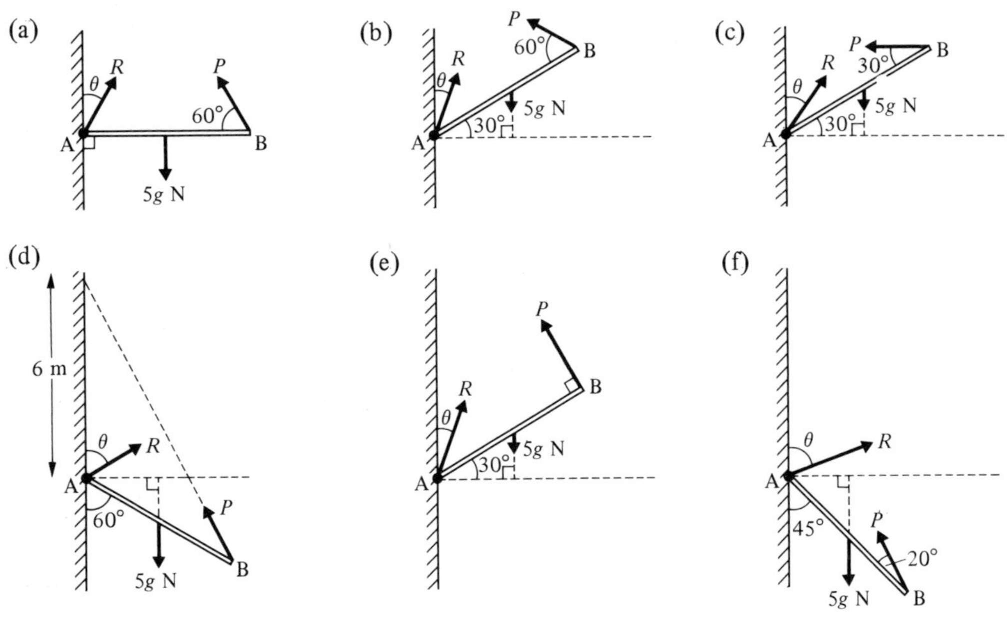

3. Each of the following diagrams shows a uniform rod AB, of mass 10 kg, with its lower end A resting on a rough horizontal floor, coefficient of friction μ. A string attached to end B keeps the rod in equilibrium. T is the tension in this string, F is the frictional force at A and R is the normal reaction at A. Find the magnitude of T, F and R and the least possible value of μ for each situation.

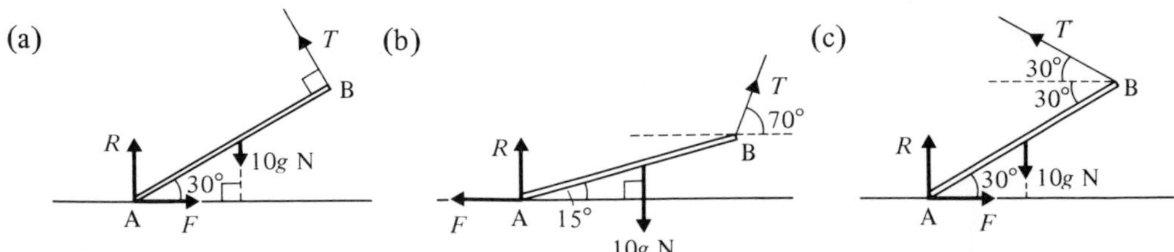

4. Each of the following diagrams shows a uniform ladder AB of mass 30 kg and length 6 m resting with its end A against a smooth vertical wall and end B on a smooth horizontal floor. The ladder is kept in equilibrium by a light horizontal string which has one end attached to B and the other end attached to the wall. R is the normal reaction at the floor, S is the normal reaction at the wall and T is the tension in the string. Find the magnitude of T for each situation.

(a)

(b)

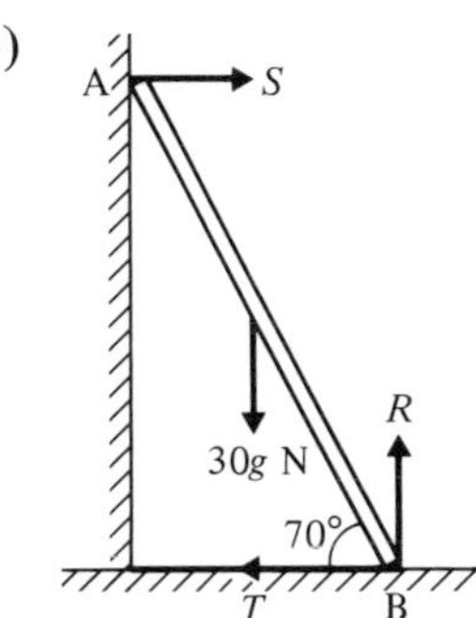

(c)

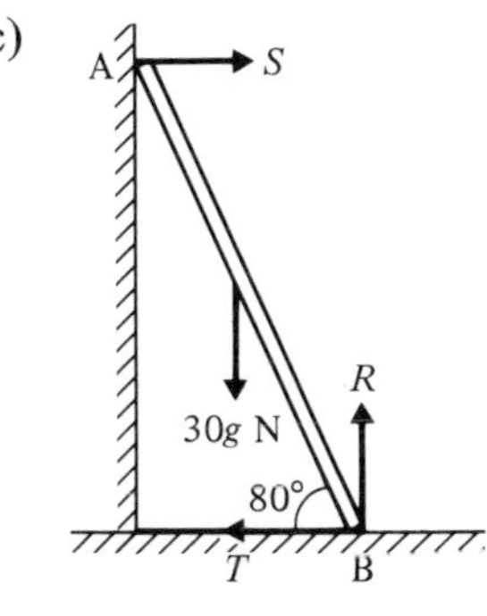

5. Each of the following diagrams shows a uniform ladder of mass 20 kg and length $2l$ resting in equilibrium with its upper end against a smooth vertical wall and its lower end on a rough horizontal floor, coefficient of friction μ. S is the normal reaction at the wall, F is the frictional force at the ground and R is the normal reaction at the ground. Find the magnitude of S, F and R, and the least possible value of μ for each situation.

(a)

(b)

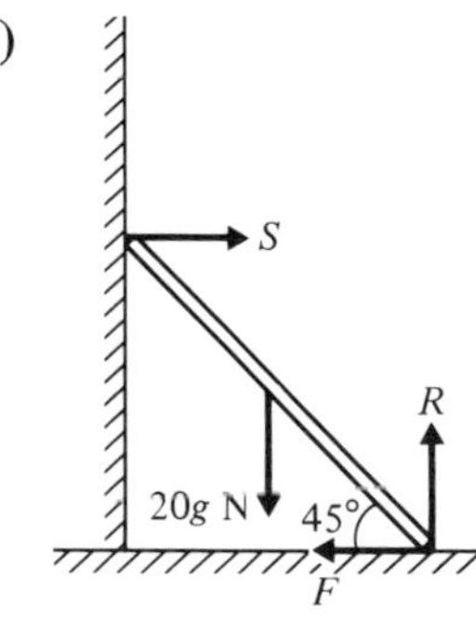

(c)

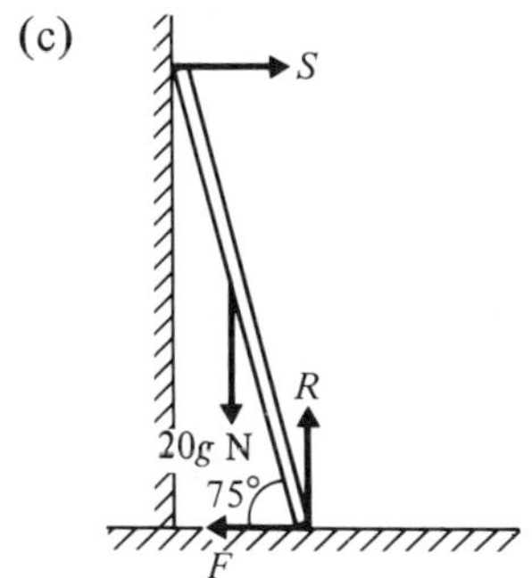

6. The diagram shows a uniform ladder AB of weight W N and length 4 m resting with its end A against a smooth vertical wall and its end B on a smooth horizontal floor. The ladder is kept in equilibrium at an angle θ to the floor by a light horizontal string attached to the wall and to a point C on the ladder.
If $\tan \theta = 2$, find the tension in the string when BC is of length
(a) 1 m, (b) 2 m, (c) 3 m.

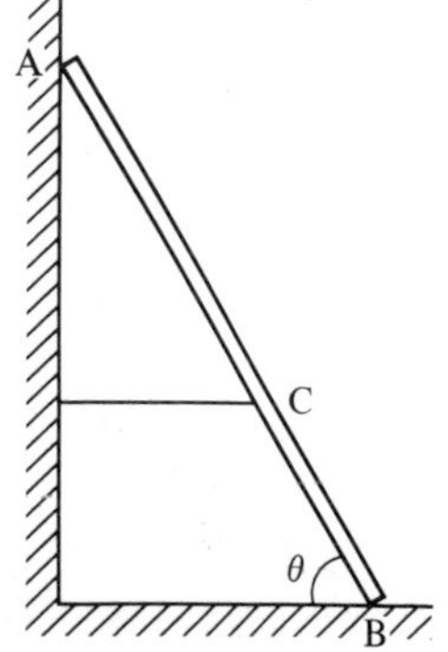

7. The diagram shows a uniform ladder of mass m and length $2l$ resting in limiting equilibrium with its upper end against a rough vertical wall (coefficient of friction μ_1) and its lower end against a rough horizontal floor (coefficient of friction μ_2). The normal reactions at the wall and the floor are S and R respectively with $\mu_1 S$ and $\mu_2 R$ the corresponding frictional forces. The ladder makes an angle θ with the floor. Find θ for each of the following cases
(a) $\mu_1 = \frac{1}{5}$ and $\mu_2 = \frac{1}{5}$,
(b) $\mu_1 = \frac{1}{5}$ and $\mu_2 = \frac{1}{3}$,
(c) $\mu_1 = \frac{1}{3}$ and $\mu_2 = \frac{1}{3}$.

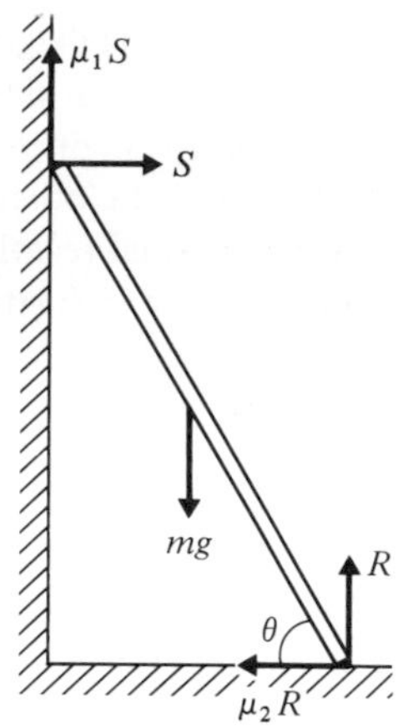

8. The diagram shows a uniform ladder AB of weight W and length $2l$ resting in equilibrium with its upper end A against a smooth vertical wall and its lower end B on a smooth inclined plane. The inclined plane makes an angle θ with the horizontal and the ladder makes an angle ϕ with the wall. Find ϕ when θ equals
(a) 10°, (b) 20°, (c) 30°.

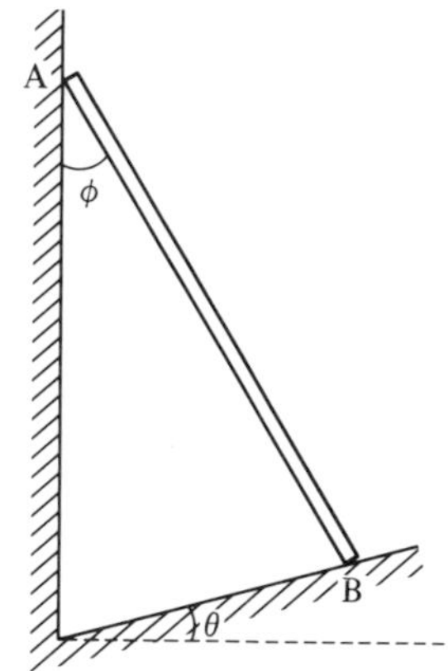

9. A uniform beam AB of mass 5 kg is freely hinged at A to a vertical wall and is maintained horizontally in equilibrium by a light string connecting B to a point on the wall, above A. The string makes an angle of 30° with BA. Find the tension in the string and the magnitude and direction of the reaction at the hinge.

10. A non-uniform beam of mass 5 kg rests horizontally in equilibrium, supported by two light strings attached to the ends of the beam. The tensions in the strings are T_1 and T_2 and the strings make angles of 30° and 40° with the beam, as shown in the diagram. Find the magnitudes of T_1 and T_2.

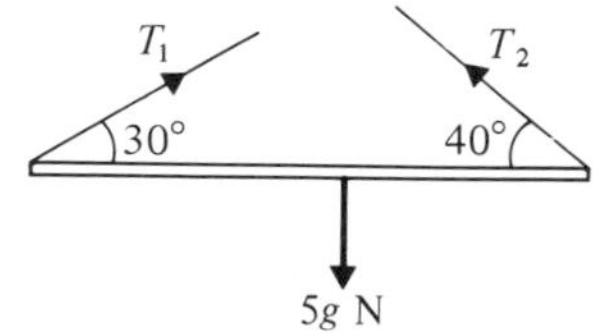

11. A non-uniform beam AB of weight 20 N and length 4 m, has end A freely hinged to a vertical wall. A light string linking B to a point on the wall above A, makes an angle of 60° with BA and allows the beam to rest horizontally in equilibrium. If the tension in the string is 12 N, find the magnitude and direction of the reaction at A and the distance from A to the centre of gravity of the beam.

12. A non-uniform beam AB is of length 8 m and its weight of 10 N acts from a point G between A and B such that AG = 6 m. The beam is supported horizontally by strings attached at A and B. The string attached to A makes an angle of 30° with AB. Find the angle that the string attached to B makes with BA, and find the tensions in the strings.

13. A uniform beam AB of length 4 m and weight 50 N is freely hinged at A to a vertical wall and is held horizontally, in equilibrium, by a string which has one end attached to B and the other end attached to a point C on the wall, 4 m above A. Find the magnitude of the reaction at A.
14. A uniform pole AB of mass 100 kg has its lower end A on rough horizontal ground and is being raised into a vertical position by a rope attached to B. The rope and the pole lie in the same vertical plane and A does not slip across the ground. Find the horizontal and vertical components of the reaction at the ground when the rope is at right angles to the pole and the pole is at 20° to the horizontal.
15. A non-uniform pole AB of mass 50 kg has its centre of gravity at the point of trisection of its length, nearer to B. The pole has its lower end A on rough horizontal ground and is being raised into a vertical position by a rope attached to B. The rope and the pole lie in the same vertical plane and A does not slip across the ground. Find the horizontal and vertical components of the reaction at the ground when the rope is at right angles to the pole and the pole is at 30° to the horizontal.
16. A uniform beam AB of length $2l$ rests with end A in contact with rough horizontal ground. A point C on the beam rests against a smooth support. AC is of length $\frac{3l}{2}$ with C higher than A, and AC makes an angle of 60° with the horizontal. If the beam is in limiting equilibrium, find the coefficient of friction between the beam and the ground.
17. A uniform ladder of weight W and length $2l$ rests with one end on a smooth horizontal floor and the other end against a smooth vertical wall. The ladder is held in this position by a light, horizontal, inextensible string of length l, which has one end attached to the bottom of the ladder and the other end fastened to a point at the base of the wall, vertically below the top of the ladder. Show that the tension in the string is $\frac{W}{2\sqrt{3}}$.
18. A uniform ladder of mass 8 kg rests in equilibrium with its base on a smooth horizontal floor and its top against a smooth vertical wall. The base of the ladder is 1 m from the wall and the top of the ladder is 2 m from the floor. The ladder is kept in equilibrium by a light string attached to the base of the ladder and to a point on the wall, vertically below the top of the ladder and 1 m above the floor. Find the tension in the string.
19. A uniform ladder of mass 25 kg rests in equilibrium with its base on a rough horizontal floor and its top against a smooth vertical wall. If the ladder makes an angle of 75° with the horizontal, find the magnitude of the normal reaction and of the frictional force at the floor, and state the minimum possible value of the coefficient of friction μ between the ladder and the floor.
20. A uniform ladder of mass 15 kg rests with its foot on a rough horizontal floor (angle of friction 15°) and its top against a smooth vertical wall. Find the minimum horizontal force that must be applied to the foot of the ladder to keep the ladder in equilibrium inclined at 60° to the horizontal.
21. A uniform ladder rests in limiting equilibrium with its base on rough horizontal ground (coefficient of friction μ) and its top against a rough vertical wall (coefficient of friction $\frac{1}{4}$). If the ladder is inclined at 30° to the vertical, find the value of μ.
22. A non-uniform ladder AB of length 6 m has its centre of gravity at a point C on the ladder such that AC = 4 m. The ladder rests in limiting equilibrium with end A on rough horizontal ground (coefficient of friction $\frac{1}{3}$) and end B against a rough vertical wall (coefficient of friction $\frac{1}{4}$). If the ladder makes an acute angle θ with the ground, show that $\tan\theta = \frac{23}{12}$.

23. A non-uniform ladder AB of length 10 m has its centre of gravity at a point C. The ladder rests in limiting equilibrium with end A on a rough horizontal floor (angle of friction 17°) and end B against a smooth vertical wall. If the ladder is inclined at an angle of 63° to the floor, find the length AC.

24. Each of the following diagrams shows a uniform rod AB of weight W, with end A freely hinged to a vertical wall. The rod is in equilibrium under the forces shown.

(a)

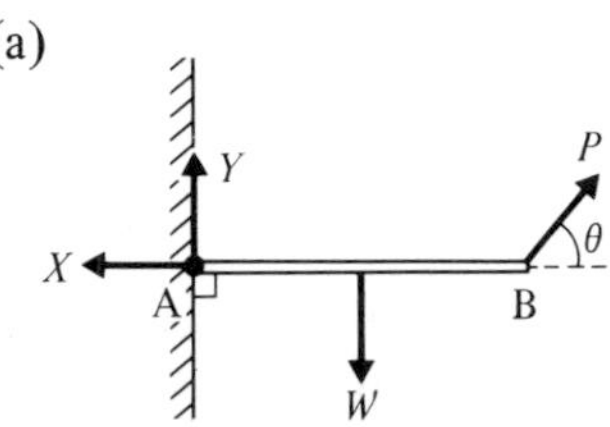

Prove: $X \tan \theta = \dfrac{W}{2}$

(b)

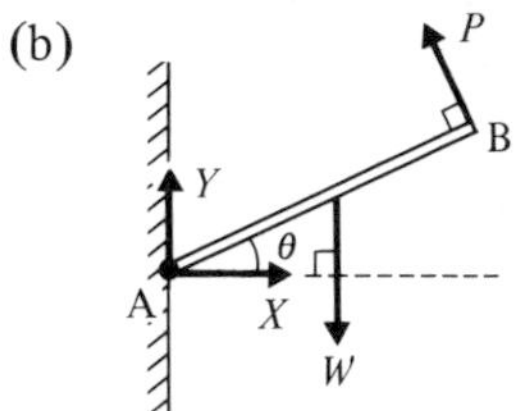

Prove: $2Y = W(1 + \sin^2 \theta)$

(c)

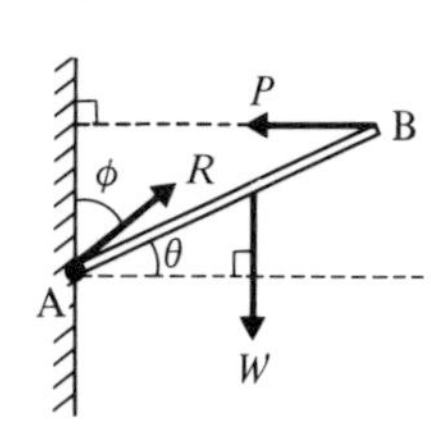

Prove: $\tan \theta \tan \phi = \frac{1}{2}$

25. Each of the following diagrams shows a uniform ladder AB of weight W, with its lower end on a horizontal floor and its top against a smooth vertical wall. The ladder is in equilibrium under the forces shown.

(a)

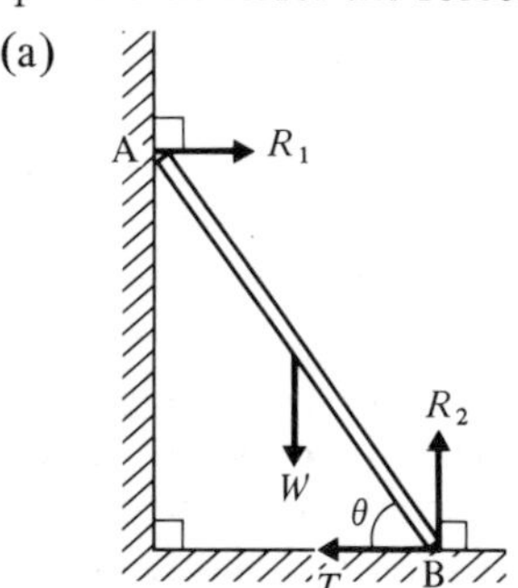

Prove: $2T \tan \theta = W$

(b)

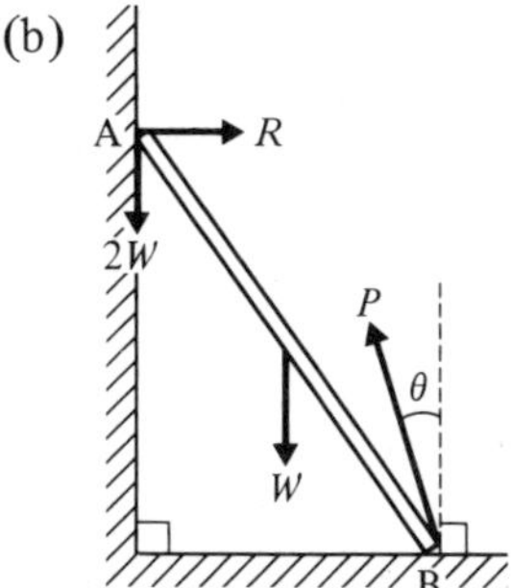

Prove: $R = 3W \tan \theta$

(c)

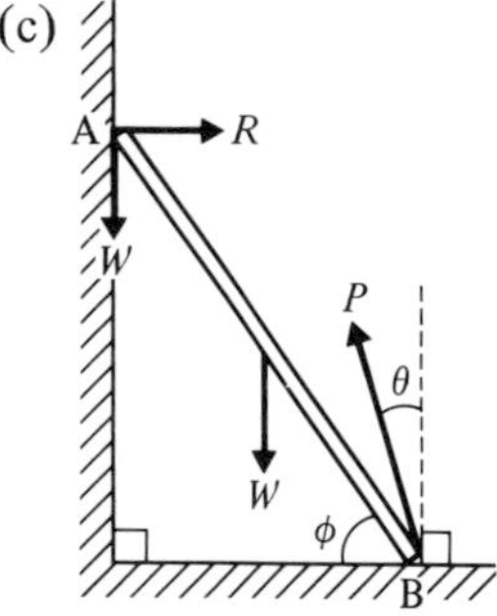

Prove: (a) $R = 2W \tan \theta$
(b) $\tan \theta \tan \phi = \frac{3}{4}$

26. A uniform ladder rests in limiting equilibrium with its top end against a smooth vertical wall and its base on a rough horizontal floor (coefficient of friction μ). If the ladder makes an angle of θ with the floor, prove that $2\mu \tan \theta = 1$.

27. A uniform ladder rests in limiting equilibrium with its top end against a rough vertical wall (coefficient of friction μ_1), and its base on a rough horizontal floor (coefficient of friction μ_2). If the ladder makes an angle of θ with the floor, prove that $\tan \theta = \dfrac{1 - \mu_1 \mu_2}{2\mu_2}$.

28. The diagram shows a uniform ladder resting in equilibrium with its top end against a smooth vertical wall and its base on a smooth inclined plane. The plane makes an angle of θ with the horizontal and the ladder makes an angle of ϕ with the wall. Prove that $\tan \phi = 2 \tan \theta$.

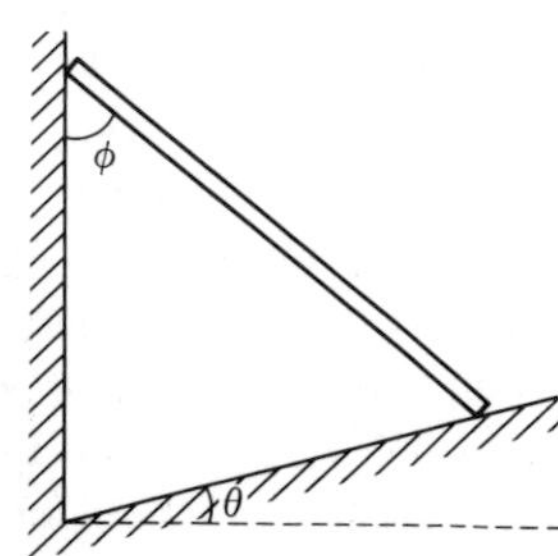

29. A uniform beam AB is supported at an angle θ to the horizontal by a light string attached to end B, and with end A resting on rough horizontal ground (angle of friction λ). The beam and the string lie in the same vertical plane and the beam rests in limiting equilibrium with the string at right angles to the beam. Prove that $\tan \lambda = \dfrac{\sin 2\theta}{3 - \cos 2\theta}$.
30. A uniform ladder of mass 30 kg is placed with its base on a rough horizontal floor (coefficient of friction $\frac{1}{4}$), and its top against a smooth vertical wall, with the ladder making an angle of 60° with the floor. Find the magnitude of the minimum horizontal force that must be applied at the base of the ladder in order to prevent slipping. What is the maximum horizontal force that could be applied at the base without slipping occurring?
31. A uniform ladder of mass 25 kg is placed with its base on a rough horizontal floor (coefficient of friction $\frac{1}{5}$), and its top against a rough vertical wall (coefficient of friction $\frac{1}{3}$), with the ladder making an angle of 61° with the floor. Find the magnitude of the minimum horizontal force that must be applied at the base of the ladder in order to prevent slipping. What is the maximum horizontal force that could be applied at the base without slipping occurring?
32. A uniform ladder of mass 10 kg and length 4 m rests with one end on a smooth horizontal floor and the other end against a smooth vertical wall. The ladder is kept in equilibrium, at an angle $\tan^{-1} 2$ to the horizontal, by a light horizontal string attached to the base of the ladder and to the base of the wall, at a point vertically below the top of the ladder. A man of mass 100 kg ascends the ladder. If the string will break when the tension exceeds 490 N, find how far up the ladder the man can go before this occurs. What tension must the string be capable of withstanding if the man is to reach the top of the ladder?
33. A uniform ladder of mass 30 kg and length 10 m has its base resting on rough horizontal ground and its top against a smooth vertical wall. The ladder rests in equilibrium, at 60° to the horizontal, with a man of mass 90 kg standing on the ladder at a point 7·5 m from its base. Find the magnitude of the normal reaction and of the frictional force at the ground. Find the minimum value for the coefficient of friction between the ladder and the ground that would enable the man to climb to the top of the ladder.
34. A uniform ladder of length 10 metres and weight W N rests with its base on a rough horizontal floor (coefficient of friction $\frac{1}{3}$), and its top against a smooth vertical wall. The ladder makes an angle θ with the horizontal, where $\tan \theta = 1{\cdot}7$. A man of weight $2W$ N starts to climb the ladder. How far up the ladder can the man climb before slipping occurs? Find, in terms of W, the magnitude of the least horizontal force that must be applied to the base of the ladder to enable the man to reach the top safely.
35. A uniform ladder AB is of weight $2W$ N and length 10 metres. It rests with end A on a rough horizontal floor and end B against a rough vertical wall. The coefficient of friction at the wall and at the floor is $\frac{1}{3}$ and the ladder makes an angle θ with the horizontal, such that $\tan \theta = \frac{16}{7}$. A man of weight $5W$ N starts to climb the ladder. How far up the ladder can the man climb before slipping occurs? When a boy of weight X N stands on the bottom rung of the ladder, i.e. at A, the man is just able to climb to the top safely. Find X in terms of W.
36. A non-uniform ladder AB of length 12 m and mass 30 kg has its centre of gravity at the point of trisection of its length, nearer to A. The ladder rests with end A on rough horizontal ground (coefficient of friction $\frac{1}{4}$), and end B against a rough vertical wall (coefficient of friction $\frac{1}{5}$). The ladder makes an angle θ with the horizontal such that

$\tan\theta = \frac{9}{4}$. A straight horizontal string connects A to a point at the base of the wall, vertically below B. A man of mass 90 kg begins to climb the ladder. How far up the ladder can he go without causing tension in the string?
What tension must the string be capable of withstanding if the man is to reach the top of the ladder safely?

Exercise 9B Examination questions

1. A uniform ladder rests with one end against a smooth vertical wall. The other end rests on rough horizontal ground and the coefficient of friction between ground and ladder is $\frac{3}{4}$. A man whose mass is equal to that of the ladder can just ascend to the top without the ladder slipping. Show that the inclination of the ladder to the vertical is $\frac{1}{4}\pi$. (Oxford)

2. A uniform rod AB of length 20 m and mass 5 kg is smoothly hinged to a fixed point A and is held inclined at 30° to the horizontal, with B above A, by a light inextensible string attached to the mid-point of the rod and to a point C vertically above A, where AC = 10 m. Find the magnitude and direction of the reaction at A. Find also the tension in the string. If a particle of mass 5 kg is now attached to the rod at B, show that the tension in the string is trebled. (S.U.J.B)

3. The diagram shows a uniform bar AB of length 36 cm and of mass 10 kg, freely hinged at A, resting on a smooth support at C at an angle of 45° with the vertical. AC = 21 cm. A body of mass 5 kg is suspended from a point D on the bar, where BD = 9 cm. Calculate the magnitude of the reaction at C.
Show that the line of action of the force at A lies along the bar and calculate its magnitude. (Cambridge)

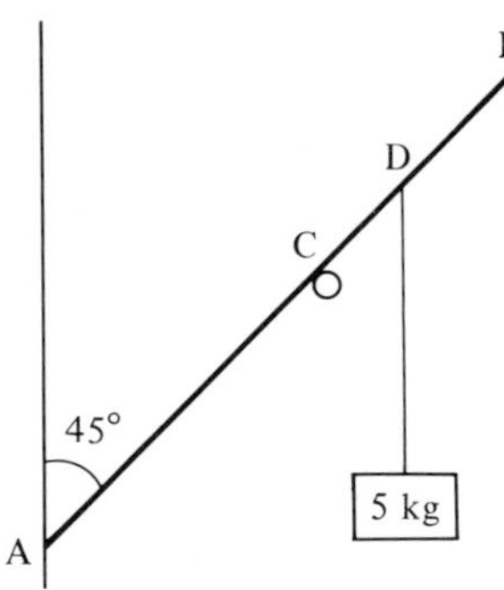

4. A uniform ladder of length $2a$ rests in limiting equilibrium in a vertical plane with its lower end on rough horizontal ground and its upper end against a smooth vertical wall. The ladder makes an angle 60° with the ground. Show that the coefficient of friction is $\sqrt{3}/6$.
The ladder is lowered in its vertical plane whilst still resting against the smooth wall and the ground to make an angle 30° with the ground. The coefficient of friction between the ladder and the ground remains at $\sqrt{3}/6$. A man whose weight is four times that of the ladder starts climbing up the ladder. Find how far he can climb up the ladder before it slips. (London)

5. Explain, with the aid of a diagram, the meaning of the expression *coefficient of friction.*
A straight uniform rod AB of weight W rests in limiting equilibrium with the end A on horizontal ground and the end B against a vertical wall. The vertical plane containing AB is perpendicular to the wall. The coefficient of friction between the rod and the ground is $\frac{4}{5}$. The coefficient of friction between the rod and the wall is $\frac{3}{5}$. Given that the inclination of AB to the horizontal is α, calculate (i) the normal reactions at A and B in terms of W, (ii) the numerical value of $\tan\alpha$. (A.E.B)

6. A uniform straight rod AB, of length $4a$ and weight W, is inclined at an angle α to the horizontal, where $\tan \alpha = \frac{1}{2}$, with the end A standing on rough horizontal ground. A point C of the rod, where $AC = 3a$, rests against a fixed smooth horizontal rail and AB is perpendicular to the rail. Given that the rod is in equilibrium, find, in terms of W, the frictional force and the normal reaction acting on the rod at A. If μ is the coefficient of friction between the rod and the ground, deduce that $\mu \geqslant \frac{4}{7}$.
Given that $\mu = \frac{3}{4}$, find, in terms of W, the weight of the heaviest particle which can be attached at B without disturbing the equilibrium. (A.E.B)

7. A uniform smooth rod of mass M and length $2l$ rests partly inside and partly outside a fixed smooth hemispherical bowl of radius a. The rim of the bowl is horizontal and one point of the rod is in contact with the rim. Prove that the inclination θ of the rod to the horizontal is given by $2a \cos 2\theta = l \cos \theta$. Find the reaction between the rod and the bowl at the rim. (Oxford)

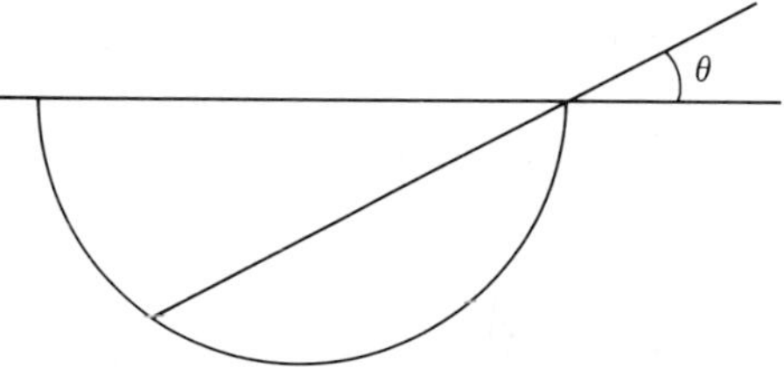

10

Resultant velocity and relative velocity

Resultant velocity

Velocity is a vector quantity since it has both magnitude and direction. Two velocities can therefore be combined by the same method as used for forces in chapter 4.

Example 1

Find, in vector form, the resultant of the following velocities: $(4\mathbf{i} - 2\mathbf{j})$ m/s, $(-7\mathbf{i} + 5\mathbf{j})$ m/s and $(8\mathbf{i} - 6\mathbf{j})$ m/s.

$$\begin{aligned}\text{Resultant velocity} &= (4\mathbf{i} - 2\mathbf{j}) + (-7\mathbf{i} + 5\mathbf{j}) + (8\mathbf{i} - 6\mathbf{j}) \\ &= (5\mathbf{i} - 3\mathbf{j}) \text{ m/s}\end{aligned}$$

The resultant velocity is $(5\mathbf{i} - 3\mathbf{j})$ m/s.

The following example shows how the magnitude and direction of the resultant velocity may be found by a scale drawing.

Example 2

Find, by scale drawing, the magnitude and direction of the resultant of the velocities 16 m/s due east and 10 m/s in a direction N 38° E.
Draw a rough sketch showing the given velocities.

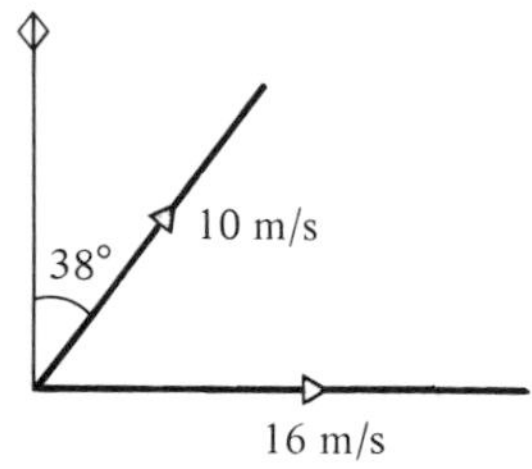

Using a scale of 1 cm ≡ 2 m/s, construct a vector triangle ABC for the given velocities

AB = 8 cm BC = 5 cm
angle ABC = 128°

The resultant velocity is represented by $\overrightarrow{AC}$.

By measurement:

AC = 11·75 cm
angle α = 20°

The resultant velocity is 23·5 m/s in a direction N 70° E.

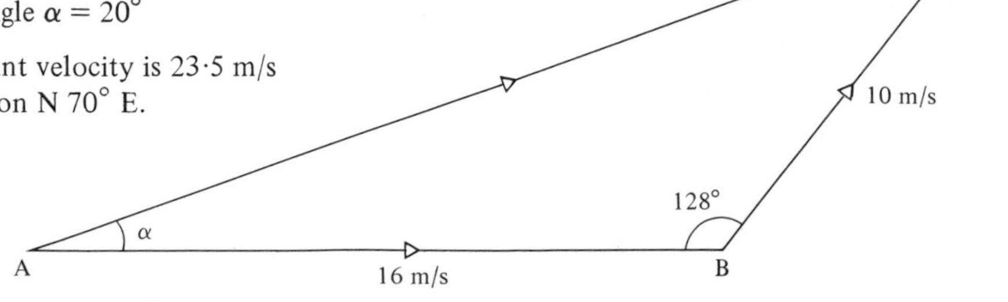

Example 3

Calculate the magnitude and the direction of the resultant of the velocities 8 km/h in a direction N 80° W and 5 km/h in a direction S 25° W.

Draw a rough sketch showing the given velocities

angle ABC $= 105°$. Let $\hat{BAC} = \alpha$

The resultant velocity is represented by $\vec{AC}$.

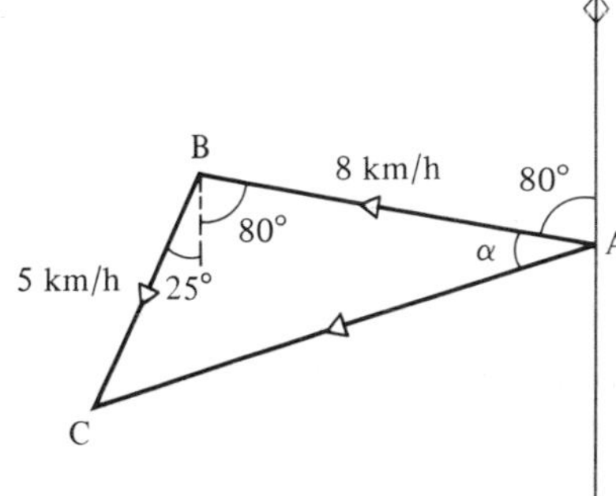

From triangle ABC, using the cosine rule:

$$AC^2 = 8^2 + 5^2 - 2(5)(8) \cos 105$$
$$= 89 + 80 \cos 75$$
$$\therefore \quad AC = 10{\cdot}47$$

From triangle ABC, using the sine rule:

$$\frac{AC}{\sin 105} = \frac{5}{\sin \alpha}$$

substituting for AC:

$$\sin \alpha = \frac{5 \sin 105}{10{\cdot}47}$$
$$\therefore \quad \alpha = 27{\cdot}47° \text{ or } 152{\cdot}53° \text{ (The obtuse angle is not applicable)}$$

The resultant velocity is 10·5 km/h in a direction S 72·53° W.

Components of velocity

It is sometimes useful to consider the components of the velocity of a body, particularly if the motion of the body is the result of the combination of two velocities.

Crossing a river by boat

Consider the problem of crossing from a point on one bank of a river to a point on the other bank. There are three cases to consider.

(i) In order to cross from a point O on one bank to a point P directly opposite to O on the other bank, the course set by the boat must be upstream. If the speed of the boat in still water is v and the speed of the current is u, then the component of v upstream must counteract u.

$$\therefore \quad v \sin \theta = u$$

The speed across the river is then $v \cos \theta$ and the crossing is made from O to P.

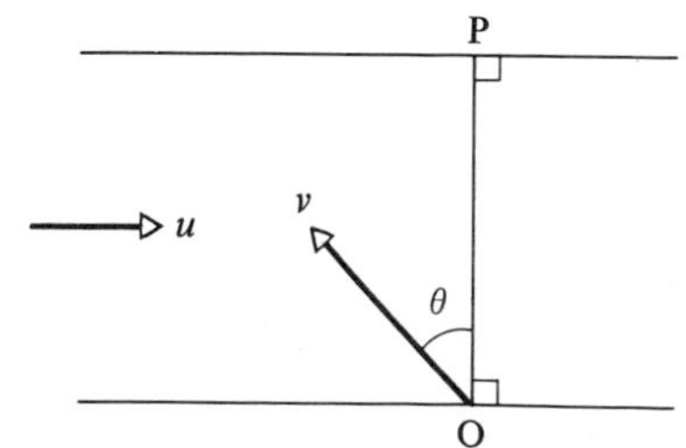

(ii) If the course set is directly across the river, then the current will carry the boat downstream. The boat has two velocities, **v** the velocity in still water, and **u** the velocity of the current downstream. The resultant velocity **V** of the boat can be found from the vector triangle. If the magnitude of **V** is written as V, then:

$$V^2 = v^2 + u^2$$

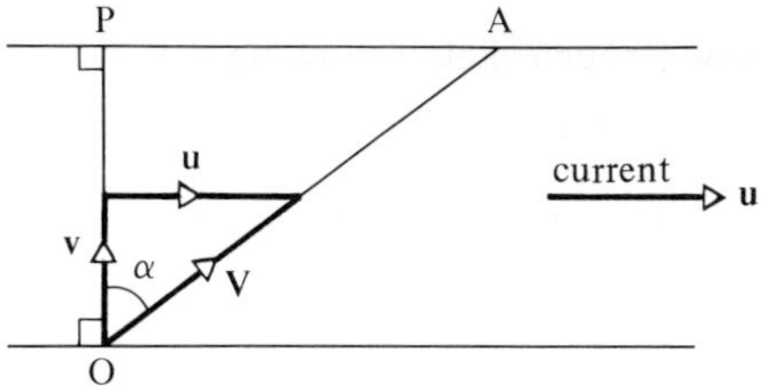

and the boat will travel at an angle α to the line OP where:

$$\tan \alpha = \frac{u}{v}$$

The time to cross the river is $t = \dfrac{\text{OP}}{v}$ and the boat is carried downstream a distance PA, where PA $= u \times t$. This will be the quickest crossing.

(iii) In order to cross the river and reach a point B on the other bank, the course set must be in such a direction that the resultant velocity of the boat is in the direction OB.

Suppose B is upstream.

From the diagram

$$\mathbf{V} = \mathbf{v} + \mathbf{u}$$

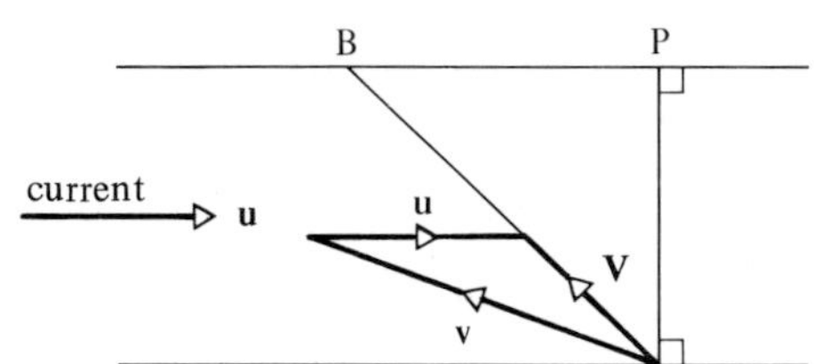

Example 4

A boat can travel at 3·5 m/s in still water. A river is 80 m wide and the current flows at 2 m/s. Calculate (a) the shortest time taken to cross the river and the distance downstream that the boat is carried, (b) the course that must be set to cross the river to a point exactly opposite the starting point and the time taken for the crossing.

(a) To cross in the shortest time, the course set is directly across the river

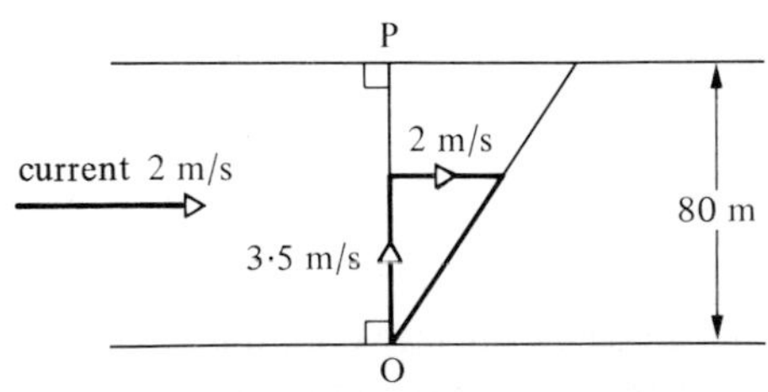

$$\text{time to cross} = \frac{\text{OP}}{3{\cdot}5}$$

$$= \frac{80}{3{\cdot}5}$$

$$= 22{\cdot}86 \text{ s}$$

$$\text{distance downstream} = 2 \times 22{\cdot}86 = 45{\cdot}72 \text{ m}$$

The time for the quickest crossing is 22·9 s and the distance downstream is 45·7 m.

(b) To cross the river directly

$$3{\cdot}5 \sin\theta = 2$$
$$\therefore \quad \sin\theta = \tfrac{4}{7}$$
$$\text{i.e.} \quad \theta = 34{\cdot}85°$$

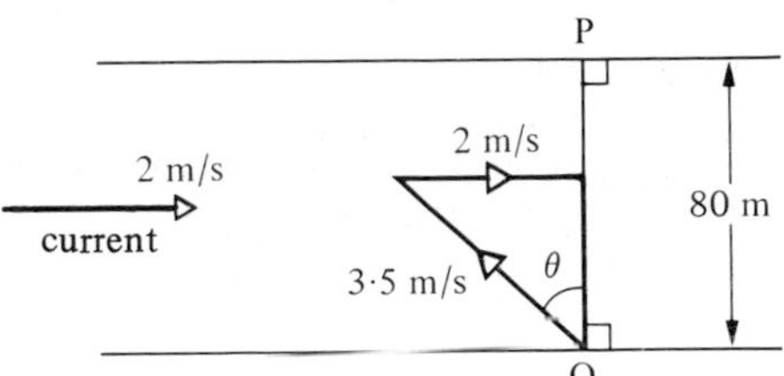

Speed across the river is $3{\cdot}5 \cos\theta$

$$\text{time to cross} = \frac{80}{3{\cdot}5\cos\theta}$$
$$= \frac{80}{3{\cdot}5\cos 34{\cdot}85°} = 27{\cdot}85 \text{ s}$$

The time taken to cross the river directly is 27·9 s and the course to be set is upstream at an angle 55·15° to the bank of the river.

Example 5

A pilot has to fly his aircraft from the point A to the point B, where B is 600 km due east of A. There is a wind of 80 km/h blowing from the north-west and the aircraft flies at 350 km/h in still air. Find the course which the pilot must set and the time taken for the flight.

This is similar to the oarsman crossing a river to a previously determined point on the opposite bank.

Draw a rough sketch in which $\overrightarrow{AC}$ represents the velocity of the wind and $\overrightarrow{CD}$ represents the velocity of the aircraft in still air.
The resultant $\overrightarrow{AD}$ of these two velocities must lie along the line AB since this is the line along which the aircraft is to travel.
The line CD gives the direction of the course to be set by the pilot.

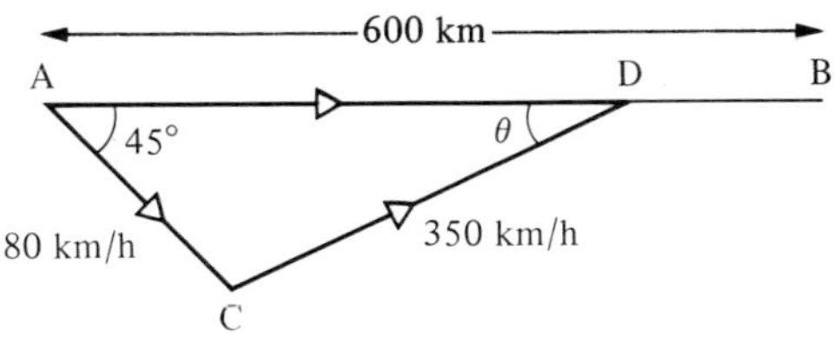

From triangle ACD, using the sine rule:

$$\frac{350}{\sin 45} = \frac{80}{\sin\theta}$$
$$\therefore \quad \sin\theta = \frac{80 \sin 45}{350}$$
$$\therefore \quad \theta = 9{\cdot}30° \text{ or } 170{\cdot}70° \text{ (The obtuse angle is not applicable)}$$
$$\therefore \quad \text{angle ACD} = 180° - 45° - 9{\cdot}30°$$
$$= 125{\cdot}70°$$

From triangle ACD, using the sine rule:

$$\frac{\text{AD}}{\sin 125{\cdot}70°} = \frac{350}{\sin 45}$$
$$\therefore \quad \text{AD} = \frac{350 \sin 125{\cdot}70°}{\sin 45}$$
$$\therefore \quad \text{AD} = 402{\cdot}0$$

$$\text{time of flight} = \tfrac{600}{402} = 1{\cdot}493 \text{ h}$$

The course to be set is 080·70° and the time taken for the flight is 1·49 h.

Exercise 10A

1. Find in vector form the resultant of each of the following sets of velocities.
   (a) $(5\mathbf{i} + 2\mathbf{j})$ m/s, $(4\mathbf{i} - 3\mathbf{j})$ m/s,
   (b) $(6\mathbf{i} + 2\mathbf{j})$ m/s, $(2\mathbf{i} + 3\mathbf{j})$ m/s, $(-\mathbf{i} + 4\mathbf{j})$ m/s,
   (c) $(2\mathbf{i} - 5\mathbf{j})$ m/s, $(3\mathbf{i} + 7\mathbf{j})$ m/s, $(-6\mathbf{i} - 8\mathbf{j})$ m/s,
   (d) $(18\mathbf{i} + 9\mathbf{j})$ km/h, $(10\mathbf{i} + 5\mathbf{j})$ m/s,
   (e) $(15\mathbf{i} - 5\mathbf{j})$ m/s, $(15\mathbf{i} + 3\mathbf{j})$ km/h, $(-6\mathbf{i} + 15\mathbf{j})$ km/h.
2. Find by scale drawing the magnitude and direction of the resultant of each of the following pairs of velocities.
   (a) 24 m/s due north, 7 m/s due east,
   (b) 5 km/h due north, 5 km/h N 60° E,
   (c) 5 m/s due north, 7 m/s S 60° E,
   (d) 10 m/s N 30° E, 8 m/s N 70° E,
   (e) 70 km/h S 35° E, 90 km/h N 25° E.
3. Find by calculation the magnitude and direction of the resultant of each of the following pairs of velocities.
   (a) 9 m/s due west, 12 m/s due north,
   (b) 6 m/s due east, 4 m/s NW,
   (c) 17 km/h due north, 15 km/h N 26° E
   (d) 10 km/h N 35° W, 15 km/h S 40° W,
   (e) 72 km/h N65° E, 20 m/s SE.
4. If the resultant of $(3\mathbf{i} + 4\mathbf{j})$ m/s and $(a\mathbf{i} + b\mathbf{j})$ m/s is $(7\mathbf{i} - \mathbf{j})$ m/s, find the values of a and b.
5. If the resultant of $(a\mathbf{i} + b\mathbf{j})$ km/h and $(b\mathbf{i} - a\mathbf{j})$ km/h is $(10\mathbf{i} - 4\mathbf{j})$ km/h, find the values of a and b.
6. The resultant of two velocities is a velocity of 10 km/h, N 30° W. If one of the velocities is 10 km/h due west, find the magnitude and direction of the other velocity.
7. The resultant of two velocities is a velocity of 6 m/s due east. If one of the velocities is 5 m/s, N 30° W, find the magnitude and direction of the other velocity.
8. The resultant of two velocities is a velocity of 19 m/s, S 60° E. If one of the velocities is 10 m/s due east, find the magnitude and direction of the other velocity.
9. A man wishes to row across a river to reach a point on the far bank, exactly opposite his starting point. The river is 100 m wide and flows at 3 m/s. In still water the man can row at 5 m/s. Find at what angle to the bank the man must steer the boat in order to complete the crossing, and the time it takes him.
10. A man wishes to row across a river to reach a point on the far bank, exactly opposite his starting point. The river is 125 m wide and flows at 1 m/s. If the man can row at 3 m/s in still water, find the direction the man must steer in order to complete the crossing, and the time it takes him.
11. A boy wishes to swim across a river, 100 m wide, as quickly as possible. The river flows at 3 km/h and the boy can swim at 4 km/h in still water. Find the time that it takes the boy to cross the river and how far downstream he travels.
12. A man who can swim at 2 m/s in still water, wishes to swim across a river, 120 m wide, as quickly as possible. If the river flows at 0·5 m/s, find the time the man takes for the crossing and how far downstream he travels.
13. A pilot has to fly his aircraft from airport A to airport B, 100 km due east of A. In still air the aircraft flies at 125 km/h. If there is a wind of 35 km/h blowing from the north, find the course that the pilot must set in order to reach B and the time the journey takes.

14. Two airfields A and B are 500 km apart with B on a bearing 060° from A. An aircraft which can travel at 200 km/h in still air, is to be flown from A to B. If there is a wind of 40 km/h blowing from the west, find the course that the pilot must set in order to reach B and find, to the nearest minute, the time taken.
15. An aircraft capable of flying at 250 km/h in still air, is to be flown from airport A to airport B, situated 300 km from A on a bearing 320°. If there is a wind of 50 km/h blowing from 030°, find the course the pilot must set and find, to the nearest minute, the time taken for the journey.
16. A man swims at 5 km/h in still water. Find the time it takes the man to swim across a river 250 m wide, flowing at 3 km/h, if he swims so as to cross the river
    (a) by the shortest route, (b) in the quickest time.
17. A man wishes to row a boat across a river to reach a point on the far bank that is 35 m downstream from his starting point. The man can row the boat at 2·5 m/s in still water. If the river is 50 m wide and flows at 3 m/s, find the two possible courses the man could set and find the respective crossing times.
18. Airfield A is 500 km due south of airfield B. A pilot, wishing to fly his aircraft from A to B, is told that there is a wind of 50 km/h blowing from N 60° E. In still air the aircraft flies at 300 km/h. What course should the pilot set in order to reach B and how long will the flight take?
    Assuming the wind does not change, how long would the return flight take?
19. Two heliports A and B are 150 km apart with B on a bearing 045° from A. A wind of 30 km/h is blowing from a direction 260°. Assuming this wind remains constant throughout, find the time required for a helicopter to fly from A to B and back to A again, if the helicopter can fly at 100 km/h in still air.
20. When swimming in a river a man finds that he has a maximum speed v when swimming downstream and u when swimming upstream.
    (a) Find an expression for his maximum speed when swimming in still water.
    (b) If the river is of width s, show that the shortest time in which the man can swim across is $\frac{2s}{v+u}$ and that such a crossing would take him a distance of $\frac{s(v-u)}{v+u}$ downstream from his starting point.
    (c) If the man wishes to swim as quickly as possible from a point on one bank to a point exactly opposite on the other bank, show that he must swim in a direction that makes an angle $\cos^{-1}\left(\frac{v-u}{v+u}\right)$ with the bank and that the crossing will take a time $\frac{s}{\sqrt{(uv)}}$.

Relative velocity

Suppose A and B are two moving bodies. The velocity of A with respect to B is the velocity of A as it appears to an observer on B and this is usually denoted by ${}_{A}\mathbf{v}_{B}$. The simplest case to consider is when A and B are moving along parallel lines.

Example 6

A man M on a train travelling due west at 25 m/s notices a second train passing him on a parallel track at 40 m/s. Calculate the velocity of the second train S relative to the first train if the directions of motion are (a) the same, (b) opposite.

(a) Draw a diagram showing the velocities.

(a) Velocity of S as it appears to M

$${}_{S}\mathbf{v}_{M} = 40 - 25$$
$$= 15 \text{ m/s due west}$$

S appears to be travelling more slowly than it is doing.

M 25 m/s
S 40 m/s

(b) Velocity of S as it appears to M

$${}_{S}\mathbf{v}_{M} = 40 + 25$$
$$= 65 \text{ m/s due east}$$

S appears to be travelling more quickly than it is doing.

M 25 m/s
S 40 m/s

Note that the observer is only concerned with the motion of the other train *relative* to himself, i.e. as though he were not moving.

Non-parallel courses

The previous example suggests the general method. Since the observer takes no account of his own motion, the velocity of A relative to B is found by reducing the observer on B to rest.

Suppose bodies A and B have velocities $\mathbf{v}_A$ and $\mathbf{v}_B$ respectively, as shown in Fig. 1. To find the velocity of A relative to B, i.e. ${}_{A}\mathbf{v}_{B}$, B must be reduced to rest. To do this, B is given a velocity equal and opposite to its own velocity. The same velocity must be applied to A, as shown in Fig. 2. The resultant velocity of A will then be the velocity of A relative to B. Thus, from Fig. 2 we have the vector equation

$${}_{A}\mathbf{v}_{B} = \mathbf{v}_A - \mathbf{v}_B \qquad \ldots [1]$$

If the velocities $\mathbf{v}_A$ and $\mathbf{v}_B$ are given in vector form, ${}_{A}\mathbf{v}_{B}$ can be found very easily from equation [1].

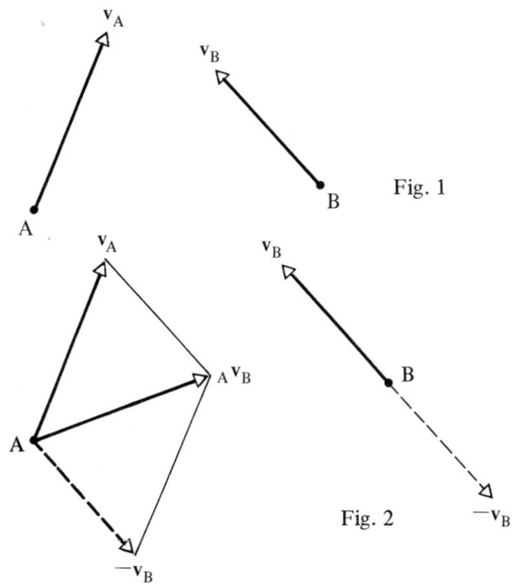

Fig. 1

Fig. 2

If $\mathbf{v}_A$ and $\mathbf{v}_B$ are not given in vector form, then the vector equation [1] can be thought of as ${}_{A}\mathbf{v}_{B} = \mathbf{v}_A + (-\mathbf{v}_B)$ and the corresponding vector triangle can be drawn. The magnitude and the direction of the relative velocity can be found in the usual way either by calculation, or by scale drawing.

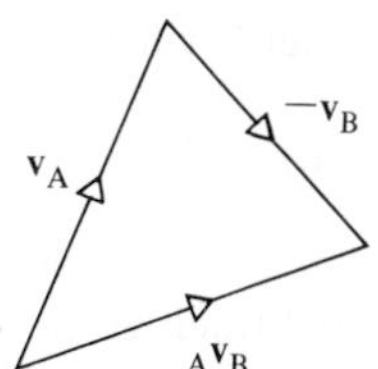

Example 7

Particle A has a velocity of $(3\mathbf{i} + 7\mathbf{j})$ m/s and particle B has a velocity of $(5\mathbf{i} + 2\mathbf{j})$ m/s. Find the velocity of A relative to B.

Given $\quad \mathbf{v}_A = (3\mathbf{i} + 7\mathbf{j})$ m/s and $\mathbf{v}_B = (5\mathbf{i} + 2\mathbf{j})$ m/s

Required $\quad {}_A\mathbf{v}_B$

$$\begin{aligned} {}_A\mathbf{v}_B &= \mathbf{v}_A - \mathbf{v}_B \\ &= (3\mathbf{i} + 7\mathbf{j}) - (5\mathbf{i} + 2\mathbf{j}) \\ \therefore \quad {}_A\mathbf{v}_B &= (-2\mathbf{i} + 5\mathbf{j}) \end{aligned}$$

The velocity of A relative to B is $(-2\mathbf{i} + 5\mathbf{j})$ m/s.

Example 8

A girl walks at 5 km/h due west and a boy runs at 12 km/h towards the south-east. Find the velocity of the boy relative to the girl.

Given $\quad \mathbf{v}_G$: ◁—— 5 km/h —— and $\mathbf{v}_B$:

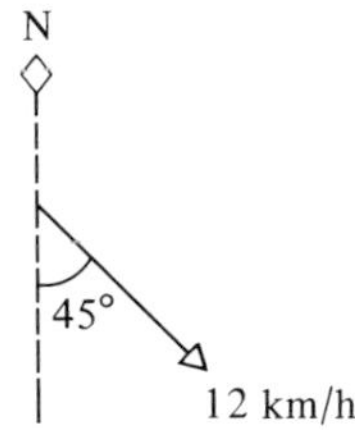

Required $\quad {}_B\mathbf{v}_G$

Using ${}_B\mathbf{v}_G = \mathbf{v}_B - \mathbf{v}_G$ in the form ${}_B\mathbf{v}_G = \mathbf{v}_B + (-\mathbf{v}_G)$, draw the vector triangle ABC.

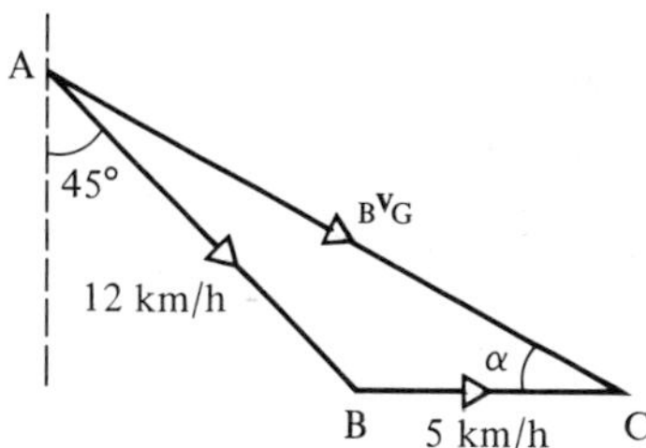

By calculation from triangle ABC, using the cosine rule

$$\begin{aligned} AC^2 &= 12^2 + 5^2 - 2(12)(5)\cos 135 \\ &= 169 + 120 \cos 45 \\ \therefore \quad AC &= 15{\cdot}93 \end{aligned}$$

From triangle ABC using the sine rule

$$\frac{12}{\sin\alpha} = \frac{15{\cdot}93}{\sin 135}$$

$$\therefore \quad \sin\alpha = \frac{12 \sin 135}{15{\cdot}93}$$

$$\therefore \quad \alpha = 32{\cdot}19^\circ$$

The velocity of the boy relative to the girl is 15·9 km/h in a direction S 57·81° E.

Measurement of an accurate scale drawing of triangle ABC could be used instead to find the magnitude and direction of the velocity of the boy relative to the girl.

True velocity

Suppose the true velocity of body A is known and the velocity of body B relative to A is also known, then the true velocity of B can be found.

Example 9

To the captain of a ship S travelling with velocity $(12\mathbf{i} - 15\mathbf{j})$ km/h, a second ship T appears to have a velocity of $(10\mathbf{i} + 20\mathbf{j})$ km/h. Find the true velocity of T.

Given $\quad \mathbf{v}_S = (12\mathbf{i} - 15\mathbf{j})$ km/h and ${}_T\mathbf{v}_S = (10\mathbf{i} + 20\mathbf{j})$ km/h

Required $\quad \mathbf{v}_T$

Using
$$ {}_T\mathbf{v}_S = \mathbf{v}_T - \mathbf{v}_S $$
$$ \therefore \quad 10\mathbf{i} + 20\mathbf{j} = \mathbf{v}_T - (12\mathbf{i} - 15\mathbf{j}) $$
$$ \therefore \quad \mathbf{v}_T = 22\mathbf{i} + 5\mathbf{j} $$

The true velocity of T is $(22\mathbf{i} + 5\mathbf{j})$ km/h.

Example 10

To a girl running at 6 m/s on a bearing of 155°, a low flying bird appears to be moving at 7 m/s on a bearing of 250°. Find the true velocity of the bird.

Given $\quad \mathbf{v}_G$:

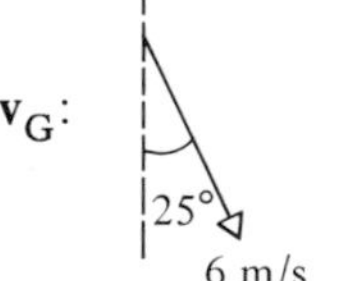

and ${}_B\mathbf{v}_G$:

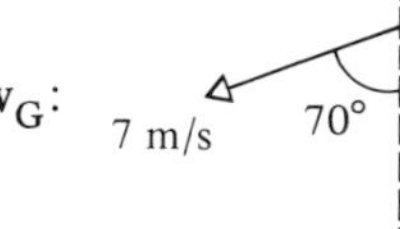

Required $\quad \mathbf{v}_B$

From the relation $\quad {}_B\mathbf{v}_G = \mathbf{v}_B - \mathbf{v}_G \quad$ it follows

that $\quad {}_B\mathbf{v}_G + \mathbf{v}_G = \mathbf{v}_B$

and thus a vector triangle can be drawn:

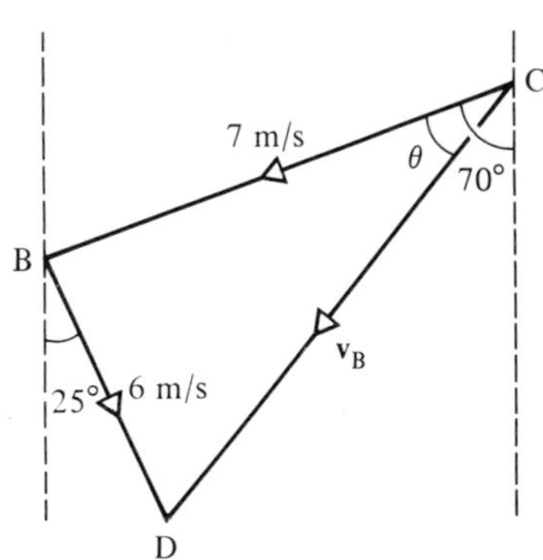

From triangle BCD, using the cosine rule
$$ CD^2 = 7^2 + 6^2 - 2(7)(6)\cos 85 $$
$$ = 85 - 84\cos 85 $$
$$ \therefore \quad CD = 8{\cdot}813 $$

From triangle BCD, using the sine rule
$$ \frac{6}{\sin\theta} = \frac{8{\cdot}813}{\sin 85} $$
$$ \therefore \quad \sin\theta = \frac{6\sin 85}{8{\cdot}813} $$
$$ \therefore \quad \theta = 42{\cdot}70^\circ $$

The true velocity of the bird is 8·81 m/s in a direction S 27·30° W.

Example 11

To a man rowing due south at 4 m/s, a boy in a boat appears to be moving due west. To a woman swimming at $1\frac{1}{2}$ m/s in a direction N 65° W, the boy appears to be moving in a direction S 10° W. Find the true magnitude and direction of the velocity of the boy.

Denoting the man by M, the woman by W and the boy by B:

Given

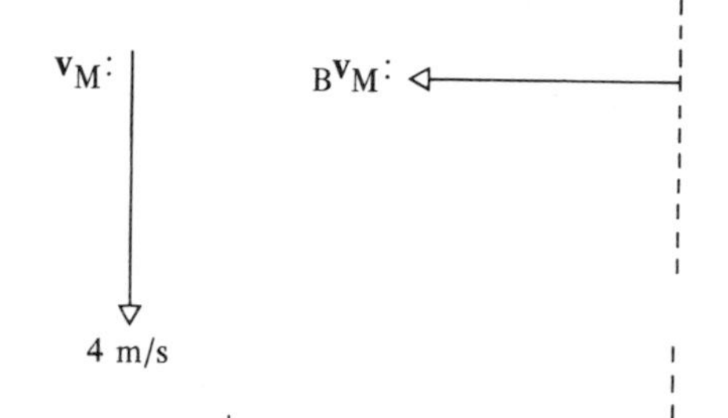

vector triangle

${}_B\mathbf{v}_M = \mathbf{v}_B - \mathbf{v}_M$

${}_B\mathbf{v}_M$

4 m/s

$\mathbf{v}_B$

and

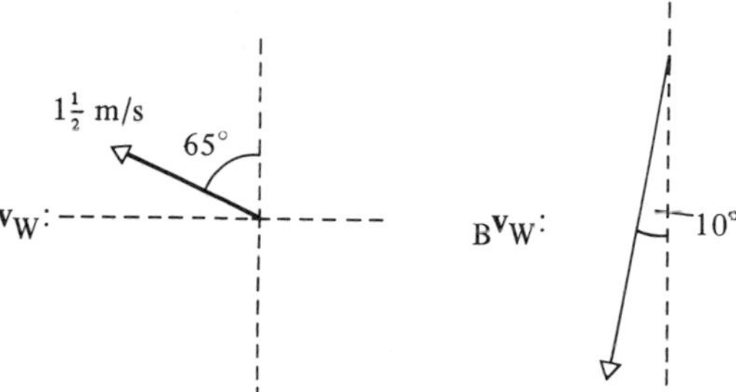

vector triangle

${}_B\mathbf{v}_W = \mathbf{v}_B - \mathbf{v}_W$

$\mathbf{v}_B$

10°

${}_B\mathbf{v}_W$

65°

$1\frac{1}{2}$ m/s

Since the two vector triangles have a common side which represents the true velocity $\mathbf{v}_B$ of the boy, they can be combined in one diagram.

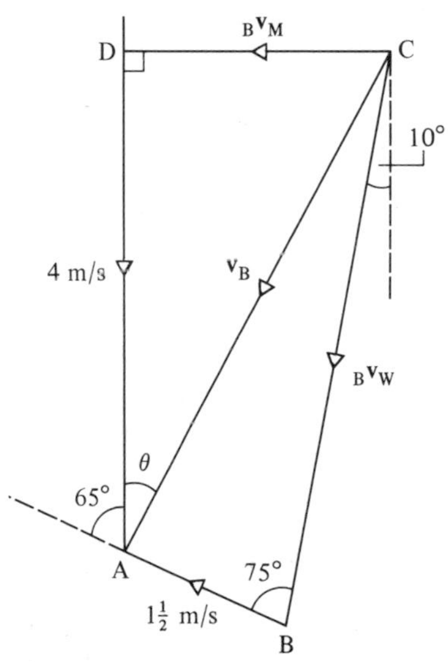

From triangle ACD, right-angled at D

$$\cos\theta = \frac{4}{|\mathbf{v}_B|} \qquad \ldots [1]$$

From triangle ACB, using the sine rule

$$\frac{|\mathbf{v}_B|}{\sin 75} = \frac{1\frac{1}{2}}{\sin(\theta - 10)} \qquad \ldots [2]$$

Substituting for $\mathbf{v}_B$ in equation [2], from [1]

$$4\sin(\theta - 10) = 1\tfrac{1}{2}\sin 75\cos\theta$$

$$\therefore\quad 4\sin\theta\cos 10 = (1\tfrac{1}{2}\sin 75 + 4\sin 10)\cos\theta$$

$$\therefore\quad \tan\theta = \frac{1\frac{1}{2}\sin 75 + 4\sin 10}{4\cos 10}$$

$$\therefore\quad \theta = 28{\cdot}55°$$

and from equation [1]

$$|\mathbf{v}_B| = \frac{4}{\cos 28{\cdot}55°} = 4{\cdot}554 \text{ m/s}$$

The true velocity of the boy is 4·55 m/s in a direction S 28·55° E.

Alternatively, the solutions could be obtained by making an accurate scale drawing of ABCD.

Exercise 10B

1. A cruiser is moving at 30 km/h due north and a battleship is moving at 20 km/h due north. Find the velocity of the cruiser relative to the battleship.
2. Particle A is moving due north at 30 m/s and particle B is moving due south at 20 m/s. Find the velocity of A relative to B.
3. A yacht and a trawler leave a harbour at 8 a.m. The yacht travels due west at 10 km/h and the trawler due east at 20 km/h. What is the velocity of the trawler relative to the yacht? How far apart are the boats at 9.30 a.m.?
4. At 10.30 a.m. a car, travelling at 25 m/s due east, overtakes a motorbike travelling at 10 m/s due east. What is the velocity of the car relative to the motorbike and how far apart are the vehicles at 10.31 a.m.?
5. Particle A has a velocity of $(12\mathbf{i} + 5\mathbf{j})$ m/s and particle B has a velocity of $(4\mathbf{i} + 3\mathbf{j})$ m/s. Find the velocity of A relative to B.
6. A pigeon is flying with velocity $(7\mathbf{i} - \mathbf{j})$ m/s and a sparrow is flying with velocity $(5\mathbf{i} + 6\mathbf{j})$ m/s. Find the velocity of the pigeon relative to the sparrow.
7. A bomber aircraft is moving with velocity $(300\mathbf{i} - 100\mathbf{j})$ km/h and a fighter aircraft is moving with velocity $(400\mathbf{i} + 500\mathbf{j})$ km/h. Find the velocity of the fighter relative to the bomber.
8. Jack rides his horse with velocity $(5\mathbf{i} + 24\mathbf{j})$ km/h whilst Jill is riding her horse with velocity $(5\mathbf{i} + 12\mathbf{j})$ km/h. Find Jack's velocity as seen by Jill. What is Jill's velocity as seen by Jack?
9. Tom walks at 4 km/h due north and Jane walks at 3 km/h due east. Find Tom's velocity relative to Jane.
10. A and B are two yachts. A has a velocity of 8 km/h due south and B has a velocity of 15 km/h due west. Find the velocity of A relative to B.
11. What is the velocity of a cruiser moving at 20 km/h due north as seen by an observer on a liner moving at 15 km/h in a direction N 30° W.
12. A car is being driven at 20 m/s on a bearing 040°. The wind is blowing from 330° with a speed of 10 m/s. Find the velocity of the wind as experienced by the driver of the car.
13. An aircraft is moving at 250 km/h in a direction N 60° E. A second aircraft is moving at 200 km/h in a direction N 20° W. Find the velocity of the first aircraft as seen by the pilot of the second aircraft.
14. Find the velocity of a crow flying at 16 m/s due north as seen by a blackbird flying at 12 m/s due east.
    What is the velocity of the blackbird as seen by the crow?
15. To a man standing on the deck of a ship which is moving with a velocity of $(-6\mathbf{i} + 8\mathbf{j})$ km/h, the wind seems to have a velocity of $(7\mathbf{i} - 5\mathbf{j})$ km/h. Find the true velocity of the wind.
16. To the pilot of a bomber aircraft travelling with velocity $(150\mathbf{i} - 200\mathbf{j})$ km/h, a fighter aircraft appears to have a velocity $(150\mathbf{i} + 440\mathbf{j})$ km/h. Find the true velocity of the fighter.
17. To a cyclist riding at 3 m/s due east, the wind appears to come from the south with speed $3\sqrt{3}$ m/s. Find the true speed and direction of the wind.
18. To the pilot of an aircraft A, travelling at 300 km/h due south, it appears that an aircraft B is travelling at 600 km/h in a direction N 60° W. Find the true speed and direction of aircraft B.
19. Jane is riding her horse at 5 km/h due north and sees Sue riding her horse apparently with velocity 4 km/h, N 60° E. Find Sue's true velocity.
20. To a person walking due east at 3 km/h, the wind appears to come from the north-east at 7 km/h. Find the true velocity of the wind.

21. To the driver of a motorboat moving at 6 km/h on a bearing 345°, a yacht appears to be moving at 18 km/h on a bearing 220°. Find the true velocity of the yacht.
22. A starling, flying at 8 m/s on a bearing 240°, sees a thrush apparently flying at 5 m/s on a bearing 300°. Find the true velocity of the thrush.
23. A train is travelling at 80 km/h in a direction N 15° E. A passenger on the train observes a plane apparently moving at 125 km/h in a direction N 50° E. Find the true velocity of the plane.
24. To a passenger on a boat which is travelling at 20 km/h on a bearing of 230°, the wind seems to be blowing from 250° at 12 km/h. Find the true velocity of the wind.
25. To a jogger jogging at 12 km/h in a direction N 10° E, the wind seems to come from a direction N 20° W at 15 km/h. Find the true velocity of the wind.
26. A, B and C are three aircraft. A has velocity $(200\mathbf{i} + 170\mathbf{j})$ m/s. To the pilot of A it appears that B has velocity $(50\mathbf{i} - 270\mathbf{j})$ m/s. To the pilot of B it appears that C has a velocity $(50\mathbf{i} + 170\mathbf{j})$ m/s. Find, in vector form, the velocities of B and C.
27. To an observer on a liner moving with velocity $(18\mathbf{i} - 17\mathbf{j})$ km/h, a yacht appears to have a velocity $(-8\mathbf{i} + 29\mathbf{j})$ km/h. To someone on the yacht the wind appears to have a velocity $(-5\mathbf{i} - 5\mathbf{j})$ km/h. Find the true velocity of the wind, giving your answer in vector form.
28. When a man cycles due north at 10 km/h, the wind appears to come from the east. When he cycles in a direction N 60° W at 8 km/h it appears to come from the south. Find the true velocity of the wind.
29. To a bird flying due east at 10 m/s, the wind seems to come from the south. When the bird alters its direction of flight to N 30° E without altering its speed, the wind seems to come from the north-west. Find the true velocity of the wind.
30. To an observer on a trawler moving at 12 km/h in a direction S 30° W, the wind appears to come from N 60° W. To an observer on a ferry moving at 15 km/h in a direction S 80° E, the wind appears to come from the north. Find the true velocity of the wind.

Interception and collision

Consider two ships A and B initially at points P and Q. Suppose A is moving with uniform velocity $\mathbf{v}_A$ and B with uniform velocity $\mathbf{v}_B$.

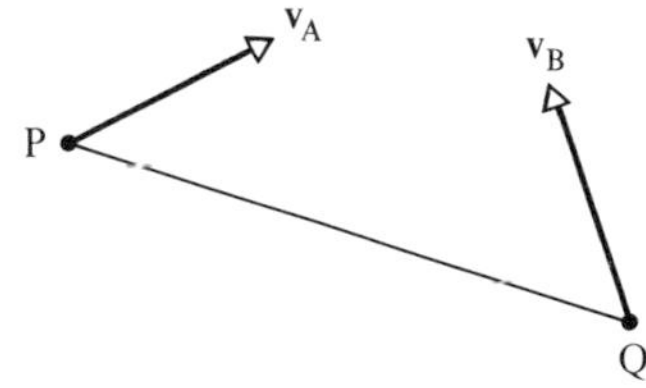

By imposing a velocity of $-\mathbf{v}_B$ on both A and B, then B can be considered to be at rest and the velocity of A is then relative to B, i.e. ${}_A\mathbf{v}_B$. Thus, if ${}_A\mathbf{v}_B$ is in the direction of PQ, then during the course of the motion A and B will meet. This may be by design (i.e. one ship intending to intercept the other) or by accident (i.e. one ship colliding with the other).
Thus for collision or interception to occur, ${}_A\mathbf{v}_B$ must be in the direction of the line joining the original position of A to that of B.

Example 12

A speedboat A and a ship B are initially 570 m apart and B is due north of A. The ship has a constant velocity of $(7\mathbf{i} + 5\mathbf{j})$ m/s and the speedboat has a constant speed of 25 m/s.
Find, in vector form, the velocity of A if it is to intercept B, and find the time taken to do so.
($\mathbf{i}$ represents a unit vector due east and $\mathbf{j}$ a unit vector due north.)

Draw a diagram showing the initial positions of A and B

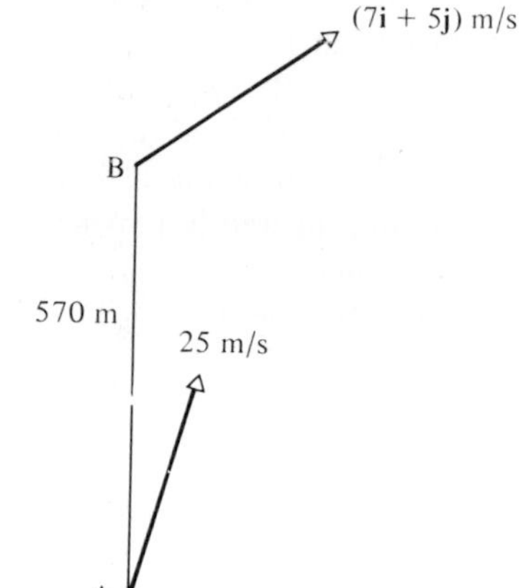

Let $\mathbf{v}_A = (a\mathbf{i} + b\mathbf{j})$ m/s

$$_A\mathbf{v}_B = \mathbf{v}_A - \mathbf{v}_B$$
$$= (a\mathbf{i} + b\mathbf{j}) - (7\mathbf{i} + 5\mathbf{j})$$
$$= (a - 7)\mathbf{i} + (b - 5)\mathbf{j}$$

For interception, $_A\mathbf{v}_B$ must be in the direction due north

Hence $_A\mathbf{v}_B = 0\mathbf{i} + (b - 5)\mathbf{j}$

and $(b - 5)$ must be positive

$\therefore \quad a - 7 = 0$ or $a = 7$

But the speed of A is 25 m/s

$$\therefore \quad \sqrt{(a^2 + b^2)} = 25$$
$$\therefore \quad a^2 + b^2 = 625$$

substituting for a

$$b^2 = 625 - 49$$

$\therefore \quad b = 24$ (negative value not applicable as $(b - 5)$ must be positive)

$$\therefore \quad \mathbf{v}_A = (7\mathbf{i} + 24\mathbf{j}) \text{ m/s}$$

Velocity of A relative to B

$$= (a - 7)\mathbf{i} + (b - 5)\mathbf{j} = 0\mathbf{i} + 19\mathbf{j}$$

$\therefore$ $_A\mathbf{v}_B$ is 19 m/s due north.

Time to intercept $= \dfrac{570}{|_A\mathbf{v}_B|} = \dfrac{570}{19} = 30$ s

Velocity of A is $(7\mathbf{i} + 24\mathbf{j})$ m/s and the time to intercept B is 30 s.

Example 13

At 8 a.m. the position vectors **r** and the velocity vectors **v** of two particles, A and B, are as follows:

$\mathbf{r}_A = (5\mathbf{i} - 3\mathbf{j})$ km $\qquad \mathbf{v}_A = (2\mathbf{i} + 5\mathbf{j})$ km/h

$\mathbf{r}_B = (7\mathbf{i} + 5\mathbf{j})$ km $\qquad \mathbf{v}_B = (-3\mathbf{i} - 15\mathbf{j})$ km/h

Show that if the velocities remain constant, a collision will occur, and find the time of the collision and the position vector of the point where it occurs.

A collision will occur if the velocity of A relative to B is in the direction of the line joining the initial position of A to that of B.

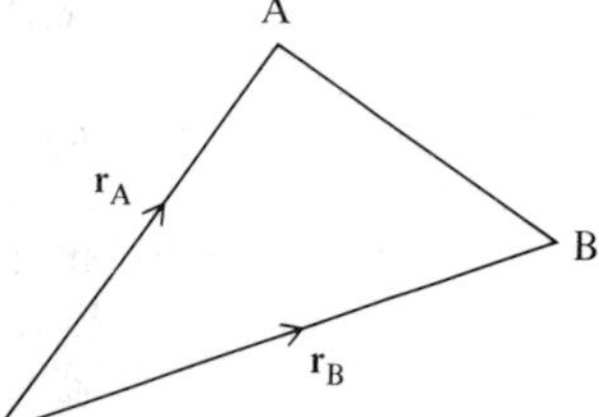

The diagram shows that $(\mathbf{r}_B - \mathbf{r}_A)$ gives the vector $\overrightarrow{AB}$ joining the initial positions of A and B and so, for collision, $_A\mathbf{v}_B$ must be parallel to $\mathbf{r}_B - \mathbf{r}_A$.

But $\quad \mathbf{r}_B - \mathbf{r}_A = 2\mathbf{i} + 8\mathbf{j}$

and $\quad _A\mathbf{v}_B = \mathbf{v}_A - \mathbf{v}_B = 5\mathbf{i} + 20\mathbf{j}$

Thus $\mathbf{r}_B - \mathbf{r}_A$ and ${}_A\mathbf{v}_B$ are parallel vectors as one is a multiple of the other. Therefore a collision will occur.

From $\mathbf{s} = \mathbf{v}t$, $\quad 2\mathbf{i} + 8\mathbf{j} = (5\mathbf{i} + 20\mathbf{j})t$

$$\text{or} \quad t = \tfrac{2}{5}\text{ h}$$

position vector of A when $t = \frac{2}{5}$

$$\mathbf{r}'_A = 5\mathbf{i} - 3\mathbf{j} + \tfrac{2}{5}(2\mathbf{i} + 5\mathbf{j})$$
$$= \tfrac{29}{5}\mathbf{i} - \mathbf{j}$$

The collision occurs at 8.24 a.m. at the point with position vector $(\frac{29}{5}\mathbf{i} - \mathbf{j})$ km.

Alternatively this problem can be solved by showing that the two particles are at the same point at the same time:

Suppose the collision occurs after t hours

Position vector of A after t hours $= \mathbf{r}_A + (\mathbf{v}_A)t$

$$= 5\mathbf{i} - 3\mathbf{j} + 2\mathbf{i}t + 5\mathbf{j}t$$
$$= (5 + 2t)\mathbf{i} + (-3 + 5t)\mathbf{j}$$

similarly, position vector of B after t hours $= 7\mathbf{i} + 5\mathbf{j} + (-3\mathbf{i} - 15\mathbf{j})t$

$$= (7 - 3t)\mathbf{i} + (5 - 15t)\mathbf{j}$$

For a collision: $\quad (5 + 2t)\mathbf{i} + (-3 + 5t)\mathbf{j} = (7 - 3t)\mathbf{i} + (5 - 15t)\mathbf{j}$

Equating coefficients of unit vectors **i** and **j**:

$$5 + 2t = 7 - 3t \quad \text{and} \quad -3 + 5t = 5 - 15t$$
$$5t = 2 \qquad\qquad 20t = 8$$
$$t = \tfrac{2}{5} \qquad\qquad t = \tfrac{2}{5}$$

So the particles have the same position vectors after $\frac{2}{5}$ h. Thus again the collision is found to occur at 8.24 a.m. and its location may be found as before.

Velocities and positions not in vector form

Similar problems involving collisions and interceptions may be posed where the data is not given in vector form.

Example 14

At a certain instant two particles P and Q are at the points A and B which are 150 cm apart with B due east of A. Particle P is travelling at $10\sqrt{3}$ cm/s due south and Q is travelling at 20 cm/s in a direction S 30° W. Show that if the velocities of P and Q remain unchanged, a collision will take place and find the time which elapses before it does so.

If a velocity of $-\mathbf{v}_P$ is imposed on both P and Q, then P will be at rest and ${}_Q\mathbf{v}_P$ must be in the direction BA if a collision is to occur.

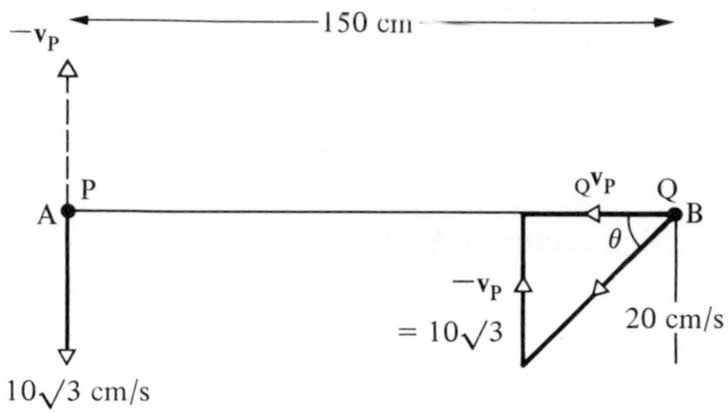

From the diagram

$$\sin\theta = \tfrac{10}{20}\sqrt{3}$$

$$= \frac{\sqrt{3}}{2}$$

hence for collision $\theta = 60°$

But it is known that the direction of motion of Q is S 30° W, so angle θ is 60°.

Collision occurs when Q, relative to P, is at the point A

From the diagram $\quad \cos 60 = \dfrac{|{}_Q\mathbf{v}_P|}{20} \quad$ and time taken $= \dfrac{AB}{|{}_Q\mathbf{v}_P|}$

$\therefore \quad |{}_Q\mathbf{v}_P| = 10 \qquad\qquad = \dfrac{150}{10} = 15$ s

A collision does take place after 15 s.

Note that if the actual position of the collision is required, then the position of either P or Q after 15 s has to be found.

After 15 s, P is $10\sqrt{3} \times 15$ cm or $150\sqrt{3}$ cm due south of A and this is where the collision occurs.

Example 15

At 8 a.m. two particles P and Q are at the points A and B, 12 km apart, with B on a bearing of 250° from A. P is moving at 4 km/h on a bearing of 320°. If the maximum speed of Q is 7 km/h, find the course on which Q should be set in order to intercept P as soon as possible, and find when the interception occurs.

Draw a sketch showing the initial positions of P and Q.
If a velocity of $-\mathbf{v}_P$ is imposed on both P and Q, then P will be at rest and, for interception to occur ${}_Q\mathbf{v}_P$ must be in the direction BA.
Draw a scale diagram in which D lies on BA (see next page)

and $\overrightarrow{BC} = -\mathbf{v}_P$

$= 4$ km/h on a bearing of 140°

and $\overrightarrow{CD} = \mathbf{v}_Q$

$= 7$ km/h

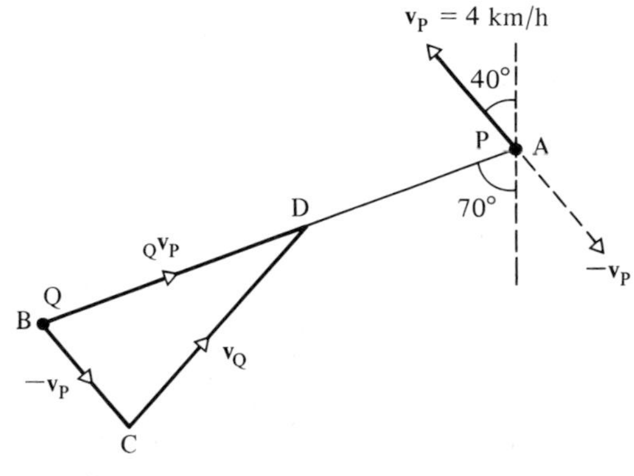

By measurement

BD = 7·3 cm
angle BCD = 78°

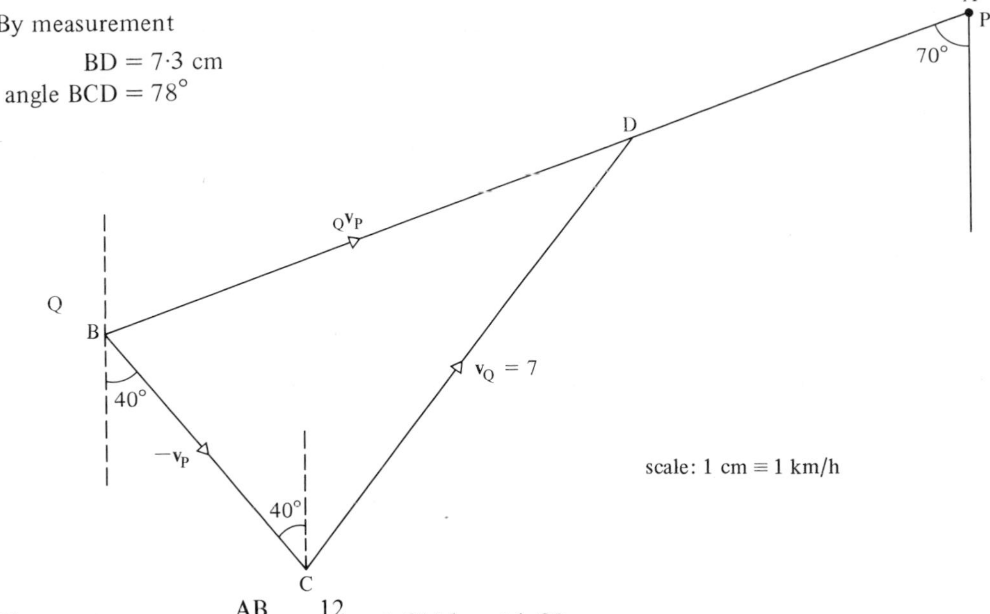

$$\text{Time to intercept} = \frac{AB}{|_{Q}\mathbf{v}_{P}|} = \frac{12}{7{\cdot}3} = 1{\cdot}644 \text{ h} = 1 \text{ h } 39 \text{ m}$$

The course to be set is 038° and interception takes place at 9.39 a.m.

Again, it should be noted that these answers could have been calculated from the triangle BCD using trigonometry.

Exercise 10C

The following questions are presented in vector form with **i** representing a unit vector due east and **j** a unit vector due north.

1. Initially two particles X and Y are 100 m apart with X due east of Y. X has a constant velocity of $(2\mathbf{i} + 3\mathbf{j})$ m/s and Y a constant speed of 5 m/s. Find, in vector form, the velocity of Y if it is to intercept X and find the time taken to do so.
2. Initially two particles X and Y are 48 m apart with Y due north of X. X has a constant velocity of $(5\mathbf{i} + 4\mathbf{j})$ m/s and Y a constant speed of 13 m/s. Find, in vector form, the velocity of Y if it is to intercept X, and find the time taken to do so.
3. At 12 noon the position vectors **r** and velocity vectors **v** of two ships A and B are as follows,

   $\mathbf{r}_A = (\mathbf{i} + 7\mathbf{j})$ km $\mathbf{v}_A = (6\mathbf{i} + 2\mathbf{j})$ km/h
   $\mathbf{r}_B = (6\mathbf{i} + 4\mathbf{j})$ km $\mathbf{v}_B = (-4\mathbf{i} + 8\mathbf{j})$ km/h

   Show that if the ships do not alter their velocities, a collision will occur and find the time at which it occurs and the position vector of its location.
4. At 12 noon the position vectors **r** and velocity vectors **v** of two ships A and B are as follows,

   $\mathbf{r}_A = (5\mathbf{i} + 2\mathbf{j})$ km $\mathbf{v}_A = (15\mathbf{i} + 10\mathbf{j})$ km/h
   $\mathbf{r}_B = (7\mathbf{i} + 7\mathbf{j})$ km $\mathbf{v}_B = (9\mathbf{i} - 5\mathbf{j})$ km/h

   Show that if the ships do not alter their velocities, a collision will occur and find the time at which it occurs and the position vector of its location.

5. At 11.30 a.m. a battleship is at a place with position vector $(-6\mathbf{i} + 12\mathbf{j})$ km and is moving with velocity $(16\mathbf{i} - 4\mathbf{j})$ km/h. At 12 noon a cruiser is at a place with position vector $(12\mathbf{i} - 15\mathbf{j})$ km and is moving with velocity $(8\mathbf{i} + 16\mathbf{j})$ km/h. Show that if these velocities are maintained the two ships will collide and find when and where the collision occurs.
6. At 11.30 a.m. a jumbo jet has a position vector $(-100\mathbf{i} + 220\mathbf{j})$ km and a velocity vector $(300\mathbf{i} + 400\mathbf{j})$ km/h. At 11.45 a.m. a cargo plane has a position vector $(-60\mathbf{i} + 355\mathbf{j})$ km and a velocity vector $(400\mathbf{i} + 300\mathbf{j})$ km/h. Show that if these velocities are maintained the planes will crash into each other and find the time and position vector of the crash.
7. At 2 p.m. the position vectors $\mathbf{r}$ and velocity vectors $\mathbf{v}$ of three ships A, B and C are as follows,

   $\mathbf{r}_A = (5\mathbf{i} + \mathbf{j})$ km $\qquad \mathbf{v}_A = (9\mathbf{i} + 18\mathbf{j})$ km/h
   $\mathbf{r}_B = (12\mathbf{i} + 5\mathbf{j})$ km $\qquad \mathbf{v}_B = (-12\mathbf{i} + 6\mathbf{j})$ km/h
   $\mathbf{r}_C = (13\mathbf{i} - 3\mathbf{j})$ km $\qquad \mathbf{v}_C = (9\mathbf{i} + 12\mathbf{j})$ km/h

   (a) Assuming that all three ships maintain these velocities, show that A and B will collide and find when and where the collision occurs.
   (b) Find the position vector of C when A and B collide and find how far C is from the collision.
   (c) When the collission occurs, C immediately changes its course, but not its speed, and steams direct to the scene. When does C arrive?
8. At 12 noon the position vectors $\mathbf{r}$ and velocity vectors $\mathbf{v}$ of three ships A, B and C are as follows,

   $\mathbf{r}_A = (10{\cdot}5\mathbf{i} + 6\mathbf{j})$ km $\qquad \mathbf{v}_A = (9\mathbf{i} + 18\mathbf{j})$ km/h
   $\mathbf{r}_B = (7\mathbf{i} + 20\mathbf{j})$ km $\qquad \mathbf{v}_B = (12\mathbf{i} + 6\mathbf{j})$ km/h
   $\mathbf{r}_C = (10\mathbf{i} + 15\mathbf{j})$ km $\qquad \mathbf{v}_C = (6\mathbf{i} + 12\mathbf{j})$ km/h

   Assuming that all three ships maintain these velocities, show that A and B will collide and find when and where the collision occurs.
   When the collision occurs, C immediately changes its course but not its speed, and steams direct to the scene. When does C arrive?
9. At certain times the position vectors $\mathbf{r}$ and velocity vectors $\mathbf{v}$ of three ships A, B and C are as follows,

   $\mathbf{r}_A = (6\mathbf{i} + 17\mathbf{j})$ km $\qquad \mathbf{v}_A = (4\mathbf{i} - 20\mathbf{j})$ km/h $\qquad$ at 11.30 a.m.
   $\mathbf{r}_B = (5\mathbf{i} - 18\mathbf{j})$ km $\qquad \mathbf{v}_B = (2\mathbf{i} + 14\mathbf{j})$ km/h $\qquad$ at 11.45 a.m.
   $\mathbf{r}_C = (2\mathbf{i} - 5\mathbf{j})$ km $\qquad \mathbf{v}_C = (12\mathbf{i} - 4\mathbf{j})$ km/h $\qquad$ at 12 noon

   Assuming that all three ships maintain these velocities, show that two of them will collide and find when and where the collision will occur.
   When the collision occurs the ship not involved immediately changes its course but not its speed, and steams direct to the scene. When will it arrive?
10. At certain times the position vectors $\mathbf{r}$ and velocity vectors $\mathbf{v}$ of three ships A, B and C are as follows,

    $\mathbf{r}_A = (8\mathbf{i} - 26\mathbf{j})$ km $\qquad \mathbf{v}_A = (7\mathbf{i} + 24\mathbf{j})$ km/h $\qquad$ at 7.48 a.m.
    $\mathbf{r}_B = (6\mathbf{i} - 10\mathbf{j})$ km $\qquad \mathbf{v}_B = (15\mathbf{i} + 5\mathbf{j})$ km/h $\qquad$ at 8.00 a.m.
    $\mathbf{r}_C = (10\mathbf{i} + 6\mathbf{j})$ km $\qquad \mathbf{v}_C = (10\mathbf{i} - 15\mathbf{j})$ km/h $\qquad$ at 8.00 a.m.

    Show that, if the ships maintain these velocities, two of the ships will collide and find when and where the collision will occur.
    When the collision occurs, the ship not involved immediately changes its course but not its speed, and steams direct to the scene of the collision. Find, in vector form, the velocity of this ship during this part of its motion and find its time of arrival.

Exercise 10D

1. At 12 noon two ships A and B are 10 km apart with B due east of A. A is travelling at 20 km/h in a direction N 60° E and B is travelling at 10 km/h due north. Show that, if the two ships maintain these velocities, they will collide and find, to the nearest minute, when the collision occurs.
2. At 11 p.m. two ships A and B are 10 km apart with B due north of A. A is travelling north-east at 18 km/h and B is travelling due east at $9\sqrt{2}$ km/h. Show that, if the two ships do not change their velocities, they will collide and find, to the nearest minute, when the collision occurs.
3. A coastguard vessel wishes to intercept a yacht suspected of smuggling. At 1 a.m. the yacht is 10 km due east of the coastguard vessel and is travelling due north at 15 km/h. If the coastguard vessel travels at 20 km/h, in what direction should it steer in order to intercept the yacht? When would this interception occur?
4. A lifeboat sets out from a harbour at 9.10 p.m. to go to the assistance of a yacht which is, at that time, 5 km due south of the harbour and is drifting due west at 8 km/h. If the lifeboat travels at 20 km/h, find the course it should set so as to reach the yacht as quickly as possible and the time when it arrives, (to the nearest half minute).
5. At 12 noon two ships A and B are 12 km apart with B on a bearing 140° from A. Ship A has a maximum speed of 30 km/h and wishes to intercept ship B, which is travelling at 20 km/h on a bearing 340°. Find the course A should set in order to intercept B as soon as possible and the time when interception occurs.
6. A helicopter sets off from its base and flies at 50 m/s to intercept a ship which, when the helicopter sets off, is at a distance of 5 km on a bearing of 335° from the base. The ship is travelling at 10 m/s on a bearing 095°. Find the course that the helicopter pilot should set if he is to intercept the ship as quickly as possible and the time interval between the helicopter taking off and it reaching the ship.
7. The driver of a speedboat travelling at 75 km/h wishes to intercept a yacht travelling at 20 km/h in a direction N 40° E. Initially the speedboat is positioned 10 km from the yacht on a bearing S 30° E. Find the course that the driver of the speedboat should set to intercept the yacht and how long the journey will take.
8. A batsman hits a ball at 15 m/s in a direction S 80° W. A fielder, 45 m and S 65° W from the batsman, runs at 6 m/s to intercept the ball. Assuming the velocities remain unchanged, find in what direction the fielder must run to intercept the ball as quickly as possible. How long does it take him, to the nearest tenth of a second?

Closest approach

If two bodies do not collide, then there will be an instant at which they are closer to each other than they are at any other instant.

Example 16

Two ships, A and B, have the following position vectors **r** and velocity vectors **v** at the times indicated:

$\mathbf{r}_A = (-\mathbf{i} + 0\mathbf{j})$ km $\quad \mathbf{v}_A = (3\mathbf{i} + \mathbf{j})$ km/h at 3 p.m.

$\mathbf{r}_B = (-\mathbf{i} - 4\mathbf{j})$ km $\quad \mathbf{v}_B = (11\mathbf{i} + 3\mathbf{j})$ km/h at 4 p.m.

Assuming the ships continue with unchanged velocities, find
(a) the position vector of A at 4 p.m.,
(b) the distance between A and B at 4 p.m.,
(c) the least distance between A and B during the motion,
(d) the time at which the ships are closest together.

(a) Position of A, t hours after 3 p.m. is $\mathbf{r}_A + (\mathbf{v}_A \times t)$
Thus position vector of A at 4 p.m. is $(-\mathbf{i} + 0\mathbf{j}) + (3\mathbf{i} + \mathbf{j}) \times 1 = (2\mathbf{i} + \mathbf{j})$ km

(b) Distance between A and B at 4 p.m. is given by the length of AB

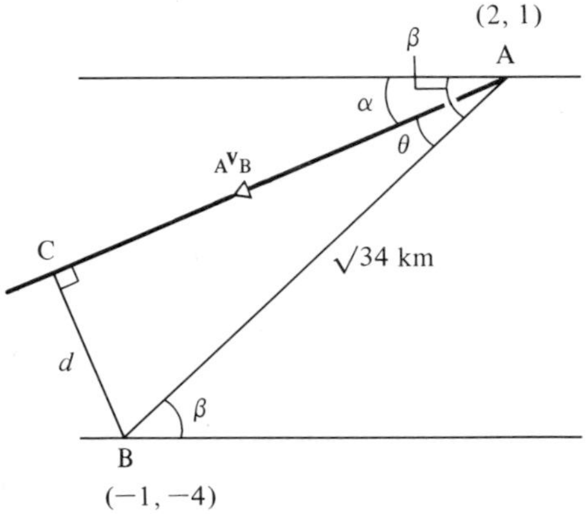

Using Pythagoras,

$$\begin{aligned} AB &= \sqrt{[(2 - -1)^2 + (1 - -4)^2]} \\ &= \sqrt{(3^2 + 5^2)} = \sqrt{34} \\ &= 5{\cdot}831 \text{ km} \end{aligned}$$

(c) Consider the motion of A relative to B

$$\begin{aligned} {}_A\mathbf{v}_B &= \mathbf{v}_A - \mathbf{v}_B \\ &= (3\mathbf{i} + \mathbf{j}) - (11\mathbf{i} + 3\mathbf{j}) \\ &= (-8\mathbf{i} - 2\mathbf{j}) \text{ km/h} \end{aligned}$$

Closest approach is given by

$$\begin{aligned} d &= AB \sin\theta \\ &= \sqrt{34} \sin\theta \quad \text{where } \theta = \beta - \alpha \end{aligned}$$

but $\tan\beta = \dfrac{1 - -4}{2 - -1} = \dfrac{5}{3}$ $\qquad \therefore \quad \beta = 59{\cdot}04°$

From ${}_A\mathbf{v}_B$ $\qquad \tan\alpha = \frac{2}{8}$ $\qquad \therefore \quad \alpha = 14{\cdot}04°$

hence $\theta = 45°$

$\therefore \quad d = \sqrt{34} \sin 45 = \sqrt{17} = 4{\cdot}123$ km

(d) Ships are closest when A is at C, relative to B

$$\begin{aligned} AC^2 &= AB^2 - CB^2 \\ &= (\sqrt{34})^2 - (\sqrt{17})^2 \\ \therefore \quad AC &= \sqrt{17} \end{aligned}$$

Time to reach point C $= \dfrac{AC}{|{}_A\mathbf{v}_B|} = \dfrac{\sqrt{17}}{\sqrt{68}} = \frac{1}{2}$ h

The position vector of A at 4 p.m. is $(2\mathbf{i} + \mathbf{j})$ km; at this time A and B are 5·83 km apart; the least distance between A and B is 4·12 km and this occurs at 4.30 p.m.

Note that these results can also be calculated by finding an expression for the square of the distance of separation of A and B at time t and then using calculus to find when this is a minimum.

Example 17

Two ships A and B are initially 20 km apart with B on a bearing of N 67° E from A. Ship A is moving at 18 km/h in a direction S 20° E and B is moving at 12 km/h due south. Assuming the velocities of A and B remain unchanged, find the least distance apart of the ships in the subsequent motion and the time at which this position is reached.

Draw a diagram showing the initial positions of the ships and consider the motion of A relative to B.

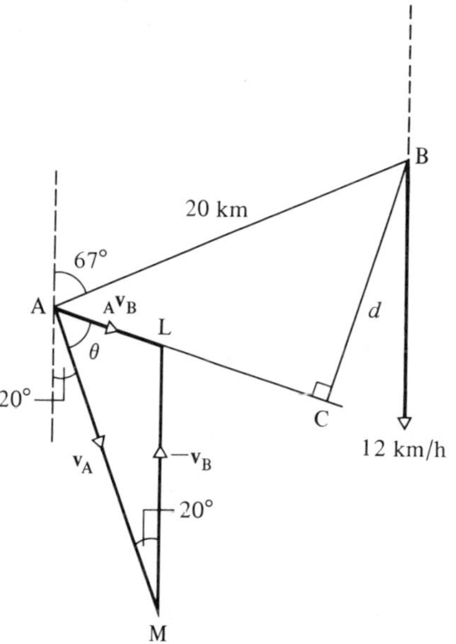

Relative to B, ship A travels along AC and

$${}_A\mathbf{v}_B = \mathbf{v}_A - \mathbf{v}_B$$

Least distance apart is d

From triangle ALM

$$|{}_A\mathbf{v}_B|^2 = 18^2 + 12^2 - 2(18)(12)\cos 20$$
$$= 468 - 432\cos 20$$
$$\therefore \quad |{}_A\mathbf{v}_B| = 7{\cdot}877$$

By the sine rule

$$\frac{|{}_A\mathbf{v}_B|}{\sin 20} = \frac{12}{\sin\theta}$$

$$\therefore \quad \sin\theta = \frac{12\sin 20}{7{\cdot}877}$$

$$\therefore \quad \theta = 31{\cdot}40^\circ$$

then angle $\text{BAC} = 180^\circ - \theta - 20^\circ - 67^\circ = 61{\cdot}60^\circ$

$$d = \text{AB}\sin \text{B}\hat{\text{A}}\text{C} = 20\sin 61{\cdot}60^\circ$$
$$= 17{\cdot}59 \text{ km}$$

$$\text{Time to reach this position} = \frac{\text{AC}}{|{}_A\mathbf{v}_B|}$$
$$= \frac{20\cos 61{\cdot}60^\circ}{7{\cdot}877} = 1{\cdot}208 \text{ h}$$

The least distance between the ships is 17·6 km and occurs after 1 hour and 12 minutes.

Exercise 10E

The following questions are presented in vector form with **i** representing a unit vector due east and **j** a unit vector due north.

1. At 12 noon the position vectors **r** and velocity vectors **v** of two ships A and B are as follows,

   $\mathbf{r}_A = (-9\mathbf{i} + 6\mathbf{j})$ km $\qquad \mathbf{v}_A = (3\mathbf{i} + 12\mathbf{j})$ km/h

   $\mathbf{r}_B = (16\mathbf{i} + 6\mathbf{j})$ km $\qquad \mathbf{v}_B = (-9\mathbf{i} + 3\mathbf{j})$ km/h

   (a) Find how far apart the ships are at 12 noon.

   (b) Assuming the velocities do not change, find the least distance between the ships in the subsequent motion.

   (c) Find when this distance of closest approach occurs and the position vectors of A and B at that time.

2. At 8 a.m. two ships A and B are 11 km apart with B due west of A. A and B travel with constant velocities of $(-4\mathbf{i} + 3\mathbf{j})$ km/h and $(2\mathbf{i} + 4\mathbf{j})$ km/h respectively. Find the least distance between the two ships in the subsequent motion and the time, to the nearest minute, at which this situation occurs.

3. At 7.30 p.m. two ships A and B are 8 km apart with B due north of A. The velocities of A and B are $12\mathbf{j}$ km/h and $-5\mathbf{i}$ km/h respectively. Assuming these velocities do not change, find the least distance between the ships in the subsequent motion and the time, to the nearest minute, at which this situation occurs.
4. A and B are two tankers and at 1300 hours B has a position vector of $(4\mathbf{i} + 8\mathbf{j})$ km relative to A. Tanker A is moving with a constant velocity of $(6\mathbf{i} + 9\mathbf{j})$ km/h and tanker B is moving with a constant velocity of $(-3\mathbf{i} + 6\mathbf{j})$ km/h. Find the least distance between the tankers in the subsequent motion and the time at which this situation occurs.
5. At certain times the position vectors $\mathbf{r}$ and velocity vectors $\mathbf{v}$ of two ships A and B are as follows,

   $\mathbf{r}_A = 20\mathbf{j}$ km $\quad \mathbf{v}_A = (9\mathbf{i} - 2\mathbf{j})$ km/h $\quad$ at 1400 hours
   $\mathbf{r}_B = (\mathbf{i} + 4\mathbf{j})$ km $\quad \mathbf{v}_B = (4\mathbf{i} + 8\mathbf{j})$ km/h $\quad$ at 1500 hours

   Assuming these velocities do not change, find
   (a) the position vector of A at 1500 hours,
   (b) the least distance between A and B,
   (c) the time at which this least separation occurs.
6. At 12 noon two ships A and B have the following position vectors $\mathbf{r}$ and velocity vectors $\mathbf{v}$,

   $\mathbf{r}_A = (5\mathbf{i} + \mathbf{j})$ km $\quad \mathbf{v}_A = (7\mathbf{i} + 3\mathbf{j})$ km/h
   $\mathbf{r}_B = (8\mathbf{i} + 7\mathbf{j})$ km $\quad \mathbf{v}_B = (2\mathbf{i} - \mathbf{j})$ km/h

   If both ships maintain these velocities, find the least distance between the ships and the time, to the nearest minute, at which this situation occurs.
7. Two ships A and B have the following position vectors $\mathbf{r}$ and velocity vectors $\mathbf{v}$ at the times indicated,

   $\mathbf{r}_A = (3\mathbf{i} + \mathbf{j})$ km $\quad \mathbf{v}_A = (2\mathbf{i} + 3\mathbf{j})$ km/h $\quad$ at 11 a.m.
   $\mathbf{r}_B = (2\mathbf{i} - \mathbf{j})$ km $\quad \mathbf{v}_B = (3\mathbf{i} + 7\mathbf{j})$ km/h $\quad$ at 12 noon

   Assuming that the ships maintain these velocities, find
   (a) the position vector of ship A at 12 noon,
   (b) the distance between A and B at 12 noon,
   (c) the least distance between A and B during the motion,
   (d) the time, to the nearest minute, when the least separation occurs.
8. A battleship B and a cruiser C have the following position vectors $\mathbf{r}$ and velocity vectors $\mathbf{v}$ at 12 noon.

   $\mathbf{r}_B = (13\mathbf{i} + 5\mathbf{j})$ km $\quad \mathbf{v}_B = (3\mathbf{i} - 10\mathbf{j})$ km/h
   $\mathbf{r}_C = (3\mathbf{i} - 5\mathbf{j})$ km $\quad \mathbf{v}_C = (15\mathbf{i} + 14\mathbf{j})$ km/h

   Assuming that the ships do not alter their velocities, find the closest distance that they come to each other.
   The battleship has guns with a range of up to 5 km. Find the length of time during which the cruiser is within range of the battleship's guns.
9. Two ships A and B have the following position vectors $\mathbf{r}$ and velocity vectors $\mathbf{v}$ at the times stated,

   $\mathbf{r}_A = (-2\mathbf{i} + 3\mathbf{j})$ km $\quad \mathbf{v}_A = (12\mathbf{i} - 4\mathbf{j})$ km/h $\quad$ at 11.45 a.m.
   $\mathbf{r}_B = (8\mathbf{i} + 7\mathbf{j})$ km $\quad \mathbf{v}_B = (2\mathbf{i} - 14\mathbf{j})$ km/h $\quad$ at 12 noon.

   Assuming that the ships do not alter their velocities, find their least distance of separation.
   If ship B has guns with a range of up to 2 km, find for what length of time A is within range.

Exercise 10F

1. Initially two ships A and B are 65 km apart with B due east of A. A is moving due east at 10 km/h and B due south at 24 km/h. The two ships continue moving with these velocities. Find the least distance between the ships in the subsequent motion and the time taken to the nearest minute for such a situation to occur.
2. Two aircraft A and B are flying at the same altitude with velocities 180 m/s due east and 240 m/s due north respectively. Initially B is 5 km due south of A. Given that the aircraft do not change their velocities, find the shortest distance between the aircraft in the subsequent motion and the time taken for such a situation to occur.
3. A road running north-south crosses a road running east-west at a junction O. Initially Paul is on the east-west road, 1·7 km west of O, and is cycling towards O at 15 km/h. At the same time Pat is at O cycling due north at 8 km/h. If Paul and Pat do not alter their velocities, find the least distance they are apart in the subsequent motion and the time taken for that situation to occur.
4. At 7.30 a.m. two ships A and B have velocities 15 km/h, N 30° E and 20 km/h due east respectively, with B 5 km due west of A. In the subsequent motion A and B do not alter their velocities. Find the distance between A and B when they are closest together and the time at which this situation occurs, to the nearest minute.
5. A road running north-south crosses a road running east-west at a junction O. John cycles towards O from the west at 3 m/s as Tom cycles towards O from the south at 4 m/s. Initially John is 600 m from O and Tom is 250 m from O. If Tom and John do not alter their velocities, find the least distance they are apart during the motion and the time taken to reach that situation.
   How far, and in what direction, are Tom and John then from O?
6. Two aircraft A and B are flying, at the same altitude, with velocities 200 m/s, N 30° E and 300 m/s, N 50° W respectively. Initially A and B are 2 km apart with B on a bearing S 70° E from A. Given that A and B do not alter their velocities find the least distance of separation between the two aircraft in the subsequent motion and the time taken to reach such a situation.
7. At 1500 hours a trawler is 10 km due east of a launch. The trawler maintains a steady 10 km/h on a bearing 180° and the launch maintains a steady 20 km/h on a bearing 071°. Find the minimum distance the boats are apart in the subsequent motion and the time at which it occurs.
   Find, to the nearest minute, the length of time for which the two boats are within 8 km of each other.
8. A battleship and a cruiser are initially 16 km apart with the battleship on a bearing N 35° E from the cruiser. The battleship travels at 14 km/h on a bearing S 29° E and the cruiser at 17 km/h on a bearing N 50° E. The guns on the battleship have a range of up to 6 km. Find
   (a) the least distance between the cruiser and the battleship in the subsequent motion,
   (b) the length of time for which the battleship has the cruiser within range of its guns.

Course for closest approach

The following example illustrates the method used to find the course which must be set in order that one body may pass as close as possible to a second body.

Example 18

A ship A is moving with a constant speed of 18 km/h in a direction N 55° E and is initially 6 km from a second ship B, the bearing of A from B being N 25° W. If B moves with a constant speed of 15 km/h, find the course B must set in order to pass as close as possible to A, the distance between the ships when they are closest together, and the time for this to occur, to the nearest minute.

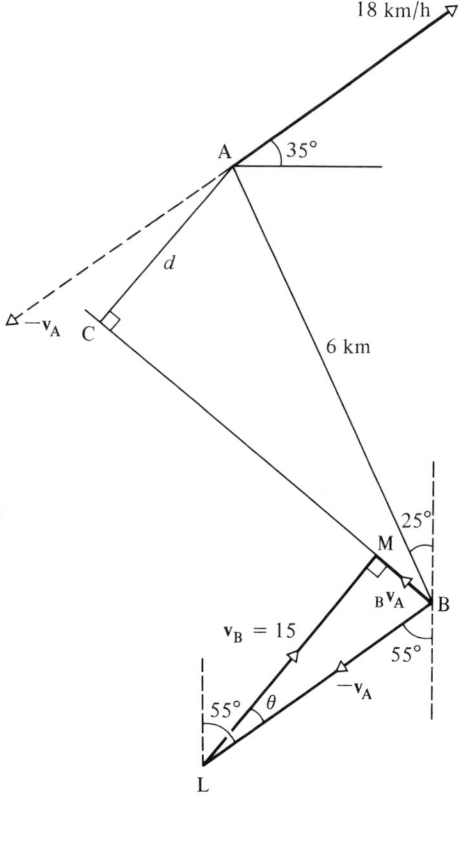

Draw a diagram showing the initial positions of A and B, and consider the motion of B relative to A. To pass as close as possible to A, B must travel relative to A along a line BC which makes as small an angle as possible with BA.
In the vector triangle BLM, LM must be perpendicular to BC.

Hence $\quad \cos\theta = \frac{15}{18}$

$\therefore \quad \theta = 33{\cdot}56°$

and course to be set is $55° - 33{\cdot}56°$
$= \text{N } 21{\cdot}44° \text{ E}$

$$|{}_B\mathbf{v}_A| = \sqrt{(18^2 - 15^2)} = 9{\cdot}950$$

$$\text{angle ABC} = 180° - 25° - 55° - (90° - \theta) = 43{\cdot}56°$$

closest approach is d

$$d = 6 \sin 43{\cdot}56°$$
$$\therefore \quad d = 4{\cdot}135 \text{ km}$$

$$\text{time taken} = \frac{\text{BC}}{|{}_B\mathbf{v}_A|} = \frac{6\cos 43{\cdot}56°}{9{\cdot}950} = 0{\cdot}437 \text{ h} \approx 26 \text{ min}$$

Course to be set by B is N 21.44° E; least distance apart is 4.14 km and this occurs after 26 minutes.

Exercise 10G

1. Motorboat B is travelling at a constant velocity of 10 m/s due east and motorboat A is travelling at a constant speed of 8 m/s. Initially A and B are 600 m apart with A due south of B. Find the course that A should set in order to get as close as possible to B. Find this closest distance and the time taken for the situation to occur.

2. At 8 a.m. two boats A and B are 5·2 km apart with A due west of B, and B travelling due north at a steady 13 km/h. If A travels with a constant speed of 12 km/h show that, for A to get as close as possible to B, A should set a course of N θ° E where $\sin\theta = \frac{5}{13}$. Find this closest distance and the time at which it occurs.
3. Two aircraft A and B are flying at the same altitude with A initially 5 km due north of B, and B flying at a steady 300 m/s on a bearing 060°. If A flies at a constant speed of 200 m/s find the course that A must set in order to fly as close as possible to B, the distance between the planes when they are closest and the time taken for this to occur.
4. A ship A is moving with a constant speed of 24 km/h in a direction N 40° E and is initially 10 km from a second ship B, the bearing of A from B being N 30° W. If B moves with a constant 22 km/h find the course that B must set in order to pass as close as possible to A, the distance between the ships when they are closest and the time taken (to the nearest minute) for this to occur.
5. At 12 noon a cruiser is 16 km due west of a destroyer. The cruiser is travelling at 40 km/h on a bearing N 30° E and the destroyer is travelling at 20 km/h. If the velocity of the cruiser and the speed of the destroyer do not change find the course that the destroyer should set to get as close as possible to the cruiser and find when this closest approch occurs. If the guns on the destroyer have a range of up to 10 km find the length of time for which the cruiser will be within range (to the nearest minute).
   If the guns on the cruiser have a range of up to 9 km find the length of time for which the battleship will be within range (to the nearest minute).
6. A fielder is positioned 40 m from a batsman and on a bearing of S 70° W from the batsman. The fielder can run at 8 m/s and the batsman hits a ball at 17 m/s in a direction N 70° W. Assuming that the fielder runs at 8 m/s from the moment the ball is hit and neglecting any change in the velocity of the ball, find the closest distance that the fielder can get to the ball and how long it takes him to get there, from the time the ball is hit.
7. A battleship commander is informed that there is a lone cruiser positioned 40 km away from him on a bearing N 70° W. The guns on the battleship have a range of up to 8 km and the top speed of the battleship is 30 km/h. The cruiser maintains a constant velocity of 50 km/h, N 60° E. Show that whatever the course the battleship sets it cannot get the cruiser within range of its guns.

Exercise 10H Examination questions

1. A boy can swim in still water at 1 m/s. He swims across a river flowing at 0·6 m/s which is 336 m wide. Find the time he takes if he travels the shortest possible distance. (London)

2. A particle A, moving with constant velocity 3**i** passes through the origin at time $t = 0$. A second particle B, moving with constant velocity (4**i** − 2**j**), passes through the origin at time $t = 1$. Find, at time $t = 3$, (a) the velocity vector of B relative to A,
   (b) the distance between A and B. (London)

3. An aircraft A is flying due east at 600 km/h and its bearing from aircraft B is 030°. If aircraft B has a speed of 1000 km/h find the direction in which B must fly in order to intercept A.
   If the aircraft are initially 50 km apart find the time taken, in minutes, for the interception to occur. (Cambridge)

4. At noon, two ships A and B have the position and velocity vectors shown, where **i** and **j** are unit vectors in the directions East and North respectively,

	Position vector	Velocity vector
Ship A	$10\mathbf{i} + 5\mathbf{j}$	$-2\mathbf{i} + 4\mathbf{j}$
Ship B	$2\mathbf{i} - \mathbf{j}$	$2\mathbf{i} + 7\mathbf{j}$

and where the speeds are measured in kilometres per hour and the distances in kilometres. If they continue on their respective courses,
   (i) find the position vector of A after a time t hours;
   (ii) show that they will collide, and give the time of the collision;
   (iii) determine how far ship A will have travelled between noon and the collision. (S.U.J.B)

5. At 12.00 h an aeroplane A is sighted 20 km due N of an airfield O travelling in a direction due E with speed 300 km h^{-1}. An aeroplane B starts immediately from O and travels with speed 400 km h^{-1} in a direction due N. By calculation or scale drawing in which the scales should be stated,
   (i) find the magnitude and direction of B's velocity relative to A, the closest distance apart and the bearing of B from A when they are closest;
   (ii) determine what course (at speed 400 km h^{-1}) B ought to have taken in order to intercept A and the time at which they would then have met. (S.U.J.B)

6. In this question distances are measured in kilometres and speeds in kilometres per hour. **i** and **j** are perpendicular unit vectors.

   A, B and C move in a plane with constant velocities and at time $t = 0$ the position vectors of A, B and C relative to an origin O are $\mathbf{i} + 3\mathbf{j}$, $9\mathbf{i} + 9\mathbf{j}$ and $6\mathbf{i} + 13\mathbf{j}$ respectively. The velocity of C relative to A is $7\mathbf{i} - 10\mathbf{j}$ and of C relative to B is $9\mathbf{i} - 12\mathbf{j}$.
   (i) Find the velocity vector of B relative to A. Show that A and B do not collide and find their shortest distance apart and the time when A and B are this distance apart.
   (ii) Show that B and C do collide and find the distance between A and C when this collision occurs. (S.U.J.B)

7. A ship A is sailing with speed u in the direction α East of North, where $0 < \alpha < 45°$, and a ship B is sailing with speed v in the direction θ East of North. The velocity of B relative to A is due Southwest. Show that $v(\cos\theta - \sin\theta) = u(\cos\alpha - \sin\alpha)$.
   The ship A changes its course to α *North of East*, keeping its speed u, while B continues with its velocity unchanged. The velocity of B relative to A is now due West. Find $\tan\theta$ in terms of α. (J.M.B)

8. Three towns A, B, C are such that B is 300 km due north of A, and C is 400 km due west of A. There is a wind blowing at 25 m s^{-1} from the direction S 30° W. An aeroplane which travels at 100 m s^{-1} in still air flies from A to B and then from B to C. Find the direction in which it steers for each part of the journey, and show that the total time taken is about $2\frac{1}{2}$ hours.
   Subsequently the aeroplane returns from C directly to A. By this time, the speed of the wind has changed slightly, but its direction is still the same. The journey from C to A takes 1 hour. Find the new speed of the wind.
   (A graphical method may be used for this question.) (Cambridge)

9. A motor launch whose speed in still water is V km h^{-1} has to travel directly from a harbour X to a buoy Y lying 5 km from X and in a direction N 30° E from X. There is a steady current of 6 km h^{-1} flowing *from* the direction N $\theta°$ W. The journey from X to Y under these conditions takes exactly one hour. Prove that $\sqrt{3}\cos\theta - \sin\theta = (V^2 - 61)/30$.
   If the return journey under the same conditions takes only half an hour, calculate the value of V and the direction of the current. (Cambridge)

11
Projectiles

Horizontal projection

In order to investigate the motion of a projectile, the horizontal and vertical motions should be considered separately.
The horizontal velocity of a projectile is constant since there is no force acting in this direction. The vertical velocity, on the other hand, is subject to the force of gravity, and the usual equations of motion $v = u + at$, $v^2 = u^2 + 2as$, $s = ut + \frac{1}{2}at^2$ and $s = \frac{(v+u)t}{2}$ are used. In some cases **i** and **j**, the unit vectors in a horizontal and vertical direction, are used in stating the position and velocity vectors.

Example 1

A particle is projected horizontally at 36 m/s from a point 122·5 m above a horizontal surface. Find the time taken by the particle to reach the surface and the horizontal distance travelled in that time.

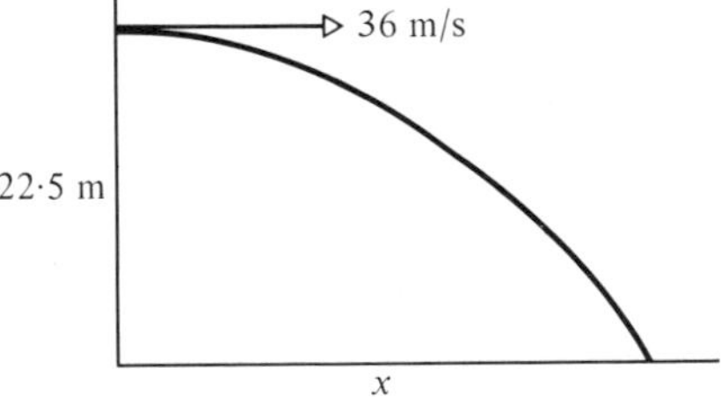

horizontal motion	vertical motion
$u = 36$ m/s	$u = 0$
$s = x$	$s = 122{\cdot}5$ m ↓
$t = t$	$t = t$
no acceleration	$a = 9{\cdot}8$ m/s^2 ↓

using $s = ut + \frac{1}{2}at^2$, for vertical motion:

$$122{\cdot}5 = 0 + \tfrac{1}{2}(9{\cdot}8)t^2$$

$$\therefore \quad t^2 = \frac{2(122{\cdot}5)}{9{\cdot}8} = 25$$

$$\therefore \quad t = 5s$$

using distance = velocity × time, for horizontal motion:

$$x = 36 \times 5$$

$$\therefore \quad x = 180 \text{ m}$$

The time taken is 5 s and the horizontal distance travelled is 180 m.

Example 2

A particle is projected horizontally with a speed of 14·7 m/s. Find the horizontal and vertical displacements of the particle from the point of projection, after 2 seconds. Find also how far the particle then is from the point of projection.

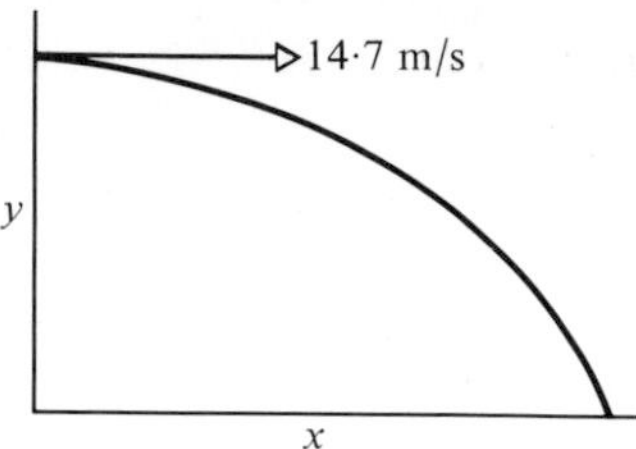

horizontal motion	vertical motion
$u = 14{\cdot}7$ m/s	$u = 0$
$s = x$	$s = y \downarrow$
$t = 2$ s	$t = 2$ s
no acceleration	$a = 9{\cdot}8$ m/s$^2 \downarrow$

using $s = ut + \frac{1}{2}at^2$, for vertical motion:

$$y = 0 + \tfrac{1}{2}(9{\cdot}8)2^2$$
$$\therefore \quad y = 19{\cdot}6 \text{ m}$$

using distance = velocity × time, for horizontal motion:

$$x = 14{\cdot}7 \times 2$$
$$= 29{\cdot}4 \text{ m}$$

Distance from point of projection $= \sqrt{(x^2 + y^2)}$
$= \sqrt{\{(29{\cdot}4)^2 + (19{\cdot}6)^2\}}$
$= 35{\cdot}33$ m

After 2 seconds the horizontal and vertical displacements are 29·4 m and 19·6 m respectively, and the displacement from the point of projection is 35·3 m.

Example 3

A particle is projected horizontally from a point 44·1 m above a horizontal plane. The particle hits the plane at a point which is, horizontally, 39 m from the point of projection. Find the initial speed of the particle.

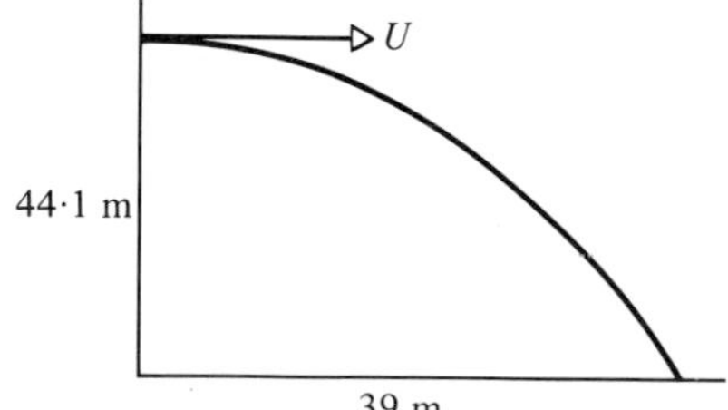

horizontal motion	vertical motion
$u = U$	$u = 0$
$s = 39$ m	$s = 44{\cdot}1$ m $\downarrow$
$t = t$	$t = t$
no acceleration	$a = 9{\cdot}8$ m/s$^2 \downarrow$

using $s = ut + \frac{1}{2}at^2$, for vertical motion:

$$44{\cdot}1 = 0 + \tfrac{1}{2}(9{\cdot}8)t^2$$
$$\therefore \quad t^2 = 9 \text{ i.e. } t = 3 \text{ s}$$

using distance = velocity × time, for horizontal motion:

$$39 = U \times 3$$
$$\therefore \quad U = 13 \text{ m/s}$$

The speed of projection is 13 m/s.

Example 4

A particle is projected horizontally with a velocity of 39·2 m/s. Find the horizontal and vertical components of the velocity of the particle 3 seconds after projection. Find, also, the speed and direction of motion of the particle at this time.

horizontal motion	vertical motion
$u = 39{\cdot}2$ m/s	$u = 0$
horizontal velocity	$v = V\downarrow$
is constant	$t = 3$ s
$\therefore\ U = 39{\cdot}2$ m/s	$a = 9{\cdot}8$ m/s$^2\downarrow$

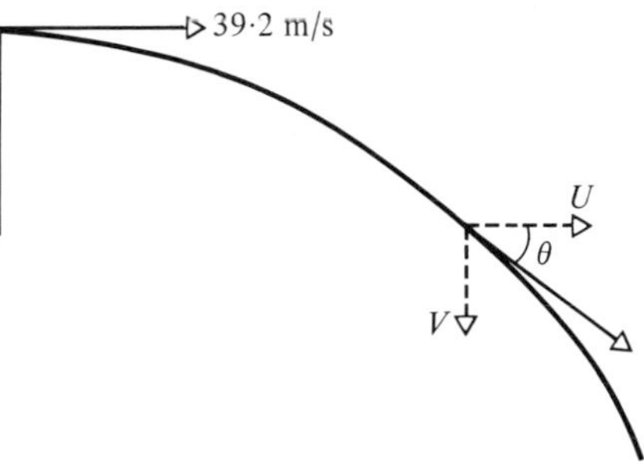

using $v = u + at$, for vertical motion:

$$V = 0 + (9{\cdot}8) \times 3$$
$$\therefore\quad V = 29{\cdot}4 \text{ m/s}$$

Speed after 3 s is given by: $\sqrt{(U^2 + V^2)}$

$$= \sqrt{\{(39{\cdot}2)^2 + (29{\cdot}4)^2\}} = 49 \text{ m/s}$$

Direction of motion is given by: $\tan\theta = \dfrac{V}{U} = \dfrac{29{\cdot}4}{39{\cdot}2}$

$$\therefore\quad \theta = 36{\cdot}87^\circ$$

The horizontal and vertical components are 39·2 m/s and 29·4 m/s respectively after 3 s; the speed of the particle is then 49 m/s and it is travelling at an angle of 36·87° below the horizontal.

Example 5

At time $t = 0$ a particle is projected with a velocity of 2**i** m/s from a point with position vector (10**i** + 90**j**) m. Find the position vector of the particle when $t = 4$ s.

Initial velocity is 2**i** m/s
∴ initial vertical velocity is zero and initial horizontal velocity is 2 m/s.

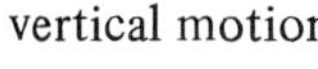

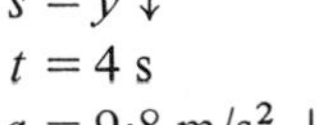

horizontal motion	vertical motion
$u = 2$ m/s	$u = 0$
$s = x$	$s = y\downarrow$
$t = 4$ s	$t = 4$ s
no acceleration	$a = 9{\cdot}8$ m/s$^2\downarrow$

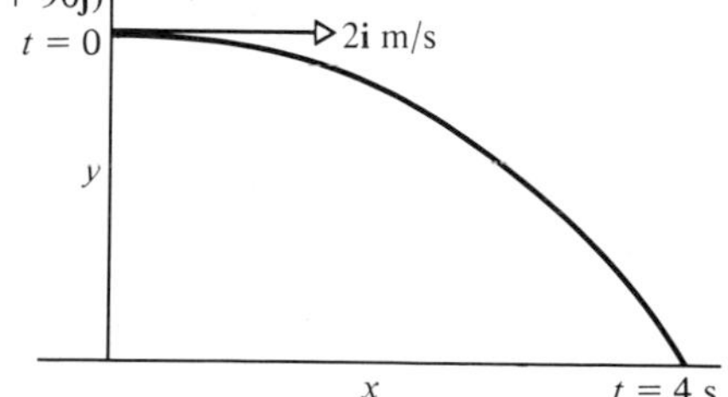

using $s = ut + \frac{1}{2}at^2$, for vertical motion:

$$y = 0 + \tfrac{1}{2}(9{\cdot}8)(4)^2$$
$$\therefore\quad y = 78{\cdot}4 \text{ m}$$

using distance = velocity × time for horizontal motion:

$$x = 2 \times 4 = 8 \text{ m}$$

Position vector after 4 seconds = 10**i** + 90**j** + 8**i** − 78·4**j** = 18**i** + 11·6**j**

The position vector after 4 seconds is (18**i** + 11·6**j**) m.

Exercise 11A

1. A particle is projected horizontally, at 20 m/s, from a point 78·4 m above a horizontal surface. Find the time taken for the particle to reach the surface and the horizontal distance travelled in that time.
2. A particle is projected horizontally with a speed of 21 m/s. Find the horizontal and vertical displacements of the particle, from the point of projection, $2\frac{6}{7}$ seconds after projection.
   Find how far the particle is then from the point of projection.
3. A particle is projected horizontally from a point 2·5 m above a horizontal surface. The particle hits the surface at a point which is, horizontally, 10 m from the point of projection. Find the initial speed of the particle.
4. At time $t = 0$ a particle is projected with velocity $5\mathbf{i}$ m/s from a point with position vector $20\mathbf{j}$ m. Find the position vector of the particle when (a) $t = 1$ s, (b) $t = 2$ s.
5. At time $t = 0$ a particle is projected with a velocity of $3\mathbf{i}$ m/s from a point with position vector $(5\mathbf{i} + 25\mathbf{j})$ m. Find the position vector of the particle when $t = 2$ s.
6. A stone is thrown horizontally at 21 m/s from the edge of a vertical cliff and falls to the sea, 40 m below. Find the horizontal distance from the foot of the cliff to the point where the stone enters the sea.
7. A batsman strikes a ball horizontally when it is 1 m above the ground. The ball is caught 10 cm above the ground by a fielder standing 12 m from the batsman. Find the speed with which the batsman hits the ball.
8. A darts player throws a dart horizontally with a speed of 14 m/s and it hits the board at a point 10 cm below the level at which it was released. Find the horizontal distance travelled by the dart.
9. A tennis ball is served horizontally with an initial speed of 21 m/s from a height of 2·8 m. By what distance does the ball clear a net 1 m high situated 12 m horizontally from the server?
10. A fielder retrieves a cricket ball and throws it horizontally with a speed of 28 m/s to the wicket-keeper standing 12 m away. If the fielder releases the ball at a height of 2 m above level ground, find the height of the ball when it reaches the wicket-keeper.
11. A gun is situated at the edge of a vertical cliff of height 90 m and has its barrel horizontal and pointing directly out to sea. The gun fires a shell at 147 m/s. How far from the base of the cliff does the shell strike the sea?
12. When an aircraft is flying horizontally at a speed of 420 km/h, it releases a bomb which, on release, has the same velocity as that of the aircraft. The bomb is released when the aircraft is a distance 2 km horizontally and h km vertically from a target. Given that the bomb hits the target, find the value of h.
13. A particle is projected horizontally at 20 m/s. Find the horizontal and vertical components of the particle's velocity two seconds after projection.
14. A particle is projected horizontally at 168 m/s. Find the magnitude and direction of the velocity of the particle five seconds after projection.
15. Initially a particle is at an origin O and is projected with a velocity of $a\mathbf{i}$ m/s. After t seconds, the particle is at the point with position vector $(30\mathbf{i} - 10\mathbf{j})$ m. Find the value of t and a.
16. Two vertical towers stand on horizontal ground and are of heights 40 m and 30 m. A ball is thrown horizontally from the top of the higher tower with a speed of 24·5 m/s and just clears the smaller tower. Find the distance (a) between the two towers, and (b) between the smaller tower and the point on the ground where the ball first lands.

17. A window in a house is situated 4·9 m above ground level. When a boy throws a ball horizontally from this window with a speed of 14 m/s, the ball just clears a vertical wall situated 10 m from the house. Find the height of the wall and how far beyond the wall the ball first hits the ground.
18. The top of a vertical tower is 20 m above ground level. When a ball is thrown horizontally from the top of this tower, it first hits the ground 24 m from the base of the tower. By how much does the ball clear a vertical wall of height 13 m situated 12 m from the tower?
19. A stone is thrown horizontally with speed u from the edge of a vertical cliff of height h. The stone hits the ground at a point which is a distance d horizontally from the base of the cliff. Show that $2hu^2 = gd^2$.
20. A vertical tower stands with its base on horizontal ground. Two particles A and B are both projected horizontally and in the same direction from the top of the tower with initial velocities of 14 m/s and 17·5 m/s respectively. If A and B hit the ground at two points 10 m apart, find the height of the tower.
21. O, A and B are three points with O on level ground and A and B respectively 3·6 metres and 40 metres vertically above O. A particle is projected horizontally from B with a speed of 21 m/s and, 2 seconds later, a particle is projected horizontally from A with a speed of 70 m/s. Show that the two particles reach the ground at the same time and at the same distance from O, and find this distance.
22. A and B are two points on level ground. A vertical tower of height $4h$ has its base at A and a vertical tower of height h has its base at B. When a stone is thrown horizontally with speed v from the top of the taller tower towards the smaller tower, it lands at a point X where $AX = \frac{3}{4}AB$. When a stone is thrown horizontally with speed u from the top of the smaller tower towards the taller tower, it also lands at the point X. Show that $3u = 2v$.

Particle projected at an angle to the horizontal

Suppose a particle is projected at an angle θ above the horizontal and with a velocity u.

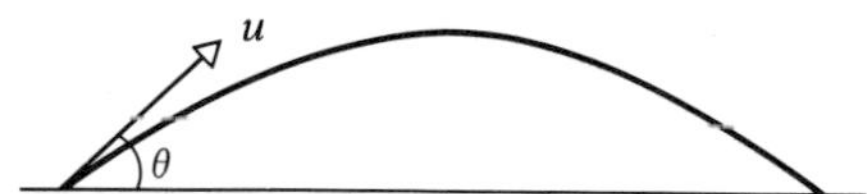

Again it is necessary to consider the motion in a horizontal and in a vertical direction separately.
The horizontal velocity remains constant and is $u \cos \theta$.
The vertical velocity, initially $u \sin \theta$, is reduced by the gravitational constant g.

Example 6

A particle is projected from a point on a horizontal plane and has an initial velocity of $28\sqrt{3}$ m/s at an angle of elevation of 60°. Find the greatest height reached by the particle and the time taken to reach this point.

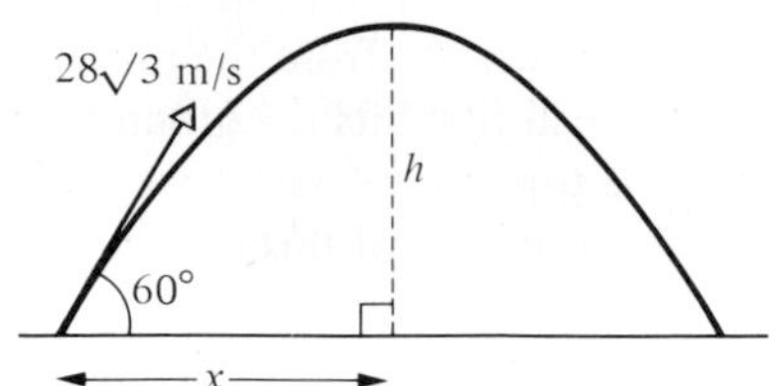

Consider the motion of the particle to the highest point of the path.

horizontal motion	vertical motion
$u = 28\sqrt{3}\cos 60$ m/s	$u = 28\sqrt{3}\sin 60$ m/s ↑
$s = x$	$s = h$ ↑
$t = t$	$v = 0$ (at highest point)
no acceleration	$a = 9{\cdot}8$ m/s² ↓ $= -9{\cdot}8$ m/s² ↑
	$t = t$

using $v^2 = u^2 + 2as$, for vertical motion: $0 = (28\sqrt{3}\sin 60)^2 + 2(-9{\cdot}8)h$

$\therefore\ h = 90$ m

using $v = u + at$, for vertical motion: $0 = 28\sqrt{3}\sin 60 + (-9{\cdot}8)t$

$\therefore\ t = 4\frac{2}{7}$ s

The greatest height reached is 90 m and the time taken is 4·29 s.

Example 7

A particle is projected from a point on a horizontal plane and has an initial velocity of 45 m/s at an angle of elevation of $\tan^{-1}\frac{3}{4}$. Find the time of flight and the range of the particle on the horizontal plane.

Consider the motion of the particle from A to B.

horizontal motion	vertical motion
$u = 45\cos\theta$ m/s	$u = 45\sin\theta$ m/s ↑
$s = R$	$s = 0$
$t = t$	$t = t$
no acceleration	$a = 9{\cdot}8$ m/s² ↓
	$= -9{\cdot}8$ m/s² ↑

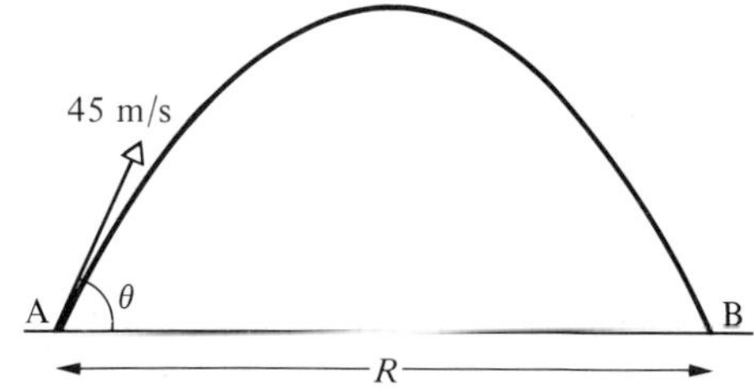

Since $\tan\theta = \frac{3}{4}$, $\sin\theta = \frac{3}{5}$ and $\cos\theta = \frac{4}{5}$.

Using $s = ut + \frac{1}{2}at^2$, for vertical motion:

$0 = 27(t) - \frac{1}{2}(9{\cdot}8)t^2$ $\quad\therefore\ t = 0$, at A or $t = \dfrac{27 \times 2}{9{\cdot}8} = 5{\cdot}51$ s, at B

Using distance = velocity × time, for horizontal motion: $R = 36 \times 5{\cdot}51 = 198{\cdot}4$ m

The time of flight is 5·51 s and the range is 198 m.

Symmetrical path

It should be noted that the trajectory, i.e. the path of a particle projected from a point A on a horizontal plane and striking the plane at some point B, is symmetrical.
The time of flight from A to B is twice the time taken by the particle to reach the highest point T. Hence the time of flight may be calculated by doubling the time taken to reach the highest point of the path.
At the highest point T, the vertical velocity is zero and the particle is instantaneously travelling horizontally.

The direction of motion at any other time is found by considering the horizontal and vertical components of the velocity. During the first half of the motion, the particle is travelling at an angle above the horizontal and in the second half, below the horizontal.

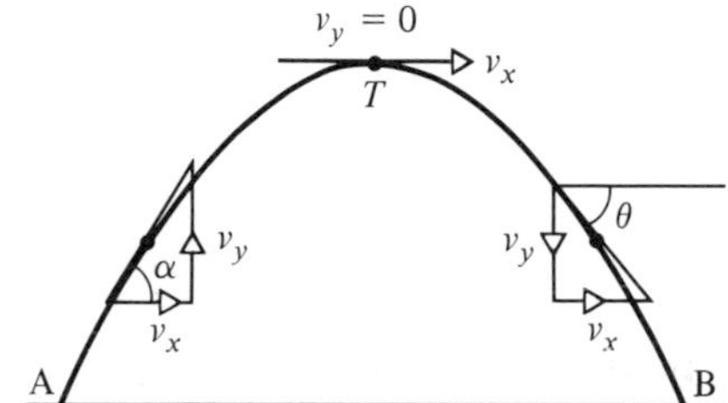

Accuracy and value of *g*

Some of the examples and questions in this chapter instruct the reader to take the value of g to be 10 m/s^2 so that the arithmetic involved is simplified. The treatment of projectiles in this chapter already ignores air resistances, so the 2% error incurred by taking g as 10 m/s^2 rather than 9·8 m/s^2 will not greatly affect the accuracy of the answers obtained. However the reader should use the more accurate value of 9·8 m/s^2 unless instructed otherwise.

Example 8

A particle is projected from an origin O with a velocity of (30**i** + 40**j**) m/s. Find the position and velocity vectors of the particle 5 seconds later.
Hence find the distance of the particle from O and the speed and direction of its motion at this time. Take $g = 10$ m/s^2.

horizontal motion	vertical motion
$u = 30$ m/s	$u = 40$ m/s ↑
$s = x$	$s = y$ ↑
$t = 5$ s	$t = 5$ s
no acceleration	$a = 10$ m/s^2 ↓
∴ $U = 30$ m/s	$= -10$ m/s^2 ↑
	$v = V$ ↑

using $s = ut + \frac{1}{2}at^2$, for vertical motion:

$$y = 40(5) - \tfrac{1}{2}(10)5^2$$
$$\therefore \quad y = 75 \text{ m}$$

using $v = u + at$, for vertical motion:

$$V = 40 - (10)5$$
$$\therefore \quad V = -10 \quad \text{i.e. } 10 \text{ m/s downwards}$$

using distance = velocity × time, for horizontal motion:

$$x = 30 \times 5$$
$$\therefore \quad x = 150 \text{ m}$$

Position vector is $(x\mathbf{i} + y\mathbf{j}) = (150\mathbf{i} + 75\mathbf{j})$ m

Velocity vector is $(U\mathbf{i} + V\mathbf{j}) = (30\mathbf{i} - 10\mathbf{j})$ m/s

Distance of particle, from O, is $\sqrt{(x^2 + y^2)} = \sqrt{(150^2 + 75^2)}$
$= 75\sqrt{5}$ m

Speed of particle is $\sqrt{(U^2 + V^2)} = \sqrt{\{(30)^2 + (-10)^2\}}$
$= 10\sqrt{10}$ m/s

Direction of motion is at an angle θ below the horizontal, given by:

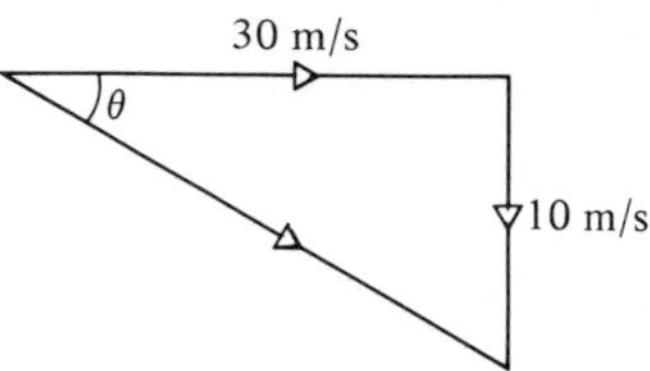

$\tan\theta = \frac{10}{30}$

$\therefore \quad \theta = 18{\cdot}43°$

The position vector is $(150\mathbf{i} + 75\mathbf{j})$ m and the velocity vector is $(30\mathbf{i} - 10\mathbf{j})$ m/s; the distance from O at this time is $75\sqrt{5}$ m and the speed of the particle is $10\sqrt{10}$ m/s at an angle of 18·43° below the horizontal.

Example 9

A stone is thrown from the edge of a vertical cliff with a velocity of 50 m/s at an angle $\tan^{-1}\frac{7}{24}$ above the horizontal. The stone strikes the sea at a point 240 m from the foot of the cliff. Find the time for which the stone is in the air and the height of the cliff.

Let θ be the angle of projection, then $\tan\theta = \frac{7}{24}$, $\sin\theta = \frac{7}{25}$ and $\cos\theta = \frac{24}{25}$.

horizontal motion	vertical motion
$u = 50\cos\theta = 48$ m/s	$u = 50\sin\theta$
	$= 14$ m/s ↑ $= -14$ m/s ↓
$s = 240$ m	$s = h$ ↓
$t = t$	$t = t$
no acceleration	$a = 9{\cdot}8$ m/s² ↓

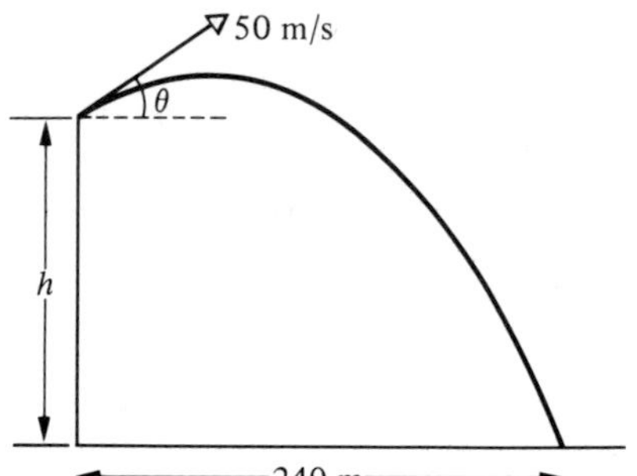

using distance = velocity × time, for horizontal motion:

$240 = 48t$

$\therefore \quad t = 5$ s

using $s = ut + \frac{1}{2}at^2$ for vertical motion: $h = -14(5) + \frac{1}{2}(9{\cdot}8)5^2$
$= 52{\cdot}5$ m

The time of flight of the stone is 5 s and the height of the cliff is 52·5 m.

Example 10

A shell fired from a gun has a horizontal range of 960 m and a time of flight of 12 s. Find the magnitude and direction of the velocity of projection. Take $g = 10\text{ m/s}^2$.

Consider the motion of the shell from A to B

horizontal motion	vertical motion
$u = U$	$u = V\uparrow$
$t = 12\text{ s}$	$t = 12\text{ s}$
$s = 960\text{ m}$	$s = 0$
no acceleration	$a = 10\text{ m/s}^2\downarrow = -10\text{ m/s}^2\uparrow$

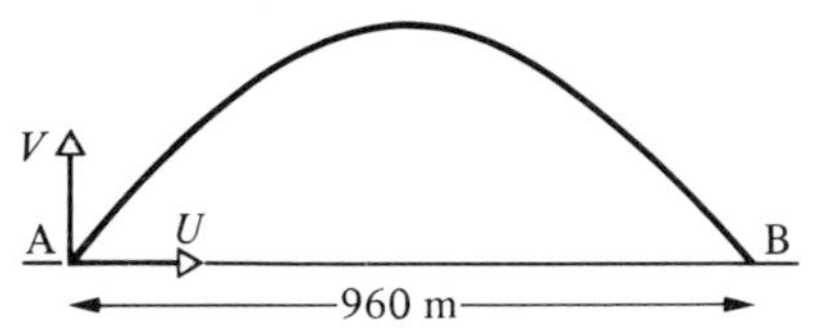

using distance = velocity × time, for horizontal motion: $960 = U \times 12$

$\therefore\quad U = 80\text{ m/s}$

using $s = ut + \frac{1}{2}at^2$, for vertical motion: $0 = (V \times 12) - \frac{1}{2}(10)12^2$

$\therefore\quad V = 60\text{ m/s}$

speed of projection is $\sqrt{(U^2 + V^2)} = \sqrt{(80^2 + 60^2)} = 100\text{ m/s}$

angle of projection is θ above the horizontal, given by

$$\tan\theta = \tfrac{60}{80}$$

or $\theta = 36{\cdot}87°$

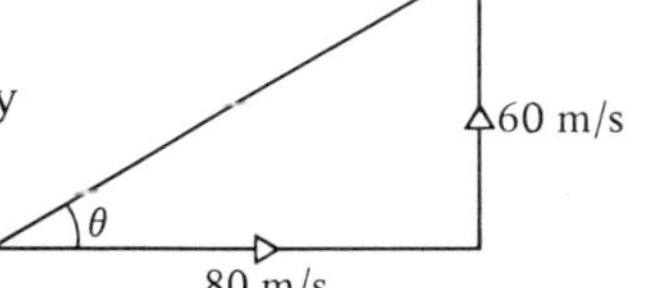

The shell is projected at 100 m/s at an angle of 36·87° above the horizontal.

Example 11

A golfer standing on level ground hits a ball with a velocity of 52 m/s at an angle α above the horizontal. If $\tan\alpha = \frac{5}{12}$ find the time for which the ball is at least 15 m above the ground. (Take $g = 10$ m/s.)

Since $\tan\alpha = \frac{5}{12}$, $\sin\alpha = \frac{5}{13}$ and $\cos\alpha = \frac{12}{13}$.

vertical motion

$u = 52\sin\alpha = 20\text{ m/s}\uparrow$

$s = 15\text{ m}\uparrow$

$t = t$

$a = 10\text{ m/s}^2\downarrow = -10\text{ m/s}^2\uparrow$

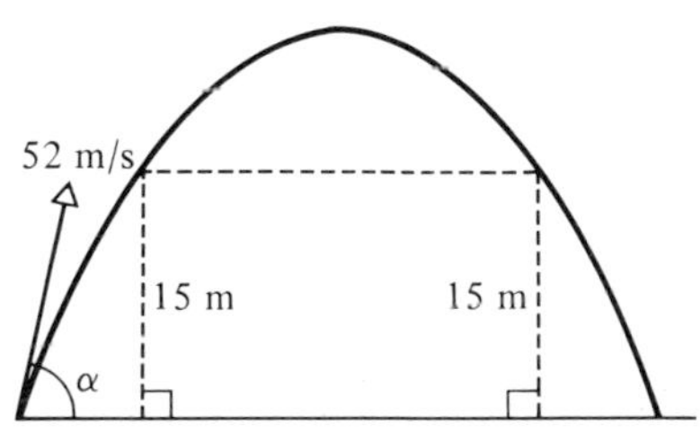

using $s = ut + \frac{1}{2}at^2$, for vertical motion:

$$15 = 20t - \tfrac{1}{2}(10)t^2$$

$$\therefore\quad t^2 - 4t + 3 = 0$$

$$\therefore\quad (t-3)(t-1) = 0 \quad \text{i.e. } t = 3 \text{ or } 1\text{ s}$$

Hence the ball is at a height of 15 m, 1 second after projection and 3 seconds after projection. The ball is at least 15 m above the ground for 2 seconds.

Example 12

A ball is projected with a velocity of 28 m/s. If the range of the ball on the horizontal plane is 64 m, find the two possible angles of projection.

Let the angle of projection be θ.

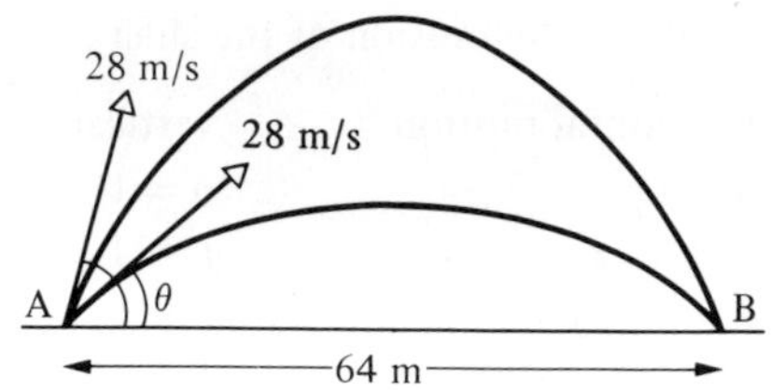

horizontal motion	vertical motion
$u = 28\cos\theta$ m/s	$u = 28\sin\theta$ m/s ↑
$s = 64$ m	$s = 0$
$t = t$	$t = t$
no acceleration	$a = 9{\cdot}8$ m/s² ↓ $= -9{\cdot}8$ m/s² ↑

using $s = ut + \frac{1}{2}at^2$, for vertical motion:

$$0 = (28\sin\theta)t - \tfrac{1}{2}(9{\cdot}8)t^2 \quad \therefore \quad t = 0 \text{ i.e. at A} \quad \text{or} \quad t = \frac{2(28\sin\theta)}{9{\cdot}8} \text{ at B}$$

using distance = velocity × time, for horizontal motion:

$$64 = 28\cos\theta \times \frac{56\sin\theta}{9{\cdot}8}$$

using $2\sin\theta\cos\theta = \sin 2\theta$, $\quad \sin 2\theta = \dfrac{64 \times 9{\cdot}8}{28 \times 28} = 0{\cdot}8 \quad \therefore \quad 2\theta = 53{\cdot}13^\circ$ or $126{\cdot}87^\circ$

$\therefore \quad \theta = 26{\cdot}57^\circ$ or $63{\cdot}44^\circ$

The two possible angles of projection are $26{\cdot}57^\circ$ and $63{\cdot}44^\circ$.

Exercise 11B

Take g as $9{\cdot}8$ m/s² unless instructed otherwise.

1. A particle is projected from a point on level ground such that its initial velocity is 56 m/s at an angle of 30° above the horizontal. Find
   (a) the time taken for the particle to reach its maximum height,
   (b) the maximum height, (c) the time of flight, (d) the horizontal range of the particle.
2. A particle is projected from a point on level ground such that its initial velocity is 60 m/s at an angle of elevation 30°. Taking g as 10 m/s², find
   (a) the time taken for the particle to reach its maximum height,
   (b) the maximum height, (c) the time of flight, (d) the horizontal range of the particle.
3. Find the horizontal range of a particle projected with a velocity of 35 m/s at an angle of $\sin^{-1}\frac{4}{5}$ above the horizontal.
4. A particle is projected from ground level with a velocity of $40\sqrt{3}$ m/s at an angle of elevation of 60°. Taking $g = 10$ m/s², find the greatest height reached by the particle and the time taken to achieve it.
5. A golfer hits a golf ball with a velocity of 44·1 m/s at an angle of $\sin^{-1}\frac{3}{5}$ above the horizontal. The ball lands on the green at a point which is level with the point of projection. Find the time for which the golf ball was in the air.
6. A gun has its barrel set at an angle of elevation of 15°. The gun fires a shell with an initial speed of 210 m/s. Find the horizontal range of the shell.

7. In a game of indoor football, a free kick is awarded whenever the ball is kicked above shoulder height. The ball is initially at rest on the floor when a player kicks it with a velocity of 14 m/s at an angle of elevation θ, with $\sin\theta = \frac{2}{5}$. Taking shoulder height as 1·5 m and assuming nothing prevents the ball from reaching its highest point, find whether a free kick is awarded or not.
8. A ball is projected from horizontal ground and has an initial velocity of 20 m/s at an angle of elevation $\tan^{-1}\frac{7}{24}$. When the ball is travelling horizontally, it strikes a vertical wall. How high above ground level does the impact occur?
9. A particle is projected from an origin O at a velocity of $(4\mathbf{i} + 13\mathbf{j})$ m/s. Find the position vector of the particle 2 seconds after projection and the distance the particle is then from O. (Take $g = 10$ m/s^2).
10. A particle is projected from the origin at a velocity of $(10\mathbf{i} + 20\mathbf{j})$ m/s. Find the position and velocity vectors of the particle 3 seconds after projection. (Take $g = 10$ m/s^2.)
11. A particle is projected at a velocity of 25 m/s at an angle of 30° above the horizontal. Find the horizontal and vertical components of the particle's velocity $2\frac{1}{2}$ seconds after projection and hence find the speed and direction of motion of the particle at that time.
12. A particle is projected from an origin O and has an initial velocity of $30\sqrt{2}$ m/s at an angle of 45° above the horizontal. Find the horizontal and vertical displacements from O of the particle 2 seconds after projection and hence find its distance from O at that time.
13. A particle is projected from an origin O at a velocity of $(4\mathbf{i} + 11\mathbf{j})$ m/s and passes through a point P which has a position vector $(8\mathbf{i} + x\mathbf{j})$ m. Taking $g = 10$ m/s^2 find the time taken for the particle to reach P from O and the value of x.
14. A stone is thrown from the edge of a vertical cliff and has an initial velocity of 26 m/s at an angle $\tan^{-1}\frac{5}{12}$ below the horizontal. The stone hits the sea at a point level with the base of the cliff and 72 m from it. Find the height of the cliff and the time for which the stone is in the air. (Take $g = 10$ m/s^2.)
15. A batsman hits a ball at a velocity of 17·5 m/s angled at $\tan^{-1}\frac{3}{4}$ above the horizontal, the ball initially being 60 cm above level ground. The ball is caught by a fielder standing 28 m from the batsman. Find the time taken for the ball to reach the fielder and the height above ground at which he takes the catch.
16. A vertical tower stands on level ground. A stone is thrown from the top of the tower and has an initial velocity of 24·5 m/s angled at $\tan^{-1}\frac{4}{3}$ above the horizontal. The stone strikes the ground at a point 73·5 m from the foot of the tower. Find the time taken for the stone to reach the ground and the height of the tower.
17. A particle projected from a point on level ground has a horizontal range of 240 m and a time of flight of 6 s. Find the magnitude and direction of the velocity of projection. (Take $g = 10$ m/s^2.)
18. A particle is projected from ground level with an initial speed of 28 m/s and during the course of its motion it must not go higher than 10 m above ground level. Find the angle of projection that would allow the particle to go as high as possible.
19. Ten seconds after its projection from the origin a particle has a position vector of $(150\mathbf{i} - 200\mathbf{j})$ m. Find, in vector form, the velocity of projection.
20. A football is kicked from a point on level ground, 15 m from a vertical wall. Three seconds later the football hits the wall at a point 6 m above ground. Find the horizontal and vertical components of the initial velocity of the ball. (Take $g = 10$ m/s^2.)
21. A football is kicked from a point O on level ground and, 2 seconds later, just clears a vertical wall of height 2·4 m. If O is 22 m from the wall, find the velocity with which the ball is kicked.

22. A particle is projected from a horizontal plane and has an initial velocity of 49 m/s at an angle of 30° above the horizontal. For how long is the particle at least 19·6 m above the level of the plane?
23. A particle is projected from a horizontal plane and has an initial velocity of 50 m/s at an angle of $\tan^{-1}\frac{4}{3}$ above the horizontal. For how long is the particle at least 60 m above the level of the plane? (Take $g = 10$ m/s^2.)
24. A particle is projected from the origin and has an initial velocity of $(7\mathbf{i} + 5\mathbf{j})$ m/s. Given that the particle passes through the point P, position vector $(x\mathbf{i} - 30\mathbf{j})$ m, find the time taken for this to occur and the value of x. (Take $g = 10$ m/s^2.)
25. A stone is thrown from the top of a vertical cliff, 100 m above sea level. The initial velocity of the stone is 13 m/s at an angle of elevation of $\tan^{-1}\frac{5}{12}$. Find the time taken for the stone to reach the sea and its horizontal distance from the cliff at that time. ($g = 10$ m/s^2.)
26. A golfer hits a golf ball with a velocity of 30 m/s at an angle of $\tan^{-1}\frac{4}{3}$ above the horizontal. The ball lands on a green, 5 m below the level from which it was struck. Find the horizontal distance travelled by the ball. (Take $g = 10$ m/s^2.)
27. A ball is thrown from the top of a vertical tower, 40 m above level ground. The initial velocity of the ball is $10\sqrt{2}$ m/s at an angle of 45° below the horizontal. Find the distance from the foot of the tower to the point where the ball first lands. (Take $g = 10$ m/s^2.)
28. A particle is projected at 84 m/s to hit a point 360 m away and on the same horizontal level as the point of projection. Find the two possible angles of projection.
29. A golfer hits a golf ball at 30 m/s and wishes it to land at a point 45 m away, on the same horizontal level as the starting point. Find the two possible angles of projection.
30. A ball is projected from horizontal ground and has an initial speed of 35 m/s. When the ball is travelling horizontally, it strikes a vertical wall. If the wall is 25 m from the point of projection, find the two possible angles of projection of the ball.
31. A particle is projected from a point O, and passes through a point A when travelling horizontally. If A is 10 m horizontally and 8 m vertically from O, find the magnitude and direction of the velocity of projection.
32. A batsman hits a ball from a point on level ground and gives the ball an initial velocity of 28 m/s at an angle of inclination of 15°. The ball lands 10 m short of the boundary. At this angle of projection, what is the least speed of projection for the ball to clear the boundary?
33. Two particles A and B are projected simultaneously, A from the top of a vertical cliff and B from the base. Particle A is projected horizontally with speed $3u$ m/s and B is projected at angle θ above the horizontal with speed $5u$ m/s. The height of the cliff is 56 m and the particles collide after 2 seconds. Find the horizontal and vertical distances from the point of collision to the base of the cliff and the values of u and θ.
34. Two stones are thrown from the top of a vertical cliff, 50 m above level ground. The stones are thrown at the same time and in the same vertical plane, one at 25 m/s and angle of elevation $\tan^{-1}\frac{3}{4}$ and the other at 25 m/s and angle of depression $\tan^{-1}\frac{3}{4}$. Find the time interval between the stones hitting the ground and the horizontal distance between their points of impact. (Take $g = 10$ m/s^2.)
35. Two particles A and B are projected simultaneously from the same point on horizontal ground and both travel in the same vertical plane; A is projected at an angle of elevation of 45° and a speed of 28 m/s and B is projected with a speed of 35 m/s. If the two particles land at points 15 metres apart find the four possible angles of projection of particle B.
36. A golfer wishes to hit a ball from a tee to a green which is at the same level as the tee. The front edge of the green is 160 m from the tee and the back edge is 185 m from the tee. If the golfer hits the ball with a speed of 49 m/s, find the possible angles of projection.

General results

Certain standard results can be established regarding the motion of a particle which is projected from a point O on a horizontal plane, with an initial velocity of U at an angle α above the horizontal.
The following examples illustrate how this may be done.

Example 13

Find the time of flight T and range R on the horizontal plane.

Consider the motion of the particle from A to B.

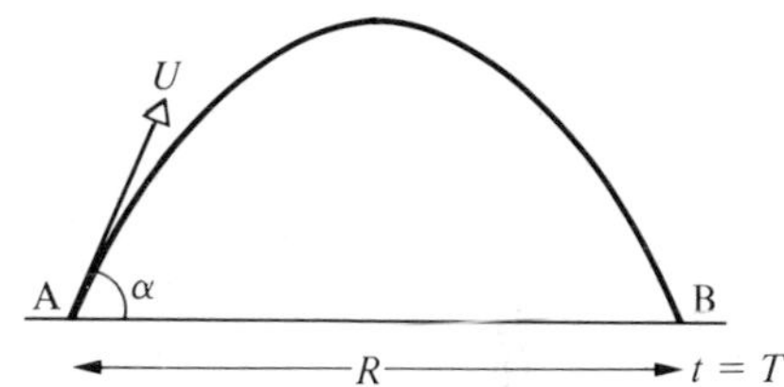

horizontal motion	vertical motion
$u = U\cos\alpha$	$u = U\sin\alpha\uparrow$
$s = R$	$s = 0$
$t = T$	$t = T$
no acceleration	$a = g\downarrow = -g\uparrow$

using $s = ut + \frac{1}{2}at^2$, for vertical motion:

$$0 = (U\sin\alpha)T - \tfrac{1}{2}gT^2$$

$$\therefore\quad T = 0 \text{ at A}$$

$$\text{or}\quad T = \frac{2U\sin\alpha}{g} \text{ at B}$$

using distance = velocity × time, for horizontal motion:

$$R = (U\cos\alpha)T$$

$$= \frac{2U^2\sin\alpha\cos\alpha}{g}$$

and using $2\sin\alpha\cos\alpha = \sin 2\alpha$

$$R = \frac{U^2\sin 2\alpha}{g}$$

The time of flight is $\dfrac{2U\sin\alpha}{g}$ and the range is $\dfrac{U^2\sin 2\alpha}{g}$.

Example 14

Find the maximum range of the particle on the horizontal plane and the angle of projection which gives this range.

From the previous example it is known that the range R is given by:

$$R = \frac{U^2\sin 2\alpha}{g}$$

For any particular value of U, the maximum value of R occurs when $\sin 2\alpha$ is a maximum;

maximum value of $2\alpha = 1$

when $\quad 2\alpha = 90^\circ$ or $\alpha = 45^\circ$

hence $\quad R_{max} = \dfrac{U^2\sin 90^\circ}{g} = \dfrac{U^2}{g}$

The maximum range is $\dfrac{U^2}{g}$ and the angle of projection to give this range is 45°.

Equation of trajectory

When a particle is projected from a point O on a horizontal plane, the equation of the trajectory may be obtained by taking x and y axes through the point of projection O, as shown in the diagram.

Consider the motion of the particle from O to the point P (x, y).

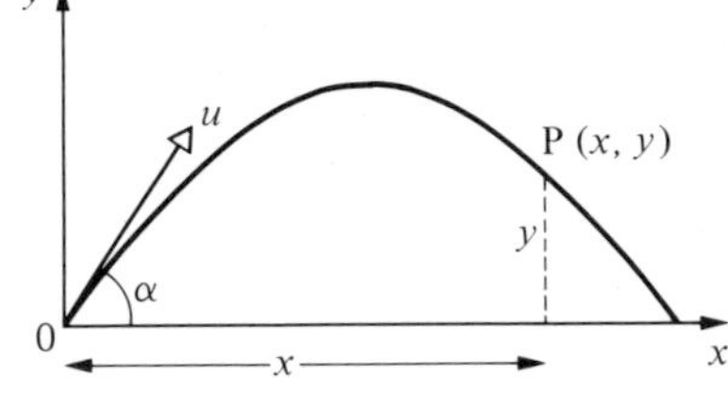

horizontal motion	vertical motion
$u = u \cos \alpha$	$u = u \sin \alpha \uparrow$
$s = x$	$s = y \uparrow$
$t = t$	$t = t$
no acceleration	$a = g \downarrow = -g \uparrow$

using $s = ut + \frac{1}{2}at^2$ for vertical motion:

$$y = (u \sin \alpha)t - \tfrac{1}{2}gt^2 \qquad \ldots [1]$$

using distance = velocity × time for horizontal motion: $x = (u \cos \alpha)t$

$$\therefore \quad t = \frac{x}{u \cos \alpha} \qquad \ldots [2]$$

substituting for t from equation [2] in equation [1]: $\therefore \quad y = \dfrac{(u \sin \alpha)x}{u \cos \alpha} - \tfrac{1}{2}g\left(\dfrac{x}{u \cos \alpha}\right)^2$

$$\therefore \quad y = x \tan \alpha - \frac{gx^2}{2u^2 \cos^2 \alpha}$$

and since $\dfrac{1}{\cos^2 \alpha} = \sec^2 \alpha = 1 + \tan^2 \alpha$

$$\therefore \quad y = x \tan \alpha - gx^2 \frac{(1 + \tan^2 \alpha)}{2u^2}$$

This is the equation of the trajectory referred to the axes shown.
There are four variables (x, y, u and α) involved in this equation and if three are known, the fourth can be determined in any particular example.

Exercise 11C

Take g as 9·8 m/s^2 unless instructed otherwise.

1. Greatest height. Time of flight. Range. Maximum range.
   A particle is projected from a point on a horizontal plane and has an initial velocity u at an angle of θ above the plane. Show that
   (a) the greatest height h reached by the particle above the plane is given by $h = \dfrac{u^2 \sin^2 \theta}{2g}$,
   (b) the time of flight T is given by $T = \dfrac{2u \sin \theta}{g}$,
   (c) the horizontal range R is given by $R = \dfrac{u^2 \sin 2\theta}{g}$,
   (d) the maximum horizontal range R_{max} is given by $R_{max} = \dfrac{u^2}{g}$ and find the value of θ for which this occurs.

The results of question **1** may be used without proof to answer questions **2** to **7**.

2. A particle is projected from a point on a horizontal plane and has an initial velocity of 140 m/s at an angle of elevation of 30°. Find the greatest height reached by the particle above the plane.
3. A particle is projected from a point on a horizontal plane and has an initial velocity of 70 m/s at an angle of 10° above the plane. Find the range of the particle on the horizontal plane.
   At this speed of projection, what is the maximum range the particle could have on the horizontal plane and for what angle of projection would it occur?
4. A particle is projected from a point on a horizontal plane and has an initial velocity of $49\sqrt{2}$ m/s at an angle of 45° above the plane. If the particle returns to the plane, find (a) the time of flight, (b) the time taken to reach the highest point.
5. A bullet fired from a gun has a maximum horizontal range of 2000 m. Find the muzzle velocity of the gun (i.e. the speed with which the bullet leaves the gun).
6. A gun is on the same horizontal level as a target 484 metres away. A bullet is fired from the gun with a muzzle velocity of 154 m/s. At what angle of elevation should the muzzle of the gun be set in order to ensure the bullet hits the target?
7. A particle projected from a point on a horizontal plane reaches a greatest height h above the plane and has a horizontal range R. If $R = 2h$, find the angle of projection.
8. Equation of trajectory.
   A particle is projected from a point O and has an initial velocity u at an angle of θ above the horizontal. In the vertical plane of projection take x and y axes as the horizontal and vertical axes respectively and O as the origin. Show that the equation of the trajectory is $y = x \tan\theta - \dfrac{gx^2}{2u^2}\sec^2\theta$.
   If the particle passes through the point with coordinates (a, b), show that $ga^2\tan^2\theta - 2u^2a\tan\theta + 2u^2b + ga^2 = 0$.

The equation of the trajectory obtained in question **8** may be used without proof to answer questions **9** to **12**.

9. A particle is projected from a point on a horizontal plane and initially travels in a direction which makes an angle of $\tan^{-1}\frac{3}{4}$ with the horizontal. In the subsequent motion the particle passes through a point above the plane which is 20 m horizontally and 10 m vertically from the point of projection. Find the speed of projection. (Take $g = 10$ m/s^2.)
10. A particle is projected from a point on a horizontal plane and has an initial speed of 28 m/s. If the particle passes through a point above the plane, 40 m horizontally and 20 m vertically from the point of projection, find the possible angles of projection.
11. A particle is projected from a point on a horizontal plane and has an initial speed of 42 m/s. If the particle passes through a point above the plane, 70 m vertically and 60 m horizontally from the point of projection, find the possible angles of projection.
12. An origin O lies on a horizontal plane and a point P lies above the plane, 50 m horizontally and 60 m vertically from O. Show that a particle projected from O with speed 35 m/s cannot pass through P.

Projectile on an inclined plane

Suppose that a particle is projected from a point O up a plane inclined at an angle α to the horizontal, and that the motion takes place in a vertical plane containing a line of greatest slope of the plane. Let the particle be projected at an angle β above the plane.

The motion of the particle is best investigated by considering the component of the initial velocity at right angles to the plane, i.e. $u \sin \beta$.

In this direction, the acceleration due to gravity is $-g \cos \alpha$.

When the particle strikes the plane at the point P, the displacement of the particle at right angles to the plane is zero.

The distance OP is called the range of the particle on the inclined plane.

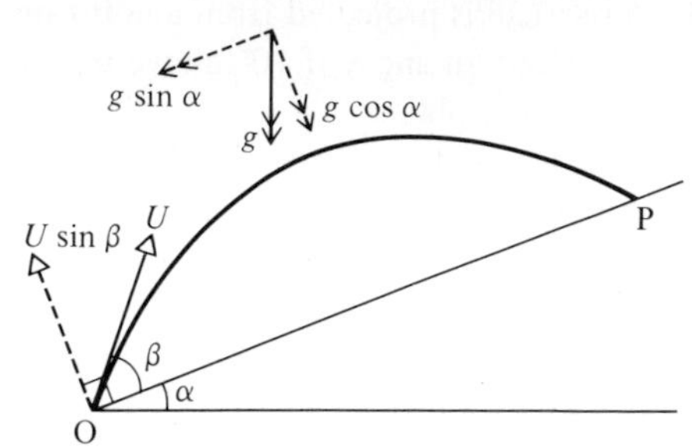

Example 15

A particle is projected up a plane which is inclined at 25° to the horizontal. The particle is projected from a point A on the plane with a velocity of 49 m/s at an angle of 50° to the plane, the vertical plane of the motion containing a line of greatest slope of the plane. Find the time taken by the particle to strike the plane and the range on the plane.

Suppose the particle strikes the plane at B.

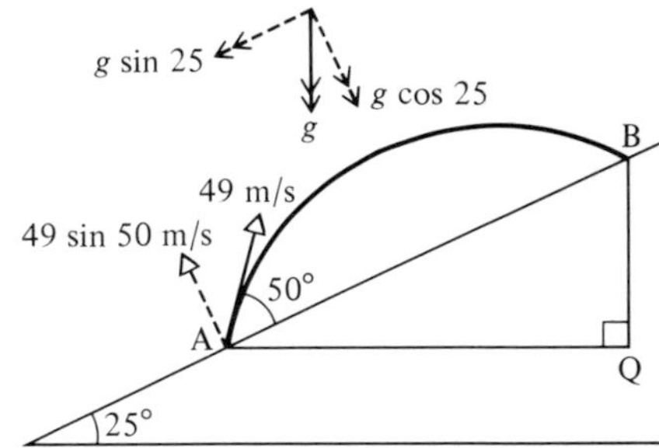

horizontal motion	motion at right angles to AB
$u = 49 \cos 75$ m/s	$u = 49 \sin 50$ m/s
$s = \text{AQ}$	$s = 0$
$t = t$	$t = t$
no acceleration	$a = -g \cos 25$ m/s^2

using $s = ut + \frac{1}{2}at^2$ for motion at right angles to AB:

$$0 = (49 \sin 50)t - \tfrac{1}{2}(g \cos 25)t^2$$

$$\therefore \quad t = 0 \text{ at A} \quad \text{or} \quad t = \frac{98 \sin 50}{g \cos 25} = 8{\cdot}452 \text{ s at B}$$

From the triangle ABQ, the range AB is given by:

$$\text{AB} = \frac{\text{AQ}}{\cos 25},$$

but AQ is the horizontal distance travelled by the particle,

$$\therefore \quad \text{AQ} = \text{horizontal velocity} \times \text{time}$$
$$= 49 \cos 75 \times 8{\cdot}452$$

$$\text{and} \quad \therefore \quad \text{AB} = \frac{49 \cos 75 \times 8{\cdot}452}{\cos 25}$$
$$= 118{\cdot}3 \text{ m}$$

The time of flight is 8·45 s and the range is 118 m.

Example 16

A particle is projected down a slope which is inclined at 30° to the horizontal. The particle is projected from a point A on the slope and has an initial velocity of 10·5 m/s at an angle θ to the slope, the vertical plane of the motion containing a line of greatest slope. If the range of the particle down the slope is 16·5 m, find the two possible values of θ.

Suppose the particle strikes the plane at P.

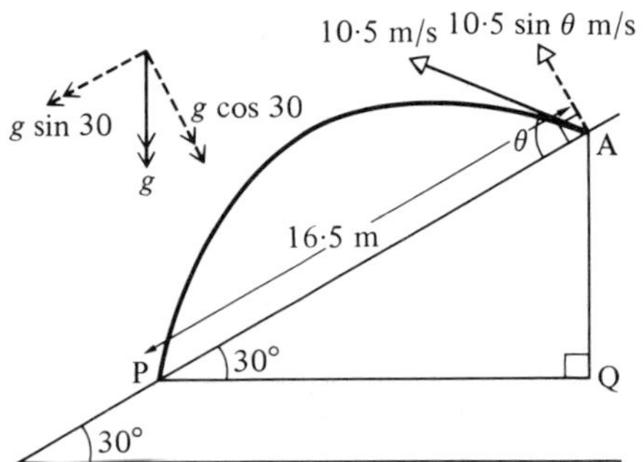

horizontal motion	motion at right angles to AP
$u = 10{\cdot}5\cos(\theta - 30)$ m/s	$u = 10{\cdot}5 \sin\theta$ m/s
$s = PQ = 16{\cdot}5\cos 30$ m	$s = 0$
$t = t$	$t = t$
no acceleration	$a = -g\cos 30$ m/s^2

using $s = ut + \frac{1}{2}at^2$ for motion at right angles to AP:

$$0 = (10{\cdot}5\sin\theta)t - \tfrac{1}{2}(g\cos 30)t^2$$

$$\therefore \quad t = 0 \text{ at A} \quad \text{or} \quad t = \frac{21\sin\theta}{g\cos 30} \text{ s at P}$$

From the triangle APQ, the range AP is given by:

$$AP = \frac{PQ}{\cos 30},$$

but PQ is the horizontal distance travelled by the particle,

$$\therefore \quad PQ = \text{horizontal velocity} \times \text{time}$$

$$= 10{\cdot}5\cos(\theta - 30) \times \frac{21\sin\theta}{g\cos 30}$$

Since the range AP is 16·5 m,

$$16{\cdot}5 = \frac{10{\cdot}5\cos(\theta - 30)}{\cos 30} \times \frac{21\sin\theta}{g\cos 30}$$

$$\therefore \quad 1{\cdot}1 = 2\sin\theta\cos(\theta - 30)$$

$$= \sin(2\theta - 30) + \sin 30$$

$$\therefore \quad 0{\cdot}6 = \sin(2\theta - 30)$$

$$\therefore \quad (2\theta - 30) = 36{\cdot}87^\circ \quad \text{or} \quad 143{\cdot}13^\circ$$

$$\therefore \quad \theta = 33{\cdot}44^\circ \quad \text{or} \quad 86{\cdot}57^\circ$$

The possible angles of projection are at $33{\cdot}44^\circ$ and $86{\cdot}57^\circ$ to the slope of the plane.

Exercise 11D

In every question of this exercise it should be assumed that the vertical plane of the motion contains a line of greatest slope, i.e. the motion is either *directly* down or *directly* up the slope.

1. A particle is projected up a plane which is inclined at 30° to the horizontal. The particle is projected from a point on the plane and has an initial velocity of $7\sqrt{3}$ m/s at 30° to the plane. Find (a) the time taken by the particle to strike the plane, (b) the range up the inclined plane.

2. A particle is projected up a plane which is inclined at 30° to the horizontal. The particle is projected from a point on the plane and has an initial velocity of 30 m/s at 20° to the plane. Find (a) the time taken by the particle to strike the plane, (b) the range up the inclined plane.
3. A particle is projected down a plane which is inclined at 30° to the horizontal. The particle is projected from a point on the plane and has an initial velocity of 14·7 m/s at 60° to the plane. Find (a) the time taken by the particle to strike the plane, (b) the range down the inclined plane.
4. A particle is projected down a plane which is inclined at 45° to the horizontal. The particle is projected from a point on the plane and has an initial velocity of $7\sqrt{2}$ m/s at 30° to the plane. Find (a) the time taken by the particle to first strike the plane, (b) the range down the inclined plane.
5. A particle is projected from the foot of a slope which is inclined at 15° to the horizontal, the initial direction of motion being at an angle of 30° to the slope. If the particle first strikes the slope 5 seconds after projection, find the initial speed of projection and the range up the inclined plane.
6. A particle is projected downwards from the top of a slope which is inclined at 30° to the horizontal, the initial direction of motion being at an angle of 45° to the slope. If the particle first strikes the slope 2 seconds after projection, find the initial speed of projection and the range down the incline.
7. A particle is projected down a slope which is inclined at 45° to the horizontal. The particle is projected from a point on the slope and has an initial velocity of $14\sqrt{2}$ m/s at an angle θ to the slope. If the particle first hits the slope 4 seconds after projection, find the value of θ.
8. A particle is projected down a slope which is inclined at 30° to the horizontal. The particle is projected from a point on the slope and has an initial velocity of 21 m/s at an angle θ to the slope. If the range of the particle down the slope is 60 m, find the two possible values of θ.
9. A particle is projected up a plane of inclination ϕ. It is projected from a point on the plane and has an initial velocity u at an angle θ to the plane.

   (a) Show that the particle will hit the plane after a time t where $t = \dfrac{2u \sin\theta}{g \cos\phi}$.

   (b) Show that the range R up the plane is given by $R = \dfrac{2u^2}{g \cos^2\phi} \sin\theta \cos(\theta + \phi)$.

   (c) If ϕ is fixed but θ can vary, show that the maximum range up the plane is given by $R_{max} = \dfrac{u^2}{g(1 + \sin\phi)}$, and find the relationship between θ and ϕ for which R_{max} occurs.
   Hint: use the identity $2 \sin A \cos B = \sin(A + B) + \sin(A - B)$.

Exercise 11E Examination questions

1. A ball is projected horizontally with speed 7 m/s from the top of a tower which is 90 m high and which stands on a horizontal plane. Find how far from the bottom of the tower the ball strikes the plane. (London)

2. An aircraft dives towards a stationary target which is at sea-level and when it is at a height of 1390 m above sea-level it launches a missile towards the target. The initial velocity of the missile is 410 m/s in a direction making an angle θ below the horizontal where $\tan\theta = \frac{9}{40}$. Calculate the time of flight of the missile from the instant it was launched until it reaches sea-level.
Show that if the target is hit then the horizontal distance of the aircraft from the target at the instant the missile was launched was 4 km. (J.M.B)

3. A particle P is projected from a point A at the top of a cliff 52 m vertically above sea-level and it moves freely under gravity until it strikes the sea at a point S. The velocity of projection of P has horizontal and upward vertical components of magnitudes 24 m/s and 7 m/s respectively. Calculate
   (i) the magnitude and angle of elevation of the velocity of projection,
   (ii) the time of flight of the particle,
   (iii) the horizontal distance between A and S,
   (iv) the speed of the particle at the instant when it strikes the sea.
   (Take the acceleration due to gravity to be 10 m/s^2.) (A.E.B)

4. The point A is at a height of 60 m vertically above the point B which is on horizontal ground. Particles P and Q are simultaneously projected from A and B respectively with the same speed, at the same angle of elevation, in the same vertical plane and in the same direction and they move freely under gravity. The particle Q strikes the ground again at C. Given that BC = 200 m and that this distance is covered in 4 seconds, calculate
   (i) the magnitudes of the horizontal and vertical components of the initial velocity of Q,
   (ii) the greatest height attained by Q above BC,
   (iii) the distance beyond C of the point D, where P strikes the ground,
   (iv) the angle made with the horizontal by the path of P at D.
   (Take the acceleration due to gravity to be 10 m/s^2.) (A.E.B)

5. A tennis player hits a ball at a point O, which is at a height of 2 m above the ground and at a horizontal distance of 4 m from the net, the initial speed being in a direction of 45° above the horizontal in a vertical plane perpendicular to the net. The ball just clears the net which is 1 m high. The acceleration due to gravity is 10 m/s^2.
   (i) Taking the horizontal and vertical through O in the plane of the motion as axes of x and y respectively, show that the equation of the path of the ball may be written in the form $y = x - \frac{5}{16}x^2$, assuming that the only force acting on the ball is that due to gravity. Find the initial speed of the ball.
   Find also,
   (ii) the distance from the net at which the ball strikes the ground;
   (iii) the magnitude and direction of the velocity with which the ball strikes the ground.
   All answers should be given correct to 2 significant figures. (S.U.J.B)

6. A particle is projected, with velocity V at an angle of elevation α, from a point O on the horizontal ground. It passes through a point P which is at a distance x horizontally from O and at a height y above the ground. Prove that $y = x\tan\alpha - \dfrac{gx^2}{2V^2}\sec^2\alpha$.
Show that, in general, there are two possible angles of projection and find these two angles in the particular case when the particle is projected with a speed of 40 m s^{-1} at a target

that lies at a horizontal distance of 160 m from, and 20 m below, the point of projection. Calculate, in the form $n : 1$, the ratio of the times of flight for the two trajectories. (Take the acceleration due to gravity to be 10 m/s^2.) (Cambridge)

7. A point O is at the foot of a plane which is inclined at an angle α to the horizontal. A particle is projected with speed V from O at angle of elevation θ to the horizontal, and moves in the vertical plane containing the line of greatest slope. It strikes the inclined plane horizontally. Express $\tan\theta$ in terms of $\tan\alpha$.

   Prove that the distance from O up the plane to the point of impact is $\dfrac{2V^2 \sec\alpha \tan\alpha}{g(1 + 4\tan^2\alpha)}$.

   (Oxford)

12
Circular motion

Radians

Angles are usually measured in degrees, minutes and seconds.
A radian is a larger unit which is sometimes used. It is defined as the angle subtended at the centre of a circle by an arc equal to the radius of the circle.

In the diagram, suppose that the radius OP is of length 1 unit and that the arc PQ is also of length 1 unit, then the angle POQ is 1 radian.
Since the circumference of this circle is 2π units, the angle at O subtended by the whole circumference is 2π radians.

hence $\quad 1 \text{ revolution} = 360^\circ = 2\pi \text{ radians}$

or $\quad \frac{1}{4} \text{ revolution} = 90^\circ = \frac{\pi}{2} \text{ radians.}$

As

$$2\pi \text{ radians} = 360^\circ$$

$$1 \text{ radian} = \frac{360^\circ}{2\pi} \text{ which is approximately } 57^\circ.$$

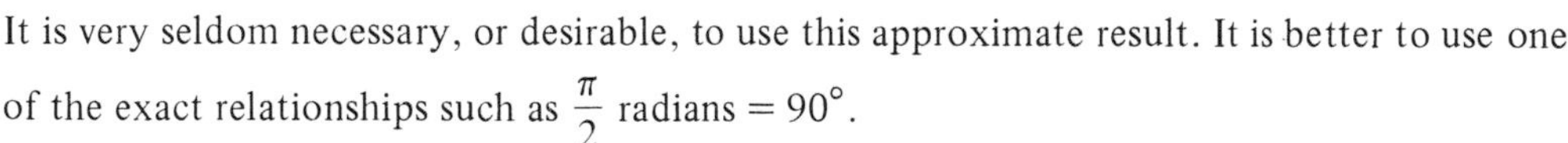

It is very seldom necessary, or desirable, to use this approximate result. It is better to use one of the exact relationships such as $\frac{\pi}{2}$ radians $= 90^\circ$.

Linear and angular speed

The linear speed of a body, i.e. its speed in a straight line, is usually measured in m/s, km/h or some similar unit.
When a body is moving on a circular path it is often useful to measure its speed as the rate of change of the angle at the centre of the circle.

Suppose a particle P moves along the circumference of a circle, centre O, at a constant speed. If the body moves along the arc from A to B in one second, then the angle AOB gives the change in angle per second.
Thus, if angle AOB is in radians, the angular speed of the body is in radians per seconds (rad/s).
If the particle makes 5 revolutions in one minute, its angular speed is 5 rev/min.

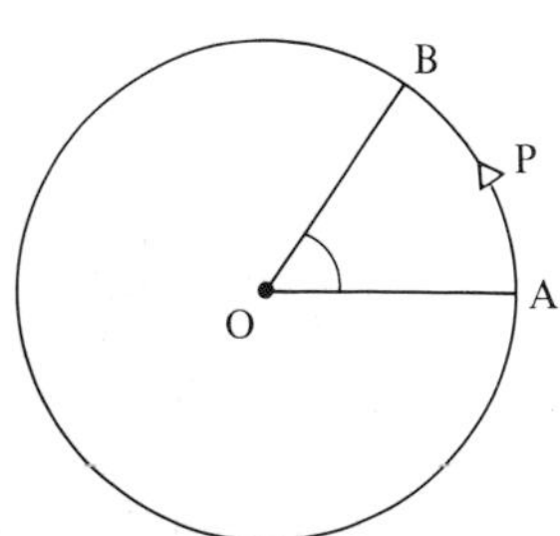

Example 1

Express an angular speed of (a) 15 rev/min in rad/min, (b) 3 rev/s in rad/min.

(a)
$$1 \text{ rev} = 2\pi \text{ rad}$$
$$\therefore \quad 15 \text{ rev/min} = 15 \times 2\pi \text{ rad/min}$$
$$= 30\pi \text{ rad/min}$$

(b)
$$1 \text{ rev} = 2\pi \text{ rad}$$
$$\therefore \quad 3 \text{ rev/s} = 3 \times 2\pi \times 60 \text{ rad/min}$$
$$= 360\pi \text{ rad/min}$$

Example 2

Express an angular speed of 8 rad/s in rev/min.

$$2\pi \text{ rad} = 1 \text{ rev}$$
$$\therefore \quad 8 \text{ rad/s} = \frac{8}{2\pi} \text{ rev/s} = \frac{240}{\pi} \text{ rev/min}$$

Relationship between linear and angular speed

When a particle P is moving along the circumference of a circle, centre O, at a constant speed, there is a relationship between the linear and angular speeds of the particle.

Suppose the linear speed is v m/s and the radius of the circle is r m. If the time taken to travel from A to B is 1 second then

$$\text{arc AB} = v \times 1 \text{ metres}$$

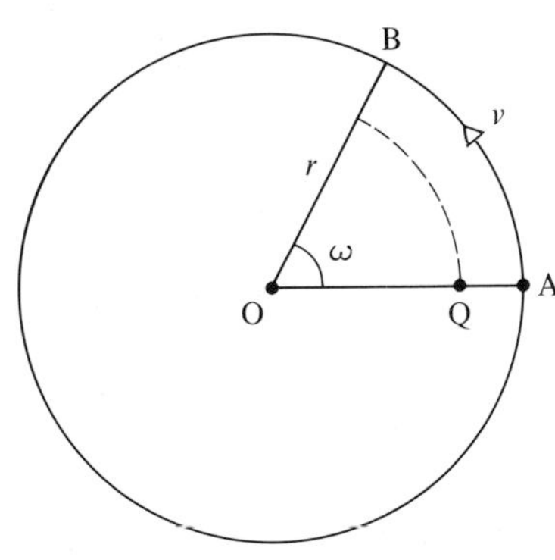

Suppose the angular speed is ω rad/s, then as the time to travel from A to B is one second, angle AOB is ω radians.

Then
$$\frac{\text{arc AB}}{2\pi r} = \frac{\omega}{2\pi}$$

substituting for arc AB,
$$\frac{v}{2\pi r} = \frac{\omega}{2\pi}$$
$$\therefore \quad \omega = \frac{v}{r} \quad \text{or} \quad v = r\omega$$

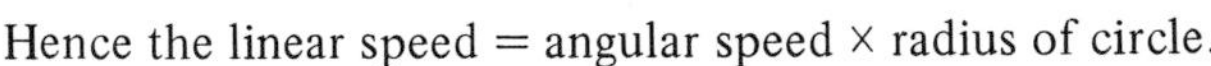

Hence the linear speed = angular speed × radius of circle.

Note that this relationship depends upon ω being measured in *radians* per unit of time.

As the particle P moves from A to B, all points on the radius OA will move to new positions on the radius OB.
The angular speed of P is equal to the angular speed of the point Q shown in the diagram, but the linear speed of P will be greater than that of Q.

Example 3

Find the speed in m/s of a particle which is moving on a circular path of radius 30 cm with an angular speed of (a) 4 rad/s, (b) 5 rev/min.

(a) $\omega = 4$ rad/s $\quad r = 30 \text{ cm} = \frac{30}{100}$ m

using $v = r\omega$

$v = \frac{30}{100} \times 4 = 1{\cdot}2$ m/s

The speed is 1·2 m/s.

(b) $\omega = 5$ rev/min $\quad r = 30 \text{ cm} = \frac{30}{100}$ m

$= 5 \times 2\pi$ rad/min

$= \dfrac{5 \times 2\pi}{60}$ rad/s

using $v = r\omega$

$\therefore \quad v = \dfrac{30}{100} \times \dfrac{5 \times 2\pi}{60} = \dfrac{\pi}{20}$ m/s

The speed is $\dfrac{\pi}{20}$ m/s.

Example 4

A particle moving on a circular path of radius 4 m has a speed of 36 m/s. Find the angular speed of the particle in (a) rad/s, (b) rev/min.

(a) $v = 36$ m/s $\quad r = 4$ m

using $v = r\omega$

$36 = 4 \times \omega$ or $\omega = 9$ rad/s

(b) 2π rad $= 1$ rev

$\therefore \quad 9 \text{ rad/s} = \dfrac{9}{2\pi} \text{ rev/s} = \dfrac{9}{2\pi} \times 60 \text{ rev/min} = \dfrac{270}{\pi}$ rev/min

The speed is 9 rad/s or $\dfrac{270}{\pi}$ rev/min.

Exercise 12A

(Leave π in the answers)

1. Express an angular speed of 100 rev/min in rad/min.
2. Express an angular speed of 90 rev/min in rad/s.
3. Express an angular speed of 10 rad/s in rev/min.
4. A uniformly rotating disc completes one revolution every ten seconds. Find the angular speed of the disc (a) in rev/min, (b) in rad/min, (c) in rad/s.
5. Find in rad/s the angular speed of a record revolving at (a) 45 rev/min, (b) $33\frac{1}{3}$ rev/min.
6. Find the speed in m/s of a particle which is moving on a circular path of radius 5 m with an angular speed of (a) 2 rad/s, (b) 300 rev/min.

7. Find the speed in m/s of a particle which is moving on a circular path of radius 50 cm with an angular speed of (a) 10 rad/s, (b) 20 rev/s.
8. A particle moving in a circular path of radius 50 cm has a speed of 12 m/s. Find the angular speed of the particle (a) in rad/s, (b) in rev/min.
9. A particle moving in a circular path of radius 20 cm has a speed of 20 cm/s. Find the angular speed of the particle (a) in rad/s, (b) in rev/min.
10. Sue and Pam stand on a playground roundabout at distances of 1 m and 1·5 m respectively from the centre of rotation. If the roundabout revolves at 12 rev/min, find the speeds with which Sue and Pam are moving.
11. Tom and John stand on a playground roundabout respectively 50 cm and 175 cm from the centre. The roundabout moves with constant angular velocity and Tom's speed is found to be 1 m/s. Find (a) the angular speed of the roundabout in rad/s, (b) the time taken for the roundabout to complete ten revolutions, (c) John's speed.

Motion on a circular path

When the speed of a car travelling in a straight line changes from 12 m/s to 15 m/s, it is said to have an acceleration because the velocity has changed.

If a body moves on a circular path, radius r, at a constant speed v, the *direction* of the motion is changing. The *velocity* of the body is therefore changing and again the body is said to have an acceleration, even though the magnitude of the velocity (i.e. its speed) is constant.

The acceleration of such a body at any instant is $\frac{v^2}{r}$ and is directed towards the centre of the circular path.

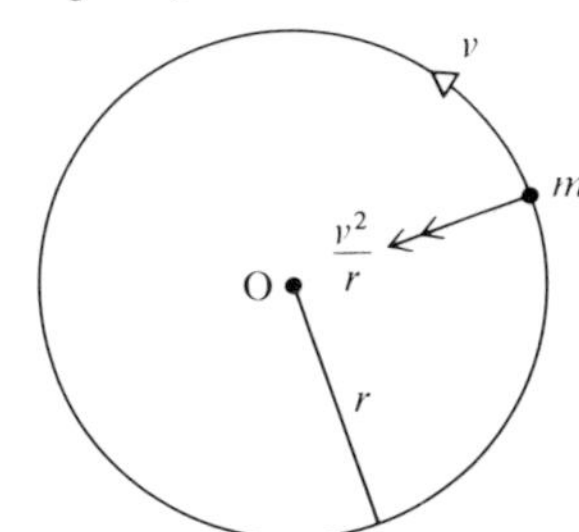

A proof of the above statement follows:

Suppose a particle is moving on a circular path of radius r with a constant speed v.

Let the time taken for the particle to travel from P to Q be δt, the angle POQ be $\delta\theta$ and the arc PQ be δs.

The velocity at P is v along the tangent PA and the velocity at Q is v along the tangent AQ.

The velocity at Q can be resolved into components of

$v \cos \delta\theta$ parallel to PA

and $v \sin \delta\theta$ perpendicular to PA.

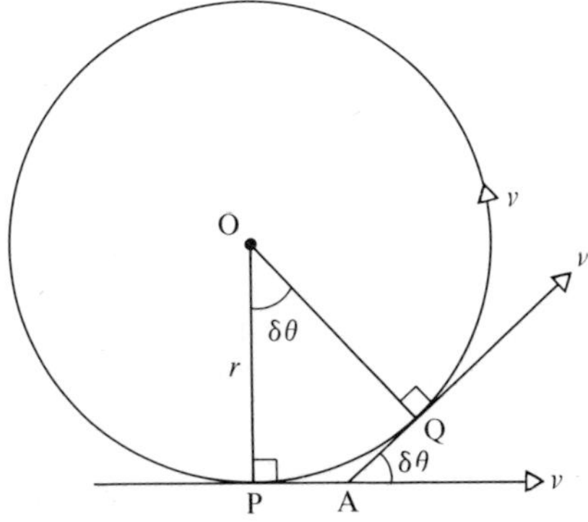

$$\text{acceleration} = \frac{\text{change in velocity}}{\text{time}}$$

For the acceleration at any instant, we must let $\delta t \to 0$ and hence $\delta\theta \to 0$

$$\text{Acceleration along PA} = \lim \left(\frac{v\cos\delta\theta - v}{\delta t}\right) = \lim \frac{v(\cos\delta\theta - 1)}{\delta t}$$

but as $\delta\theta \to 0$, $(\cos\delta\theta - 1) \to 0$

$\therefore$ acceleration along PA $= 0$

$$\text{Acceleration along PO} = \lim \left(\frac{v\sin\delta\theta - 0}{\delta t}\right) = \lim \frac{v\sin\delta\theta}{\delta t}$$

but as $\delta\theta \to 0$, $\sin\delta\theta \to \delta\theta$

$\therefore$ acceleration along PO$=$ $\lim \dfrac{v\delta\theta}{\delta t} = v\omega$

but $v = r\omega$ $\quad\therefore\quad$ acceleration $= \dfrac{v^2}{r}$ and is towards the centre of the circular path.

Example 5

A particle describes a horizontal circle of radius 2 metres at a speed of 3 m/s. Find the acceleration of the particle.

Acceleration towards centre of circle $= \dfrac{v^2}{r}$

Given $\quad v = 3$ m/s and $r = 2$ m

$\therefore$ acceleration $= \dfrac{(3)^2}{2} = 4{\cdot}5\ \text{m/s}^2$

The acceleration of the particle is $4{\cdot}5\ \text{m/s}^2$ towards the centre of the circle.

Central force

For a body to have an acceleration, there has to be a force acting on the body. In the case of a body following a circular path, since the acceleration is directed towards the centre of the circle, the force must also be in this direction.

The magnitude of the force will be

mass $\times$ acceleration $= m\dfrac{v^2}{r}$, where m is the mass of the body.

Example 6

A body of mass 250 g moves with constant angular speed of 4 rad/s in a horizontal circle of radius 3·5 m. Find the force that must act on the body towards the centre of the circle.

$$\text{Force} = \text{mass} \times \text{acceleration}$$

$$= m\frac{v^2}{r} = mr\omega^2 \qquad \text{since } v = r\omega$$

$$\therefore \quad \text{Force} = \tfrac{250}{1000} \times 3{\cdot}5 \times 4^2 = 14\ \text{N}$$

The force acting on the body towards the centre of the circle is 14 N.

The force which acts upon a body so that it follows a circular path may be provided in various ways.

(i) Particle on a string

A particle is attached to one end of a string, the other end of the string being attached at a point O on a smooth horizontal surface. If the particle describes circles on the surface, the necessary force towards the centre of the circle is provided by the tension in the string.

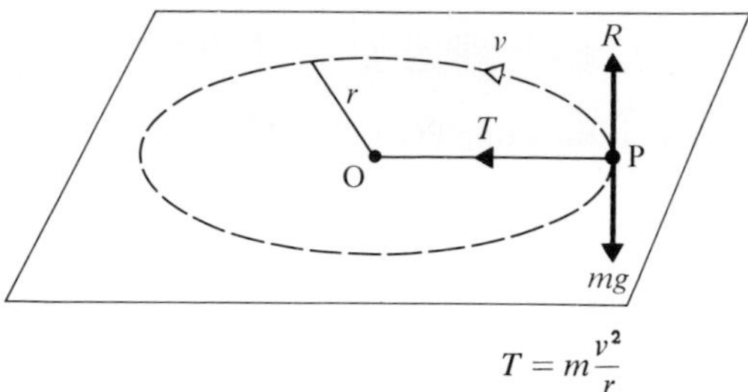

$$T = m\frac{v^2}{r}$$

(ii) Bead on a circular wire

If a bead is threaded on a smooth horizontal circular wire and moves at a speed v, the necessary force towards the centre of the circular wire is provided by the force between the bead and the wire. In addition, the wire supports the weight of the bead, and the vertical reaction R equals mg.

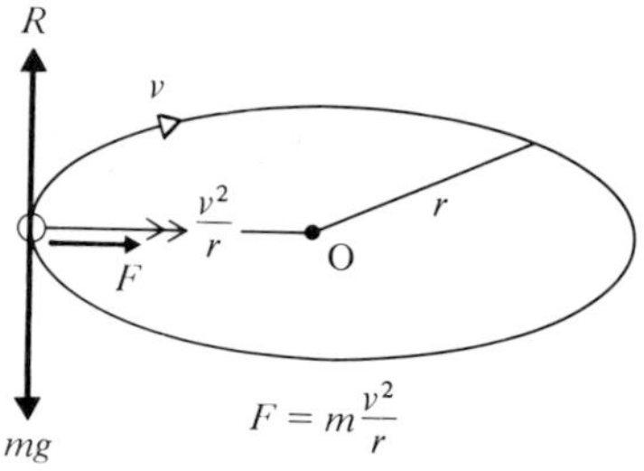

$$F = m\frac{v^2}{r}$$

(iii) Particle on a rotating disc

If a particle rests on the surface of a rotating horizontal disc, the only horizontal force acting on the particle is the frictional force between the particle and the surface of the disc. This frictional force provides the necessary force towards the centre of rotation.

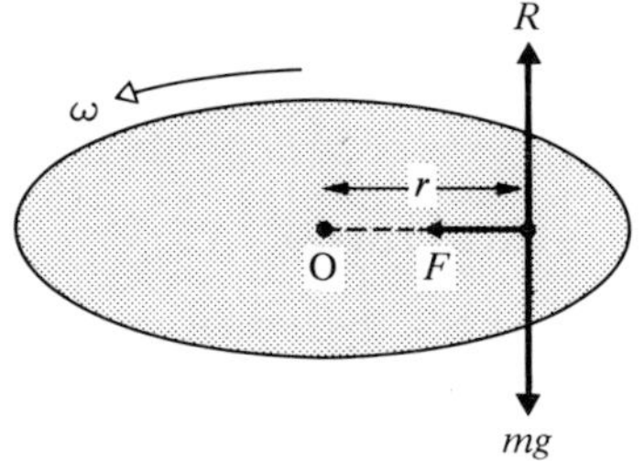

For any particular surface, there will be a maximum value of F, i.e. $F_{max} = \mu R$, and then the particle will be on the point of slipping.

(iv) Car on circular path

Again, the necessary force towards the centre of the circular path is provided by the frictional force between the tyres of the car and the surface of the road.

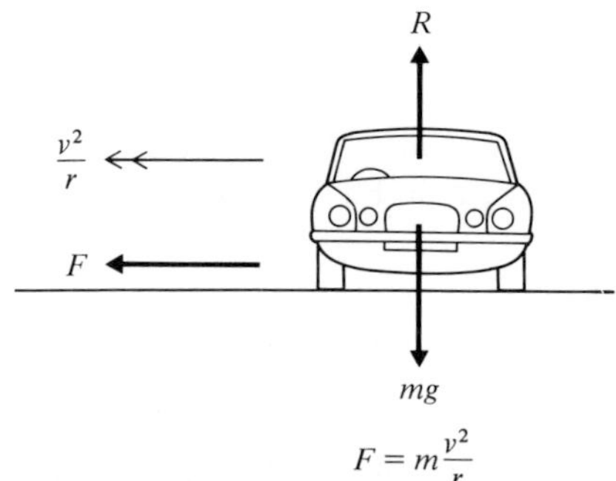

$$F = m\frac{v^2}{r}$$

Example 7

A particle of mass 300 g is attached to one end of a light inextensible string of length 40 cm, the other end of the string being fixed at O on a smooth horizontal surface. If the particle describes circles, centre O, find the tension in the string when (a) the speed of the particle is $2\sqrt{2}$ m/s, (b) the angular speed of the particle is 5 rad/s.

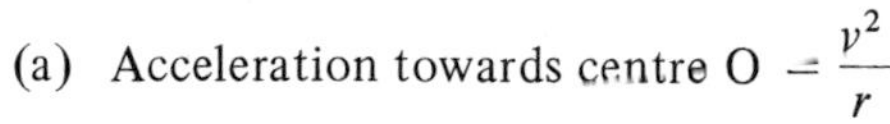

(a) Acceleration towards centre O $= \dfrac{v^2}{r}$

$\therefore$ force towards O, i.e. the tension in the string,

$$= \frac{mv^2}{r}$$

$$\therefore \quad \text{tension} = \tfrac{300}{1000} \times \frac{(2\sqrt{2})^2}{0{\cdot}4}$$

$$= 6\text{ N}$$

The tension in the string is 6 N.

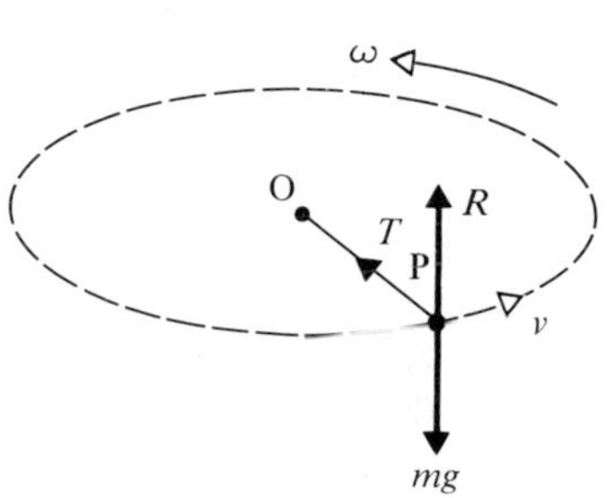

(b) Again, tension $= m\dfrac{v^2}{r} = mr\omega^2$

$$\therefore \quad \text{tension} = \tfrac{300}{1000} \times 0{\cdot}4 \times (5)^2 = 3\text{ N}$$

The tension in the string is 3 N.

Example 8

A car travels along a horizontal road which is an arc of a circle of radius 125 m. The greatest speed at which the car can travel without slipping is 42 km/h. Find the coefficient of friction between the tyres of the car and the surface of the road.

Let the mass of the car be m kg

Resolving vertically $\qquad R = mg \qquad \ldots [1]$

Friction provides the force towards the centre of the circle and since the car is on the point of slipping,

$$F_{max} = \mu R = m\frac{v^2}{r} \qquad \ldots [2]$$

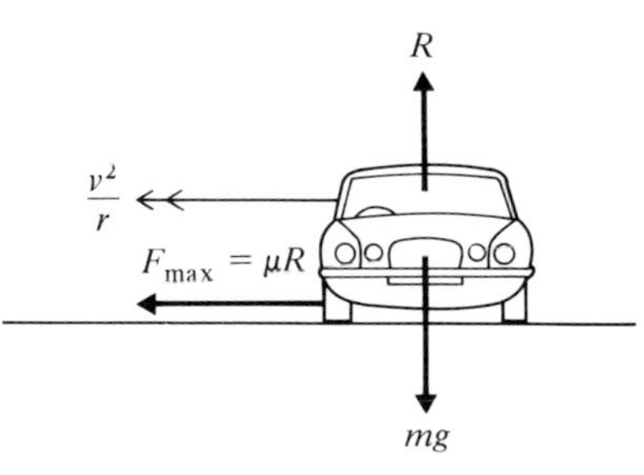

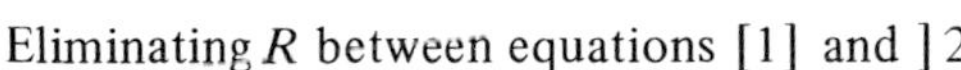

Eliminating R between equations [1] and]2

$$\mu mg = m\frac{v^2}{r} \quad \text{or} \quad \mu = \frac{v^2}{rg}$$

but $r = 125$ m and $v = 42$ km/h $= \dfrac{42 \times 1000}{60 \times 60}$ m/s

$$\therefore \quad \mu = \left(\frac{42 \times 1000}{60 \times 60}\right)^2 \times \frac{1}{125(9{\cdot}8)} = \frac{1}{9}$$

The coefficient of friction between the tyres and the road is $\frac{1}{9}$.

Example 9

A small box is placed on the surface of a horizontal disc at a point 5 cm from the centre of the disc. The box is on the point of slipping when the disc rotates at $1\frac{2}{5}$ rev/s. Find the coefficient of friction between the box and the surface of the disc.

Let the mass of the box be m kg

Resolving vertically $\qquad R = mg \qquad \ldots [1]$

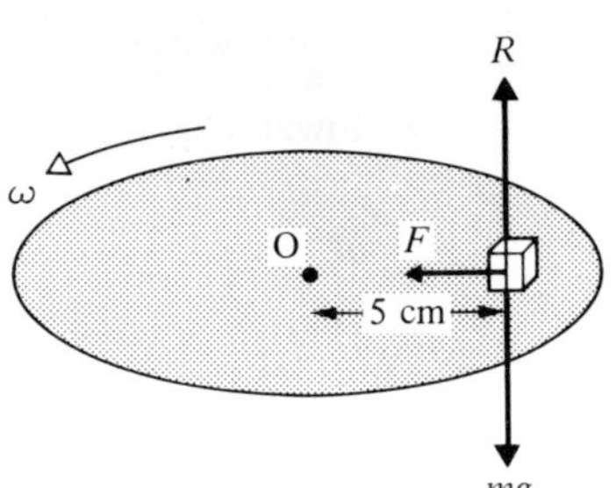

Friction provides the force towards O and since the box is on the point of slipping,

$$F_{\text{max}} = \mu R = m\frac{v^2}{r}$$

$$\text{or} \quad \mu R = mr\omega^2 \qquad \ldots [2]$$

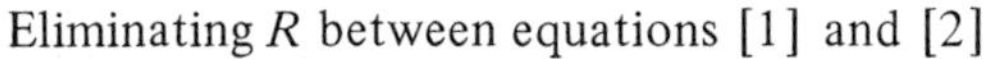

Eliminating R between equations [1] and [2]

$$\mu = \frac{r\omega^2}{g}$$

but $r = 5$ cm $= \frac{5}{100}$ m and $\omega = 1\frac{2}{5}$ rev/s $= \frac{7}{5} \times 2\pi$ rad/s

$$\therefore \quad \mu = \left(\frac{5}{100}\right) \times \left(\frac{14\pi}{5}\right)^2 \times \left(\frac{1}{9{\cdot}8}\right) = 0{\cdot}395$$

The coefficient of friction between the box and the disc is 0·395.

Exercise 12B

1. A particle moves with a constant angular speed of 2 rad/s around a circular path of radius 1·5 m. Find the acceleration of the particle.
2. A particle moves with a constant speed of 3 m/s around a circular path of radius 50 cm. Find the acceleration of the particle.
3. A record is revolving at 45 rev/min. Points A, B and C lie on the record at 5 cm, 10 cm and 20 cm from the centre respectively. Find the accelerations of A, B and C.
4. A body of mass 2 kg moves with a constant angular speed of 5 rad/s around a horizontal circle of radius 10 cm. Find the magnitude of the horizontal force that must be acting on the body towards the centre of the circle.
5. A body of mass 0·5 kg moves with a constant speed of 4 m/s around a horizontal circle of radius 1 m. Find the magnitude of the horizontal force that must be acting on the body towards the centre of the circle.
6. A body of mass 1·5 kg moves with a constant angular speed of 600 rev/min around a horizontal circle of radius 10 cm. Find the magnitude of the horizontal force that must be acting on the body towards the centre of the circle.
7. A body of mass 100 g moves with a constant angular speed around a horizontal circle of radius 50 cm. If the body completes 50 revolutions in 3 minutes, find the magnitude of the horizontal force that must be acting on the body towards the centre of the circle.
8. A particle of mass 250 g lies on a smooth horizontal surface and is connected to a point O on the surface by a light inextensible string of length 20 cm. With the string taut, the particle describes a horizontal circle, centre O, with constant angular speed ω. Find the tension in the string when ω is (a) 10 rad/s, (b) 20 rad/s, (c) 100 rev/min.

9. A particle of mass 125 g lies on a smooth horizontal surface and is connected to a point O on the surface by a light inextensible string of length 50 cm. With the string taut, the particle describes a horizontal circle, centre O, with constant speed v. Find the tension in the string when v is (a) 2 m/s, (b) 10 m/s, (c) 20 m/s.
10. A body of mass 2 kg lies on a smooth horizontal surface and is connected to a point O on the surface by a light inextensible string of length 25 cm. With the string taut, the body describes a horizontal circle with centre O. If the tension in the string is 18 N, find the angular speed of the body.
11. A body of mass 3 kg lies on a smooth horizontal surface and is connected to a point O on the surface by a light inextensible string of length 1·5 m. With the string taut, the body describes a horizontal circle with centre O. If the tension in the string is 32 N, find the speed of the body.
    Find the greatest possible speed of the body around the circle given that the string will break if the tension exceeds 98 N.
12. A smooth wire is in the form of a circle of radius 20 cm and centre O. The wire has a bead of mass 100 g threaded on it. The wire is held horizontally and the bead moves along the wire with a constant speed of 3 m/s. Find
    (a) the vertical force experienced by the wire due to the bead. (State whether up or down.)
    (b) the vertical force experienced by the bead due to the wire. (State whether up or down.)
    (c) the horizontal force experienced by the wire due to the bead. (State whether towards or away from O.)
    (d) the horizontal force experienced by the bead due to the wire. (State whether towards or away from O.)
13. A bend in a level road forms a circular arc of radius 50 m. A car travels around the bend at a speed of 14 m/s without slipping occurring. What information does this give about the value of the coefficient of friction between the tyres of the car and the road surface?
14. A bend in a level road forms a circular arc of radius 75 m. The greatest speed at which a car can travel around the bend without slipping occurring is 63 km/h. Find the coefficient of friction between the tyres of the car and the road surface.
15. A bend in a level road forms a circular arc of radius 54 m. Find the greatest speed at which a car can travel around the bend without slipping occurring if the coefficient of friction between the tyres of the car and the road surface is 0·3.
16. A body of mass 200 g rests without slipping on a horizontal disc which is rotating at 5·6 rad/s. The coefficient of friction between the body and the disc is 0·4. Find the greatest possible distance between the body and the centre of rotation.
    Repeat the question for a body of mass 400 g.
17. A body is placed on a horizontal disc at a point which is 15 cm from the centre of the disc. When the disc rotates at 30 rev/min, the body is just on the point of slipping. Find the coefficient of friction between the body and the surface of the disc.
18. A level road is to be constructed; one design requirement is that any vehicle travelling around any bend in the road, can travel at any speed up to 21 m/s without slipping occurring. It is calculated that under the worst conditions, the coefficient of friction between a vehicle's tyres and the surface of the road could be as low as 0·125. Find the minimum radius that any bend in the road may have.
19. A level railway track is in the form of a circular arc of radius 200 m. Find the horizontal force acting on the rails when a train of mass 15 000 kg travels around the bend with a speed of 20 m/s.

20. Two particles A and B, of masses 200 g and 50 g respectively, are connected by a light inextensible string. Particle A, at one end of the string lies on a smooth horizontal table. The string passes smoothly through a small hole O in the table and particle B hangs freely at the other end of the string. When particle A describes a horizontal circle about O with angular speed 3·5 rad/s, particle B hangs 30 cm below the level of the surface of the table. Find the length of the string.

21. Two particles A and B, of masses 100 g and 250 g respectively, are connected by a light inextensible string of length 20 cm. Particle A at one end of the string lies on a smooth horizontal table. The string passes smoothly through a small hole O in the table and particle B hangs freely at the other end of the string. Find the speed (in m/s) with which A must be made to perform horizontal circles about O in order to keep B in equilibrium at a position 7·5 cm below the level of the surface of the table.

22. A light inextensible string of length l has one end fixed at a point on a smooth horizontal surface, and the other end attached to a body of mass m lying on the surface. With the string taut, the body is given an initial speed of $2\sqrt{(gl)}$ in a direction parallel to the plane of the surface and perpendicular to the string. Show that the tension in the string during the ensuing circular motion will be four times the weight of the particle.

23. A car is just on the point of slipping when travelling on level ground at a speed v around a bend of radius r. Under the same road surface conditions, the car is just on the point of slipping when travelling on level ground at a speed $2v$ around a bend of radius R. Show that $R = 4r$.

24. Two particles A and B of masses m_1 and m_2 respectively are connected by a light inextensible string of length l. Particle A, at one end of the string, lies on a smooth horizontal table. The string passes smoothly through a small hole O in the table and particle B hangs freely at the other end of the string. If A follows a horizontal circular path with centre O and angular speed ω, show that particle B will rest in equilibrium at a point which is a distance x below the level of the surface of the table, where

$$x = \frac{lm_1\omega^2 - m_2 g}{m_1\omega^2}.$$

Conical pendulum

A particle is attached to the lower end of a light inextensible string, the upper end of which is fixed. When the particle describes a horizontal circle, the string describes the curved surface of a cone. This arrangement is known as a *conical pendulum.*

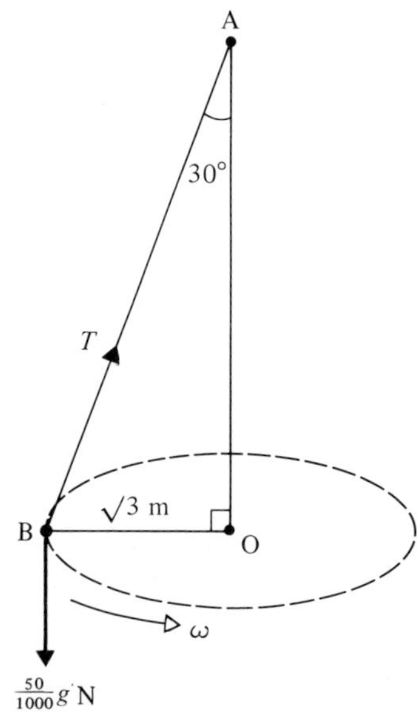

Example 10

A conical pendulum consists of a light inextensible string AB, fixed at A and carrying a particle of mass 50 g at B. The particle moves in a horizontal circle of radius $\sqrt{3}$ m and centre vertically below A. If the angle between the string and the vertical is 30°, find the tension in the string and the angular speed of the particle.

Resolving vertically $\quad T \cos 30 = \frac{50}{1000} g \quad \ldots [1]$

$\therefore \quad T = 0{\cdot}5658$ N

The horizontal component of the tension provides the force towards the centre O of the circle.

Using $F = ma$ $\qquad T\cos 60 = mr\omega^2$

$$= \frac{50}{1000} \times \sqrt{3} \times \omega^2$$

and using $T = \dfrac{50}{1000}g \times \dfrac{1}{\cos 30}$ from equation [1]

$$\frac{50}{1000}g \times \frac{\cos 60}{\cos 30} = \frac{50}{1000} \times \sqrt{3} \times \omega^2$$

$$\therefore \quad \omega = 1{\cdot}807 \text{ rad/s}$$

The tension in the string is 0·566 N and the angular speed of the particle is 1·81 rad/s.

Motion of a car rounding a banked curve

When a car travels along a circular arc on a horizontal road, the frictional force between the tyres and the road provides the necessary horizontal force F towards the centre of the circular arc.
If the speed of the car exceeds a particular value, then the force required (i.e. $m\dfrac{v^2}{r}$) may be greater than can be provided by the maximum frictional force, and the car will slip.

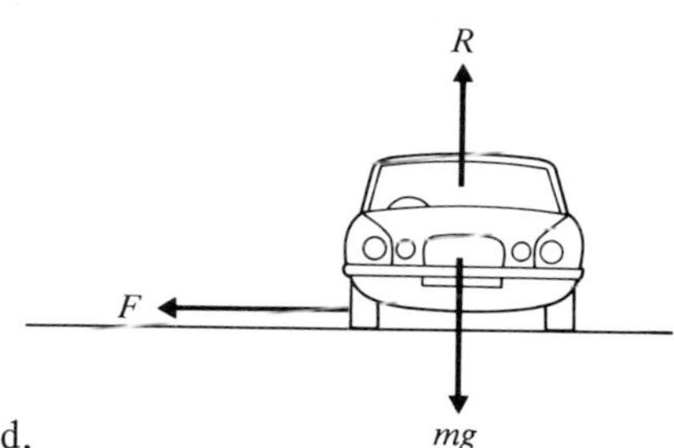

In practice, a bend in a road or race track may be banked, i.e. the level of the road on the inside of the bend is lower than the level of the road on the outside of the bend. The normal reaction R between the car and the road is then no longer vertical. The horizontal component of R can therefore provide or help to provide the necessary force, directed towards the centre.

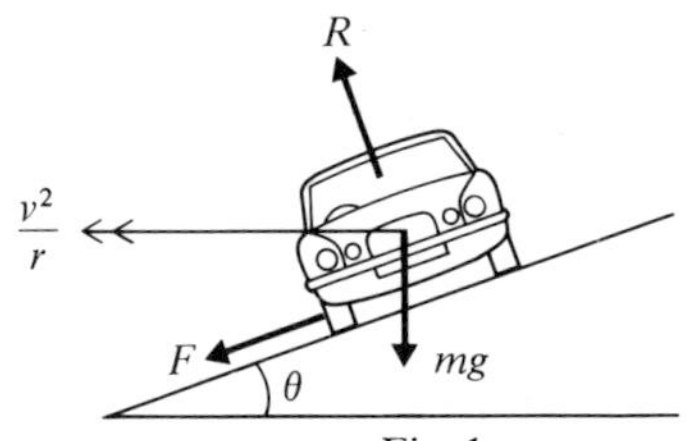

Fig. 1

If F is the frictional force between the tyres and the road surface, Fig. 1,

$$F\cos\theta + R\sin\theta = m\frac{v^2}{r}$$

If the car travels too slowly there will be a tendency for the car to slip down the slope. To prevent this, a frictional force F' will act, Fig. 2,

Fig. 2

$$R\sin\theta - F'\cos\theta = m\frac{v^2}{r}$$

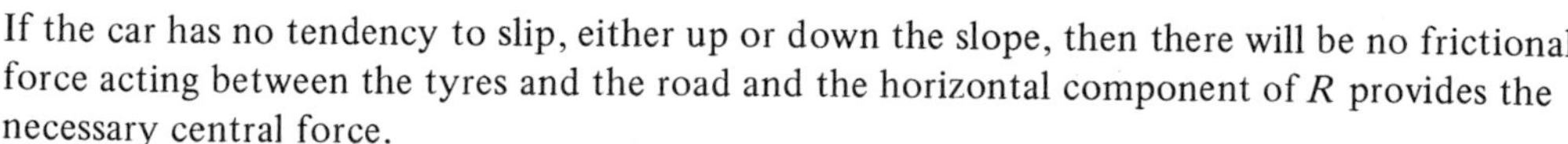
If the car has no tendency to slip, either up or down the slope, then there will be no frictional force acting between the tyres and the road and the horizontal component of R provides the necessary central force.

In this case $\qquad R\sin\theta = m\dfrac{v^2}{r}$

Example 11

A car travels around a bend of radius 400 m on a road which is banked at an angle θ to the horizontal. If the car has no tendency to slip when travelling at 35 m/s, find θ.

Since there is no tendency to slip, there is no frictional force between the tyres and the road.

Resolving vertically

$$R\cos\theta = mg \qquad \ldots [1]$$

Using $F = ma$, the force towards the centre of the bend is provided by a component of R

$$\therefore \quad R\sin\theta = m\frac{v^2}{r} \qquad \ldots [2]$$

Dividing equation [2] by equation [1]:

$$\tan\theta = \frac{v^2}{rg}$$

Substituting for r and v, $\quad \tan\theta = \dfrac{(35)^2}{400 \times 9{\cdot}8} = \frac{5}{16}$

$$\therefore \quad \theta = 17{\cdot}35^\circ$$

The angle θ is $17{\cdot}35^\circ$.

Example 12

A car travels around a bend in a road which is a circular arc of radius 62·5 m. The road is banked at an angle $\tan^{-1}\frac{5}{12}$ to the horizontal. If the coefficient of friction between the tyres of the car and the road surface is 0·4, find (a) the greatest speed, (b) the least speed at which the car can be driven around the bend without slipping occuring.

(a) When the car is travelling as fast as possible, the maximum frictional force μR acts so as to prevent the car slipping up the slope.

Resolving vertically

$$R\cos\theta = mg + \mu R\sin\theta \qquad \ldots [1]$$

Using $F = ma$ horizontally

$$R\sin\theta + \mu R\cos\theta = m\frac{v^2}{r} \qquad \ldots [2]$$

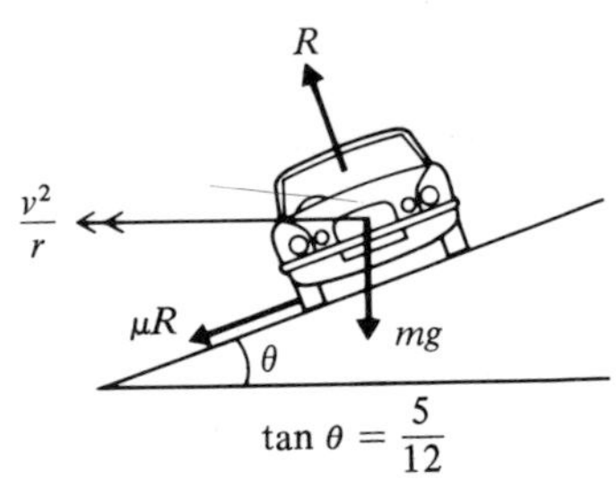

Eliminating R between these equations

$$\frac{\cos\theta - \mu\sin\theta}{\sin\theta + \mu\cos\theta} = \frac{rg}{v^2} \quad \text{or} \quad \frac{1 - \mu\tan\theta}{\tan\theta + \mu} = \frac{rg}{v^2}$$

Substituting for μ, $\tan\theta$, r and g $\quad \dfrac{1 - (0{\cdot}4) \times \frac{5}{12}}{\frac{5}{12} + 0{\cdot}4} = \dfrac{62{\cdot}5 \times 9{\cdot}8}{v^2}$

and $\quad \therefore \quad v = 24{\cdot}5$ m/s

(b) When the car is travelling as slowly as possible, the maximum frictional force μR acts so as to prevent the car slipping down the slope.

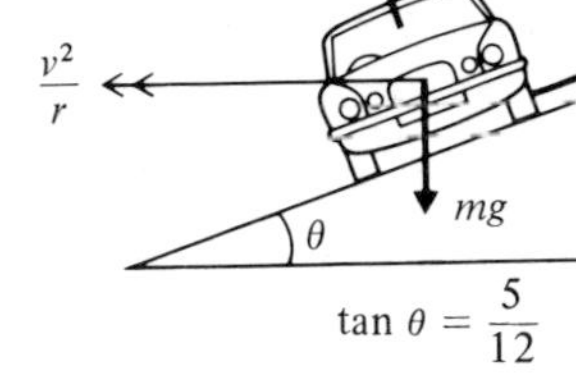

Resolving vertically

$$R\cos\theta + \mu R\sin\theta = mg \qquad \ldots [1]$$

Using $F = ma$, horizontally

$$R\sin\theta - \mu R\cos\theta = m\frac{v^2}{r} \qquad \ldots [2]$$

Eliminating R between these equations

$$\frac{\cos\theta + \mu\sin\theta}{\sin\theta - \mu\cos\theta} = \frac{rg}{v^2} \quad \text{or} \quad \frac{1 + \mu\tan\theta}{\tan\theta - \mu} = \frac{rg}{v^2}$$

Substituting for μ, $\tan\theta$, r and g

$$\frac{1 + (0{\cdot}4) \times \frac{5}{12}}{\frac{5}{12} - 0{\cdot}4} = \frac{62{\cdot}5 \times 9{\cdot}8}{v^2}$$

and $\quad \therefore \quad v = 2{\cdot}958$ m/s

The greatest and least speeds at which the car can travel without slipping are 24·5 m/s and 2·96 m/s respectively.

Motion of a train rounding a banked curve

If the outer rail is raised above the level of the inner rail, this will have a similar effect to the banking on a road. The normal reaction again has a horizontal component $R\sin\theta$ acting towards the centre of the circular arc.

If the train travels at the speed v for which the banking was designed, then the force $R\sin\theta$ will provide the necessary central force towards the centre of the curve. If the train travels at a speed greater than v, then the force $R\sin\theta$ will be insufficient to provide the necessary force. In such a situation there will be a reaction between the outer rail and the flange of wheel B, and the horizontal component of this reaction will, together with $R\sin\theta$, provide the necessary force.

If the train travels at a speed less than v, then $R\sin\theta$ will be greater than the central force required, and the reaction between the inner rail and the flange of wheel A will compensate.

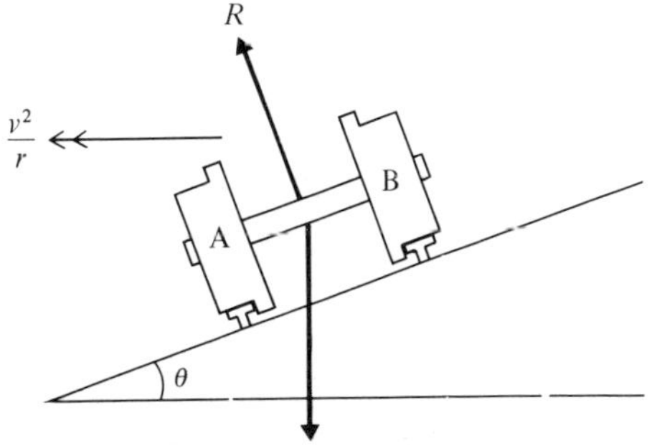

Example 13

A train of mass 50 tonnes travels around a bend of radius 900 m. The outer rail of the track is raised 7 cm above the inner rail and the distance between the rails is 1·4 m. Calculate the speed at which the train should travel so that there is no force acting between the flanges on the wheels and the rails.

Let the angle of the banking be θ, then

$$\sin\theta = \tfrac{7}{140} = \tfrac{1}{20} \approx \tan\theta$$

At v m/s, there is no reaction between the wheel flanges and the rails.

Resolving vertically

$$R\cos\theta = 50\,000g \qquad \ldots [1]$$

Using $F = ma$, force towards the centre of curve provided by component of normal reaction R is

$$R\sin\theta = 50\,000 \times \frac{v^2}{900} \qquad \ldots [2]$$

Dividing equation [2] by equation [1] $\quad \tan\theta = \dfrac{v^2}{900g}$

$$\therefore \quad \tfrac{900}{20}g = v^2 \quad \text{substituting for } \tan\theta$$

$$\therefore \quad v = 21 \text{ m/s}$$

The train should travel at 21 m/s for there to be no force between the flanges and the rails.

Example 14

A light inextensible string AB of length 33 cm has a particle of mass 50 g attached to it at a point P, 13 cm from the end A. The ends of the string are attached at two fixed points in the same vertical line with A 21 cm above B. The particle moves in a horizontal circle, 5 cm vertically below A with both parts of the string taut, at a constant speed of $2\sqrt{3}$ m/s. Find the tensions in the two parts of the string.

Let O be the centre of the horizontal circle,

then $\quad OP^2 = 13^2 - 5^2 = 144$

$\therefore$ radius of circle $= 12$ cm $= 0{\cdot}12$ m

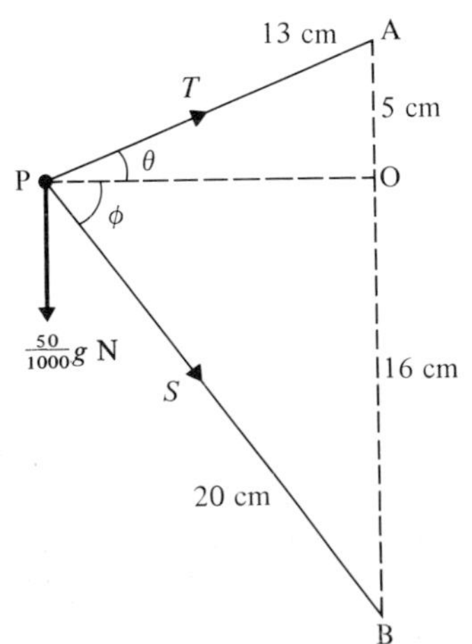

Resolving vertically

$$T\sin\theta = \tfrac{50}{1000}g + S\sin\phi$$

$$\text{or} \quad \tfrac{5}{13}T = \frac{g}{20} + \tfrac{16}{20}S \qquad \ldots [1]$$

The horizontal components of the tensions in the two parts of the string provide the necessary central force, towards O.

Using $F = ma$

$$T\cos\theta + S\cos\phi = \frac{50}{1000} \times \frac{v^2}{\text{OP}}$$

$$\text{or} \quad \frac{12}{13}T + \frac{12}{20}S = \frac{(2\sqrt{3})^2}{20 \times 0{\cdot}12}$$

$$\therefore \quad \frac{T}{13} + \frac{S}{20} = \frac{5}{12} \qquad \ldots [2]$$

Eliminating T between equations [1] and [2]:

$$5\left(\frac{5}{12} - \frac{S}{20}\right) = \frac{9{\cdot}8}{20} + \frac{16}{20}S \qquad \therefore \quad S = 1{\cdot}517 \text{ N}$$

substituting in equation [2] for S $\quad T = 4{\cdot}431$ N

The tensions in the upper and lower parts of the string are 4·43 N and 1·52 N respectively.

Exercise 12C

1. Each of the following diagrams shows a conical pendulum consisting of a light inextensible string AB fixed at A and carrying a bob of mass 10 kg at B. The bob moves in a horizontal circle, centre vertically below A, with a constant angular speed ω. The tension in the string is T.

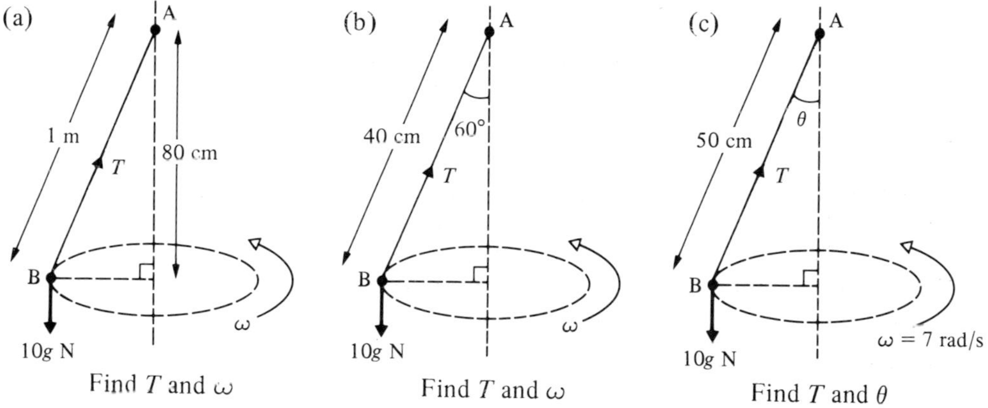

2. A car moves in a horizontal circular path of radius 50 m around a banked track. If the car has no tendency to slip when driven around the track at a speed of 14 m/s, find the angle at which the track is banked.
3. A car moves in a horizontal circular path of radius 75 m around a bend which is banked at $\tan^{-1}\frac{5}{12}$ to the horizontal. At what speed should the car be driven if it is to have no tendency to slip?
4. A particle moves in a horizontal circular path, of radius 20 cm, on a smooth surface that is banked at an angle θ to the horizontal. When the particle moves at a speed of 70 cm/s it does not slip; find θ.
5. A car moves in a horizontal circle of radius 75 m around a bend which is banked at an angle of 20° to the horizontal. At what speed should the car be driven if it is to have no tendency to slip?

6. The bend in a road is to be constructed of radius 150 m and such that, under any road conditions, a car can be driven in a horizontal circular path around the bend at 63 km/h without slipping occuring. At what angle to the horizontal should the road be banked?
7. A conical pendulum consists of a light inextensible string AB of length 50 cm fixed at A and carrying a bob of mass 2 kg at B. The bob describes a horizontal circle about the vertical through A with a constant angular speed of 5 rad/s. Find the tension in the string.
8. A particle is attached to one end of a light inextensible string which has its other end attached to a fixed point A. With the string taut, the particle describes a horizontal circle with constant angular speed 2·8 rad/s, the centre of the circle being at a point O vertically below A. Find the distance OA.
9. A bend in a railway track forms a horizontal circular arc of radius 1·25 km. A train of mass 40 tonnes travels round the bend at a constant speed of 63 km/h. The distance between the rails is 1·5 m.

   (a) With the rails set at the same horizontal level (see Fig. 1), find the force exerted on the side of the rails.

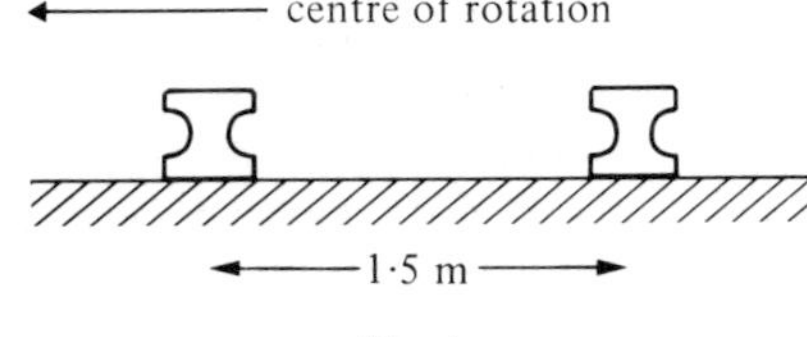

Fig. 1

   (b) With the outer rail raised a distance x above the inner rail (see Fig. 2), the force on the side of the rails is reduced to zero. Find x.

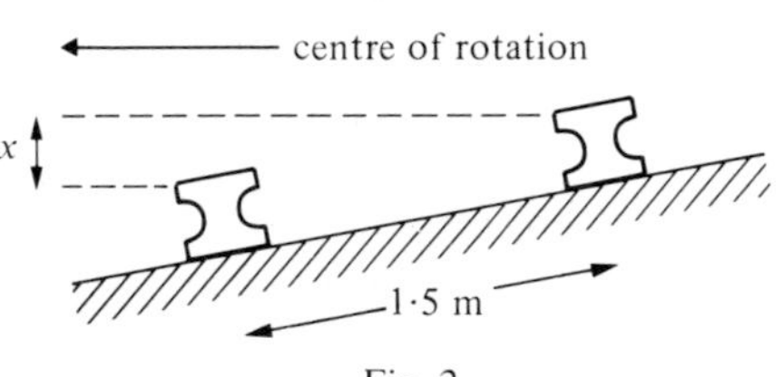

Fig. 2

10. Repeat question **9** for a train of mass 50 tonnes travelling at 50·4 km/h around a bend of radius 2 km.
11. A car moves in a horizontal circular path of radius 60 m around a bend that is banked at an angle $\tan^{-1}\frac{7}{24}$ to the horizontal. If the coefficient of friction between the tyres of the car and the road surface is $\frac{3}{7}$, find the greatest speed with which the car can be driven around the bend without slipping occurring.
12. A car moves in a horizontal circular path of radius 60 m around a bend that is banked at an angle $\tan^{-1} 0{\cdot}5$ to the horizontal. Without slipping occurring, the maximum speed with which the car can travel around the bend is 28 m/s. Find the coefficient of friction between the tyres of the car and the road surface.
13. A film stuntman has to drive a car in a horizontal circular path of radius 105 m around a bend that is banked at 45° to the horizontal. The stuntman finds that he must drive with a speed of at least 21 m/s if he is to avoid slipping sideways down the slope. Find the coefficient of friction between the tyres of the car and the road surface.
14. A car moves in a horizontal circular path of radius 140 m around a banked corner of a race track. The greatest speed with which the car can be driven around the corner without slipping occurring is 42 m/s. If the coefficient of friction between the tyres of the car and the surface of the track is $\frac{1}{3}$, find the angle of banking.
15. A bend in a road is in the form of a horizontal circular arc of radius r, with the road surface banked at an angle θ to the horizontal. Show that a car will have no tendency to slip when driven around the bend with speed $\sqrt{(rg \tan \theta)}$.

16. A light inextensible string AB has a particle attached at end B and A is fixed. With the string taut, the particle describes a horizontal circle with constant angular speed ω.
If the centre of the circle is at a point which is a distance x vertically below A show that $\omega^2 x = g$.

17. A light inextensible string AB of length l has end A fixed and carries a particle of mass m at B. With the string taut, the particle describes a horizontal circle about the vertical axis through A, with constant angular speed ω. Show that the tension T in the string is given by $T = ml\omega^2$.

18. A vehicle is just on the point of slipping when parked on a bend that is banked at an angle of 20° to the horizontal. Find the coefficient of friction between the vehicle's tyres and the surface of the road.
If the vehicle were driven around this bend in a horizontal circular path of radius 60 m, find the greatest speed it could attain without slipping occurring.

19. One corner of a race track is banked at 45° to the horizontal. A car is to be driven around the corner in a horizontal circular path of radius 60 m. The coefficient of friction between the tyres of the car and the surface of the track is 0·5. Find the greatest and least speeds with which the car can travel without slipping.

20. A point A lies 25 cm above a smooth horizontal surface. A light inextensible string of length 50 cm has one end fixed at A and carries a body B of mass 0·5 kg at its other end. With the string taut, B moves in a circular path on the surface, the centre of the circle being vertically below A. If B moves with constant angular speed of 4 rad/s, find the magnitude of (a) the tension in the string, (b) the reaction force between B and the surface.

21. A smooth bead of mass 100 g is threaded on a light inextensible string of length 70 cm. The string has one end attached to a fixed point A and the other to a fixed point B 50 cm vertically below A. The bead moves in a horizontal circle about the line AB with a constant angular speed, of ω rad/s and the string taut. If the bead is at a point C on the string with AC = 40 cm, find the value of ω and the tension in the string.

22. A smooth hemispherical shell of radius r is fixed with its rim horizontal. A small ball bearing of mass m lies inside the shell and performs horizontal circles on the inner surface of the shell, the plane of these circles lying 5 cm below the level of the rim. Find the angular speed of the ball bearing in rad/s.

23. A light inextensible string AB of length 90 cm has a particle of mass 600g fastened to it at a point C. The ends A and B are attached to two fixed points in the same vertical line as each other, with A 60 cm above B. The particle moves in a horizontal circle at a constant angular speed of 5 rad/s with both parts of the string taut and CB horizontal. Find the tensions in the two parts of the string.

24. A light inextensible string AB of length $2l$ has a particle attached to its mid-point C. The ends A and B of the string are fastened to two fixed points with A distance l vertically above B. With both parts of the string taut, the particle describes a horizontal circle about the line AB with constant angular speed ω. If the tension in CA is three times that in CB, show that $\omega = 2\sqrt{\dfrac{g}{l}}$.

25. At any speed greater than v, a car will slip when driven in a horizontal circular path of radius r around a track that is banked at an angle θ to the horizontal. If μ is the coefficient of friction between the tyres of the car and the surface of the track,

show that $\mu = \dfrac{v^2 - rg\tan\theta}{rg + v^2\tan\theta}$.

Exercise 12D Examination questions

1. The coefficient of friction between a parcel and the horizontal floor of the delivery truck on which it rests is $\frac{5}{8}$. The truck is being driven round a circular bend in a horizontal road at a uniform speed of 14 m/s. Show that the parcel will not slide across the floor of the truck if the radius of the bend is greater than 32 m. (J.M.B)

2. In a conical pendulum, the inelastic string is of length 50 cm and the mass attached to the end of the string is 400 g. The mass describes a horizontal circle of radius 30 cm at a constant speed of v m/s. Calculate (i) the tension in the string, (ii) the numerical value of v. (A.E.B)

3. A small stone of mass 125 g is attached to one end of a light inextensible string, $1\frac{1}{2}$ m in length. The other end of the string is fixed at a point on the surface of a smooth horizontal table. The stone travels in a circle on the table at a constant speed of 12 m/s with the string taut. Calculate the tension in the string.
If the string would break were the tension to exceed $33\frac{1}{3}$ N, calculate the maximum possible speed of the stone. (A.E.B)

4. A car undergoing trials is moving on a horizontal surface around a circular bend of radius 50 m at a steady speed of 14 m/s. Calculate the least value of the coefficient of friction between the tyres of the car and the surface.
Find the angle to the horizontal at which this bend should be banked in order that the car can move in a horizontal circle of radius 50 m around it at 14 m/s without any tendency to side-slip.
Another section of the test area is circular and is banked at 30° to the horizontal. The coefficient of friction between the tyres of the car and the surface of this test area is 0·6. Calculate the greatest speed at which the car can move in a horizontal circle of radius 70 m around this banked test area.
(Take the acceleration due to gravity to be 10 m/s^2.) (A.E.B)

5. A small bead, of mass m, is suspended from a fixed point O by a light inelastic string. The point O is at a height $4a$ above a smooth horizontal table and the bead moves on the table in a horizontal circle of radius $3a$ with constant speed $(2ga)^{\frac{1}{2}}$. Find the reaction between the bead and the table.
Find also the reaction in the case when the bead moves in the same circle with the same speed as before, but is, instead, threaded on a smooth string of length $8a$ whose ends are fastened at O and at the point N on the table vertically below O. (Cambridge)

6. Prove that a particle moving in a circle of radius r with uniform speed v has an acceleration of $\dfrac{v^2}{r}$ directed towards the centre of the circle.

(a) Two particles with masses 0·1 kg and 1 kg are connected by a light inextensible string of length 20 cm passing through a hole in a smooth table. The first particle is on the table and the second particle hangs at rest. If the first particle is rotating at 300 revolutions per minute, what is the depth of the second particle below the table? (Use $g = 10$ m/s^2.)

(b) A particle of mass m kg is describing horizontal circles at the rate of 100 revolutions per minute inside a smooth hemispherical bowl of radius r metres which is fixed with its rim horizontal. What is the depth of these circles below the rim of the hemisphere? (Use $g = 10$ m/s^2.) (S.U.J.B.)

7. A particle is moving at constant speed u in a horizontal circle of radius a and centre O. Prove that the acceleration of the particle is of magnitude u^2/a and is directed towards O. A smooth ring R of mass m is threaded on to a smooth fixed vertical pole on which it is free to move. One end of a light inextensible string of length $2l$ is tied to the pole at the uppermost point A and the other end is tied to R. A particle P of mass m is attached to the string at its mid-point. With each half of the string taut, P moves in a horizontal circle at constant speed v, and the distance RA $= 6l/5$. It may be assumed that, in this motion, the string does not wrap itself around the pole and that, at any instant, the triangle APR is vertical.
(i) Calculate the tensions, in terms of mg, in the two halves of the string.
(ii) Show that $v^2 = 16gl/5$.
(iii) Find the time, in terms of l and g, for P to complete one revolution. (A.E.B)

8. A circular cone of semi-vertical angle α is fixed with its axis vertical and its vertex, A, lowest, as shown in the diagram. A particle P of mass m moves on the inner surface of the cone, which is smooth. The particle is joined to A by a light inextensible string AP of length l. The particle moves in a horizontal circle with constant speed v and with the string taut. Find the reaction exerted on P by the cone.
Find the tension in the string and show that the motion is possible only if $v^2 \geqslant gl \cos \alpha$. (J.M.B)

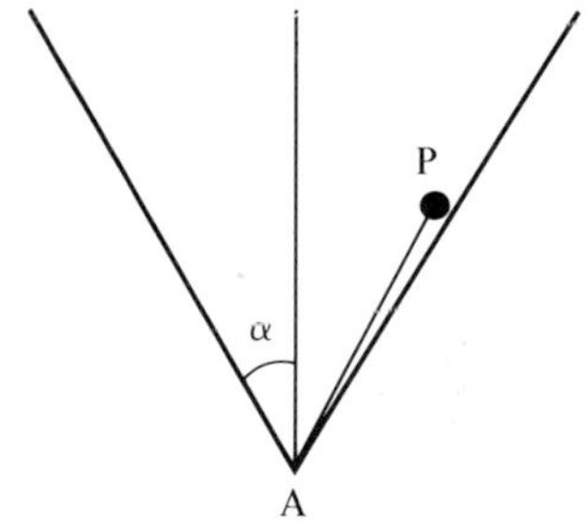

Note Some questions involving motion in a vertical circle are included in exercise 13F.

13
Work, energy and power

Work

If the point of application of a force of P newtons moves through a distance s metres in the direction of the force, then the work done is

$$P \times s \text{ joules}$$

The unit of work is the joule, abbreviated J.

Work done against gravity

In order to raise a mass of m kg vertically at a constant speed, a force of mg N must be applied vertically upwards to the mass. In raising the mass a distance s metres, the work done against gravity will be

$$mgs \text{ joules}$$

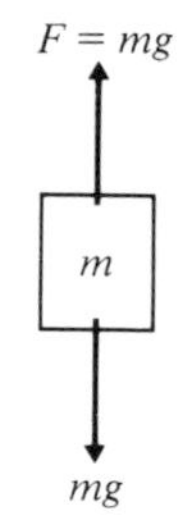

Example 1

Find the work done against gravity when a body of mass 3·5 kg is raised through a vertical distance of 6 m.

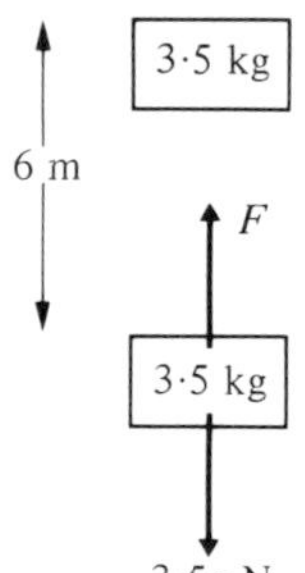

Vertical force required is F

$$F = 3{\cdot}5g$$

Work done $= F \times s$

$= 3{\cdot}5g \times 6 = 205{\cdot}8$ J

The work done against gravity is 205·8 J.

General motion at constant speed

In order to move a body at a constant speed, a force equal in magnitude to the forces of resistance acting on the body, has to be applied to the body.

Example 2

A block of wood is pulled a distance of 4 m across a horizontal surface against resistances totalling 7·5 N. If the block moves at a constant velocity, find the work done against the resistances.

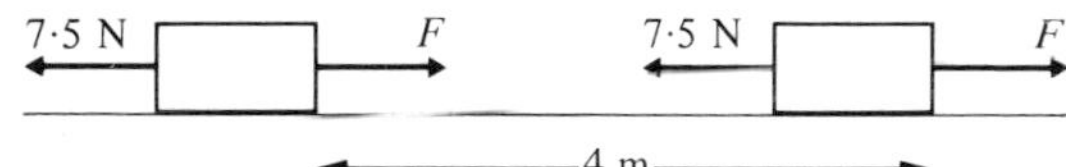

Let the pulling force be F

Horizontally $\quad F = 7{\cdot}5$ N

Work done against resistances

$$= 7{\cdot}5 \times \text{horizontal distance moved}$$
$$= 7{\cdot}5 \times 4 = 30 \text{ J}$$

The work done against the resistances is 30 J.

Example 3

A horizontal force pulls a mass of 2·25 kg a distance of 8 m across a rough horizontal surface, coefficient of friction $\frac{1}{3}$. The body moves with constant velocity and the only resisting force is that due to friction. Find the work done against friction.

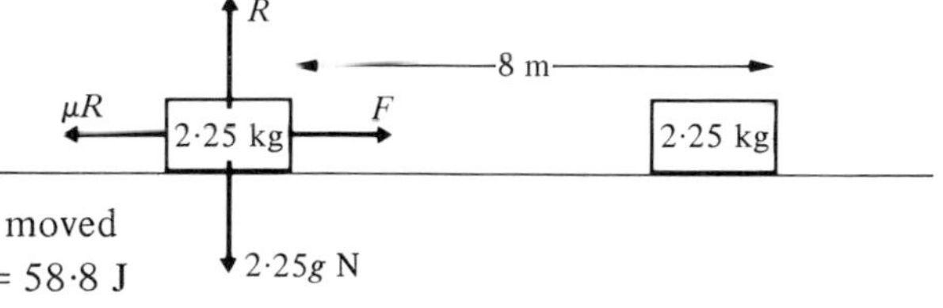

The maximum frictional force is μR

Resolving vertically $\quad R = 2{\cdot}25g$

Work done against friction

$$= \mu R \times \text{distance moved}$$
$$= \tfrac{1}{3}(2{\cdot}25g) \times 8 = 58{\cdot}8 \text{ J}$$

The work done against friction is 58·8 J.

Work done against gravity and friction

When a body is pulled at a uniform speed up the surface of a rough inclined plane, work is done both against gravity and against the frictional force which is acting on the body due to the contact with the rough surface of the plane.

Example 4

A rough surface is inclined at $\tan^{-1}\frac{7}{24}$ to the horizontal. A body of mass 5 kg lies on the surface and is pulled at a uniform speed a distance of 75 cm up the surface by a force acting along a line of greatest slope. The coefficient of friction between the body and the surface is $\frac{5}{12}$. Find (a) the work done against gravity, (b) the work done against friction.

Given $\tan\theta = \frac{7}{24}$, $\sin\theta = \frac{7}{25}$, $\cos\theta = \frac{24}{25}$.

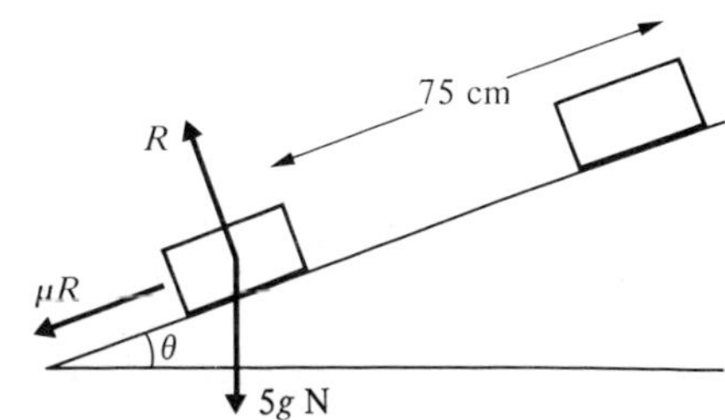

(a) Work done against gravity

$$= \text{force} \times \text{vertical distance moved}$$
$$= 5g \times \frac{75}{100}\sin\theta = 5g \times \frac{75}{100} \times \frac{7}{25}$$
$$= 10{\cdot}29 \text{ J}$$

The work done against gravity is 10·3 J.

(b) Resolving perpendicular to the plane

$$R = 5g \cos\theta = 5g \times \tfrac{24}{25} = \tfrac{24}{5}g$$

Maximum frictional force $= \mu R$

$= \frac{5}{12} \times \frac{24}{5}g = 19{\cdot}6$ N

Work done against friction $= 19{\cdot}6 \times \frac{75}{100} = 14{\cdot}7$ J

The work done against friction is 14·7 J.

Exercise 13A

1. Find the work done against gravity when a body of mass 5 kg is raised through a vertical distance of 2 m.
2. Find the work done against gravity when a body of mass 1 kg is raised through a vertical distance of 3 m.
3. Find the work done against gravity when a person of mass 80 kg climbs a vertical distance of 25 m.
4. A man building a wall lifts 50 bricks through a vertical distance of 3 m. If each brick has a mass of 4 kg, how much work does the man do against gravity?
5. A body of mass 2 kg is moved vertically upwards at a constant speed of 5 m/s. Find how much work is done against gravity in each second.
6. A body of mass 200 g is moved vertically upwards at a constant speed of 2 m/s. Find how much work is done against gravity in each second.
7. A body of mass 10 kg is pulled a distance of 20 m across a horizontal surface against resistances totalling 40 N. If the body moves with uniform velocity, find the work done against the resistances.
8. A box is pulled a distance of 8 m across a horizontal surface against resistances totalling 7 N. If the box moves with uniform velocity, find the work done against the resistances.
9. A horizontal force pulls a body of mass 5 kg a distance of 8 m across a rough horizontal surface, coefficient of friction 0·25. The body moves with uniform velocity and the only resisting force is that due to friction. Find the work done.
10. A horizontal force pulls a body of mass 3 kg a distance of 20 m across a rough horizontal surface, coefficient of friction $\frac{2}{7}$. The body moves with uniform velocity and the only resisting force is that due to friction. Find the work done.
11. A horizontal force drags a body of mass 4 kg a distance of 10 m across a rough horizontal floor at a constant speed. The work done against friction is 49 J. Find the coefficient of friction between the body and the surface.
12. A smooth surface is inclined at 30° to the horizontal. A body of mass 15 kg lies on the surface and is pulled at a uniform speed a distance of 10 m up a line of greatest slope. Find the work done against gravity.
13. A surface is inclined at $\tan^{-1}\frac{3}{4}$ to the horizontal. A body of mass 50 kg lies on the surface and is pulled at a uniform speed a distance of 5 m up a line of greatest slope against resistances totalling 50 N. Find (a) the work done against gravity, (b) the work done against the resistances.
14. A rough surface is inclined at $\tan^{-1}\frac{5}{12}$ to the horizontal. A body of mass 130 kg is pulled at a uniform speed a distance of 50 m up the surface by a force acting along a line of greatest slope. The coefficient of friction between the body and the surface is $\frac{2}{7}$. Find

(a) the frictional force acting, (b) the work done against friction,
(c) the work done against gravity.

15. A rough surface is inclined at 30° to the horizontal. A body of mass 100 kg is pulled at a uniform speed a distance of 20 m up the surface by a force acting along a line of greatest slope. The coefficient of friction between the body and the surface is 0·1. Find
(a) the work done against friction, (b) the work done against gravity.

16. A rough surface is inclined at $\tan^{-1}\frac{3}{4}$ to the horizontal. Find the total work done when a body of mass 50 kg is pulled at a uniform speed a distance of 15 m up the surface by a force acting along a line of greatest slope. The coefficient of friction between the body and the surface is $\frac{1}{3}$ and the only resistances to motion are those due to gravity and friction.

17. A rough surface is inclined at an angle θ to the horizontal. A body of mass m is pulled at a uniform speed a distance x up the surface by a force acting along a line of greatest slope. The coefficient of friction between the body and the plane is μ. If the only resistances to motion are those due to gravity and friction, show that the total work done on the body is $mgx(\sin\theta + \mu\cos\theta)$.

Energy

The energy of a body is a measure of the capacity which the body has to do work.
When a force does work on a body it changes the energy of the body.
Energy can exist in a number of different forms, but we shall consider two main types: kinetic energy and potential energy.

Kinetic energy

The kinetic energy of a body is that energy which it possesses by virtue of its motion.
When a force does work on a body so as to increase its speed, then the work done is a measure of the increase in the kinetic energy of the body.

Suppose a constant force F acts on a body of mass m, which is initially at rest on a smooth horizontal surface, and after moving a distance s across the surface the body has a speed v.

$$\text{Work done on the body} = F \times s$$

$$\text{But} \quad F = \text{mass} \times \text{acceleration, and acceleration} = \frac{v^2 - 0}{2s}$$

$$\therefore \quad F = \frac{mv^2}{2s}$$

$$\therefore \quad \text{work done on the body} = \frac{mv^2}{2s} \times s = \frac{mv^2}{2}$$

The quantity $\frac{mv^2}{2}$ is defined as the kinetic energy of mass m moving with velocity v. A body at rest therefore has zero kinetic energy.

Example 5

Find the kinetic energy of a particle of mass 250 g moving with a speed of $4\sqrt{2}$ m/s.

$$\text{Kinetic energy} = \tfrac{1}{2}mv^2$$
$$= \tfrac{1}{2}(0{\cdot}25)(4\sqrt{2})^2 = 4 \text{ J}$$

The kinetic energy of the particle is 4 J.

Potential energy

The potential energy of a body is that energy it possesses by virtue of its position.
When a body of mass m kg is raised vertically a distance h metres, the work done against gravity is mgh joules. The work done against gravity is a measure of the *increase* in the potential energy of the body, i.e. the capacity of the body to do work is increased.
When a body is lowered vertically its potential energy is decreased.
There is no zero of potential energy, although an arbitrary level may be used from which *changes* in the potential energy of a body may be measured.

Example 6

Find the change in the potential energy of a child of mass 48 kg when (a) ascending, (b) descending a vertical distance of 2 m.

(a) Change in potential energy $= \text{work done against gravity}$
$= mgh$
$= 48g \times 2 = 940{\cdot}8 \text{ J}$

The change in potential energy is 940·8 J.

(b) When descending, the child is *losing* potential energy, i.e. losing its potential to do work

Loss in potential energy $= mgh$
$= 48g \times 2 = 940{\cdot}8 \text{ J}$

The loss in potential energy is 940·8 J.

Example 7

A cricket ball of mass 400 g moving at 3 m/s and a golf ball of mass 100 g have equal kinetic energies. Find the speed at which the golf ball is moving.

Kinetic energy of cricket ball $= \frac{1}{2}mv^2$
$= \frac{1}{2}\left(\frac{400}{1000}\right)(3)^2 = 1{\cdot}8 \text{ J}$

Kinetic energy of golf ball $= \frac{1}{2}\left(\frac{100}{1000}\right)v^2$

$$\therefore \quad 1{\cdot}8 = \frac{v^2}{20}$$

$$\therefore \quad v^2 = 36 \text{ or } v = 6 \text{ m/s}$$

The golf ball is moving at 6 m/s.

Example 8

A body of mass 4 kg decreases its kinetic energy by 32 J. If initially it had a speed of 5 m/s, find its final speed.

Initial kinetic energy $= \frac{1}{2}mv^2$

$= \frac{1}{2}(4)(5)^2 = 50$ J

$\therefore$ final kinetic energy $= 50 - 32 = 18$ J

Let final speed of body be v $\therefore$ $18 = \frac{1}{2}(4)v^2$ or $v = 3$ m/s

The final speed of the body is 3 m/s.

Exercise 13B

1. Find the kinetic energy of
   (a) a body of mass 5 kg moving with speed 4 m/s,
   (b) a body of mass 2 kg moving with speed 3 m/s,
   (c) a car of mass 800 kg moving with speed 10 m/s,
   (d) a particle of mass 100 g moving with speed 20 m/s,
   (e) a bullet of mass 10 g moving with speed 400 m/s.
2. Find the potential energy gained by
   (a) a body of mass 5 kg raised through a vertical distance of 10 m,
   (b) a man of mass 60 kg ascending a vertical distance of 5 m,
   (c) a lift of mass 1 tonne ascending a vertical distance of 20 m.
3. Find the potential energy lost by
   (a) a body of mass 20 kg falling through a vertical distance of 2 m,
   (b) a man of mass 80 kg descending a vertical distance of 10 m,
   (c) a lift of mass 500 kg descending a vertical distance of 20 m.
4. Find the gain in kinetic energy when
   (a) a car of mass 1 tonne increases its speed from 5 m/s to 6 m/s,
   (b) a body of mass 5 g increases its speed from 200 m/s to 300 m/s.
5. Find the loss in kinetic energy when
   (a) a body of mass 2 kg decreases its speed from 2 m/s to 1 m/s,
   (b) a car of mass 800 kg decreases its speed from 18 km/h to rest.
6. A body of mass 5 kg, initially moving with speed 2 m/s, increases its kinetic energy by 30 J. Find the final speed of the body.
7. A body of mass 5 kg, initially moving with speed 3 m/s, increases its kinetic energy by 40 J. Find the final speed of the body.
8. Find the gain in kinetic energy when a car of mass 900 kg accelerates at 2 m/s^2 for 5 s if initially it is at rest.
9. A body of mass 6 kg, initially moving with speed 12 m/s, experiences a constant retarding force of 10 N for 3 s. Find the kinetic energy of the body at the end of this time.

Changes in the energy possessed by a body

Gain of energy due to work done on a body

When a force acts upon a body so as to increase the kinetic (or potential) energy of that body, the work done by the force is equal to the increase in the energy of the body.

Example 9

A constant force pushes a body of mass 500 g in a straight line across a smooth horizontal surface. The body passes through a point A with a speed of 3 m/s and then through a point B with a speed of 5 m/s, B being 3 m from A. For the motion of the body from A to B, find
(a) the increase in the kinetic energy of the body,
(b) the work done by the force,
(c) the magnitude of the force.

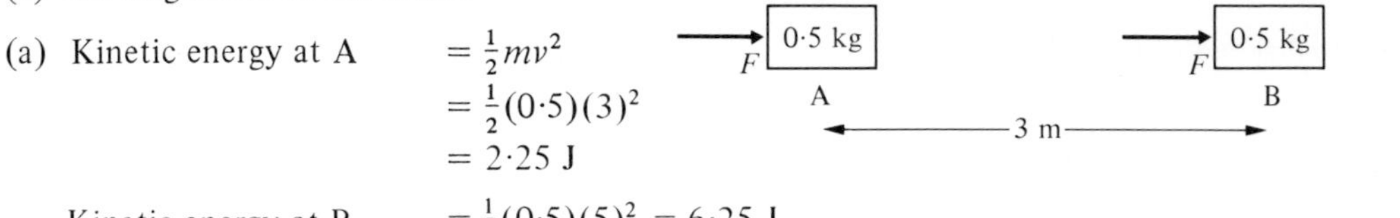

(a) Kinetic energy at A $= \frac{1}{2}mv^2$
$= \frac{1}{2}(0{\cdot}5)(3)^2$
$= 2{\cdot}25$ J

Kinetic energy at B $= \frac{1}{2}(0{\cdot}5)(5)^2 = 6{\cdot}25$ J

The increase in kinetic energy is 4 J.

(b) There is no change in potential energy and there is no friction to overcome.
$\therefore$ work done by force = increase in kinetic energy
= 4 J

Work done by force is 4 J.

(c) Work done by force = force × distance moved
$= F \times 3$
$\therefore \quad 3F = 4$ or $F = 1\frac{1}{3}$ N

The magnitude of the force is $1\frac{1}{3}$ N.

Loss of energy due to work done against a resistance

If a body has to move against some form of resistance, the body has to use some of its energy to overcome this resistance. The decrease in the energy of the body is equal to the work done against this resistance.

Example 10

A and B are two points 6 m apart on a horizontal surface. A particle of mass 400 g passes through the point A with a speed of 8 m/s, and through the point B with a speed of 5 m/s. The resistance against which the particle moves is constant in magnitude.

For the motion of the particle from A to B, find
(a) the loss in the kinetic energy of the particle,
(b) the work done against the resistances,
(c) the magnitude of the resistances.

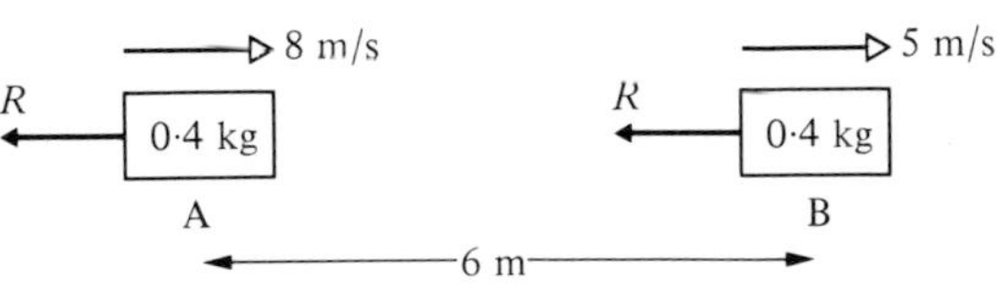

(a) Loss in kinetic energy $= \frac{1}{2}mu^2 - \frac{1}{2}mv^2$
$= \frac{1}{2}(0{\cdot}4)8^2 - \frac{1}{2}(0{\cdot}4)5^2 = 7{\cdot}8$ J

The loss in kinetic energy is 7·8 J.

(b) Work done against resistances = loss in kinetic energy

$$= 7{\cdot}8 \text{ J}$$

The work done against resistances is 7·8 J.

(c) Work done against resistances R is (force × distance moved)

$$\therefore \quad 7{\cdot}8 = R \times 6$$
$$\text{or} \quad R = 1{\cdot}3 \text{ N}$$

The magnitude of the resistances is 1·3 N.

Loss or gain in energy

In all questions involving the loss or gain of energy, the work done against or by external forces must be considered. An equation can then be written down, balancing the work done and the change in energy.

Example 11

The point A is 4 m vertically above the point B. A body of mass 200 g is projected from A vertically downwards with a speed of 3 m/s. For the motion of the body from A to B, neglecting resistances, find

(a) the loss in the potential energy of the body,
(b) the gain in the kinetic energy of the body,
(c) the kinetic energy of the body when at B.

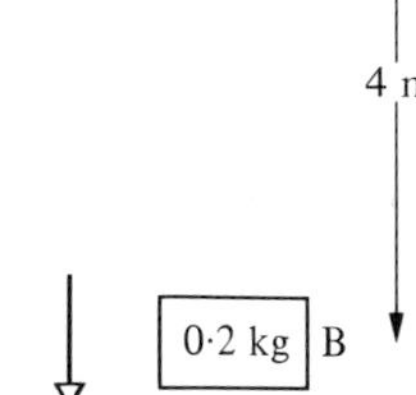

(a) Loss in potential energy $= mgh$

$$= 0{\cdot}2(9{\cdot}8)(4) = 7{\cdot}84 \text{ J}$$

The loss in potential energy is 7·84 J.

(b) Since no work is done against resistance

gain in kinetic energy = loss in potential energy

$$= 7{\cdot}84 \text{ J}$$

Note that this answer could be obtained by using the velocity formulae:

$$v^2 = u^2 + 2gs$$
$$= (3)^2 + 2(9{\cdot}8)(4) = 87{\cdot}4$$

$$\text{gain in kinetic energy} = \tfrac{1}{2}mv^2 - \tfrac{1}{2}mu^2$$
$$= \tfrac{1}{2}(0{\cdot}2)(87{\cdot}4 - 3^2) = 7{\cdot}84 \text{ J}$$

The gain in kinetic energy is 7·84 J.

(c) Kinetic energy at A $= \tfrac{1}{2}(0{\cdot}2)(3)^2 = 0{\cdot}9$ J

In moving from A to B, the body gains 7·84 J of kinetic energy

$$\therefore \quad \text{kinetic energy at B} = 0{\cdot}9 + 7{\cdot}84 = 8{\cdot}74 \text{ J}$$

The kinetic energy at B is 8·74 J.

Example 12

A body of mass 2 kg falls vertically, passing through two points A and B. The speeds of the body as it passes A and B are 1 m/s and 4 m/s respectively. The resistance against which the body falls is 9·6 N. Find the distance AB by using energy considerations.

Let the distance AB $= s$

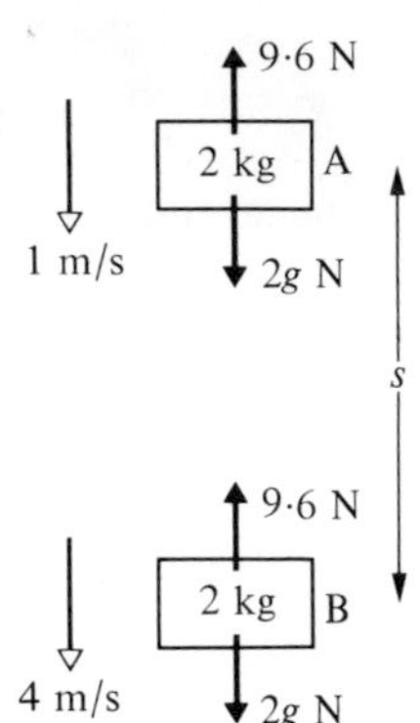

In falling from A to B

$$\text{loss in potential energy} = mgs = 2(9{\cdot}8)s = 19{\cdot}6s \text{ J}$$

$$\text{gain in kinetic energy} = \tfrac{1}{2}m(v^2 - u^2) = \tfrac{1}{2}(2)(4^2 - 1^2) = 15 \text{ J}$$

But, loss in P.E. = gain in K.E. + work done against resistance

$$\therefore \quad 19{\cdot}6s = 15 + (9{\cdot}6 \times s)$$

$$\therefore \quad 10s = 15 \quad \text{or} \quad s = 1{\cdot}5 \text{ m}$$

The distance AB is 1·5 m.

Example 13

From the point A situated at the bottom of a rough inclined plane, a body is projected with a speed of 5·6 m/s along and up a line of greatest slope. The plane is inclined at $\tan^{-1}\frac{4}{3}$ to the horizontal. The coefficient of friction between the body and the plane is $\frac{4}{7}$ and the body first comes to rest at a point B. By energy considerations, find the distance AB.

$\tan\theta = \frac{4}{3}$, $\sin\theta = \frac{4}{5}$, $\cos\theta = \frac{3}{5}$.

Let AB $= s$ metres and the mass of the body be m kg.

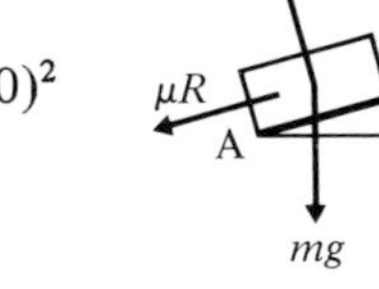

In moving from A to B

$$\text{loss in kinetic energy} = \tfrac{1}{2}m(5{\cdot}6)^2 - \tfrac{1}{2}m(0)^2 = \tfrac{1}{2}m(5{\cdot}6)^2 = 15{\cdot}68m \text{ J}$$

$$\text{gain in potential energy} = mgh = m(9{\cdot}8)(s\sin\theta) = m(9{\cdot}8)s(\tfrac{4}{5}) = 7{\cdot}84ms \text{ J}$$

Resolving perpendicular to the plane, $R = mg\cos\theta$

$$\therefore \quad \text{frictional force } \mu R = \mu mg\cos\theta = \tfrac{4}{7}m(9{\cdot}8)(\tfrac{3}{5}) = 3{\cdot}36m \text{ N}$$

$$\text{work done against friction} = \mu R \times s = 3{\cdot}36ms \text{ J}$$

But, loss in K.E. = gain in P.E. + work done against friction

$$\therefore \quad 15{\cdot}68m = 7{\cdot}84ms + 3{\cdot}36ms$$

$$\therefore \quad 15{\cdot}68 = 11{\cdot}2s \quad \text{or} \quad s = 1{\cdot}4 \text{ metres}$$

The distance AB is 1·4 metres.

Exercise 13C

1. A and B are two points 4 m apart on a smooth horizontal surface. A body of mass 2 kg is initially at rest at A and is pushed by a force of constant magnitude acting in a direction $\overrightarrow{AB}$. The body reaches B with a speed of 4 m/s. For the motion of the body from A to B find (a) the increase in the kinetic energy of the body, (b) the work done by the force, (c) the magnitude of the force.
2. A constant force pushes a body of mass 4 kg in a straight line across a smooth horizontal surface. The body passes through a point A with a speed of 5 m/s and then through a point B with a speed of 8 m/s, B being 6 m from A. For the motion of the body from A to B find (a) the increase in the kinetic energy of the body, (b) the work done by the force, (c) the magnitude of the force.
3. A and B are two points 3 m apart on a smooth horizontal surface. A body of mass 6 kg is initially at rest at A and is pushed towards B by a constant force of 9 N. For the motion of the body from A to B, find (a) the work done by the force, (b) the increase in the kinetic energy of the body, (c) the final velocity of the body.
4. A constant force of magnitude 8 N pushes a body of mass 4 kg in a straight line across a smooth horizontal surface. The body passes through a point A with a speed of 4 m/s and then through a point B, 5 m from A. For the motion of the body from A to B, find (a) the work done by the force, (b) the increase in the kinetic energy of the body, (c) the final velocity of the body.
5. A particle of mass 100 g moves in a straight line across a horizontal surface against resistances of constant magnitude. The particle passes through a point A with a speed of 7 m/s and then through a point B with a speed of 3 m/s, B being 2 m from A. For the motion of the particle from A to B, find (a) the loss in the kinetic energy of the body, (b) the work done against the resistances, (c) the magnitude of the resistances.
6. A body of mass 5 kg moves in a straight line across a horizontal surface against a constant resistance of magnitude 10 N. The body passes through a point A and then comes to rest at a point B, 9 m from A. For the motion of the body from A to B, find (a) the work done against the resistance, (b) the loss in kinetic energy of the body, (c) the speed of the body when at A.
7. A body of mass 5 kg slides across a rough horizontal surface. In sliding 5 m the speed of the body decreases from 8 m/s to 6 m/s. Find (a) the loss in the kinetic energy of the body, (b) the work done against friction, (c) the coefficient of friction between the body and the surface.
8. A body of mass 2 kg, initially at rest, falls a vertical distance of 10 m. Neglecting resistances find (a) the loss in the potential energy of the body, (b) the gain in the kinetic energy of the body, (c) the final speed of the body.
9. A body of mass 3 kg is projected vertically downwards from a point A with speed 4 m/s. The body passes through a point B, 5 m below A. For the motion of the body from A to B, neglect resistances and find (a) the loss in the potential energy of the body, (b) the gain in the kinetic energy of the body, (c) the kinetic energy of the body when at B.
10. A and B are two points in the same vertical line with A above B. A body of mass 4 kg is released from rest at A and falls vertically, passing through B with speed 7 m/s. For the motion of the body from A to B, neglect resistances and find (a) the gain in the kinetic energy of the body, (b) the loss in the potential energy of the body, (c) the distance AB.

11. A ball of mass 500 g is thrown vertically upwards from ground level and has an initial speed of 10 m/s. The ball passes through a point B, directly above the point of projection, with speed 4 m/s. For the motion of the ball from A to B, neglect resistances and find
(a) the loss in the kinetic energy of the ball,
(b) the gain in the potential energy of the ball,
(c) the height of B above the point of projection.

12. A and B are two points 15 m apart in the same vertical line, with A above B. A body of mass 5 kg is released from rest at A and falls vertically against constant resistances of 25 N. For the motion of the body from A to B, find
(a) the loss in the potential energy possessed by the body,
(b) the work done against the resistances, (c) the gain in the kinetic energy of the body,
(d) the speed of the body when passing through B.

13. A body of mass 5 kg falls vertically against constant resistances of X N. The body passes through two points A and B, 2·5 m apart with A above B, when travelling with speeds of 2 m/s and 6 m/s respectively. By energy considerations, find the value of X.

14. A body of mass 2 kg falls vertically against constant resistances of 14 N. The body passes through two points A and B when travelling with speeds of 3 m/s and 10 m/s respectively. Using energy considerations, find the distance AB.

15. A bullet of mass 15 g is fired towards a fixed wooden block. The bullet enters the block when travelling horizontally at 400 m/s and comes to rest after penetrating a distance of 25 cm. Find (a) the loss in the kinetic energy of the bullet,
(b) the work done against the resistance of the wood,
(c) the force of resistance experienced by the bullet (assumed constant throughout).

16. A bullet of mass 8 g is fired towards a fixed wooden block and enters the block when travelling horizontally with a speed of 300 m/s. If the wood provides a constant resistance to motion of 1800 N, find how far into the block the bullet will penetrate.

17. A smooth slope is inclined at $\tan^{-1}\frac{3}{4}$ to the horizontal. A body of mass 4 kg is released from rest at the top of the slope and reaches the bottom with speed 7 m/s. Find
(a) the gain in the kinetic energy of the body,
(b) the loss in the potential energy of the body, (c) the length of the slope.

18. Point A is situated at the base of a smooth plane which is inclined at 30° to the horizontal. A body of mass 2 kg is projected from A along and up a line of greatest slope. The body first comes to rest at a point B, 5 m from A. By energy considerations, find the speed at which the body was projected.

19. Point A is situated at the base of a smooth plane which is inclined at $\tan^{-1}\frac{5}{12}$ to the horizontal. A body is projected from A with a speed of 14 m/s along and up a line of greatest slope, and first comes to rest at a point B. By energy considerations find the distance AB.

20. A rough slope of length 5 m is inclined at an angle of 30° to the horizontal. A body of mass 2 kg is released from rest at the top of the slope and travels down the slope against constant resistances of X N. If the body reaches the bottom of the slope with speed 2 m/s, find (a) the loss in the potential energy of the body,
(b) the gain in the kinetic energy of the body,
(c) the work done against the resistances, (d) the value of X.

21. A rough slope of length 10 m is inclined at an angle of $\tan^{-1}\frac{3}{4}$ to the horizontal. A block of mass 50 kg is released from rest at the top of the slope and travels down the slope reaching the bottom with speed 8 m/s. Find (a) the work done against friction,
(b) the coefficient of friction between the block and the surface.

22. Point A is situated at the bottom of a rough plane which is inclined at $\tan^{-1}\frac{3}{4}$ to the horizontal. A body is projected from A with a speed of 14 m/s along and up a line of greatest slope. The coefficient of friction between the body and the plane is 0·25 and the body first comes to rest at a point B. By energy considerations, find the distance AB.
23. Point A is situated at the bottom of a rough plane which is inclined at 45° to the horizontal. A body of mass 500 g is projected from A along and up a line of greatest slope. The coefficient of friction between the body and the plane is $\frac{3}{7}$. The body first comes to rest at a point B, $4\sqrt{2}$ m from A, before returning to A. Find
    (a) the work done against friction when the body moves from A to B,
    (b) the initial speed of the body,
    (c) the work done against friction when the body moves from A to B and back to A,
    (d) the speed of the body on return to A.
24. Point A is situated at the bottom of a rough plane which is inclined at an angle θ to the horizontal. A body of mass m is projected from A, along and up a line of greatest slope. The coefficient of friction between the body and the plane is μ. The body first comes to rest at a point B, a distance x from A, before returning to A. Show that
    (a) the work done against friction when the body moves from A to B and back to A is given by $2\mu mgx \cos\theta$,
    (b) the initial speed of projection of the body is $\sqrt{\{2gx(\sin\theta + \mu\cos\theta)\}}$,
    (c) the speed of the body on its return to A is $\sqrt{\{2gx(\sin\theta - \mu\cos\theta)\}}$.

Power

Power is a measure of the rate at which work is being done.
If 1 joule of work is done in 1 second, the rate of working is said to be 1 watt (W).
Hence, a machine which does 1000 joules of work per second is working at a rate of 1000 watts or 1 kilowatt (i.e. 1000 W = 1 kW).

Example 14

Find the rate at which work is being done when a mass of 20 kg is lifted vertically at a constant speed of 5 m/s.

$$\begin{aligned}\text{Work done} &= \text{increase in potential energy}\\ &= 20gh \text{ where } h \text{ is height lifted vertically}\\ \therefore\quad \text{work done in one second} &= 20g(5)\text{ J} = 980\text{ J}\\ \therefore\quad \text{rate of working} &= 980\text{ J per second} \quad\text{or}\quad 980\text{ W}\end{aligned}$$

Rate of working is 980 W.

Example 15

A pump ejects 12 000 kg of water at a speed of 4 m/s in 40 seconds. Find the average rate at which the pump is working.

$$\begin{aligned}\text{kinetic energy given to water} &= \tfrac{1}{2}mv^2\\ &= \tfrac{1}{2}(12\,000)(4)^2\text{ J in 40 s}\\ &= \tfrac{1}{2}\left(\tfrac{12\,000}{40}\right)(4)^2\text{ J in 1 s} = 2400\text{ J in 1 s}\end{aligned}$$

The average rate of working is 2400 W or 2·4 kW.

Pump raising and ejecting water

If a pump is used not only to raise water, but also to eject it at a given speed, then the total work being done is the sum of the potential energy and the kinetic energy given to the water each second.

Example 16

A pump draws water from a tank and issues it from the end of a hose which is 2·5 m vertically above the level from which the water is drawn. The cross-sectional area of the hose is 10 cm^2, and the water leaves the end of the hose at a speed of 5 m/s. Find the rate at which the pump is working. (Take the density of water to be 1000 kg/m^3.)

$$\text{Volume of water raised and issued per second} = 5 \times \frac{10}{(100)^2}\ \text{m}^3$$

$$\text{mass of water raised and issued per second} = \frac{50}{(100)^2} \times 1000\ \text{kg} = 5\ \text{kg}$$

$$\begin{aligned}\text{potential energy given to water per second} &= mgh \\ &= 5(9{\cdot}8)(2{\cdot}5)\ \text{J}\end{aligned}$$

$$\begin{aligned}\text{kinetic energy given to water per second} &= \tfrac{1}{2}mv^2 \\ &= \tfrac{1}{2}(5)5^2\ \text{J}\end{aligned}$$

$$\text{work done per second, by the pump} = 5(9{\cdot}8)(2{\cdot}5) + \tfrac{1}{2}(5)5^2\ \text{J} = 185\ \text{J}$$

The rate at which the pump is working is 185 W.

Exercise 13D

Take the density of water as 1000 kg/m^3.

1. What is the average rate at which work must be done in lifting a mass of 100 kg a vertical distance of 5 m in 7 seconds?
2. What is the rate at which work must be done in lifting a mass of 500 kg vertically at a constant speed of 3 m/s?
3. In every minute a machine pumps 300 kg of water along a horizontal hose from rest at one end to eject it at a speed of 4 m/s at the other. Find the average rate at which the machine is working.
4. A bowling machine bowls 6 balls every minute. Each ball has a mass of 150 g and leaves the machine with a speed of 20 m/s. Find the average rate at which the machine is working.
5. Find the average rate at which a climber of mass 80 kg must work when climbing a vertical distance of 30 m in 2 minutes.
6. In building a section of a wall, a man has to lift 500 bricks a vertical distance of 150 cm. Each brick has a mass of 2 kg and the man completes the section in 5 minutes. Find the man's average rate of working.
7. A pump raises 75 kg of water a vertical distance of 20 m in 14 seconds. Find the average rate at which the pump is working.
8. In every minute a pump draws 6 m^3 of water from a well and issues it at a speed of 5 m/s from a nozzle situated 4 m above the level from which the water was drawn. Find the average rate at which the pump is working.

9. In every minute a pump draws 5 m^3 of water from a well and issues it at a speed of 6 m/s from a nozzle situated 6 m above the level from which the water was drawn. Find the average rate at which the pump is working.
10. A pump draws water from a tank and issues it at a speed of 8 m/s from the end of a pipe of cross-sectional area 0·01 m^2, situated 10 m above the level from which the water is drawn. Find (a) the mass of water issued from the pipe in each second,
(b) the rate at which the pump is working.
11. A pump draws water from a tank and issues it at a speed of 10 m/s from the end of a hose of cross-sectional area 5 cm^2, situated 4 m above the level from which the water is drawn. Find the rate at which the pump is working.
12. In each minute a pump draws 2·4 m^3 of water from a well 5 m below ground, and issues it at ground level through a pipe of cross-sectional area 50 cm^2. Find
(a) the speed with which the water leaves the pipe,
(b) the rate at which the pump is working.
If in fact the pump is only 75% efficient (i.e. 25% of the power is lost in the running of the pump), find the rate at which it must work.
13. In each minute a pump, working at 3·48 kW, raises 1·5 m^3 of water from an underground tank and issues it from the end of a pipe situated at ground level. The water leaves the pipe with speed 10 m/s and the pump is 50% efficient (i.e. 50% of the power is lost in the running of the pump). Find (a) the area of cross section of the pipe,
(b) the depth below ground level from which the water is drawn.
14. In each minute a pump, working at a constant rate of 0·825 kW, draws 0·3 m^3 of water from a well and issues it through a nozzle situated 5 m above the level from which the water was drawn. If the pump is 60% efficient (i.e. 40% of the power is lost in the running of the pump), find the velocity with which the water is ejected.
Find also the area of cross section of the nozzle.

Vehicles in motion

Consider a car being driven along a road. The forward force F N which propels the car is supplied by the engine. Suppose the engine is working at a constant rate of P watts,

then
$$\begin{aligned} P &= \text{work done per second} \\ &= \text{force} \times \text{distance moved per second} \\ &= F \times \text{speed} = Fv \end{aligned}$$

thus $F = \frac{P}{v}$ gives the forward force exerted by the engine working at P watts, when the car is travelling at v m/s.

The expression $\frac{P}{v}$ is frequently referred to as the *tractive force* being exerted by the engine at the instant when the vehicle is travelling at v m/s.

For problems involving vehicles in motion, if the power or rate of working is involved, draw a diagram showing the forces acting, including $\frac{P}{v}$ the tractive force. Then apply $F = ma$ in the direction of motion; if there is no acceleration, use the fact that the resultant force acting on the vehicle must be zero.

Example 17

The engine of a car is working at a steady rate of 5 kW. The car of mass 1200 kg is being driven along a level road against a constant resistance to motion of 325 N. Find
(a) the acceleration of the car when its speed is 8 m/s,
(b) the maximum speed of the car.

(a) Power $P = 5\text{ kW} = 5000\text{ W}$

$v = 8\text{ m/s}$

Using $F = ma$, $\quad \dfrac{P}{v} - 325 = 1200a$

$\therefore \quad \dfrac{5000}{8} - 325 = 1200a$

$\therefore \quad a = 0{\cdot}25\text{ m/s}^2$

The acceleration is $0{\cdot}25\text{ m/s}^2$.

(b) To find the maximum speed of the car, consider the situation when the acceleration is zero

then $\quad \dfrac{P}{v} = 325$

$\therefore \quad \dfrac{5000}{v} = 325 \quad$ or $\quad v = 15{\cdot}38\text{ m/s}$

The maximum speed of the car is 15·4 m/s.

Example 18

The greatest speed at which a cyclist can travel is 9 m/s when on a level road and cycling against a constant resistance of 40 N. Find the maximum rate at which the cyclist can work.

Let the rate of working be P;
at maximum speed there is no acceleration.

$\therefore \quad \dfrac{P}{v} = 40$

but $\quad v = 9\text{ m/s} \qquad \therefore \quad P = 360\text{ W}$

The cyclist can work at 360 W.

Example 19

A car travels along a level road against a constant resistance to motion of 500 N. The mass of the car is 1500 kg and the maximum speed it can attain is 40 m/s. Find the rate at which the engine of the car is working.
If the engine of the car works at the same rate and the resistances remain unchanged, find the maximum speed of the car when ascending an incline of $\sin^{-1}\frac{1}{49}$.

On the level, let the rate of working be P;
at maximum speed there is no acceleration.

$\therefore \quad \dfrac{P}{40} = 500$

$\therefore \quad P = 20\,000\text{ W}$

The car is working at 20 kW.

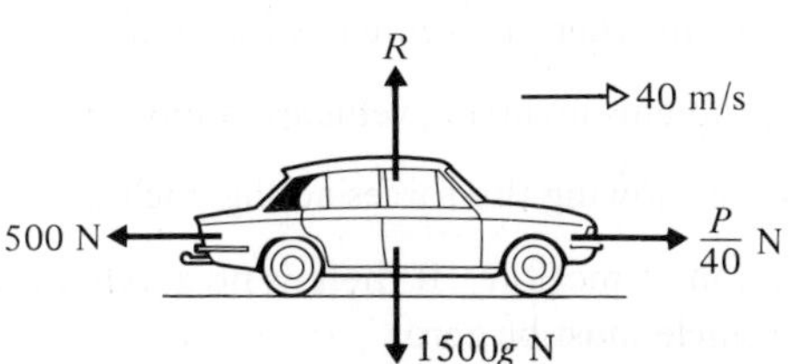

On the incline, let the maximum speed be v; at maximum speed there is no acceleration.

$$\therefore \quad \frac{P}{v} = 500 + 1500g \sin\theta$$

$$\therefore \quad \frac{20\,000}{v} = 500 + 1500(9{\cdot}8)\tfrac{1}{49}$$

$$\therefore \quad v = 25 \text{ m/s}$$

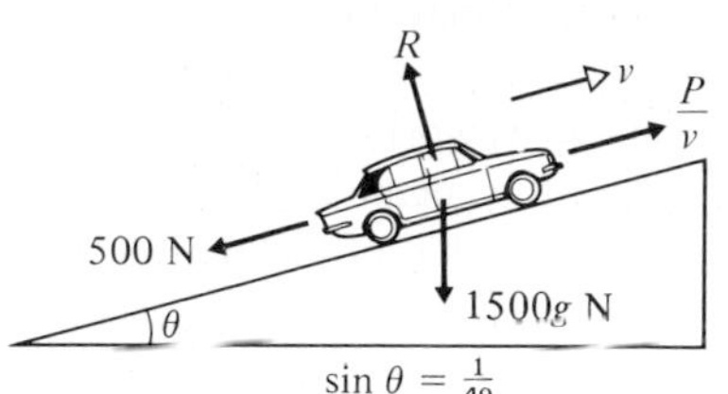

The maximum speed of the car on the incline is 25 m/s.

Exercise 13E

1. A car is driven along a level road against a constant resistance to motion of 400 N. Find the maximum speed at which the car can move when its engine works at a steady rate of
(a) 4 kW, (b) 6 kW, (c) 8·8 kW
2. A car of mass 1 tonne is driven along a level road against a constant resistance to motion of 200 N. With the engine of the car working at a steady rate of 8 kW find
(a) the acceleration of the car when its speed is 5 m/s,
(b) the acceleration of the car when its speed is 10 m/s, (c) the maximum speed of the car.
3. A car of mass 900 kg is driven along a level road against a constant resistance to motion of 300 N. With the engine of the car working at a steady rate of 12 kW, find
(a) the acceleration of the car when its speed is 4 m/s,
(b) the acceleration of the car when its speed is 10 m/s,
(c) the maximum speed of the car.
4. With its engines working at a steady 14 kW, the maximum speed that a car can attain along a level road is 35 m/s. Find the magnitude of the resistances experienced by the car.
5. A cyclist travels along a level road at a constant speed of 8 m/s. If the resistances to motion total 50 N, find the rate at which the cyclist is working.
6. A cyclist finds that when cycling along a level road against a resistance to motion of 20 N, the greatest speed that he can attain is 10 m/s. Find the maximum rate at which the cyclist can work.
7. A train of mass 100 tonnes travels along level track with its engines developing a constant 60 kW of power. If the greatest speed that the train can attain along the track is 108 km/h, find the magnitude of the resistance to motion.
Assuming this resistance remains unchanged, find the acceleration of the train when it travels along the track at 54 km/h with its engines working at the same rate as before.
8. A cyclist and his bike have a combined mass of 75 kg. Find the maximum speed that the cyclist can attain on level ground when working at a constant rate of 210 W against resistances totalling 21 N.
With the resistances and the rate of working unchanged, the cyclist ascends a slope of inclination $\sin^{-1}\frac{1}{15}$. Find the maximum speed of the cyclist up the slope.
9. With its engine working at a constant rate of 15 kW, a car of mass 800 kg ascends a hill of 1 in 98 against a constant resistance to motion of 420 N. Find
(a) the acceleration of the car up the hill when travelling with a speed of 10 m/s,
(b) the maximum speed of the car up the hill.
10. With its engines working at a constant rate of 369 kW, a train of mass 100 tonnes ascends a hill of 1 in 50 at a constant speed of 54 km/h. Find the magnitude of the resistance to motion experienced by the train.

11. A car of mass 900 kg can attain a maximum speed of 48 m/s when travelling along a level road against a constant resistance to motion of 350 N. Find the rate at which the car engine is working.
With the engine of the car working at the same rate and the resistances unchanged, the car ascends a hill of inclination $\sin^{-1}\frac{1}{18}$. Find the maximum speed of the car up the hill.
12. A cyclist and his bike have a combined mass of 75 kg and the maximum rate at which the cyclist can work is 0·392 kW. If the greatest speed with which the cyclist can ride along a level road is 8 m/s, find the magnitude of the constant resistance to motion.
With this resistance unchanged, find the greatest speed at which the cyclist can ascend a hill of inclination $\sin^{-1}\frac{1}{15}$.
13. (a) With its engine working at a steady rate of 32 kW, a car of mass 1 tonne travels at a constant speed of 40 m/s along a level road. Find the magnitude of the resistance to motion experienced by the car.
(b) Given that the resistance to motion is directly proportional to the speed at which the car is travelling, find the magnitude of the resistance experienced by the car when travelling at 30 m/s.
(c) Find the rate at which the engine must work for the car to ascend a slope of 1 in 98 at a constant speed of 20 m/s, the resistance to motion still obeying the same rule of proportionality.
14. When a car of mass 900 kg has its engine working at a constant rate of 7·35 kW, the car can ascend a hill of 1 in 63 at a constant speed of 15 m/s. Find
(a) the resistance to motion experienced by the car,
(b) the maximum speed of the car when travelling down the same slope with its engine working at the same rate as before and the resistance to motion unchanged.
15. With its engine working at a constant rate of 9·8 kW, a car of mass 800 kg can descend a slope of 1 in 56 at twice the steady speed that it can ascend the same slope, the resistances to motion remaining the same throughout. Find the magnitude of the resistance and the speed of ascent.

Exercise 13F Examination questions

1. A force, acting vertically upwards on a body of mass 10 kg, moves the body vertically from rest to a height 5 m above its starting point and gives it a speed of 6 m/s. Find the work done by the force. [Take g as 10 m/s^2.] (London)

2. In a downhill ski race the competitors descend from the start at a point 1000 m above sea-level to the finish at a point 500 m above sea-level. A particular competitor when fully equipped has a total mass of 70 kg. He starts from rest and crosses the finishing line at a speed of 36 km/h. For this competitor, calculate (i) his loss of potential energy, (ii) his gain of kinetic energy.
Hence calculate the total mechanical work done in overcoming the resistances to his motion. (J.M.B)

3. A motor operates a machine which conveys 120 ball bearings a minute through a vertical height of 2 m and discharges each at a speed of 4 m/s. The mass of each ball bearing is 25 g. Given that 30% of the work done by the motor is lost calculate the power developed by the motor. [Take g as 10 m/s^2.] (Cambridge)

4. A block of mass 6·5 kg is projected with a velocity of 4 m/s up a line of greatest slope of a rough plane. Calculate the initial kinetic energy of the block.
The coefficient of friction between the block and the plane is $\frac{2}{3}$ and the plane makes an angle θ with the horizontal where $\sin\theta = \frac{5}{13}$. The block travels a distance of d m up the plane before coming instantaneously to rest. Express in terms of d
 (i) the potential energy gained by the block in coming to rest,
 (ii) the work done against friction by the block in coming to rest.
Hence calculate the value of d. [Take g as 10 m/s^2.] (Cambridge)

5. At all speeds the resistance to the motion of a lorry of mass 5000 kg is proportional to the speed of the lorry. With the engine working at 84 kW, the lorry can attain a maximum speed of 12 m/s up a straight road which is inclined at $\sin^{-1}(\frac{1}{20})$ to the horizontal.
 (i) Show that the resistance is 4500 N at this speed.
 (ii) Find the acceleration at an instant when the lorry is moving at 12 m/s along a straight horizontal road with the engine working at 84 kW.
 (iii) Calculate the greatest steady speed at which the lorry could descend the straight road which is inclined at $\sin^{-1}(\frac{1}{20})$ to the horizontal with the engine working at 84 kW.
(Take the acceleration due to gravity to be 10 m/s^2.) (A.E.B)

6. The forces resisting the motion of a car are constant at all speeds and total 480 N. When the engine of the car is working at a rate of K kilowatts, the maximum speed of the car on a level road is 37·5 m/s.
 (i) Calculate the value of K.
The car is moving at this maximum speed when the rate of working of the engine is suddenly increased to $(K + 12)$ kilowatts and the resulting initial acceleration of the car is 0·256 m/s^2.
 (ii) Calculate the mass of the car.
The car moves down a straight road whose inclination to the horizontal is $\sin^{-1}(\frac{1}{5})$, at a constant speed of 8 m/s in low gear with the engine of the car working at a rate of H kilowatts to provide a constant braking force.
 (iii) Calculate the value of H. (Take the acceleration due to gravity to be 10 m/s^2.) (A.E.B)

7. A particle of mass m is attached to one end, A of a light inextensible string, AB, of length a. The particle is projected horizontally with speed V so that it moves in a vertical circular arc about B as centre. Find the tension in the string when it makes an angle θ with the downward vertical.
If $V = \sqrt{(2ag)}$, where does the string go slack? (S.U.J.B)

8. A heavy particle connected to a fixed point O by a light inelastic string of length a is moving in a vertical circle about O. Its speed when at the lowest point of the circle is $(\frac{7}{2}ga)^{\frac{1}{2}}$. Find the inclination of the string to the vertical when it becomes slack, and show that the speed of the particle is then $(\frac{1}{2}ga)^{\frac{1}{2}}$. Find also the maximum height above O reached by the particle. (Oxford)

9. A particle P is projected horizontally with speed u from the lowest point A of the smooth inside surface of a fixed hollow sphere of internal radius a.
 (i) In the case when $u^2 = ga$ show that P does not leave the surface of the sphere. Show also that, when P has moved halfway along its path from A towards the point at which it first comes to rest, its speed is $\sqrt{(ga\{\sqrt{3} - 1\})}$.
 (ii) Find u^2 in terms of ga in the case when P leaves the surface at a height $\frac{3}{2}a$ above A, and find, in terms of a and g, the speed of P as it leaves the surface. (J.M.B)

14

Momentum and impulse

Momentum

The momentum of a body of mass m, having a velocity v, is mv. If the units of mass and velocity are kg and m/s respectively, then the units of momentum are newton-seconds (N s). There is no named unit for momentum in the way that there is for force (newton) and energy (joule).
Since the momentum of a body depends upon the *velocity* with which the body is moving, momentum is a vector quantity.

Example 1

Find the magnitude of the momentum of
(a) a cricket ball of mass 420 g thrown at 20 m/s,
(b) a steam-roller of mass 6 tonnes moving at 0·4 m/s.

(a)
$$\text{Momentum} = \text{mass} \times \text{velocity}$$
$$\text{Magnitude of momentum} = \tfrac{420}{1000} \times 20 = 8{\cdot}4 \text{ N s}$$

The magnitude of the momentum is 8·4 N s.

(b)
$$\text{Momentum} = \text{mass} \times \text{velocity}$$
$$\text{Magnitude of momentum} = (6 \times 1000) \times \tfrac{4}{10} = 2400 \text{ N s}$$

The magnitude of the momentum is 2400 N s.

Changes in momentum

If the velocity of a body changes from u to v, then its momentum also changes.
The change in momentum can be found by considering the initial momentum mu and the final momentum mv.

Example 2

Find the change in the momentum of a body of mass 2 kg when its speed changes from
(a) 6 m/s to 15 m/s in the same direction,
(b) 5 m/s to 3 m/s in the opposite direction.

(a) Draw two diagrams

initial → 6 m/s [2 kg]   final → 15 m/s [2 kg]

Taking velocities to the right as positive

$$\text{initial momentum} = 2 \times 6 = 12 \text{ N s}$$
$$\text{final momentum} = 2 \times 15 = 30 \text{ N s}$$

Thus the change in the momentum is 18 N s.

(b) Draw two diagrams

Taking velocities to the right as positive

initial momentum $= 2 \times 5 = 10$ N s
final momentum $= 2 \times (-3) = -6$ N s

Thus the change in momentum is 16 N s.

Impulse

The impulse of a constant force F is defined as $F \times t$, where t is the time for which the force is acting.

$$\text{Impulse} = F \times t$$

but $F = m \times a$

$$\therefore \quad \text{impulse} = ma \times t$$
$$= mv - mu \qquad \text{using } v = u + at$$
$$= \text{change in momentum}$$

Thus, impulse of a force = change in momentum produced.

It should be noted that in the case of a hammer hitting a nail, or a batsman hitting a ball, the force involved is large but it only acts for a short time. In other cases the force may continue to act for a much greater length of time.

Example 3

A body of mass 4 kg is initially at rest on a smooth horizontal surface. A horizontal force of 3·5 N acts on the body for 8 seconds. Find
(a) the magnitude of the impulse given to the body,
(b) the magnitude of the final momentum of the body,
(c) the final speed of the body.

(a) Draw two diagrams

impulse = force × time
$= 3{\cdot}5 \times 8 = 28$ N s

The impulse given to the body is 28 N s.

(b) Body initially at rest, so initial momentum is zero.

impulse = change in momentum
= final momentum − initial momentum
∴ 28 = final momentum − 0

The final momentum of the body is 28 N s.

(c) Final momentum = 28 N s
$\therefore \quad 4 \times v = 28 \quad \text{or} \quad v = 7$ m/s

The final speed of the body is 7 m/s.

Example 4

A body of mass 3 kg is initially moving with a constant velocity of $15\mathbf{i}$ m/s. It experiences a force of $8\mathbf{i}$ N for 6 seconds. Find
(a) the impulse given to the body by the force,
(b) the velocity of the body when the force ceases to act.

(a)

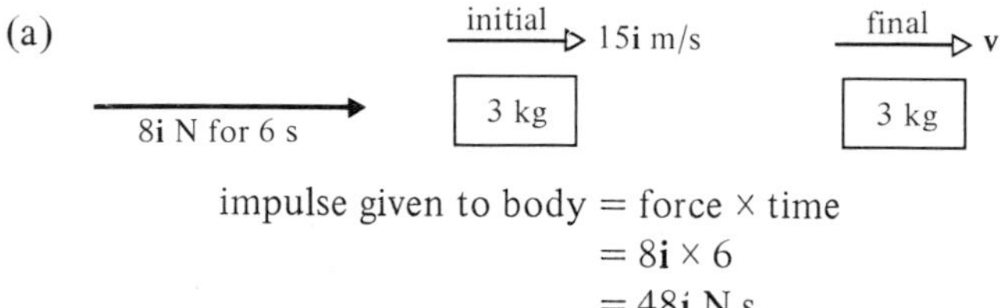

$$\begin{aligned}\text{impulse given to body} &= \text{force} \times \text{time}\\ &= 8\mathbf{i} \times 6\\ &= 48\mathbf{i}\text{ N s}\end{aligned}$$

The impulse given to the body is $48\mathbf{i}$ N s.

(b)

$$\begin{aligned}\text{impulse} &= \text{change in momentum}\\ &= m\mathbf{v} - m\mathbf{u}\\ \therefore\quad 48\mathbf{i} &= 3\mathbf{v} - (3 \times 15\mathbf{i})\\ 93\mathbf{i} &= 3\mathbf{v}\\ \therefore\quad \mathbf{v} &= 31\mathbf{i}\text{ m/s}\end{aligned}$$

The final velocity of the body is $31\mathbf{i}$ m/s.

Example 5

A force of $(2\mathbf{i} + 7\mathbf{j})$ N acts on a body of mass 5 kg for 10 seconds. The body was initially moving with constant velocity of $(\mathbf{i} - 2\mathbf{j})$ m/s. Find the final velocity of the body in vector form, and hence obtain its final speed.

Let the final velocity of the body be $(a\mathbf{i} + b\mathbf{j})$ m/s

$$\begin{aligned}\text{impulse} &= \text{change in momentum}\\ \therefore\quad (2\mathbf{i} + 7\mathbf{j}) \times 10 &= 5 \times (a\mathbf{i} + b\mathbf{j}) - 5 \times (\mathbf{i} - 2\mathbf{j})\\ \therefore\quad 20\mathbf{i} &= 5a\mathbf{i} - 5\mathbf{i} \text{ or } a = 5\\ \text{and}\quad 70\mathbf{j} &= 5b\mathbf{j} + 10\mathbf{j} \text{ or } b = 12\end{aligned}$$

Final velocity is $(5\mathbf{i} + 12\mathbf{j})$ m/s

$$\begin{aligned}\text{speed of body} &= \sqrt{(5^2 + 12^2)}\\ &= 13\text{ m/s}\end{aligned}$$

The final velocity is $(5\mathbf{i} + 12\mathbf{j})$ m/s and the final speed is 13 m/s.

Exercise 14A

1. Find the magnitude of the momentum possessed by
   (a) a car of mass 700 kg moving with a speed of 20 m/s,
   (b) a ball of mass 200 g moving with a speed of 15 m/s,
   (c) a man of mass 75 kg running with a speed of 4 m/s,
   (d) a train of mass 200 tonnes moving with a speed of 18 m/s,
   (e) a bullet of mass 15 g moving with a speed of 400 m/s.

2. Find the change in the momentum of a body of mass 5 kg when
   (a) its speed changes from 4 m/s to 7 m/s without its direction of motion changing,
   (b) its speed changes from 2 m/s to 12 m/s without its direction of motion changing,
   (c) its speed changes from 4 m/s to 7 m/s and its direction of motion is reversed,
   (d) its speed changes from 2 m/s to 12 m/s and its direction of motion is reversed.
3. A body of mass 2 kg is initially at rest on a smooth horizontal surface. A horizontal force of 6 N acts on the body for 3 seconds. Find
   (a) the magnitude of the impulse given to the body,
   (b) the magnitude of the final momentum of the body,
   (c) the final speed of the body.
4. A body is initially at rest on a smooth horizontal surface when a horizontal force of 7 N acts on it for 5 seconds. Find the magnitude of the final momentum of the body.
5. A body of mass 12 kg is initially moving with a constant velocity of 2**i** m/s when it is acted upon by a force of 6**i** N for 10 seconds. Find (a) the impulse given to the body by the force, (b) the velocity of the body when the force ceases to act.
6. A body of mass 5 kg is initially moving with a constant velocity of 8**i** m/s when it experiences a force of −10**i** N for 2 seconds. Find (a) the impulse given to the body by the force, (b) the velocity of the body when the force ceases to act.
7. A body of mass 5 kg is initially moving with a constant velocity of 2**i** m/s when it experiences a force of −10**i** N for 2 seconds. Find (a) the impulse given to the body by the force, (b) the velocity of the body when the force ceases to act.
8. A body of mass 4 kg is initially moving with a constant velocity of 2**i** m/s. After a force of a**i** N has acted on the body for 5 seconds, the body has velocity 12**i** m/s. Find
   (a) the initial momentum of the body,
   (b) the momentum of the body at the end of the 5 seconds,
   (c) the value of a.
9. A body of mass 6 kg is initially moving with a constant velocity of 9**i** m/s. After a force of b**i** N has acted on the body for 9 seconds, the body has velocity −3**i** m/s. Find the value of b.
10. A body of mass 5 kg is initially moving with a constant velocity of 7**i** m/s. After a force of 10**i** N has acted on the body for t seconds, the body has a velocity of 13**i** m/s. Find the value of t.
11. A body of mass 500 g is initially moving with a velocity of 8**i** m/s. After a force of −2**i** N has acted on the body for t seconds, the body has the same speed as before but its direction of motion has been reversed. Find the value of t.
12. A body of mass 4 kg, initially at rest on a smooth horizontal surface, experiences a force of (5**i** − 12**j**) N for 2 seconds. Find the final velocity of the body in vector form and hence obtain its final speed.
13. A body of mass 10 kg is initially moving with constant velocity (2**i** − 7**j**) m/s. The body experiences a force of (5**i** + 10**j**) N for 4 seconds. Find the final velocity of the body in vector form and hence obtain its final speed.
14. A body of mass 750 g is initially moving with constant velocity (4**i** + 6**j**) m/s. After a force of (a**i** + b**j**) N has acted on the body for 6 seconds, the body has a velocity of (12**i** − 10**j**) m/s. Find the values of a and b.

15. Each part of this question involves a smooth sphere colliding normally with a fixed vertical wall. The diagrams show the situations before and after collision. For each part take **i** as a unit vector in the direction → and find in vector form,
(i) the impulse given to the ball by the wall, (ii) the impulse given to the wall by the ball.

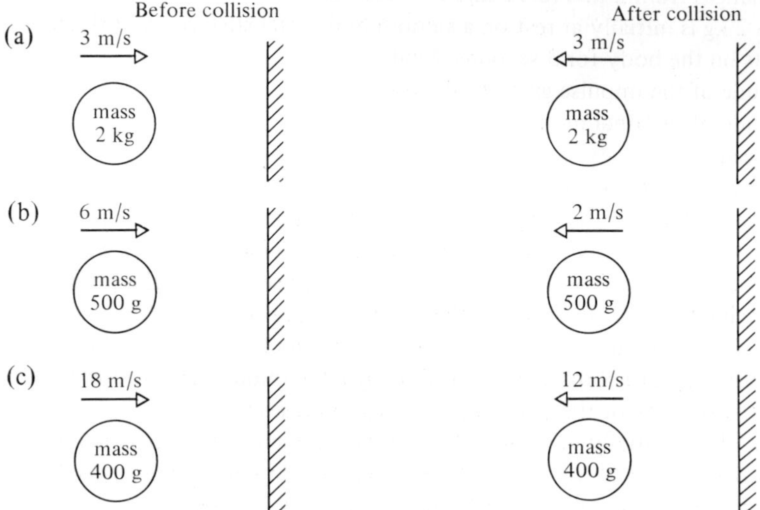

Force on a surface

When a jet of water strikes a surface and does not rebound, the surface destroys the momentum of the jet of water. The loss in momentum is equal to the impulse of the force exerted on the water by the surface. By Newton's Third Law, this force is equal in magnitude but opposite in direction to the force exerted on the surface by the water.
If the amount of water hitting the surface per second is known, then the momentum destroyed per second can be calculated.

$$\begin{aligned} \text{Since impulse} &= \text{momentum destroyed} \\ \text{and also} \quad \text{impulse} &= \text{force} \times \text{time} \\ \therefore \quad \text{force} \times 1 &= \text{momentum destroyed per second} \\ \text{or} \quad \text{force} &= \text{momentum destroyed per second} \end{aligned}$$

If the water does rebound from the surface, the direction of motion of the jet is reversed and therefore the change in momentum is increased.

Example 6

Water issues horizontally from a pipe and strikes a nearby wall without rebounding. The cross-sectional area of the pipe is 6 cm^2 and the water travels at 4 m/s. Find the magnitude of the force acting on the wall. (Take density of water as 1000 kg/m^3.)

$$\text{Volume of water hitting wall in 1 second} = \frac{6}{(100)^2} \times 4 \text{ m}^3$$

$$\text{Mass of water hitting wall in 1 second} = \text{volume of water} \times \text{density}$$

Since the wall is nearby, it may be assumed that there is no significant difference between the velocity of the water as it leaves the pipe and when it hits the wall.

$$\text{Momentum destroyed per second} = \frac{6}{(100)^2} \times 4 \times 1000 \times 4 = 9{\cdot}6 \text{ N s}$$

$$\text{Impulse} = \text{force} \times \text{time} = \text{momentum destroyed}$$

$$\therefore \quad F \times 1 = 9{\cdot}6$$

$$\therefore \quad F = 9{\cdot}6 \text{ N}$$

The force exerted on the wall by the water is 9·6 N.

Exercise 14B

In each of the following questions assume that the surface being struck by the fluid is sufficiently close to the source to allow any change in the velocity of the fluid between emission and impact to be neglected.

1. Water issues horizontally from a hose of cross-sectional area 5 cm^2 at a speed of 10 m/s. The water hits a vertical wall without rebounding. Find the magnitude of the force acting on the wall. (Take the density of water as 1000 kg/m^3.)
2. Water issues horizontally from a pipe of cross-sectional area 0·1 m^2 at a speed of 8 m/s. The water hits a vertical wall without rebounding. Find the magnitude of the force acting on the wall. (Take the density of water as 1000 kg/m^3.)
3. A water cannon emits water at a speed of 16 m/s through a pipe of cross-sectional area 0·015 m^2. If the water hits a nearby vertical surface, find the force on the surface, assuming that the water does not rebound after impact. (Take the density of water as 1000 kg/m^3.)
4. During a water pistol fight, Mike squirts water into the face of Jack. The water is emitted through a nozzle of cross-sectional area 1 mm^2 and travels horizontally with a speed of 8 m/s to hit Jack's face at right angles without rebounding. Find the magnitude of the force acting on Jack's face. (Take the density of water as 1000 kg/m^3.)
5. A factory waste pipe emits fluid of density 1200 kg/m^3 through a pipe of cross-sectional area 25 cm^2. The fluid leaves the pipe horizontally with a speed of 5 m/s and strikes a vertical wall without rebounding. Find the magnitude of the force acting on the wall.
6. In order to wash a wound in a person's arm a nurse squirts saline solution, density 1·03 g/cm^3, onto the wound from a syringe. The needle of the syringe emits the solution horizontally with a speed of 4 m/s through a hole of cross-sectional area 2·5 mm^2. The horizontal jet of solution hits the vertical surface of the wound and does not rebound. Find the magnitude of the force acting on the surface of the wound.

Impact

When a collision occurs between two bodies A and B, the force exerted on B by A is equal and opposite to the force exerted on A by B. This is another example of Newton's Third Law.
In the absence of any other forces acting on the two bodies, the changes in the momenta of A and B will be equal in magnitude, but opposite in direction. The gain in the momentum of one body will equal the loss in momentum of the other body; hence, the sum of the momenta of A and B before the impact will be equal to the sum of their momenta after the impact. This is referred to as the *Principle of Conservation of Linear Momentum.*

In dealing with examples on collision it is advisable to draw two diagrams; one showing the situation before the collision and the other showing the situation after the collision.

Since momentum is a vector quantity, the direction of motion of each body must be carefully and clearly indicated so that the correct sign may be attached to the momentum. In some instances after colliding, two bodies are said to coalesce. In this case the bodies do not rebound from each other, but they have a common velocity after the collision and move as a single body.

Example 7

A body of mass 2 kg moving on a smooth horizontal surface at 3 m/s, collides with a second body of mass 1 kg which is at rest. After the collision the bodies coalesce. Find the common speed of the bodies after impact.

Let v be the common speed of the bodies after the collision.

Draw two diagrams.

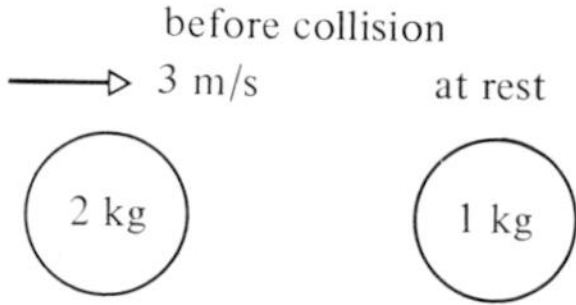

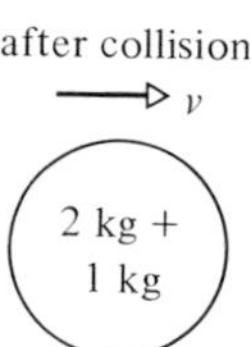

Taking velocities to the right as positive

$$\text{Momentum before collision} = (2 \times 3) + (1 \times 0)$$
$$\text{Momentum after collision} = (2 + 1) \times v$$

By conservation of momentum $6 + 0 = 3v$ or $v = 2$ m/s

The common speed of the two bodies after the impact is 2 m/s.

Example 8

The two bodies shown collide on a smooth horizontal surface. Find the value of v, the speed of the lighter body after impact.

Taking velocities to the right as positive

$$\text{Momentum before collision} = (4 \times 3) + (3 \times 2)$$
$$\text{Momentum after collision} = (4 \times 2{\cdot}5) + (3 \times v)$$

By conservation of momentum $12 + 6 = 10 + 3v$ or $v = 2\frac{2}{3}$ m/s

After the impact the lighter body has a speed of $2\frac{2}{3}$ m/s.

Example 9

The two bodies shown collide on a horizontal surface. Find the speed v of the lighter body after impact.

Taking velocities to the right as positive

$$\text{Momentum before collision} = (2 \times 6) + (5 \times (-4))$$
$$\text{Momentum after collision} = (2 \times (-v)) + (5 \times (-1))$$

By conservation of momentum $\quad 12 - 20 = -2v - 5 \;\text{ or }\; 2v = 3$

$$v = 1\tfrac{1}{2} \text{ m/s}$$

The speed of the lighter body after the impact is $1\frac{1}{2}$ m/s, and its direction of motion is reversed.

Loss of energy

When bodies collide, there is no loss of momentum but there is a loss of kinetic energy. Some of the kinetic energy possessed by the bodies is transformed into other forms of energy at the impact, e.g. heat and sound energy.

Example 10

Two smooth spheres A and B, of masses 150 g and 350 g, are travelling towards each other along the same horizontal line with speeds of 4 m/s and 2 m/s respectively. After the collision, the direction of motion of B has been reversed and it is travelling at a speed of 1 m/s. Find the speed of A after the collision and the loss of kinetic energy due to the collision.

Draw two diagrams showing the situation before and after the collision.

Taking velocities to the right as positive

$$\text{Momentum before collision} = 0{\cdot}15(4) + 0{\cdot}35(-2)$$
$$\text{Momentum after collision} = 0{\cdot}15(v) + 0{\cdot}35(1)$$

By conservation of momentum $\quad 0{\cdot}6 - 0{\cdot}7 = 0{\cdot}15v + 0{\cdot}35 \;\text{ or }\; v = -3$ m/s

Thus the speed of A after the impact is 3 m/s. The negative sign indicates that the 'after collision' diagram shows the incorrect direction of motion of sphere A.

Loss of kinetic energy $= \frac{1}{2}(0{\cdot}15)4^2 + \frac{1}{2}(0{\cdot}35)2^2 - [\frac{1}{2}(0{\cdot}15)3^2 + \frac{1}{2}(0{\cdot}35)1^2] = 1{\cdot}05$ J

The speed of A after impact is 3 m/s and the loss of kinetic energy is 1·05 J.

Recoil of a gun

When a shot is fired from a gun, an explosion occurs in the barrel of the gun. The explosion takes the form of expanding gases which exert a force on the shot and an equal and opposite force on the gun.
Initially, the shot and the gun are at rest. The gain in momentum of the shot after the explosion will be equal and opposite to the gain in momentum of the gun.
In a horizontal direction therefore, the forward momentum of the shot is equal to the backward momentum of the gun.

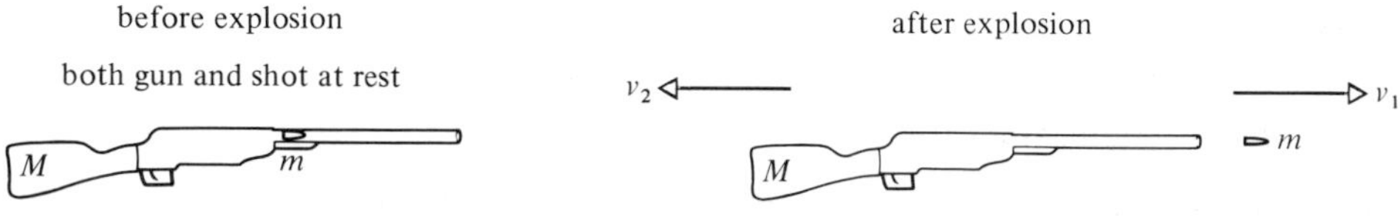

$$\text{Momentum before explosion} = M(0) + m(0)$$
$$\text{Momentum after explosion} = m(v_1) + M(-v_2)$$

By conservation of momentum $\quad 0 = mv_1 - Mv_2 \quad \text{or} \quad Mv_2 = mv_1$

Example 11

A bullet is fired from a gun with a horizontal velocity of 400 m/s. The mass of the gun is 3 kg and the mass of the bullet is 60 g. Find the initial speed of recoil of the gun and the gain in the kinetic energy of the system.

Draw two diagrams.

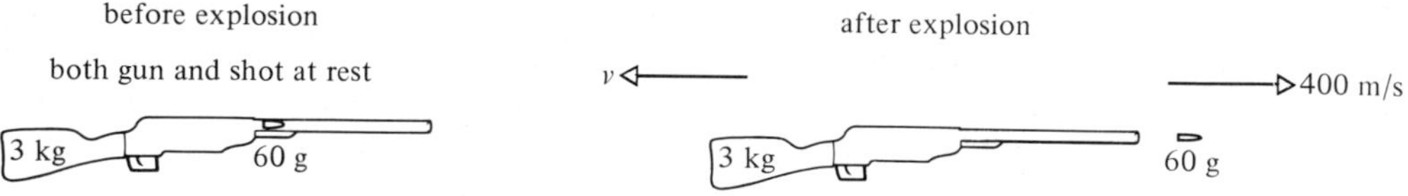

Taking velocities to the right as positive

$$\text{Momentum before explosion} = 3(0) + (0{\cdot}06)(0)$$
$$\text{Momentum after explosion} = 3(-v) + (0{\cdot}06)(400)$$

By conservation of momentum $\quad 0 = -3v + 24$

$$\therefore \quad v = 8 \text{ m/s}$$

$$\text{Initial kinetic energy} = \tfrac{1}{2}(3)(0)^2 + \tfrac{1}{2}(0{\cdot}06)(0)^2$$
$$\text{Final kinetic energy} = \tfrac{1}{2}(3)(8)^2 + \tfrac{1}{2}(0{\cdot}06)(400)^2$$
$$\text{Gain in kinetic energy} = \tfrac{1}{2}(192) + \tfrac{1}{2}(9600) - 0 = 4896 \text{ J}$$

The initial speed of recoil of the gun is 8 m/s and the gain in kinetic energy due to the explosion is 4896 J.

Exercise 14C

In questions **1** to **7**, the two diagrams show the situation before and after a collision between two bodies A and B moving along the same straight line on a smooth horizontal surface. Find the speed v in each case.

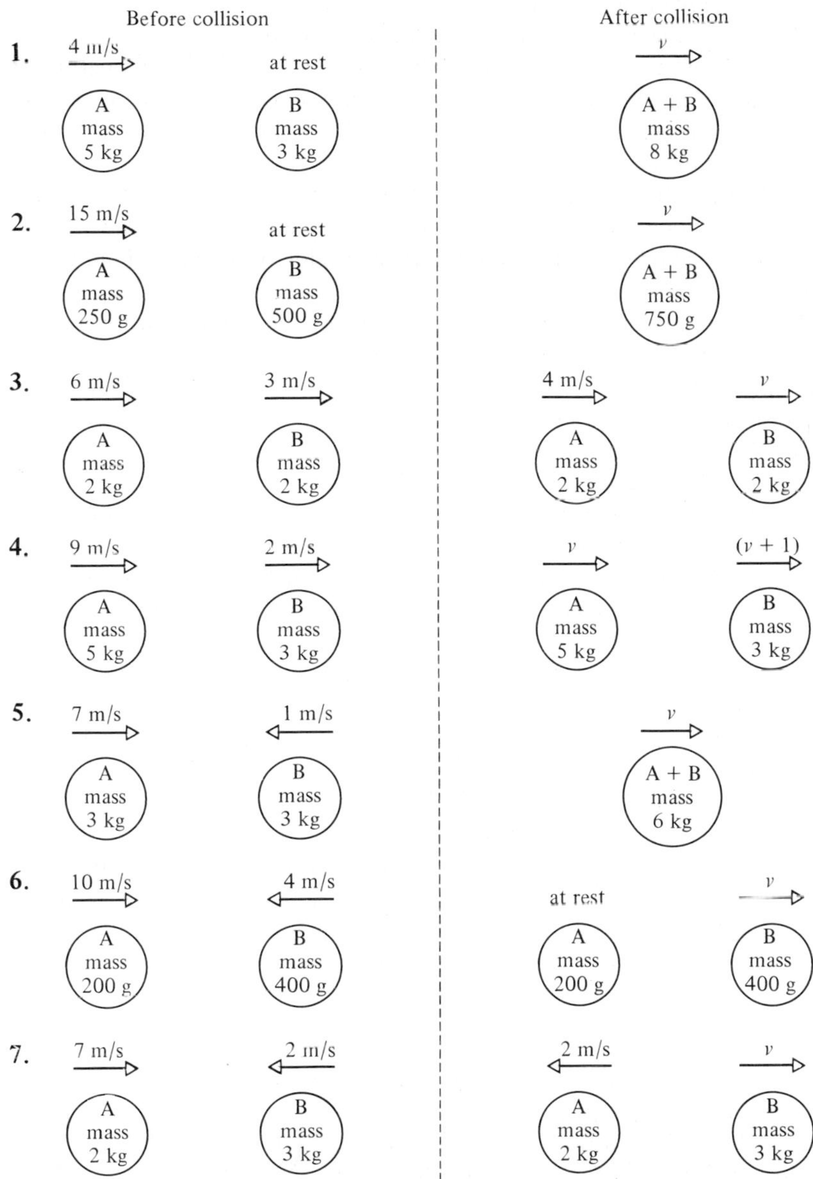

8. Two identical railway trucks are travelling in the same direction along the same straight piece of track with constant speeds of 6 m/s and 2 m/s. The faster truck catches up with the other one and on collision, the two trucks automatically couple together. Find the common speed of the trucks after collision.

9. A particle A of mass 300 g lies at rest on a smooth horizontal surface. A second particle B of mass 200 g is projected along the surface with speed 6 m/s and collides directly with A. If the collision reduces B to rest, find
   (a) the speed with which A moves after the collision,
   (b) the initial kinetic energy of the system,
   (c) the final kinetic energy of the system,
   (d) the loss in the kinetic energy of the system during the collision.
10. A particle A of mass 150 g lies at rest on a smooth horizontal surface. A second particle B of mass 100 g is projected along the surface with speed u m/s and collides directly with A. On collision the masses coalesce and move on with speed 4 m/s. Find the value of u and the loss in the kinetic energy of the system during impact.
11. Two smooth spheres A and B, of equal radii and masses 3 kg and $1\frac{1}{2}$ kg respectively, are travelling along the same horizontal line. The velocities of A and B are $6\mathbf{i}$ m/s and $-2\mathbf{i}$ m/s respectively. The spheres collide and after collision B has a velocity of $4\mathbf{i}$ m/s. Find the velocity of A after the collision.
12. Two smooth spheres A and B, of equal radii and masses 180 g and 100 g respectively, are travelling directly towards each other along a horizontal path. The initial speeds of A and B are 2 m/s and 6 m/s respectively. After collision both spheres have reversed their original directions of motion and B now has a speed of 3 m/s. Find the speed of A after impact and the loss in the kinetic energy of the system.
13. A bullet of mass 20 g is fired from a gun of mass 2·5 kg. The bullet leaves the gun with a speed of 500 m/s. Find the initial speed of recoil of the gun and the gain in the kinetic energy of the system.
14. A shell of mass 5 kg is fired from a gun of mass 2000 kg. The shell leaves the gun with a speed of 400 m/s. Find the initial speed of recoil of the gun and the gain in the kinetic energy of the system.
15. A wooden stake of mass 4 kg is to be driven vertically downwards into the ground using a mallet of mass 6 kg. The speed of the mallet just prior to impact is 10 m/s. After impact the mallet remains in contact with the stake (i.e. the weight of both mallet and stake aid penetration). Find the speed with which the stake begins to enter the ground.
    If the ground offers a constant resistance to motion of 1000 N, how far will the stake penetrate on each blow? (Take $g = 10$ m/s^2.)
16. Each part of this question involves three smooth spheres A, B and C, of equal radii, moving along the same straight line. A collides with B and then B collides with C. The diagrams show the situations before any collision, after A has collided with B, and after B has collided with C. Find the unknown speeds u and v in each case.

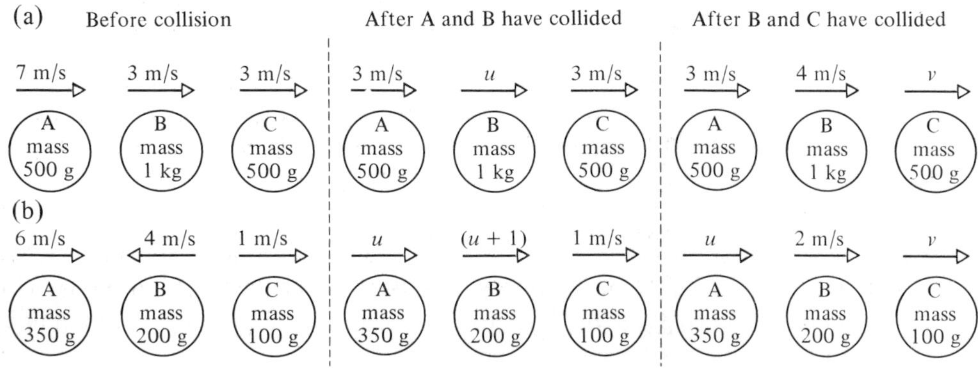

17. Two smooth spheres A and B, of equal radii and masses of 750 g and 1 kg respectively, are initially at rest on a smooth horizontal surface. A is projected directly towards B with speed 5 m/s and after collision A has not changed its direction of motion but has a speed of 1 m/s. The collision sets B into motion and it goes on to strike a fixed wall at right angles, the impact reversing B's direction of motion and halving its speed. B then collides again with A and this collision reduces B to rest. Find the final speed of A and the total loss in kinetic energy due to the collisions.
18. Two smooth spheres A and B, of equal radii and masses m_1 and m_2 respectively, are moving towards each other along the same horizontal line each with speed u. After collision both spheres have reversed their original directions of motion and A now travels with speed $\frac{u}{2}$. Show that $3m_1 > 2m_2$.

    Sphere B then strikes a fixed wall at right angles, the impact reversing the direction of motion of B and halving its speed. Show that B will again collide with A provided $3m_1 > 4m_2$.
19. Three smooth spheres A, B and C, of equal radii and masses m_1, m_2 and m_3 respectively, lie at rest in a straight line on a horizontal surface with B between A and C. Sphere B is projected towards C with speed u and C is projected towards B with speed $3u$. The collision between B and C reverses the direction of motion of sphere C which then travels with speed $2u$. Show that B will collide with A provided $m_3 > \frac{m_2}{5}$.

    What would be the necessary relationship between m_2 and m_3 for this second collision to occur had A been given an initial speed of u away from the other spheres?

Exercise 14D Examination questions

1. A ball A of mass 0·2 kg is moving at 4 m/s when it collides directly with a stationary ball B of mass 0·1 kg. After the collision, A moves with velocity 2·5 m/s in the same direction as before. Calculate (a) the speed of B after the collision, (b) the total loss of kinetic energy, in J, due to the collision. (London)
2. A bullet is fired horizontally with a speed of 600 m/s into a block of wood of mass 0·245 kg, resting on a smooth horizontal plane, and becomes embedded in the block. Given that the block begins to move with a speed of 12 m/s, find, in kg, the mass of the bullet. (London)
3. Two trucks P and Q, of mass 150 g and 250 g respectively, are free to move on a straight horizontal track of a model railway. Initially Q is at rest and P is moving towards Q with a velocity of 40 cm/s. Immediately after the impact Q has a velocity of 8 cm/s *relative to* P. Calculate (i) the velocities of each truck after the collision,
   (ii) the impulse imparted to each truck by the impact. (A.E.B)
4. A particle of mass m moving with speed v strikes a particle of mass $3m$ at rest and coalesces with it. Express the final kinetic energy as a fraction of the original kinetic energy. (London)

5. A jet of water is played horizontally upon a vertical wall. The water impinges at 20 m s^{-1} and the area of the cross-section of the jet is 50 cm^2. Assuming that the mass of 1 m^3 of water is 1 tonne and that the water drops vertically after impact, calculate the thrust upon the wall.
If the speed of the jet is reduced so that the thrust upon the wall becomes 1 kN, find the speed with which the jet now impinges. (Cambridge)

6. A railway wagon of mass 12 tonnes moving freely at 4 m s^{-1} collides with an empty stationary wagon of mass 4 tonnes at the base of a slope inclined at an angle α to the horizontal, where $\sin\alpha = \frac{1}{50}$. The wagons move together up the slope against a constant frictional resistance of 50 newtons per tonne. Calculate the distance they travel up the slope before coming to instantaneous rest.
The empty wagon is held at rest and the heavier one returns down the slope against the resistance, which remains constant at 50 newtons per tonne. Calculate the kinetic energy of the moving wagon at the moment it reaches the original point and show that this is $\frac{27}{80}$ of the kinetic energy before the impact took place. [Take g as 10 m/s^2.] (Cambridge)

7. Two particles, A and B of masses $2m$ and $3m$ respectively, are attached to the ends of a light inextensible string of length c and are placed close together on a horizontal table. The particle A is projected vertically upwards with speed $\sqrt{(6gc)}$.
   (i) Show that, at the instant immediately after the string tightens, B is moving with velocity $\frac{4}{5}\sqrt{(gc)}$.
   (ii) State the impulse of the tension in the string.
   (iii) Find the height to which A rises above the table before it comes to instantaneous rest.
   (iv) Calculate the loss in kinetic energy due to the tightening of the string. (A.E.B)

15

Elasticity

Elastic strings

In the situations in the preceding chapters the strings connecting bodies, passing over pulleys or maintaining equilibrium etc. have been said to be inextensible.
In this chapter the strings are said to be elastic which implies that they can be stretched and will regain their natural length once the stretching force has been removed.

Hooke's Law

This law states that the tension in a stretched string is proportional to the extension of the string from its natural (or unstretched) length.

tension $\propto$ extension

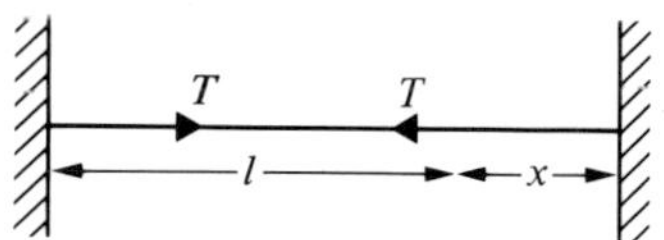

This is usually written in the form

$$T = \lambda\frac{x}{l}$$

where T is the tension in the string, x is the extension and l is the natural length of the string. The constant λ is called the modulus of the string; by considering the units in the above equation, it is seen that the units of λ are those of force, i.e. newtons.

Example 1

An elastic string is of natural length 3 m and modulus 15 N. Find (a) the tension in the string when the extension is 40 cm, (b) the extension of the string when the tension is 3 N.

(a)

extension = 40 cm = 0·4 m
natural length = 3 m

modulus = 15 N
tension = T

Hooke's Law gives $T = \lambda\frac{x}{l}$

$$\therefore \quad T = 15 \times \frac{(0{\cdot}4)}{3} = 2 \text{ N}$$

The tension in the string is 2 N.

(b)

extension = x
natural length = 3 m

modulus = 15 N
tension = 3 N

Using Hooke's Law $T = \lambda\frac{x}{l}$

$$\therefore \quad 3 = 15 \times \frac{x}{3}$$

$$\therefore \quad x = 0{\cdot}6 \text{ m}$$

The extension of the string is 60 cm.

Elastic springs

Hooke's Law also applies to an elastic spring which is either stretched or compressed. When a spring is compressed Hooke's Law gives the *thrust* in the spring due to its compression to a length which is less than its natural length.

Example 2

A spring is of natural length 1·5 m and modulus 25 N. Find the thrust in the spring when it is compressed to a length of 1·2 m.

natural length = 1·5 m   modulus = 25 N

compression = 1·5 − 1·2 = 0·3 m   thrust = T

Using Hooke's Law $T = \lambda \frac{x}{l}$

$\therefore \quad T = 25 \times \frac{0{\cdot}3}{1{\cdot}5}$

$\therefore \quad T = 5 \text{ N}$

The thrust in the spring is 5 N.

Equilibrium of a suspended body

When an elastic string has one end fixed and a mass attached to its other end so that the mass is suspended in equilibrium, then the string is stretched by the force due to the mass.

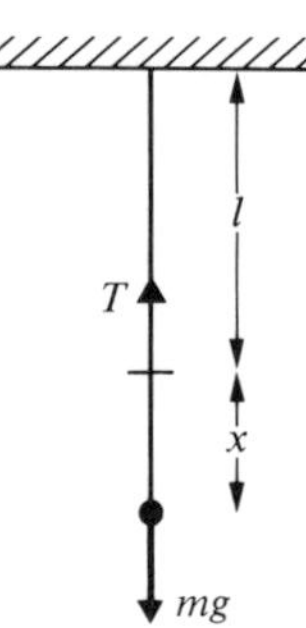

Resolving vertically $T = mg$

Using Hooke's Law $T = \lambda \frac{x}{l}$

hence $mg = \lambda \frac{x}{l}$

Example 3

A light elastic string of natural length 75 cm has one end fixed and a mass of 800 g freely suspended from the other end. Find the modulus of the string if the total length of the string in the equilibrium position is 95 cm.

In the equilibrium position

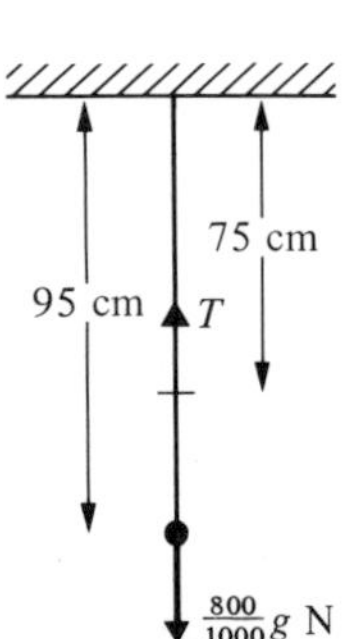

Resolving vertically $T = 0{\cdot}8g$

Using Hooke's Law $T = \lambda \frac{x}{l}$

$\therefore \quad 0{\cdot}8g = \lambda \frac{(0{\cdot}95 - 0{\cdot}75)}{0{\cdot}75}$

$\therefore \quad \lambda = 3g \text{ N}$

The modulus of the string is $3g$ N.

Example 4

A light elastic spring has its upper end A fixed and a body of mass 0·6 kg attached to its other end B. If the modulus of the spring is $4{\cdot}5g$ N and its natural length 1·5 m, find the extension of the spring when the body hangs in equilibrium.
The end B of the spring is pulled vertically downwards to C, where BC = 10 cm. Find the initial acceleration of the body when it is released from this position.

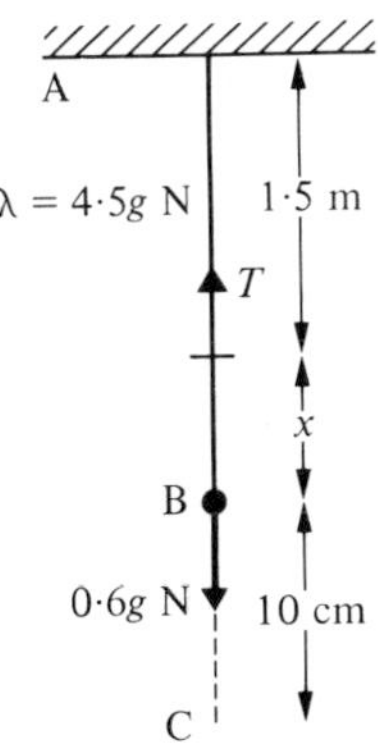

In the equilibrium position

Resolving vertically $\quad T = 0{\cdot}6g$

Using Hooke's Law $\quad T = \lambda\dfrac{x}{l}$

$$\therefore \quad 0{\cdot}6g = 4{\cdot}5g \times \frac{x}{1{\cdot}5}$$

$$\therefore \quad x = 0{\cdot}2 \text{ m}$$

When B is pulled down 10 cm, the total extension is then 30 cm.

Using Hooke's Law $\quad T = 4{\cdot}5g \times \dfrac{0{\cdot}3}{1{\cdot}5}$

$$= 0{\cdot}9g$$

Using $F = ma$

$$0{\cdot}9g - 0{\cdot}6g = 0{\cdot}6a$$

$$\therefore \quad a = \frac{g}{2} \text{ or } 4{\cdot}9 \text{ m/s}^2$$

The extension in the equilibrium position is 20 cm and the initial acceleration when the body is pulled down and released is 4·9 m/s^2.

Example 5

A body of mass M kg lies on a smooth horizontal surface and is connected to a point O on the surface by a light elastic string of natural length 50 cm and modulus 70 N. When the body moves in a horizontal circular path about O with constant speed of 3·5 m/s, the extension in the string is 20 cm. Find the mass of the body.

String

natural length = 50 cm = 0·5 m
extension = 20 cm = 0·2 m
modulus = 70 N
tension = T

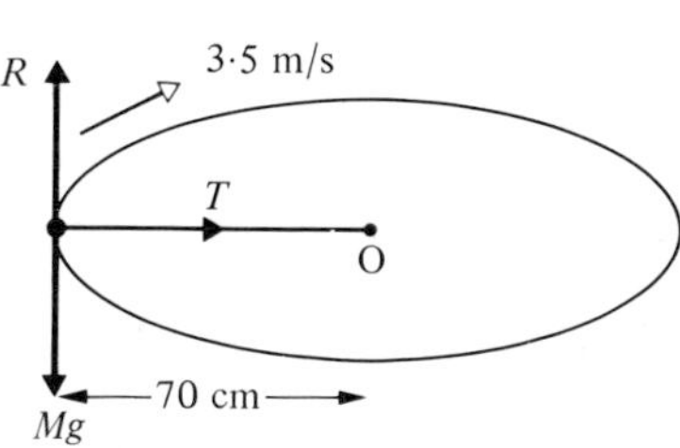

Using Hooke's Law

$$T = \lambda\frac{x}{l}$$

$$\therefore \quad T = 70 \times \frac{0{\cdot}2}{0{\cdot}5} = 28 \text{ N} \qquad \ldots [1]$$

But the tension T provides the force towards O necessary for circular motion

$$\therefore \quad T = \frac{mv^2}{r}$$

$$\therefore \quad T = M \times \frac{(3{\cdot}5)^2}{0{\cdot}7} \qquad \ldots [2]$$

From equations [1] and [2]

$$28 = M \times \frac{(3{\cdot}5)^2}{0{\cdot}7}$$

$$\therefore \quad M = 1{\cdot}6$$

The mass of the body is 1·6 kg.

Exercise 15A

1. An elastic string is of natural length 1 m and modulus 20 N. Find the tension in the string when the extension is 20 cm.
2. A spring is of natural length 50 cm and modulus 10 N. Find the thrust in the spring when it is compressed to a length of 40 cm.
3. When the tension in a spring is 5 N, the length of the spring is 25 cm longer than its natural length of 75 cm. Find the modulus of the spring.
4. When the length of a spring is 60% of its unstretched length, the thrust in the spring is 10 N. Find the modulus of the spring.
5. Find the extension of an elastic string of natural length 60 cm and modulus 18 N when the tension in the string is 6 N.
6. The tension in an elastic string is 8 N when the extension in the string is 25 cm. If the modulus of the string is 8 N, find the unstretched length of the string.
7. When an elastic string is stretched to a length of 18 cm, the tension in the string is 14 N. If the modulus of the string is 28 N, find the natural length of the string.
8. A light elastic string of natural length 20 cm and modulus $2g$ N has one end fixed and a mass of 500 g freely suspended from the other end. Find the total length of the string.
9. When a mass of 5 kg is freely suspended from one end of a light elastic string, the other end of which is fixed, the string extends to twice its natural length. Find the modulus of the string.
10. A light spring of natural length 60 cm and modulus $3g$ N has one end fixed and a body of mass 2 kg freely suspended from its other end. Find the extension of the spring. What mass would cause the same length of extension when suspended from a spring of natural length 50 cm and modulus $2g$ N?
11. An elastic string is of natural length 24 cm and modulus 18 N. One end of the string is attached to a fixed point O on a smooth horizontal surface and to the other end is attached a body of mass 2 kg. The body is held at rest on the surface at a point 32 cm away from O and then released. Find the initial acceleration of the body.
12. An elastic string is of natural length 45 cm and modulus 27 N. One end of the string is attached to a fixed point O on a rough horizontal surface and to the other end is attached a body of mass 5 kg. The body is held at rest on the surface at a point 60 cm away from O and then released. If the coefficient of friction between the body and the surface is $\frac{1}{7}$, find the initial acceleration of the body.

13. The diagram shows a body of mass 10 kg freely suspended from a light elastic spring of natural length 50 cm and modulus $25g$ N. Find the extension in the spring when
    (a) the body is at rest,
    (b) the top of the spring is moved upwards with constant velocity,
    (c) the top of the spring is moved upwards with a constant acceleration of 4·9 m/s^2,
    (d) the top of the spring is moved downwards with a constant acceleration of 4·9 m/s^2.

14. A smooth surface is inclined at an angle of 30° to the horizontal. A body A of mass 2 kg is held at rest on the surface by a light elastic string which has one end attached to A and the other to a point on the surface 1·5 m away from A up a line of greatest slope. If the modulus of the string is $2g$ N, find its natural length.
15. A particle of mass 5 kg is attached to one end of a light elastic string of natural length 1 m and modulus $4g$ N. The other end of the string is fastened to a fixed point O at the top of a smooth slope that is inclined at $\tan^{-1}\frac{3}{4}$ to the horizontal. The particle is held on the slope at a point that is 2·5 m from O down a line of greatest slope. If the particle is released from rest, find its initial acceleration towards O.
    What would the acceleration have been had the slope been rough, coefficient of friction 0·25?
16. A light elastic string has one end fixed and a body of mass $\sqrt{3}$ kg freely suspended from its other end. With a horizontal force of X N acting on the body, the system is in equilibrium with the string extended to twice its natural length and making an angle of 30° with the downward vertical. Find the modulus of the string and the value of X.
17. A body of mass 4 kg lies on a smooth horizontal surface and is connected to a point O on the surface by a light elastic string of natural length 64 cm and modulus 25 N. When the body moves with constant speed v m/s in a horizontal circle with centre O, the extension in the string is 36 cm. Find v.
18. A body of mass 5 kg lies on a smooth horizontal surface and is connected to a point O on the surface by a light elastic string of natural length 2 m and modulus 30 N. Find the extension in the string when the body moves in a horizontal circle with centre O at a constant speed of 3 m/s.
19. A and B are two fixed points on the same horizontal level and distance 48 cm apart. A light elastic string of natural length 40 cm has one end attached to A and the other to B. A body of mass 200 g is attached to the mid-point of the string and hangs in equilibrium at a point 7 cm below the level of A and B. Find the modulus of the string.
20. A body of mass m lies on a smooth horizontal surface and is connected to a point O on the surface by a light elastic string of natural length l and modulus λ. When the body moves with constant speed v around a horizontal circular path, centre at O, the extension in the string is $\frac{1}{4}l$. Show that $\lambda = \dfrac{16mv^2}{5l}$.

Energy stored in an elastic string

Since work is done in stretching an elastic string, it follows that a stretched string possesses energy equal in magnitude to the work done in stretching the string.

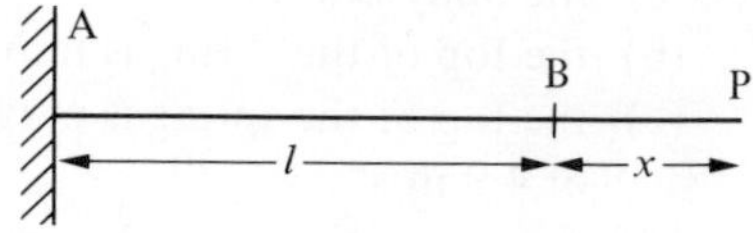

An elastic string of natural length l is stretched to a length of $(l + x)$ and the modulus of elasticity for the string is λ.
The end A of the string is fixed.

Initially the other end of the string is at B, so that AB $= l$, and subsequently this end of the string is pulled to the point P so that AP $= (l + x)$.

Let T_B and T_P be the tensions in the string when the end of the string is at B and P respectively.

Energy stored in the string due to stretching from B to P

$$\begin{aligned}
&= \text{work done in stretching from B to P} \\
&= \text{average force} \times \text{distance} \\
&= \frac{T_B + T_P}{2} \times \text{BP} \\
&= \tfrac{1}{2}\left[\frac{\lambda \times 0}{l} + \frac{\lambda \times x}{l}\right] x \\
&= \frac{\lambda x^2}{2l}
\end{aligned}$$

Example 6

An elastic string is of natural length 1·8 m and modulus 30 N. Find the energy stored in the string when it is extended to a length 2·1 m.

modulus $= 30$ N
natural length $= 1{\cdot}8$ m

$$\text{extension} = 2{\cdot}1 - 1{\cdot}8 = 0{\cdot}3 \text{ m}$$

$$\text{energy} = \frac{\lambda x^2}{2l} = \frac{30(0{\cdot}3)^2}{2(1{\cdot}8)} = 0{\cdot}75 \text{ J}$$

The energy stored in the string is 0·75 J.

It should be noted that in the example above, an *un*stretched string is stretched.
If an already stretched string is stretched by an extra amount, the additional energy stored in the string can be calculated as the difference between the initial and final energies of the string.

Example 7

An elastic string is of natural length 3·6 m and modulus 45 N. Find the work done in stretching it from a length of 4 m to a length of 4·2 m.

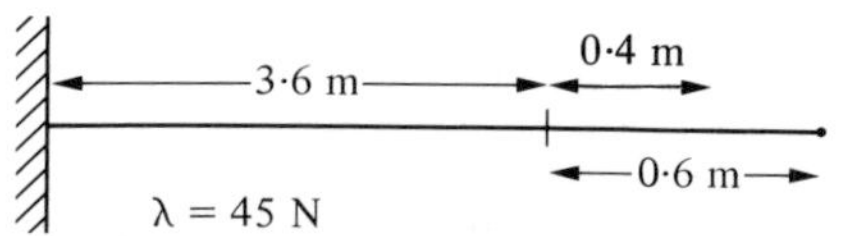

When string is of length 4 m, the extension is 0·4 m

$$\therefore \quad \text{energy stored} = \frac{45(0{\cdot}4)^2}{2(3{\cdot}6)} = 1 \text{ J}$$

When string is of length 4·2 m, the extension is 0·6 m

$$\therefore \quad \text{energy stored} = \frac{45(0{\cdot}6)^2}{2(3{\cdot}6)} = 2{\cdot}25 \text{ J}$$

$\therefore$ work done in stretching the string from 4 m to 4·2 m is 1·25 J.

Example 8

A light elastic string is of natural length 2 m and modulus $15g$ N. One end of the string is attached to a fixed point and a body of mass 3 kg hangs in equilibrium from the other end. The body is pulled down 10 cm and then released. Find
(a) the extension of the string in the equilibrium position,
(b) the energy stored in the string just before the body is released,
(c) the speed of the body as it passes through the equilibrium position.

(a) In the equilibrium position

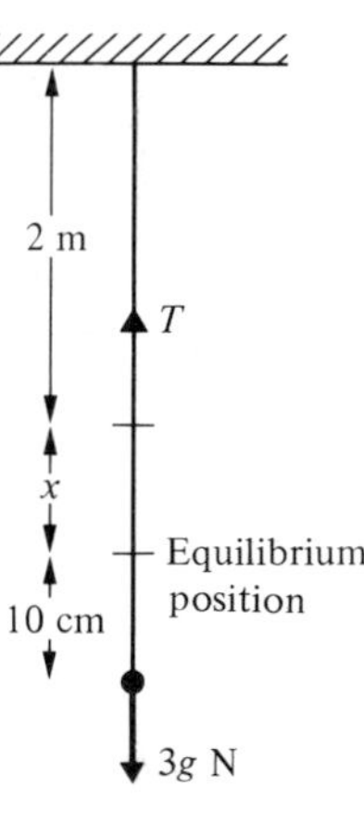

Resolving vertically $\quad T = 3g$

Using Hooke's Law $\quad T = \dfrac{15g \times x}{2}$

$$\therefore \quad \frac{15gx}{2} = 3g$$

$$\therefore \quad x = 0{\cdot}4 \text{ m or } 40 \text{ cm}$$

The extension in the equilibrium position is 40 cm.

(b) When string is extended a further 10 cm

$$\text{total extension} = 50 \text{ cm} = 0{\cdot}5 \text{ m}$$

$$\text{Energy of string} = \frac{\lambda x^2}{2l}$$

$$= 15g\frac{(0{\cdot}5)^2}{2 \times 2} = \frac{15g}{16} \text{ J}$$

The energy stored in the string just before the body is released is $\dfrac{15g}{16}$ J.

(c) When the body reaches the equilibrium position, it has gained kinetic energy and gained potential energy, and the string has lost some of the energy that was stored in it.

Energy stored in string when body is in the equilibrium position

$$= \frac{\lambda x^2}{2l}$$

$$= 15g\frac{(0{\cdot}4)^2}{2 \times 2} = \frac{3g}{5} \text{ J}$$

By the conservation of energy

gain in P.E. of body + gain in K.E. of body = loss of energy stored in string

$$\therefore \quad 3g\left(\frac{10}{100}\right) + \tfrac{1}{2}(3)v^2 = \frac{15g}{16} - \frac{3g}{5}$$

$$\therefore \quad v^2 = \frac{g}{40} \quad \text{or} \quad v = 0{\cdot}495 \text{ m/s}$$

The speed of the body when it passes through the equilibrium position is 0·495 m/s.

Energy stored in a spring

When an elastic *spring* is extended or compressed by an amount x, from its natural length l, the result already stated for an elastic string applies, i.e.

$$\text{Energy stored in a spring} = \frac{\lambda x^2}{2l}.$$

Exercise 15B

1. An elastic string is of natural length 2 m and modulus 10 N. Find the energy stored in the string when it is extended to a length of 3 m.
2. An elastic string is of natural length 1 m and modulus 20 N. Find the energy stored in the string when it is extended to a length of 1·3 m.
3. Find the work that must be done to stretch an elastic string of modulus 200 N from its natural length of 2 m to a stretched length of 2·5 m.
4. Find the work that must be done to compress a spring of modulus 500 N from its natural length of 10 cm to a shortened length of 8 cm.
5. A spring is of natural length 50 cm and modulus 60 N. How much energy is released when the length of the spring is reduced from 1·5 m to 1 m?
6. An elastic string is of natural length 4 m and modulus 24 N. Find the work that must be done to stretch the string from a length of 5 m to a length of 6 m.
7. A light elastic string is of natural length 50 cm and modulus 147 N. One end of the string is attached to a fixed point and a body of mass 3 kg is freely suspended from the other end. Find (a) the extension of the string in the equilibrium position, (b) the energy stored in the string.
8. A body lies on a smooth horizontal table and is connected to a point O on the table by a light elastic string of natural length 1·5 m and modulus 24 N. Initially the body lies at a point P 1·5 m from O. The body is pulled directly away from O and held at a point Q, 2 m from O, and then released. Find
   (a) the initial energy stored in the string when the body is at P,
   (b) the energy stored in the string when the body is held at Q,
   (c) the kinetic energy of the body as it passes through P after release from Q.
9. The diagram shows a body of mass 2 kg freely suspended from a spring of natural length 75 cm and modulus $6g$ N, the other end of which is fixed to a point A. The body initially hangs freely in equilibrium at a point B. It is then pulled down a further distance of 25 cm to a point C and released from rest. Find
   (a) the distance AB,
   (b) the energy stored in the spring when the body rests at B,
   (c) the energy stored in the spring when the body is held at C,
   (d) the kinetic energy of the body when it passes through B after release from C.

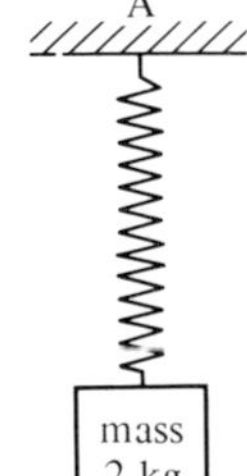

10. The diagram shows a body A of mass 480 g projected along a smooth horizontal surface with speed 2 m/s to collide directly with a body B of mass 320 g, initially at rest. B is attached to a fixed wall 20 cm away by a spring of natural length 20 cm and modulus 36 N. After the collision, A and B move on

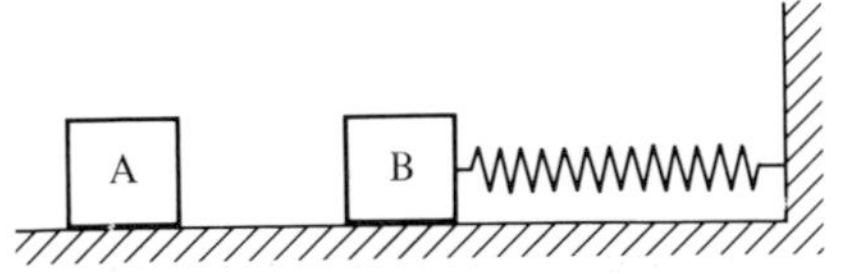

together and the thrust in the spring brings them momentarily to rest before accelerating them away from the wall. Find

(a) the common velocity of A and B immediately after collision,

(b) the length of the spring when A and B are momentarily at rest.

Elastic impact

In the last chapter the collision of two elastic bodies in motion was considered, using the principle of conservation of linear momentum.

There is another law which involves the coefficient of restitution of the bodies and their relative speeds before and after impact. This is known as *Newton's Experimental Law*.

The coefficient of restitution e for any two bodies is a measure of their elasticity.

It depends upon both of the bodies and the material of which they are made.

Newton's Experimental Law states that

$$e = \frac{\text{speed of separation of the bodies}}{\text{speed of approach of the bodies}}$$

With the law stated in this form, both the numerator and the denominator of the fraction are positive quantities.

The following diagrams illustrate how these speeds of approach and separation are obtained in different situations:

before impact | after impact

(a)

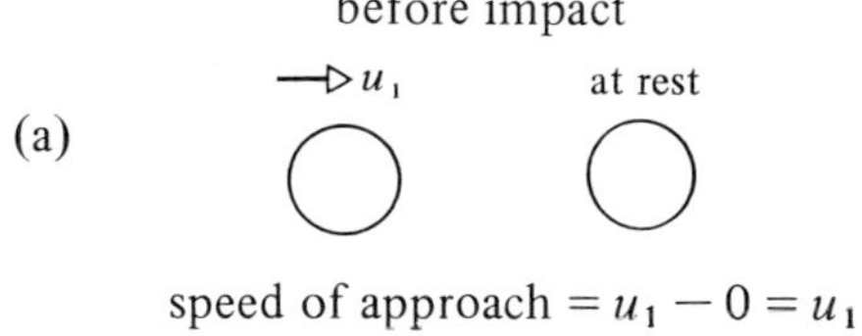

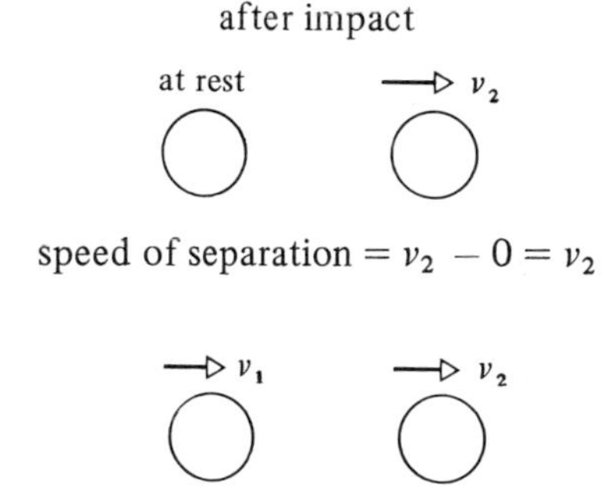

speed of approach $= u_1 - 0 = u_1$ | speed of separation $= v_2 - 0 = v_2$

(b)

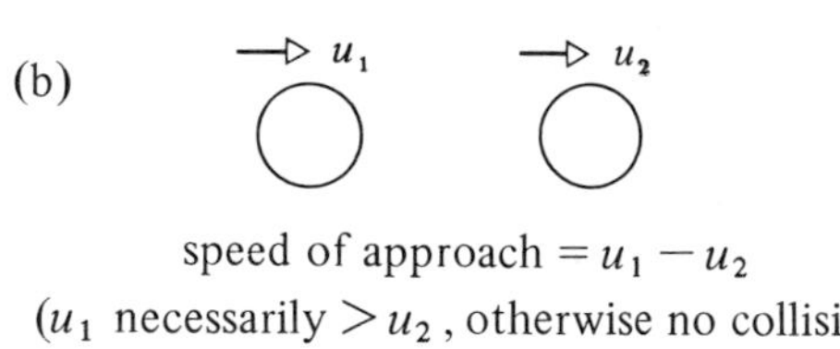

speed of approach $= u_1 - u_2$

(u_1 necessarily $> u_2$, otherwise no collision)

speed of separation $= v_2 - v_1$

($v_2 > v_1$ or bodies do not separate)

(c)

u_1 u_2

speed of approach $= u_1 + u_2$

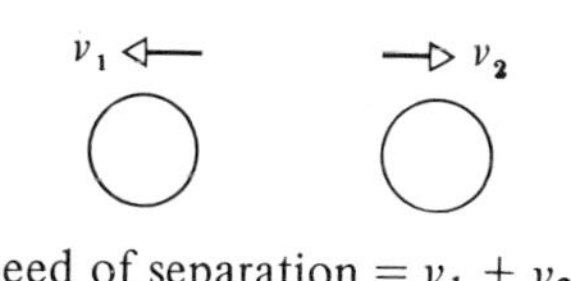

speed of separation $= v_1 + v_2$

Value of coefficient of restitution

For two perfectly elastic bodies in collision, $e = 1$. For two inelastic bodies, i.e. ones which coalesce on impact, $e = 0$. In all other cases e will be a fraction of one.

Example 9

Two spheres collide when moving along the same straight line on a horizontal surface. The speeds of the spheres before and after the collision are as shown in the diagram. Find the value of e in each case.

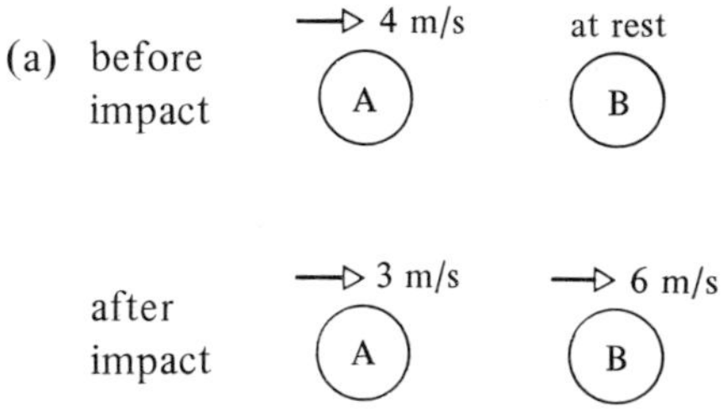

(a) before impact: A → 4 m/s; B at rest

after impact: A → 3 m/s; B → 6 m/s

$$e = \frac{\text{speed of separation}}{\text{speed of approach}} = \frac{6-3}{4-0} = \frac{3}{4}$$

The coefficient of restitution is $\frac{3}{4}$

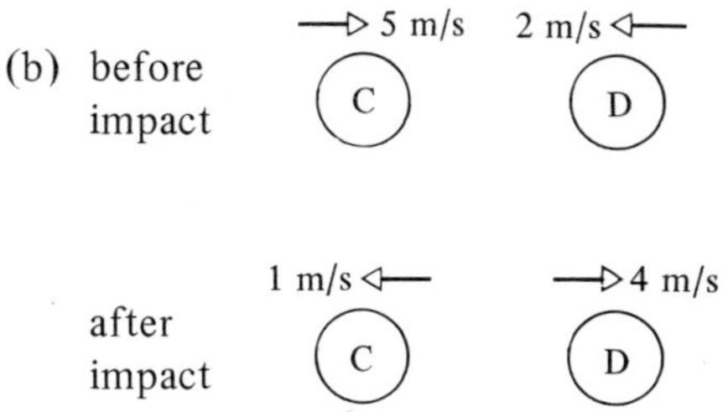

(b) before impact: C → 5 m/s; D 2 m/s ←

after impact: C 1 m/s ←; D → 4 m/s

$$e = \frac{\text{speed of separation}}{\text{speed of approach}} = \frac{4+1}{5+2} = \frac{5}{7}$$

The coefficient of restitution is $\frac{5}{7}$

Example 10

Find the unknown speed for the given spheres if their speeds before and after impact and their coefficient of restitution e are as shown.

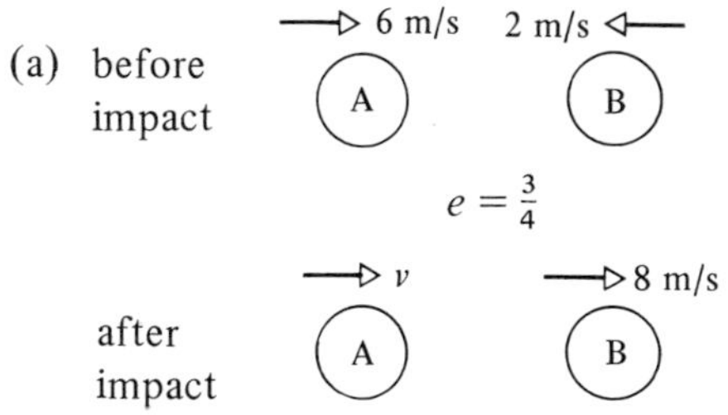

(a) before impact: A → 6 m/s; B 2 m/s ←

$e = \frac{3}{4}$

after impact: A → v; B → 8 m/s

$$e = \frac{\text{speed of separation}}{\text{speed of approach}} \qquad \therefore \quad \frac{3}{4} = \frac{8-v}{6+2}$$

$$3(8) = 4(8-v)$$

$$\therefore \quad v = 2 \text{ m/s}$$

The unknown speed after impact is 2 m/s in the direction shown

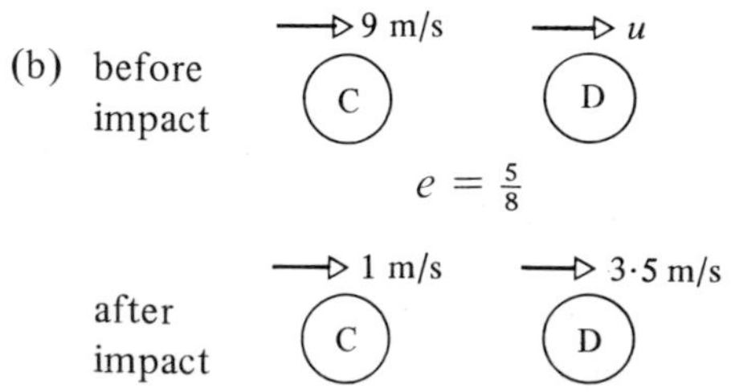

(b) before impact: C → 9 m/s; D → u

$e = \frac{5}{8}$

after impact: C → 1 m/s; D → 3·5 m/s

$$e = \frac{\text{speed of separation}}{\text{speed of approach}} \qquad \therefore \quad \frac{5}{8} = \frac{3{\cdot}5-1}{9-u}$$

$$5(9-u) = 8(3{\cdot}5-1)$$

$$\therefore \quad u = 5 \text{ m/s}$$

The unknown speed before impact is 5 m/s in the direction shown

It should be noted that when the diagrams giving the 'before' and 'after' impact situations are not given, care must be taken to translate accurately the given data on to the appropriate diagram.
In some cases where there is more than one unknown quantity, both the law of conservation of linear momentum and Newton's Experimental Law need to be used.

Example 11

Two spheres, A and B, are of equal radii and masses 2 kg and $1\frac{1}{2}$ kg respectively. The spheres A and B move towards each other along the same straight line on a smooth horizontal surface with speeds of 4 m/s and 6 m/s respectively. If the coefficient of restitution between the spheres is $\frac{2}{5}$, find their speeds and directions after impact.

Draw two diagrams.

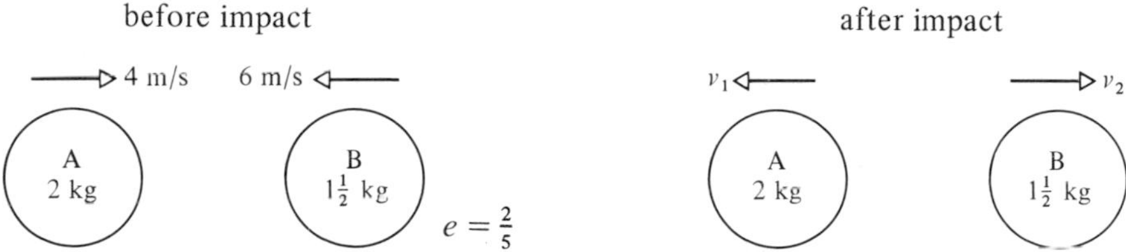

Let v_1 and v_2 (in the directions shown) be the speeds of the spheres after impact.

Newton's Law gives
$$\frac{2}{5} = \frac{\text{speed of separation}}{\text{speed of approach}}$$
$$= \frac{v_1 + v_2}{4 + 6}$$
$$\therefore \quad 4 = v_1 + v_2 \qquad \ldots [1]$$

Conservation of momentum (taking speeds to the left as positive)
$$2(-4) + 1\tfrac{1}{2}(6) = 2v_1 + 1\tfrac{1}{2}(-v_2)$$
$$\therefore \quad 1 = 2v_1 - \tfrac{3}{2}v_2 \qquad \ldots [2]$$

Subtracting equation [2] from twice equation [1]
$$7 = \tfrac{7}{2}v_2 \quad \text{or} \quad v_2 = 2 \text{ m/s}$$

Substituting in equation [1]
$$4 = v_1 + 2 \quad \text{or} \quad v_1 = 2 \text{ m/s}$$

The speeds of A and B after the collision are 2 m/s in the directions shown.

Example 12

Three particles A, B and C, of equal mass m are (in that order) travelling along the same straight line on a smooth horizontal surface. Initially, A and B have speeds of 3 m/s and 2 m/s respectively and are moving towards each other. The initial speed of C is 1·5 m/s and it is moving away from A and B. The coefficient of restitution between any two particles is $\frac{4}{5}$. Find the final speeds of A, B and C.

Draw two diagrams for the first impact between A and B.

before impact

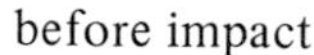

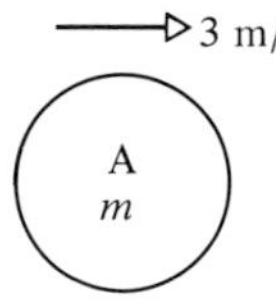

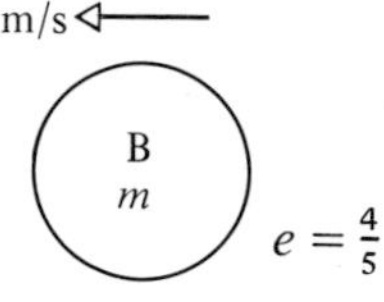

after impact

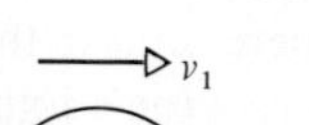

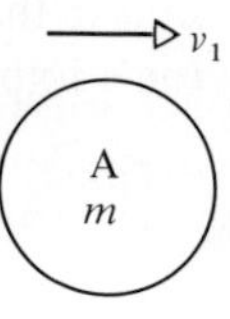

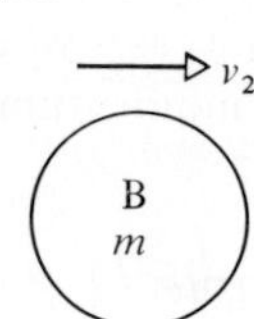

Newton's Law gives
$$\frac{4}{5} = \frac{v_2 - v_1}{3 + 2}$$
$$\therefore \quad 4 = v_2 - v_1 \qquad \ldots [1]$$

Conservation of momentum (taking velocities to the right as positive)
$$3m - 2m = mv_1 + mv_2$$
$$\therefore \quad 1 = v_1 + v_2 \qquad \ldots [2]$$

Adding equations [1] and [2] gives
$$v_2 = 2\tfrac{1}{2} \text{ m/s}$$
and by substitution
$$v_1 = -1\tfrac{1}{2} \text{ m/s}$$

Hence after this impact, A is moving to the left in the diagram at $1\frac{1}{2}$ m/s and B is moving towards C at $2\frac{1}{2}$ m/s. A second impact will occur between B and C since the velocity of B is greater than that of C.

before second impact

$\longrightarrow 2\frac{1}{2}$ m/s $\qquad \longrightarrow 1\frac{1}{2}$ m/s

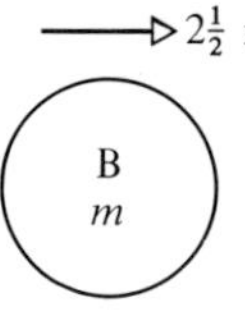

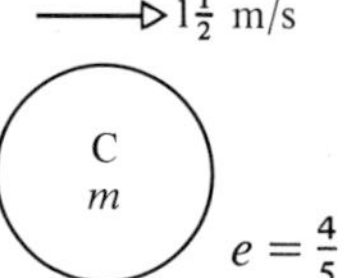

after second impact

$\longrightarrow v_3 \qquad \longrightarrow v_4$

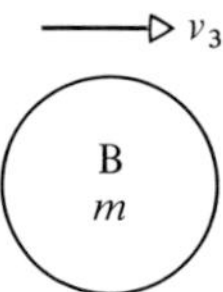

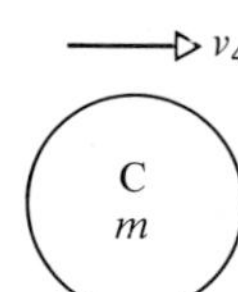

Newton's Law gives
$$\frac{4}{5} = \frac{v_4 - v_3}{2\frac{1}{2} - 1\frac{1}{2}}$$
$$\therefore \quad 4 = 5v_4 - 5v_3 \qquad \ldots [3]$$

Conservation of momentum
$$2\tfrac{1}{2}m + 1\tfrac{1}{2}m = mv_3 + mv_4$$
$$\therefore \quad 4 = v_3 + v_4 \qquad \ldots [4]$$

Solving equations [3] and [4] simultaneously
$$v_3 = 1{\cdot}6 \text{ m/s} \text{ and } v_4 = 2{\cdot}4 \text{ m/s}$$

From the values of v_1, v_3 and v_4, it can be seen that no further collisions take place.

After the two impacts which occur the speeds of A, B and C are 1·5, 1·6 and 2·4 m/s respectively.

Exercise 15C

1. In each part of this question the two diagrams show the situations before and after a collision between two spheres A and B of equal radii moving along the same straight line on a smooth horizontal surface.
For (a), (b) and (c), find the value of the coefficient of restitution e.

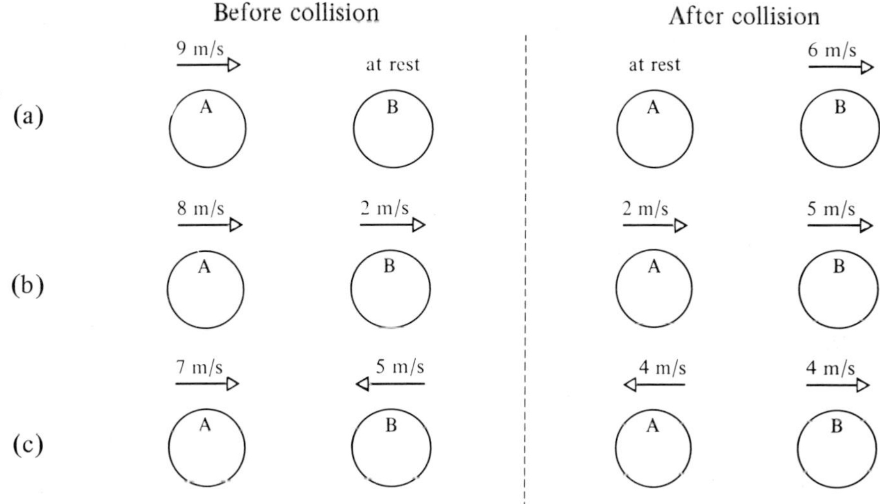

For (d) to (h), find the speeds u and v (e is given for each part).

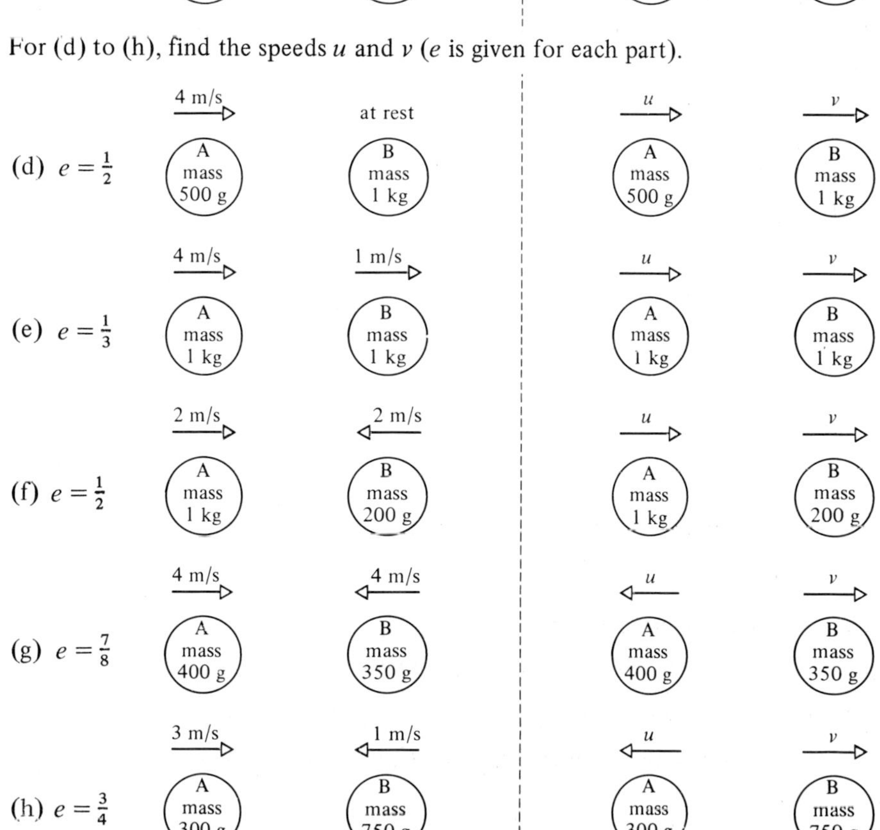

2. Two particles A and B of masses 200 g and 500 g respectively, are travelling along the same straight line on a smooth horizontal surface. Particle A, initially travelling at 6 m/s, catches up and collides with B which was initially travelling at 2 m/s. After the collision, A has not changed its direction of motion but now has a speed of 1 m/s. Find the speed of B after collision and the coefficient of restitution for the particles.
3. Two spheres A and B are of equal radii and masses 1 kg and 1·5 kg respectively. A and B move towards each other along the same straight line on a smooth horizontal surface with velocities $2\mathbf{i}$ m/s and $-\mathbf{i}$ m/s respectively. If the coefficient of restitution between A and B is $\frac{2}{3}$, find the velocities of the spheres after collision.
4. Two spheres of equal radii and masses 250 g and 150 g are travelling towards each other along a straight line on a smooth horizontal surface. Initially, the 250 g sphere has a speed of 3 m/s and the 150 g sphere a speed of 2 m/s. The spheres collide and the collision reduces the 250 g sphere to rest. Find the coefficient of restitution between the spheres and the kinetic energy lost in the collision.
5. Two particles A and B of masses $2m$ and m respectively, are travelling directly towards each other each with speed u. If the coefficient of restitution between the spheres is $\frac{1}{2}$, show that after collision, A is at rest and B has speed u. Find the loss in kinetic energy due to the collision.
6. Each part of this question involves three smooth spheres A, B and C, of equal radii, moving along the same straight line. A collides with B, and then B collides with C. The diagrams show the situations before any collision, after A has collided with B, and after B has collided with C. Find the speeds u, v, w and x.

(a) $e = \frac{3}{4}$ between any two spheres in collision

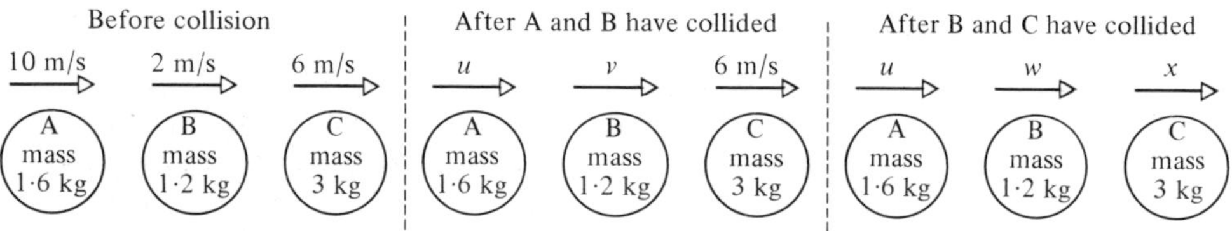

(b) Between A and B $e = \frac{2}{3}$

Between B and C $e = \frac{1}{2}$

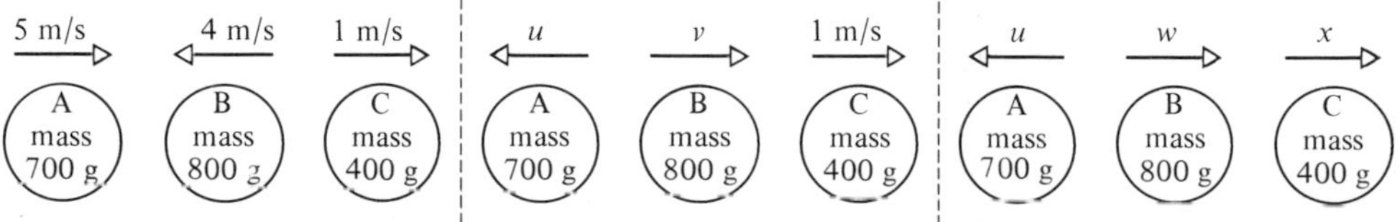

7. Three particles A, B and C of masses 80 g, 200 g and 500 g respectively, are all travelling in the same direction along the same straight line on a smooth horizontal surface. Initially the speeds of A, B and C are 6 m/s, 2 m/s and 2 m/s respectively, and B lies between A and C; A collides with B and after this collision, B collides with C. If the coefficient of restitution between any two particles colliding is $\frac{3}{4}$, find the final speeds of A, B and C.

8. A, B and C are three spheres of equal radii and masses of 750 g, 500 g and 1 kg respectively. The spheres are travelling along the same straight line on a smooth horizontal surface with B between A and C. Initially the velocities of A, B and C are $5\mathbf{i}$ m/s, $-3\mathbf{i}$ m/s and $4\mathbf{i}$ m/s respectively, where $\mathbf{i}$ is a unit vector in the direction ABC. The coefficient of restitution between A and B is $\frac{7}{8}$ and between B and C is $\frac{1}{2}$. Find the velocities of A, B and C after all collisions have taken place.
9. Two particles A and B are travelling in the same direction along the same straight line on a smooth horizontal surface with speeds of $2u$ and u respectively. Particle A catches up and collides with B, coefficient of restitution e. If the mass of A is twice that of B, find expressions for the speeds of A and B after collision.
10. Two identical spheres each of mass m are projected directly towards each other on a smooth horizontal surface. Each sphere is given an initial speed of u and the coefficient of restitution between the spheres is e. Show that the collision between the two spheres causes a loss in kinetic energy of $mu^2(1-e^2)$.
11. Two identical particles each of mass m are projected towards each other with speeds of $3u$ and u. Show that after collision, the directions of motion of both particles will be reversed, provided $e > \frac{1}{2}$.
    Find an expression for the loss in the kinetic energy of the particles due to this impact.
12. Three identical spheres A, B and C, each of mass m, lie in a straight line on a smooth horizontal surface with B between A and C. Spheres A and B are projected directly towards each other with speeds $3u$ and $2u$ respectively and C is projected directly away from A and B with speed $2u$. If the coefficient of restitution between any two spheres in collision is e, show that B will only collide with C provided $e > \frac{3}{5}$.

Impact with a surface

Consider an elastic sphere travelling with speed v and striking a fixed surface, the plane of which is perpendicular to the direction of motion of the sphere. By applying Newton's Law, it is found that after impact the direction of motion of the sphere has been reversed; its speed is now ev where e is the coefficient of restitution for the impact.

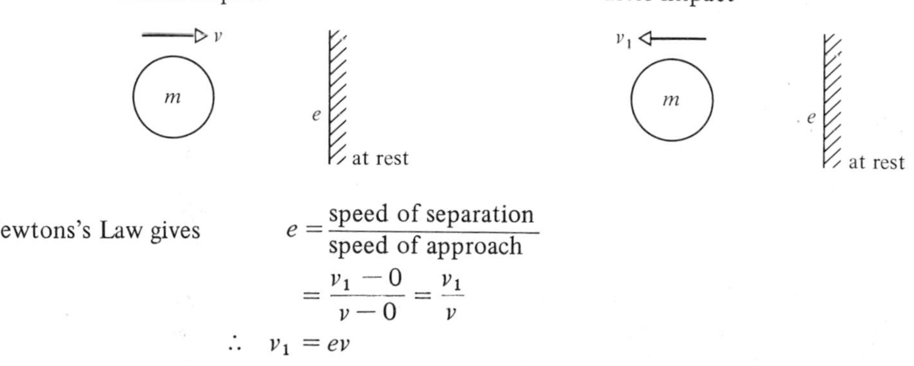

Newtons's Law gives
$$e = \frac{\text{speed of separation}}{\text{speed of approach}}$$
$$= \frac{v_1 - 0}{v - 0} = \frac{v_1}{v}$$
$$\therefore \quad v_1 = ev$$

A body is said to collide *normally* with a fixed surface if its direction of motion prior to the impact is at right angles to the surface.

Example 13

(a) A smooth sphere collides normally with a fixed vertical wall. From the information in the diagram find the coefficient of restitution e.

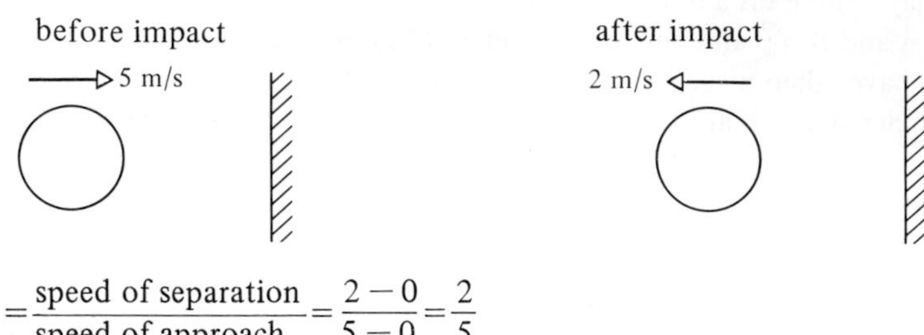

$$e = \frac{\text{speed of separation}}{\text{speed of approach}} = \frac{2-0}{5-0} = \frac{2}{5}$$

The coefficient of restitution is $\frac{2}{5}$.

(b) A smooth sphere collides normally with a fixed vertical wall. From the information in the diagram find the speed v of the sphere after impact.

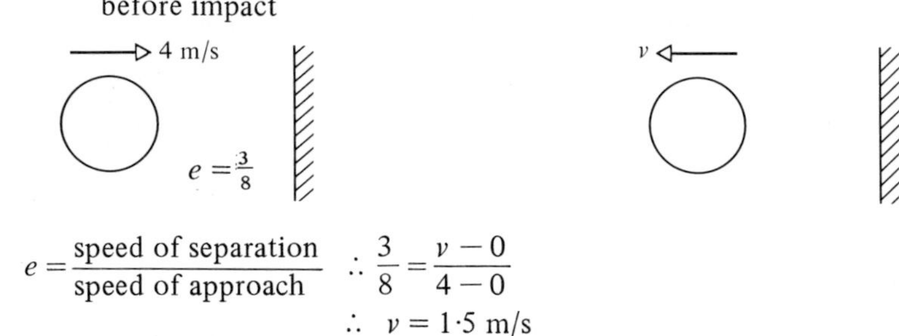

$$e = \frac{\text{speed of separation}}{\text{speed of approach}} \quad \therefore \frac{3}{8} = \frac{v-0}{4-0}$$

$$\therefore \quad v = 1{\cdot}5 \text{ m/s}$$

The speed after impact is 1·5 m/s.

Example 14

Two spheres A and B of equal radii are initially at rest on a smooth horizontal surface; they have masses of 0·5 kg and 0·4 kg respectively. They are projected towards each other with speeds of 3 m/s and 2 m/s respectively, the coefficient of restitution being $\frac{4}{5}$. After collision, B collides normally with a fixed vertical wall, the coefficient of restitution between B and the wall being $\frac{2}{3}$. Find the velocities of A and B after the first impact between them and after the second impact between them.

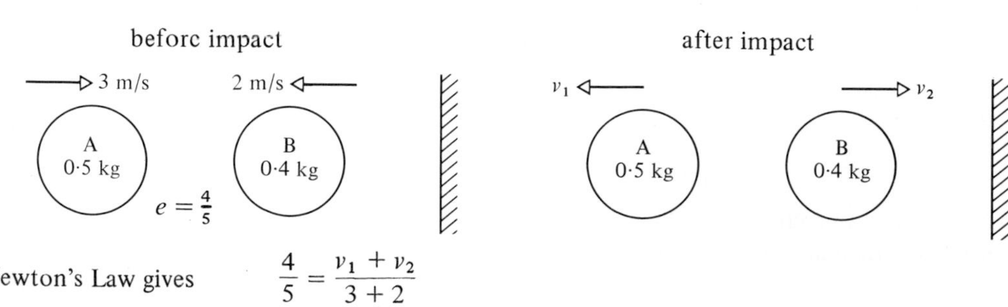

Newton's Law gives
$$\frac{4}{5} = \frac{v_1 + v_2}{3+2}$$

$$\therefore \quad 4 = v_1 + v_2 \qquad \ldots [1]$$

Conservation of momentum (taking velocities to the right as positive)

$$0{\cdot}5(3) + 0{\cdot}4(-2) = 0{\cdot}5(-v_1) + 0{\cdot}4(v_2)$$

$$\therefore \quad 7 = -5v_1 + 4v_2 \qquad \ldots [2]$$

Solving equations [1] and [2] simultaneously gives

$$v_2 = 3 \text{ m/s} \quad \text{and} \quad v_1 = 1 \text{ m/s}$$

Hence B now approaches the vertical wall with a speed of 3 m/s.

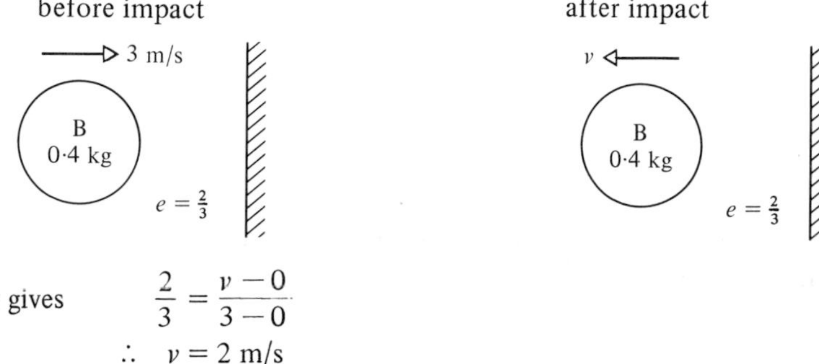

Newton's Law gives
$$\frac{2}{3} = \frac{v - 0}{3 - 0}$$

$$\therefore \quad v = 2 \text{ m/s}$$

Hence B now leaves the wall with a speed of 2 m/s and therefore collides again with A which has a speed of 1 m/s.

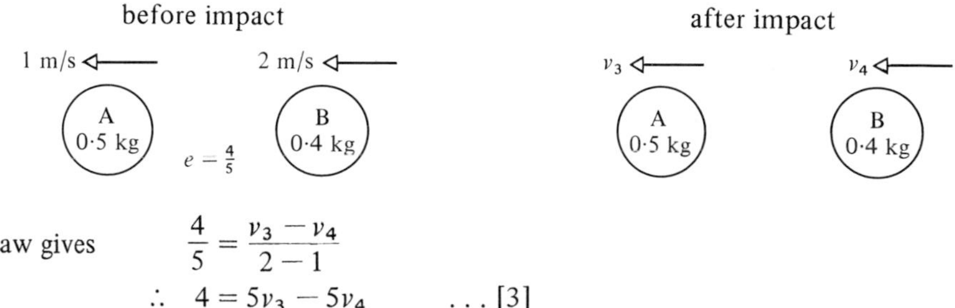

Newton's Law gives
$$\frac{4}{5} = \frac{v_3 - v_4}{2 - 1}$$

$$\therefore \quad 4 = 5v_3 - 5v_4 \qquad \ldots [3]$$

Conservation of momentum (taking velocities to the left as positive)

$$0{\cdot}5(1) + 0{\cdot}4(2) = 0{\cdot}5v_3 + 0{\cdot}4v_4$$

$$\therefore \quad 13 = 5v_3 + 4v_4 \qquad \ldots [4]$$

Solving equations [3] and [4] simultaneously gives

$$v_3 = 1{\cdot}8 \text{ m/s} \quad \text{and} \quad v_4 = 1 \text{ m/s}$$

The velocities of A and B after the first impact between them are 1 m/s and 3 m/s, and after the second impact between them are 1·8 m/s and 1 m/s respectively, in the directions indicated in the diagrams.

Exercise 15D

1. Each part of this question involves a smooth sphere colliding normally with a fixed vertical wall. The diagrams show the situations before and after collision.
For (a) and (b), find the coefficient of restitution e.

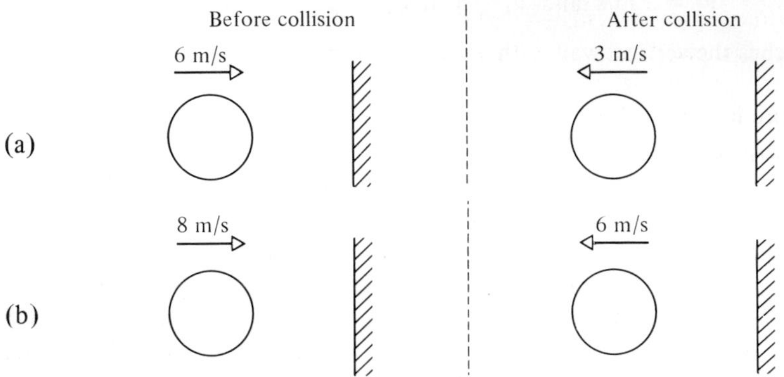

For (c) and (d), find the speed v after impact (e is given for each part).

10 m/s $\quad v$

(c) $e = \frac{3}{5}$

2 m/s $\quad v$

(d) $e = \frac{3}{4}$

For (e) and (f), find the speed u before impact (e is given for each part).

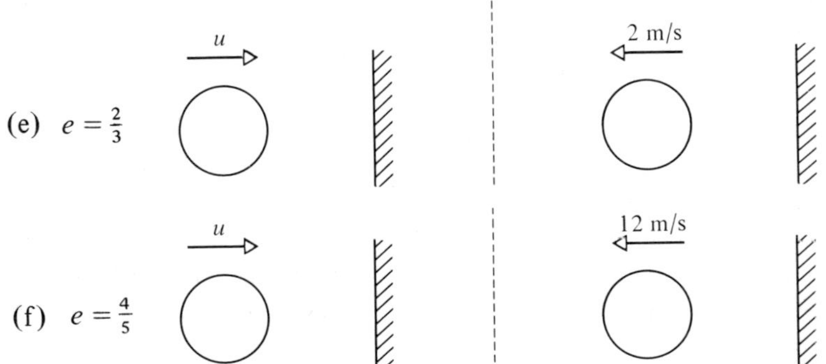

2. A body moves horizontally with velocity $6\mathbf{i}$ m/s and strikes a vertical wall normally. If the coefficient of restitution between the body and the wall is 0·75, find the velocity of the body after impact.

3. A body is moving horizontally with velocity $5\mathbf{i}$ m/s when it strikes a vertical wall normally and rebounds with velocity $-4\mathbf{i}$ m/s. Find the coefficient of restitution between the body and the wall.
4. A particle is initially at a point A on a smooth horizontal surface, midway between two vertical walls which are parallel to each other and 2 metres apart. The particle is projected from A with a speed of 2 m/s in a direction perpendicular to the walls. If the coefficient of restitution between the particle and each wall is $\frac{1}{2}$, find the time taken for the body to return to A having touched each wall once and once only.
5. Two spheres A and B of equal radii are initially at rest on a smooth horizontal surface. A is projected directly towards B with a speed of 10 m/s and the collision between the two spheres reduces A to rest. B continues after collision to strike a fixed vertical surface at right angles and rebounds to hit A again. The coefficient of restitution between A and B is $\frac{4}{5}$ and between B and the wall is $\frac{5}{8}$. If A has a mass of 200 g, find the mass of B and the speeds of the spheres after the second collision between them.
6. A and B are two spheres of equal radii and masses 750 g and 600 g respectively. The spheres are moving directly towards each other along a smooth horizontal surface with A travelling with a speed of 6 m/s and B with speed of 4 m/s. The spheres collide (coefficient of restitution $\frac{4}{5}$) and the collision reverses the direction of motion of sphere B which then strikes a fixed vertical wall at right angles. If the coefficient of restitution between B and the wall is $\frac{2}{3}$, show that B will collide again with A and find the speeds of A and B after this second collision.
7. A particle of mass m is travelling in a straight line with speed u along a smooth horizontal surface. The particle strikes a fixed vertical wall, the plane of the wall being at right angles to the direction of motion of the particle. If the kinetic energy lost by the particle due to the impact is E, show that the coefficient of restitution between the particle and the wall is given by $\sqrt{\left(\dfrac{mu^2 - 2E}{mu^2}\right)}$.
8. Two spheres A and B, of equal radii and masses of $2m$ and m respectively, are each travelling with speed u towards each other on a smooth horizontal surface. If the coefficient of restitution between A and B is e_1, show that (a) if $e_1 = \frac{1}{2}$, then A is reduced to rest by the collision, (b) if $e_1 > \frac{1}{2}$, both spheres will have their original directions of motion reversed by the collision.
   With $e_1 > \frac{1}{2}$ and B going on after collision to hit a fixed vertical wall at right angles, show that B will collide again with A provided $e_2 > \dfrac{2e_1 - 1}{1 + 4e_1}$, where e_2 is the coefficient of restitution between B and the wall.
9. Two spheres A and B of equal radii and masses 500 g and 200 g respectively, are initially travelling along the same straight line towards each other on a smooth horizontal surface. Sphere A has an initial speed of 5 m/s and sphere B an initial speed of 7 m/s. The spheres collide and the coefficient of restitution between them is 0·75. The collision reverses the direction of motion of B which then strikes a wall normally, coefficient of restitution 0·5. The first collision between A and B takes place 3 m from the wall.
   (a) Show that there will be a second collision between A and B.
   (b) Find the time interval between the first and second collisions between the spheres.
   (c) Find how far from the wall the second collision between A and B occurs.
   (d) Find the speeds of A and B after the second collision between them.

Exercise 15E Examination questions

1. An elastic string of natural length 1 m obeys Hooke's law. When it is stretched to 1·2 m the energy stored in it is 16 J. Find the energy stored in the string when it is stretched to 1·5 m. (London)

2. Two identical spheres, moving in opposite directions, collide directly. As a result of the impact one of the spheres is brought to rest. The coefficient of restitution between the spheres is $\frac{1}{2}$. Show that the ratio of the speeds of the spheres before the impact is 3 : 1. (London)

3. A light elastic string, of natural length c and modulus of elasticity $4mg$, is attached at one end to a fixed point A. A particle of mass m is tied to the other end B of the string.
   (i) If the particle hangs in equilibrium, calculate the length AB of the extended string.
   (ii) If the particle is held at A and allowed to fall vertically, use the principle of conservation of energy to find the greatest distance between A and B in the ensuing motion.
   (iii) If the particle moves in a horizontal circle, centre O, where O is at a depth d vertically below A, show that the angular velocity of the particle is $\sqrt{(g/d)}$. (A.E.B)

4. A small sphere A, of mass m moving with velocity $2u$ on a smooth horizontal plane, impinges directly on a small sphere B, of equal radius and mass $2m$ moving with velocity u in the same direction on the plane. Given that after impact B moves with velocity $3u/2$, calculate (i) the coefficient of restitution between A and B, (ii) the loss in kinetic energy due to the impact.
   The sphere B continues to move with velocity $3u/2$ until it hits a vertical wall from which it rebounds and is then brought to rest by a second impact with the approaching sphere A. Calculate (iii) the coefficient of restitution between B and the wall, (iv) the final speed of the sphere A. (A.E.B)

5. Three smooth spheres, A, B, C, of equal radii and with masses m, $2m$, $2m$ respectively, lie on a horizontal plane with their centres in a straight line and with B between A and C. The coefficient of restitution between any pair of spheres is e. Initially B and C are at rest, and A is projected towards B with speed u. It is observed that A rebounds from B with speed $\frac{1}{12}u$. Prove that $e = \frac{5}{8}$, and find the velocity of C after the second impact. (Oxford)

6. A light elastic string of modulus λ and natural length $3a$ is fixed to two points on the same horizontal level at a distance $4a$ apart. Two particles of weight W are attached, one at each point of trisection. If the sloping string make an angle θ with the horizontal in equilibrium, prove that (i) the central portion of the string extends by $\frac{1}{3}a(3-2\cos\theta)$;
   (ii) $\tan\theta = \dfrac{3W}{\lambda(3-2\cos\theta)}$. (S.U.J.B)

7. Two particles A and B of masses $2m$ and $3m$ respectively are placed on a smooth horizontal plane. The coefficient of restitution between A and B is $\frac{1}{2}$. The particle A is made to move with speed u directly towards B which is at rest. Calculate the speeds of A and B after their collision, the impulse of the force transmitted from A to B and the loss in kinetic energy due to the collision.
   The particles A and B are now connected by a light inextensible string and are placed side by side on the smooth horizontal plane. The particle A is given a horizontal velocity v directly away from B. Calculate the impulse of the tension in the string at the instant when the string tightens. Calculate also the resulting common velocity of the two particles and show that the loss in kinetic energy due to the tightening of the string is $0{\cdot}6\,mv^2$. (A.E.B)

16

Variable acceleration

In chapter 2 the motion of a body in a straight line with uniform or constant acceleration was considered. The four uniform acceleration formulae could be used:

$$v = u + at \qquad s = ut + \tfrac{1}{2}at^2$$

$$v^2 = u^2 + 2as \qquad s = \frac{(u+v)t}{2}$$

If the acceleration is non-uniform, differential equations and calculus must be used in place of these equations.

velocity = rate of change of distance

or $v = \dfrac{ds}{dt}$

The dot notation is used to show differentiation with respect to time, for example,

$$\text{velocity} = \frac{ds}{dt} = \dot{s}$$

In a similar way, $\ddot{s}$ stands for the second differential of s with respect to time.

acceleration = rate of change of velocity

or $a = \dfrac{dv}{dt} = \dot{v}$

and $a = \dfrac{d}{dt}\left(\dfrac{ds}{dt}\right) = \dfrac{d^2s}{dt^2} = \ddot{s}$

In all the worked examples in this chapter, the letters a, v, s and t have their usual meaning and SI units are used throughout.

Example 1

A body moves in a straight line such that $s = 4t^2 - t^3$.
Find (a) s when $t = 3$, (b) v when $t = 2$, (c) a when $t = 1$.

(a) Given $s = 4t^2 - t^3$
when $t = 3$, $s = 4(3)^2 - (3)^3 = 36 - 27 = 9$ m

The displacement s is 9 m when $t = 3$ s.

(b) Given $s = 4t^2 - t^3$

$$v = \frac{ds}{dt} = 8t - 3t^2$$

when $t = 2$, $v = 8(2) - 3(2)^2 = 4$ m/s

The velocity v is 4 m/s when $t = 2$ s.

(c) Given $s = 4t^2 - t^3$

$$v = \frac{ds}{dt} = 8t - 3t^2$$

and $a = \dfrac{d^2s}{dt^2} = 8 - 6t$

when $t = 1$, $a = 8 - 6(1) = 2\ \text{m/s}^2$

The acceleration a is $2\ \text{m/s}^2$ when $t = 1$ s.

Acceleration as a function of time

In order to obtain an expression for the velocity when the acceleration is given as a function of time, the process of integration has to be employed:

$$a = \frac{dv}{dt} = f(t)$$

$$\therefore \int dv = \int f(t)dt$$

or $v = \int f(t)dt + \text{constant}$

The value of the constant can only be determined if more information is given in the question.

Example 2

A body moves in a straight line such that $a = 4t$, and initially, i.e. when $t = 0$, the velocity of the body is 3 m/s. Find (a) a when $t = 5$ s, (b) v when $t = 2$ s.

(a) Given $a = 4t$

when $t = 5$, $a = 4(5) = 20\ \text{m/s}^2$

The acceleration a is $20\ \text{m/s}^2$ when $t = 5$ s.

(b) Given $a = \dfrac{dv}{dt} = 4t$

$$\int dv = \int 4t\, dt$$

or $v = 2t^2 + C$

but when $t = 0$, $v = 3$

thus $3 = 2(0)^2 + C$ $\therefore C = 3$

$\therefore$ $v = 2t^2 + 3$ substituting for C

when $t = 2$, $v = 2(2)^2 + 3 = 11$ m/s

The velocity v is 11 m/s when $t = 2$ s.

The relationships between displacement s, velocity v and acceleration a may be shown as follows.

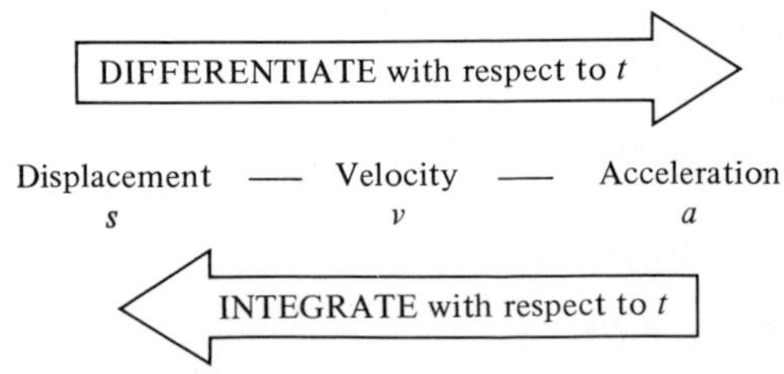

Thus if s is given as a function of time, differentiation will give an expression for velocity, and a second differentiation will give an expression for acceleration.
On the other hand, if a is given as a function of time, integration will give an expression for velocity, and a second integration will give an expression for displacement.
Again, it should be remembered that a constant must be introduced at each integration.

Example 3

A body moves in a straight line such that $v = 2t^2 - 11t + 14$. Initially, i.e. when $t = 0$, the displacement of the body from some fixed point O on the line is 50 m. Find

(a) the initial velocity of the body,
(b) the values of t when the body is at rest,
(c) the acceleration of the body when $t = 5$ s,
(d) the displacement of the body from O when $t = 6$ s.

(a) Given $v = 2t^2 - 11t + 14$

when $t = 0$, $v = 2(0) - 11(0) + 14 = 14$

$\therefore$ $v = 14$ m/s

The initial velocity of the body is 14 m/s.

(b) Given $v = 2t^2 - 11t + 14$

The body is at rest when $v = 0$

$\therefore$ $0 = 2t^2 - 11t + 14$

giving $t = 3{\cdot}5$ or 2 s

The body is at rest when $t = 3{\cdot}5$ s and when $t = 2$ s.

(c) Given $v = 2t^2 - 11t + 14$

acceleration $a = \frac{dv}{dt} = \dot{v}$

$= 4t - 11$

when $t = 5$, $a = 4(5) - 11$

$\therefore$ $a = 9$ m/s^2

When $t = 5$ s the acceleration of the body is 9 m/s^2.

(d) Given $v = 2t^2 - 11t + 14$

by integration, $s = \frac{2t^3}{3} - \frac{11t^2}{2} + 14t + C$

but $s = 50$ when $t = 0$,

$\therefore$ $50 = \frac{2}{3}(0) - \frac{11}{2}(0) + 14(0) + C$ or $C = 50$

hence $s = \frac{2}{3}t^3 - \frac{11}{2}t^2 + 14t + 50$

when $t = 6$, $s = \frac{2}{3}(6)^3 - \frac{11}{2}(6)^2 + 14(6) + 50$

$= 144 - 198 + 84 + 50$

$\therefore$ $s = 80$ m

When $t = 6$ s, the displacement of the body is 80 m.

Motion in the i-j plane

If motion, with non-uniform acceleration, takes place in the **i–j** plane and vector notation is employed, the same methods as in the previous examples may be used and each component can be differentiated or integrated separately, i.e.

$$\text{if} \quad \mathbf{s} = f(t)\mathbf{i} + g(t)\mathbf{j}$$

$$\text{then} \quad \mathbf{v} = \frac{d\mathbf{s}}{dt} = \frac{d}{dt}\Big(f(t)\Big)\mathbf{i} + \frac{d}{dt}\Big(g(t)\Big)\mathbf{j}$$

Example 4

If $\mathbf{v} = t^2\mathbf{i} + 3t\mathbf{j}$ and, when $t = 0$, $\mathbf{s} = 18\mathbf{i} - 24\mathbf{j}$, find
(a) **a** when $t = 2$ s, (b) **s** when $t = 6$ s.

(a) Given $\mathbf{v} = t^2\mathbf{i} + 3t\mathbf{j}$

$$\mathbf{a} = \frac{d\mathbf{v}}{dt} = 2t\mathbf{i} + 3\mathbf{j}$$

$$\text{when } t = 2, \quad \mathbf{a} = 2(2)\mathbf{i} + 3\mathbf{j} = 4\mathbf{i} + 3\mathbf{j}$$

When $t = 2$ s, the acceleration is $(4\mathbf{i} + 3\mathbf{j})$ m/s^2.

(b) Given $\mathbf{v} = t^2\mathbf{i} + 3t\mathbf{j}$

$$\frac{d\mathbf{s}}{dt} = t^2\mathbf{i} + 3t\mathbf{j}$$

$$\therefore \int d\mathbf{s} = \int (t^2\mathbf{i} + 3t\mathbf{j})\,dt$$

$$\therefore \quad \mathbf{s} = \frac{t^3}{3}\mathbf{i} + 3\frac{t^2}{2}\mathbf{j} + A\mathbf{i} + B\mathbf{j}$$

but $\mathbf{s} = 18\mathbf{i} - 24\mathbf{j}$ when $t = 0$,

$$\therefore \quad 18\mathbf{i} - 24\mathbf{j} = 0\mathbf{i} + 0\mathbf{j} + A\mathbf{i} + B\mathbf{j}$$

$$\text{or} \quad 18\mathbf{i} - 24\mathbf{j} = A\mathbf{i} + B\mathbf{j}$$

$$\therefore \quad A = 18 \text{ and } B = -24$$

$$\mathbf{s} = \left(\frac{t^3}{3} + 18\right)\mathbf{i} + \left(\tfrac{3}{2}t^2 - 24\right)\mathbf{j}$$

$$\text{when } t = 6 \text{ s} \quad \mathbf{s} = \left(\frac{6^3}{3} + 18\right)\mathbf{i} + \left(\tfrac{3}{2}(6)^2 - 24\right)\mathbf{j}$$

$$\text{or} \quad \mathbf{s} = 90\mathbf{i} + 30\mathbf{j}$$

When $t = 6$ s, the displacement, **s**, is $(90\mathbf{i} + 30\mathbf{j})$ m.

Exercise 16A

All units in this exercise are in SI units.
Questions **1** to **14** involve motion along a straight line. The letter s represents the displacement of the body at time t from a fixed point O on the line. The letters v and a represent the velocity and the acceleration of the body at time t.

1. If $s = 5t^3$, find s when $t = 2$s.
2. If $s = t^3 + t$, find v when $t = 3$ s.

3. If $s = 5t^2 - t^3$, find a when $t = 1$ s.
4. If $v = 6t^2$, find v when $t = 2$ s.
5. If $v = t^3$, find a when $t = 2$ s.
6. If $v = 4t + 5$ and $s = 10$ m when $t = 1$ s, find s when $t = 2$ s.
7. If $a = 6t$, find a when $t = 5$ s.
8. If $a = 6t$ and the body is initially at rest, find v when $t = 4$ s.
9. If $a = \frac{2}{3}t$, find s when $t = 6$ s given that $v = 4$ m/s, and $s = 10$ m when $t = 3$ s.
10. If $s = t^2 - 3$ find (a) s when $t = 2$ s, (b) an expression for the velocity of the body at time t, (c) the velocity when $t = 2$ s, (d) the value of t when velocity is 8 m/s, (e) the displacement of the body from O when $v = 8$ m/s.
11. If $s = 2t^3 - 21t^2 + 60t$, find (a) s when $t = 3$ s,
    (b) the values of t when the body is at rest, (c) the initial velocity of the body,
    (d) an expression for the acceleration of the body at time t,
    (e) the initial acceleration of the body.
12. If $v = 8t - 3t^2$ and the body is initially at O, find (a) v when $t = 2$ s,
    (b) an expression for the acceleration of the body at time t,
    (c) the acceleration when $t = 3$ s,
    (d) an expression for the displacement of the body from O at time t,
    (e) how far the body is from O when $t = 3$ s.
13. If $v = (3t - 2)(t - 4)$ and $s = 8$ m when $t = 1$ s, find (a) the initial velocity of the body,
    (b) the values of t when the body is at rest, (c) the acceleration of the body when $t = 3$ s,
    (d) the distance the body is from O when $t = 2$ s.
14. If $a = 2t$ and initially the body is at rest at O, find the velocity of the body when $t = 3$ s and the distance the body is then from O.

Questions **15** to **20** involve motion in the **i**–**j** plane with **s**, **v** and **a** representing position, velocity and acceleration vectors at time t.

15. If $\mathbf{s} = t^3\mathbf{i} + 2t^2\mathbf{j}$, find **a** when $t = 1$ s.
16. If $\mathbf{s} = t^3\mathbf{i} + 9t\mathbf{j}$, find the speed of the body when $t = 2$ s.
17. If $\mathbf{v} = 2t^2\mathbf{i} + 6\mathbf{j}$, find **a** when $t = 3$ s.
18. If $\mathbf{v} = 3t^2\mathbf{i} + 10t\mathbf{j}$, find the distance the body is from the origin when $t = 2$ s given that $\mathbf{s} = (4\mathbf{i} - 4\mathbf{j})$ m when $t = 0$.
19. If $\mathbf{a} = (4t + 2)\mathbf{i} - 3\mathbf{j}$ and the body is initially at rest, find **v** when $t = 3$ s.
20. If $\mathbf{a} = 6t\mathbf{i} + 2\mathbf{j}$, find **s** when $t = 3$ s given that $\mathbf{v} = (4\mathbf{i} - \mathbf{j})$ m/s and $\mathbf{s} = (2\mathbf{i} + 3\mathbf{j})$ m when $t = 1$ s.
21. The acceleration of a particle moving in a straight line is governed by the relationship $a = \dfrac{3t^2}{2}$ where t is the time in seconds. When $t = 2$ s the particle passes through the point O with velocity 1 m/s. Find the velocity of the particle when $t = 4$ s and the distance the particle is then from O.
22. A particle moves in a straight line with its acceleration a obeying the rule $a = 2t$, where t is the time in seconds. When $t = 0$ both s and v are equal to zero, where s is the distance of the particle from a point O and v is the velocity of the particle. For what value of t are s and v again numerically equal?
23. A particle moves along a straight line and t seconds after passing through an origin O the velocity of the body is given by $v = kt^2 - ct$ where k and c are constants. When $t = 2$ the body is again at O and has an acceleration of 6 m/s^2. Find the values of k and c.

24. A body moves along a straight line with acceleration given by $a = 2t$ where t is the time in seconds. When $t = 0$ the body is at rest at an origin O. The acceleration continues until $t = 3$, whereupon it ceases and the body is uniformly retarded to rest. During this retardation the acceleration is given by $a = -2\text{ m/s}^2$. Find the value of t when the body comes to rest and the displacement of the body from O at that time.

25. A body moves along a straight line with acceleration given by $a = \frac{7t}{36}$ where t is the time in seconds. When $t = 0$ the body is at rest at an origin O. The acceleration continues until $t = 6$, whereupon it ceases and the body is retarded to rest. During this retardation the acceleration is given by $a = -\frac{t}{4}$. Find the value of t when the body comes to rest and the displacement of the body from O at that time.

Velocity as a function of displacement

In the examples so far considered s, v and a have always been given as functions of the time t. This may not always be the case; dependent upon the information given and what is required, one of the following should be used.

$$a = \frac{v\,dv}{ds} \quad \text{or} \quad v = \frac{ds}{dt}$$

The first of these relationships may be obtained as follows:

$$a = \frac{dv}{dt} = \frac{dv}{ds} \times \frac{ds}{dt}$$
$$= \frac{dv}{ds} \times v = v\frac{dv}{ds}$$

Example 5

If $v = 3s^2 - 4s$, find (a) v when $s = 2$ m, (b) a when $s = 2$ m.

(a) Given $\quad v = 3s^2 - 4s$

when $s = 2$ m, $\quad v = 3(2)^2 - 4(2) = 4$ m/s

When $s = 2$ m, the velocity, v, is 4 m/s.

(b) We are given $v = 3s^2 - 4s$. To find a we use $a = v\frac{dv}{ds}$.

now $\quad \frac{dv}{ds} = 6s - 4$

$$\therefore \quad a = v\frac{dv}{ds} = (3s^2 - 4s)(6s - 4)$$

when $s = 2$ m, $\quad a = (12 - 8)(12 - 4) = 32\text{ m/s}^2$

When $s = 2$ m, the acceleration a is 32 m/s^2.

Example 6

If $v = \dfrac{20}{3s-2}$, find (a) v when $s = 4$ m, (b) s when $v = 5$ m/s, (c) t when $s = 20$ m, given that $s = 0$ when $t = 0$.

(a) Given
$$v = \frac{20}{3s-2}$$
when $s = 4$ m,
$$v = \frac{20}{3(4)-2} = \frac{20}{10}$$
$$\therefore \quad v = 2 \text{ m/s}$$

When $s = 4$ m, the velocity, v, is 2 m/s.

(b) Given
$$v = \frac{20}{3s-2}$$
when $v = 5$ m/s,
$$5 = \frac{20}{3s-2}$$
$$\therefore \quad 3s - 2 = 4 \quad \text{or} \quad s = 2 \text{ m}$$

When $v = 5$ m/s, the displacement, s, is 2 m.

(c) To find t, given v as a function of s, we must use $v = \dfrac{ds}{dt}$ as this expression involves t.

hence
$$\frac{ds}{dt} = \frac{20}{3s-2}$$
$$\therefore \quad \int (3s-2)ds = \int 20\, dt$$
$$\text{or} \quad \tfrac{3}{2}s^2 - 2s = 20t + C$$

but $s = 0$ when $t = 0$ $\therefore$ $0 = 0 + C$ or $C = 0$
$$\therefore \quad \tfrac{3}{2}s^2 - 2s = 20t$$

when $s = 20$ m,
$$\tfrac{3}{2}(20)^2 - 2(20) = 20t$$
$$\therefore \quad t = 28 \text{ s}$$

When the displacement s is 20 m, then $t = 28$ s.

Acceleration as a function of displacement

In this case, an expression for the velocity v is found by again using $a = v\dfrac{dv}{ds}$, as shown below.

$$\text{acceleration} \quad a = f(s)$$
$$\therefore \quad v\frac{dv}{ds} = f(s)$$
$$\therefore \quad \int v\,dv = \int f(s)\, ds$$
$$\therefore \quad \frac{v^2}{2} = \int f(s)\, ds + \text{constant}$$

Example 7

If $a = 3s + 5$ and initially $v = 1$ m/s when $s = 1$ m, find v when $s = 2$ m.

Given
$$a = 3s + 5$$
$$\therefore \quad v\frac{dv}{ds} = 3s + 5$$
$$\text{or} \quad \int v\,dv = \int (3s + 5)\,ds$$
$$\therefore \quad \frac{v^2}{2} = \tfrac{3}{2}s^2 + 5s + C$$

but $v = 1$ m/s when $s = 1$ m,
$$\therefore \quad \tfrac{1}{2} = \tfrac{3}{2} + 5 + C \quad \text{or} \quad C = -6$$
$$\therefore \quad \frac{v^2}{2} = \frac{3s^2}{2} + 5s - 6$$
$$\text{and when } s = 2\text{ m,} \quad \frac{v^2}{2} = \tfrac{3}{2}(4) + 5(2) - 6$$
$$\therefore \quad v^2 = 20 \quad \text{or} \quad v = 2\sqrt{5}\text{ m/s}$$

When the displacement s is 2 m, the velocity is $2\sqrt{5}$ m/s.

Having found v in terms of s, it is then possible to find s in terms of t by substituting $\frac{ds}{dt}$ for v and integrating.

Example 8

If $a = s + 5$ and initially $s = 0$ and $v = 5$ m/s, find
(a) an expression for v as a function of s, (b) an expression for t as a function of s.

(a)
$$a = s + 5$$
$$\therefore \quad v\frac{dv}{ds} = s + 5$$
$$\int v\,dv = \int (s + 5)\,ds$$
$$\therefore \quad \frac{v^2}{2} = \frac{s^2}{2} + 5s + C$$
$$v = 5 \text{ when } s = 0 \text{ gives } C = \tfrac{25}{2}$$
$$\therefore \quad v^2 = s^2 + 10s + 25 = (s + 5)^2$$
$$\therefore \quad v = s + 5 \qquad \text{The negative root is not applicable}$$

The expression for v as a function of s is $v = s + 5$.

(b) Since
$$v = s + 5$$
$$\therefore \quad \frac{ds}{dt} = s + 5$$
$$\int \frac{ds}{s + 5} = \int dt$$
$$\therefore \quad \ln(s + 5) = t + C'$$
$$s = 0 \text{ when } t = 0 \text{ gives} \quad C' = \ln 5$$
$$\text{thus} \quad \ln\left(\frac{s + 5}{5}\right) = t$$

The expression for t as a function of s is $t = \ln\left(\frac{s + 5}{5}\right)$.

Acceleration as a function of velocity

When the acceleration is given as a function of the velocity there are two possibilities:

(i) use $a = v\dfrac{dv}{ds}$ which will lead to a relationship between v and s,

or (ii) use $a = \dfrac{dv}{dt}$ which will lead to a relationship between v and t.

The choice will depend on the further information given and also on what relationship is required.

Example 9

If $a = \dfrac{6}{v^2}$, find s when $v = 4$ m/s given that $s = 0{\cdot}5$ m when $v = 2$ m/s.

Given $$a = \frac{6}{v^2}$$

Since the data given and the answer required involve s, use $a = v\dfrac{dv}{ds}$

$$\text{thus} \quad v\frac{dv}{ds} = \frac{6}{v^2}$$

$$\therefore \quad \int v^3\,dv = \int 6ds \quad \text{or} \quad \frac{v^4}{4} = 6s + C$$

but $v = 2$ m/s when $s = 0{\cdot}5$ m

$$\therefore \quad C = 1$$

$$\text{thus} \quad v^4 = 4(6s + 1)$$

$$\text{when } v = 4, \quad 4^4 = 4(6s + 1)$$

$$\therefore \quad 64 = 6s + 1 \quad \text{or} \quad s = 10{\cdot}5 \text{ m}$$

When $v = 4$ m/s the displacement s is $10{\cdot}5$ m.

Example 10

If $a = \dfrac{4}{v^3}$, find t when $v = 2$ m/s given that when $t = 0$, $v = 0$.

Given $$a = \frac{4}{v^3}$$

Since the data given and the answer required involve t, use $a = \dfrac{dv}{dt}$.

$$\text{thus} \quad \frac{dv}{dt} = \frac{4}{v^3}$$

$$\int v^3\,dv = \int 4\,dt$$

$$\therefore \quad \frac{v^4}{4} = 4t + C$$

but $v = 0$ when $t = 0$ $\quad\therefore \quad C = 0$

$$\therefore \quad \frac{v^4}{4} = 4t$$

$$\text{and when } v = 2 \text{ m/s}, \quad \frac{2^4}{4} = 4t \quad \text{or} \quad t = 1 \text{ s}$$

When $v = 2$ m/s, $t = 1$ s.

Example 11

If $a = 4 + 3v$, find s when $v = 2$ m/s given that $s = 0$ when $v = 0$.

Since the data given and the answer required involve s, use $a = \frac{vdv}{ds}$.

$$\text{thus} \quad v\frac{dv}{ds} = 4 + 3v$$

$$\int ds = \int \frac{vdv}{4+3v}$$

$$= \tfrac{1}{3}\int \frac{4 + 3v - 4}{4+3v}\,dv$$

This rearrangement is necessary to produce an expression which can be integrated.

$$= \tfrac{1}{3}\int \left(1 - \frac{4}{4+3v}\right) dv$$

$$s = \frac{v}{3} - \tfrac{4}{9}\ln(4 + 3v) + C$$

$s = 0$ when $v = 0$ gives $\quad C = \tfrac{4}{9}\ln 4$

$$\text{hence} \quad s = \frac{v}{3} - \tfrac{4}{9}\ln(4 + 3v) + \tfrac{4}{9}\ln 4$$

$$\text{or} \quad s = \frac{v}{3} + \tfrac{4}{9}\ln\left(\frac{4}{4+3v}\right)$$

$$\text{when } v = 2\text{ m/s}, \quad s = \tfrac{2}{3} + \tfrac{4}{9}\ln\left(\frac{4}{4+6}\right) = 0{\cdot}2594\text{ m}$$

When $v = 2$ m/s, the displacement s is 0·259 m.

Exercise 16B

Questions **1** to **18** involve bodies moving along a straight line. The letter s represents the displacement of the body at time t from a fixed point O on the line. The letters v and a represent the velocity and the acceleration of the body at time t. All units are in SI units.

v given as a function of s.

1. If $v = \frac{4}{1+s}$, find v when $s = 1$ m.
2. If $v = 2s - 3$, find a when $s = 4$ m.
3. If $v = \frac{s^2}{15}$, find t when $s = 10$ m given that $s = 3$ m when $t = 0$.
4. If $v = 4s - 2$, find s when $v = 8$ m/s.
5. If $v = s - 2$, find s when $a = 3$ m/s^2.
6. If $v = 2s + 3$, find t when $s = 3$ m given that $s = 0$ when $t = 0$.

a given as a function of s.

7. If $a = \frac{1}{2s-1}$, find a when $s = 3$ m.
8. If $a = \frac{40}{s^2}$, find v when $s = 20$ m, given that initially $v = 0$ and $s = 10$ m.
9. If $a = 2s - 3$, find s when $a = 1$ m/s^2.
10. If $a = \frac{1}{s+2}$, find s when $v = 1{\cdot}5$ m/s, given that initially $v = 0$ and $s = 0$.

11. If $a = s + 2$ and initially $s = 0$ and $v = 2$ m/s, find
    (a) an expression for v as a function of s
    (b) an expression for t as a function of s
12. If $a = 4s + 2$ and initially $s = 0$ and $v = 1$ m/s, find
    (a) the value of v when $s = 3$ m
    (b) the value of t when $s = 3$ m

a given as a function of v.

13. If $a = 10 - v^2$, find a when $v = 3$ m/s.
14. If $a = \frac{7}{v}$, find s when $v = 6$ m/s, given that $v = 3$ m/s when $s = 0$.
15. If $a = \frac{2}{3v^2}$, find t when $v = 4$ m/s given that $v = 0$ when $t = 0$.
16. If $a = 2 + v$, find t when $v = 12$ m/s given that $v = 0$ when $t = 0$.
17. If $a = \frac{v^2}{5}$, find s when $v = 6$ m/s given that initially $s = 0$ and $v = 1$ m/s.
18. If $a = 5 + 2v$, find s when $v = 5$ m/s given that initially $s = 0$ and $v = 0$.

Miscellaneous

19. Four points O, A, B and C lie in that order on a straight line with OA = 2 m, OB = 4 m and OC = 8 m. Initially a body is at A and moves in the direction OC such that its velocity v at any instant is given by $v = \frac{3}{s}$ m/s where s is the distance in metres that the body is from O at that instant. Find the time taken for the body to go from B to C.
20. OAB is a straight line with A between O and B and OA = 2 m. A body is initially at rest at A and moves towards B with its acceleration at any instant given by $a = \frac{k}{s}$ m/s^2, where s metres is the distance that the body is from O at that instant and k is a constant. Show that if v m/s is the velocity of the body, then $v^2 = 2k \ln \frac{s}{2}$.
21. A body is initially at rest at an origin O. It subsequently moves in a straight line away from O with acceleration at any instant given by $a = k(1 + v)$, where v is the velocity of the particle at that instant and k is a constant. If s is the displacement of the particle from O at any instant, show that $s = \frac{1}{k}[v - \ln(1 + v)]$.
22. Three points O, A and B lie in that order on a straight line with OA = d m and AB = 4 m. A particle travels along the line from O towards B with its velocity v at any instant given by $v = \frac{8}{3 + s}$ m/s, where s metres is the distance that the particle is from O at that instant. If the particle takes three seconds to travel from A to B, find the value of d.
23. A particle moves along a straight line with its acceleration at any instant given by $a = \alpha + \beta v^2$, where v is the velocity of the particle at that instant and α and β are constants. Initially the particle is at rest at a point O on the line. If the displacement of the particle from O during the motion is s, show that $v^2 = \frac{\alpha}{\beta}(e^{2\beta s} - 1)$.

24. A body moves along a straight line with its acceleration at any time t given by $a = 5 - 2v$, where v is the velocity of the body at time t. (All units are SI units). When $t = 0$, the body is at rest. Show that $t = \frac{1}{2} \ln\left(\frac{5}{5-2v}\right)$ s, and hence obtain an expression for v as a function of t.
Show that as the motion continues, the velocity of the body approaches a maximum value and find this maximum value.
25. O and A are two points on a straight line with OA $= d$. A particle moves along the line with its acceleration a at any time t given by $a = -k^2 s$, where s is the displacement of the particle from O at time t and k is a constant. When the particle is at A, its speed v is zero. Show that
(a) $v = k\sqrt{(d^2 - s^2)}$, (b) the maximum speed is kd and occurs when the particle is at O.

Simple harmonic motion (S.H.M.)

If a particle moves along a straight line so that its acceleration is proportional to its distance from a fixed point O on that line and is always directed towards O, then the motion of the particle is said to be simple harmonic.
The effect of the acceleration always being directed towards the fixed point O is to make the particle oscillate about O as shown below:

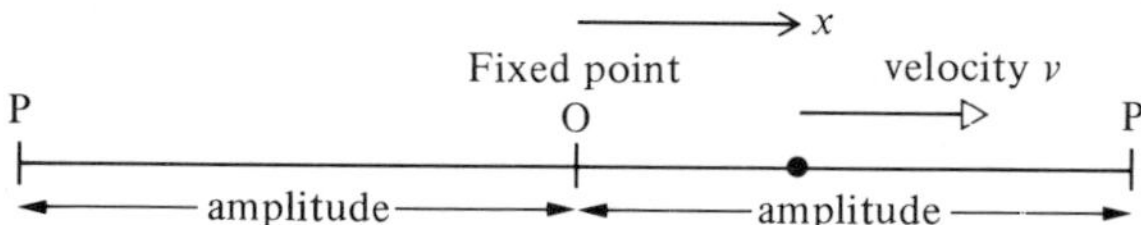

The particle is shown at a point between O and P, distance x from the fixed point O. Under simple harmonic motion the particle will move to P from the position shown; it then oscillates repeatedly from P through O to P′ and back. The particle just reaches P and P′; at these points it will have zero velocity. The distance OP′ equals the distance OP and is called the *amplitude* of the motion. The particle is said to complete one *cycle* in travelling from P through P′ back to P.

According to the definition of S.H.M. acceleration $\propto x$

Introducing the constant of proportionality n^2, this can be written as:

$$\ddot{x} = -n^2 x \qquad \ldots [1]$$

A squared constant is used to ease later integration, and the negative sign appears because the acceleration is towards the point O, i.e. in the direction of decreasing x.
Equation [1] gives the acceleration as a function of displacement and so, using the techniques developed earlier in this chapter, we use acceleration $= v\frac{dv}{dx}$.

$$\ddot{x} = -n^2 x$$

$$\text{thus} \quad v\frac{dv}{dx} = -n^2 x$$

$$\therefore \quad \int v\,dv = \int -n^2 x\,dx$$

$$\therefore \quad \frac{v^2}{2} = \frac{-n^2 x^2}{2} + C$$

but $v = 0$ when the particle is at P or P$'$, i.e. when $x = \pm a$

$$\text{thus} \quad 0 = \frac{-n^2 a^2}{2} + C \text{ or } C = \frac{n^2 a^2}{2}$$

$$\therefore \quad v^2 = n^2(a^2 - x^2) \qquad \ldots [2]$$

From equation [2] it can be seen that the maximum speed of the particle occurs when $x = 0$ and then $v = na$.

Thus $\qquad v_{max} = na$ occurs when the particle is at the point O $\qquad \ldots [3]$

Again from equation [2] $\qquad v = \pm n\sqrt{(a^2 - x^2)}$

Taking the positive root, $\qquad \dfrac{dx}{dt} = n\sqrt{(a^2 - x^2)}$

$$\therefore \quad \int \frac{dx}{\sqrt{(a^2 - x^2)}} = \int n\, dt$$

or $\quad \sin^{-1} \dfrac{x}{a} = nt + \epsilon$ where ϵ is the constant of integration

$$\therefore \quad x = a \sin(nt + \epsilon) \qquad \ldots [4]$$

Differentiating [4] with respect to t gives

$$v = an \cos(nt + \epsilon) \qquad \ldots [5]$$

It can also be shown that taking the negative root will lead to this same result.

The constant ϵ is called the *phase angle* and depends on the position of the particle when timing commences, i.e. when $t = 0$. This is often arbitrary and it is wise to choose to commence timing when the particle is at a position which gives ϵ a convenient value.
If $t = 0$ when the particle is at O, then $x = 0$ when $t = 0$.
Thus equation [4] becomes $0 = a \sin \epsilon$, allowing the value of ϵ to be taken as zero:

i.e. equation [4] becomes $x = a \sin nt$
equation [5] becomes $v = an \cos nt$

If $t = 0$ when the particle is at P, then $x = a$ when $t = 0$.

In this case equation [4] becomes $a = a \sin \epsilon$, allowing the value of ϵ to be taken as $\dfrac{\pi}{2}$ radians.

i.e. equation [4] becomes $x = a \cos nt$
equation [5] becomes $v = -an \sin nt$.

Suppose timing commences when the particle is at O. In order to find the value of t when the particle is at P, i.e. when $x = a$, use

$$x = a \sin nt$$

when $x = a$, $\qquad a = a \sin nt$ or $\sin nt = 1$

$$\therefore \quad nt = \frac{\pi}{2} \text{ or } \frac{5\pi}{2} \text{ or } \frac{9\pi}{2} \quad \text{and so } t = \frac{\pi}{2n} \text{ or } \frac{5\pi}{2n} \text{ or } \frac{9\pi}{2n} \ldots$$

Thus the particle returns to P after a period of time $\dfrac{4\pi}{2n} = \dfrac{2\pi}{n}$.

The period T (or periodic time) of the motion is the time for one complete cycle.

$$\therefore \quad T = \frac{2\pi}{n} \qquad \ldots [6]$$

Summary

For a particle moving with S.H.M. the following equations apply:

1. $\ddot{x} = -n^2 x$
2. $v^2 = n^2(a^2 - x^2)$
3. $v_{max} = na$
4. $x = a \sin(nt + \epsilon)$
5. $v = an \cos(nt + \epsilon)$
6. $T = \dfrac{2\pi}{n}$

These are important equations and they should be memorized

Example 12

Find the period of the S.H.M. defined by the equation $\ddot{x} = -25x$.

comparing $\ddot{x} = -25x$ with $\ddot{x} = -n^2 x$ gives $n = 5$

$\therefore$ using $T = \dfrac{2\pi}{n}$ gives $T = \dfrac{2\pi}{5}$

The period is $\dfrac{2\pi}{5}$ s.

Example 13

Find the maximum speed and the maximum acceleration of a particle moving with S.H.M. of period $\dfrac{\pi}{4}$ s and amplitude 25 cm.

Given the time period $\dfrac{\pi}{4} = \dfrac{2\pi}{n}$

$\therefore \quad n = 8; \qquad$ also $a = \frac{1}{4}$ m

since $v_{max} = na$

$v_{max} = 8(\frac{1}{4}) = 2$ m/s

$\ddot{x} = -n^2 x$ This is a maximum when $x = -a$

$\therefore \quad \ddot{x}_{max} = -(8)^2(-\frac{1}{4}) = 16$ m/s^2

The maximum speed is 2 m/s and the maximum acceleration is 16 m/s^2.

Example 14

A particle moves with S.H.M. about a mean position O. When the particle is 50 cm from O its speed is 3·6 m/s and when it is 120 cm from O its speed is 1·5 m/s. Find the amplitude and the periodic time of the motion.

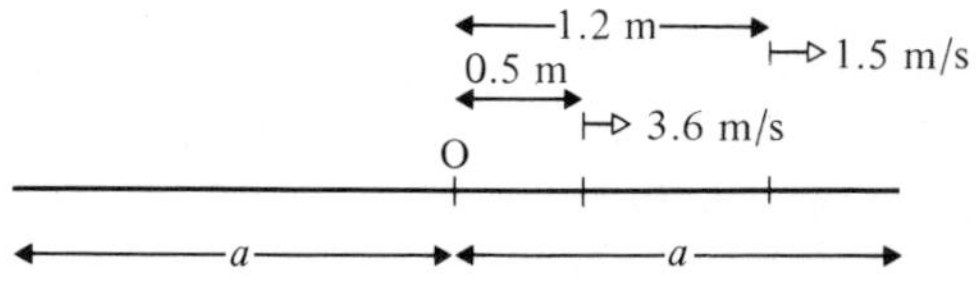

$v = 3{\cdot}6$ m/s when $x = 0{\cdot}5$ m,
so using $v^2 = n^2(a^2 - x^2)$ gives

$$3{\cdot}6^2 = n^2(a^2 - 0{\cdot}5^2) \qquad \ldots [1]$$

$v = 1{\cdot}5$ m/s when $x = 1{\cdot}2$ m, so using $v^2 = n^2(a^2 - x^2)$ gives

$$1{\cdot}5^2 = n^2(a^2 - 1{\cdot}2^2) \qquad \ldots [2]$$

Dividing equation [1] by equation [2] $\quad \dfrac{3{\cdot}6^2}{1{\cdot}5^2} = \dfrac{a^2 - 0{\cdot}5^2}{a^2 - 1{\cdot}2^2}$

and solving $\quad a = 1{\cdot}3$ m

substituting in equation [2] $\quad 1{\cdot}5^2 = n^2(1{\cdot}3^2 - 1{\cdot}2^2)$

$$\therefore \quad n = 3$$

$$\text{Periodic time} = \frac{2\pi}{n} = \frac{2\pi}{3}\text{ s}$$

The amplitude of the motion is $1{\cdot}3$ m and the periodic time is $\dfrac{2\pi}{3}$ s.

Example 15

A particle is projected from a point O at time $t = 0$ and performs S.H.M. with O as the centre of oscillation. The motion is of amplitude 20 cm and time period 4 s. Find
(a) the speed of projection,
(b) the speed of the particle when $t = 1{\cdot}5$ s,
(c) the value of t when the particle is first at a point 10 cm from O.

(a) amplitude $= 0{\cdot}2$ m $\qquad$ period $T = \dfrac{2\pi}{n} = 4$ s

$t = 0$ → v
O
←0·2 m→

$$\therefore \quad n = \frac{\pi}{2}$$

At O, $\qquad$ velocity $v = v_{max}$

$$= na$$

$$\therefore \quad v_{max} = \frac{\pi}{2}(0{\cdot}2) = \frac{\pi}{10}$$

The speed of projection is $\dfrac{\pi}{10}$ m/s.

(b) To find v when $t = 1{\cdot}5$ s, use $v = an\cos(nt + \epsilon)$

since timing starts, i.e. $t = 0$, when the particle is at O, $\epsilon = 0$

$$\text{thus} \quad v = an\cos nt$$

$$= (0{\cdot}2)\left(\frac{\pi}{2}\right)\cos\left(\frac{\pi}{2} \times \frac{3}{2}\right) = \frac{\pi}{10}\cos\frac{3\pi}{4}$$

$$= \frac{\pi}{10}\left(\frac{-1}{\sqrt{2}}\right)$$

When $t = 1{\cdot}5$ s, the speed of the particle is $\dfrac{\pi\sqrt{2}}{20}$ m/s.

(c) To find t when $x = 0{\cdot}1$ m, use $x = a \sin(nt + \epsilon)$

since timing starts when the particle is at O, $\epsilon = 0$

$$\text{thus} \quad x = a \sin nt$$

$$\therefore \quad 0{\cdot}1 = (0{\cdot}2) \sin \frac{\pi t}{2}$$

$$\text{or} \quad \sin \frac{\pi t}{2} = \tfrac{1}{2}$$

$$\therefore \quad \frac{\pi t}{2} = \frac{\pi}{6}, \frac{5\pi}{6}, \ldots$$

$$\text{or} \quad t = \tfrac{1}{3}, \tfrac{5}{3}, \ldots$$

The particle is first at a point 10 cm from O when $t = \frac{1}{3}$ s.

Relationship between circular motion and S.H.M.

Consider a particle P, mass m, moving in a circle with angular velocity ω, the radius of the circle being r.

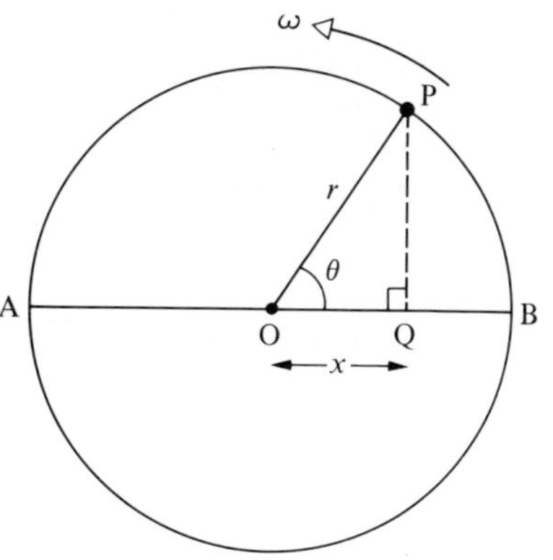

Let Q be the projection of P on the line AOB at any time and let OQ $= x$, and $\text{P}\hat{\text{O}}\text{Q} = \theta$.

Acceleration of P towards centre of motion O

$$= r\omega^2$$

Acceleration of Q towards O

$$= r\omega^2 \cos\theta$$
$$= \omega^2 x \quad \text{because } x = r\cos\theta$$

Since this acceleration is directed towards O

$$\ddot{x} = -\omega^2 x$$

Hence as P moves with constant speed around the circle, the point Q moves along the line BOA executing S.H.M. and $\omega = n$

$$\text{hence period } T \text{ of S.H.M.} = \frac{2\pi}{\omega}.$$

Exercise 16C

1. Find the periodic time of the simple harmonic motion governed by each of the following equations. (All the equations use SI units).
(a) $\ddot{x} = -x$, (b) $\ddot{x} = -4x$, (c) $\ddot{x} = -9x$.
2. A particle moves with S.H.M. about a mean position O with a periodic time of $\frac{\pi}{2}$ s. Find the magnitude of the acceleration of the particle when 1 metre from O.
3. A particle moves with S.H.M. about a mean position O. When the particle is 25 cm from O, it is accelerating at 1 m/s^2 towards O. Find the periodic time of the motion and the magnitude of the acceleration of the particle when 20 cm from O.

4. A particle moves with S.H.M. of time period $\frac{\pi}{2}$ s and has a maximum speed of 3 m/s. Find the maximum acceleration experienced by the particle.
5. A particle moves with S.H.M. of time period $\frac{\pi}{2}$ s and amplitude 2 m. Find the maximum speed of the particle.
6. A particle moves with S.H.M. about a mean position O. The amplitude of the motion is 5 m and the period is 8π s. Find the maximum speed of the particle and its speed when 3 m from O.
7. A particle moves with S.H.M. about a mean position O. The amplitude of the motion is 65 cm and the time period is $\frac{\pi}{4}$ s. Find how far the particle is from O when its speed is 2 m/s.
8. A particle moves with S.H.M. about a mean position O. The particle has zero velocity at a point which is 50 cm from O and a speed of 3 m/s at O. Find
   (a) the maximum speed of the particle, (b) the amplitude of the motion,
   (c) the periodic time of the motion.
9. A particle moves with S.H.M. about a mean position O. When the particle is 60 cm from O, its speed is 1·6 m/s, and when it is 80 cm from O, its speed is 1·2 m/s. Find the amplitude and period of the motion.
10. The effect of the waves on an empty oil drum floating in the sea is to make it bob up and down with S.H.M. If the drum encounters 20 waves every minute and for each wave the vertical distance from peak to trough is 80 cm find the amplitude and period of the motion and the maximum speed of the drum.
11. A particle moves with S.H.M. about a mean position O. When passing through two points which are 2 m and 2·4 m from O the particle has speeds of 3 m/s and 1·4 m/s respectively. Find the amplitude of the motion and the greatest speed attained by the particle.
12. A particle moves with S.H.M. about a mean position O. The particle is initially projected from O with speed $\frac{\pi}{6}$ m/s and just reaches a point A, 2 m from O. Find how far the particle is from O three seconds after projection.
    How many seconds after projection is the particle a distance of 1 m from O
    (a) for the first time, (b) for the second time, (c) for the third time?
13. A particle is released from rest at a point A, 1 m from a second point O. The particle accelerates towards O and moves with simple harmonic motion of time period 12 s and O as the centre of oscillation. Find how far the particle is from O one second after release.
    How many seconds after release is the particle at the mid-point of OA
    (a) for the first time, (b) for the second time?
14. A particle is projected from a point A at time $t = 0$ and performs S.H.M. with A as the centre of oscillation. The amplitude of the motion is 50 cm and the periodic time is 3 s. Find (a) the speed of projection, (b) the speed of the particle when $t = 1$ s,
    (c) the speed of the particle when $t = 2$ s, (d) the distance the particle is from A when $t = 2$ s.
15. A particle is released from rest at a point A at time $t = 0$ and performs S.H.M. about a mean position B. The particle just returns to A during each oscillation and AB $= 2\sqrt{2}$ m. If the particle passes through B with speed $\pi\sqrt{2}$ m/s, find the value of t when the particle is first travelling with a speed of π m/s. How far from B is the particle then?
16. The head of a piston moves with S.H.M. of amplitude $\frac{\sqrt{3}}{10}$ m about a mean position O. How far from O is the head of the piston when it is travelling with a speed equal to half of its maximum speed?

17. A particle is fastened to the mid-point of a stretched spring lying on a smooth horizontal surface. The particle is set in motion so that it moves with S.H.M. about a mean position O. If one metre is the greatest distance the particle is from O during the motion, find how far from O the particle is when it is travelling with a speed equal to four-fifths of its greatest speed.
18. AOBCA$'$ is a straight line with AA$' = 4$ m, AO $= 2$ m, AB $= 3$ m and OC $= \sqrt{2}$ m. A particle performing S.H.M. of time period 4·8 s and with O as the centre of oscillation, just reaches the points A and A$'$. Timing commences when the particle is at O and is travelling in the direction OA$'$. Find the time and the velocity of the particle when it first reaches B and when it first reaches C.
19. A particle is projected from a point O with speed 3 m/s and performs S.H.M. about O with an amplitude of 1 m. Find the time that elapses between the first and second occasions that the particle reaches a point B, 80 cm from O.
20. A body moves with S.H.M. of period 12 s and amplitude 2 m. The displacement of the body from the centre of the oscillation at time t is given by $x = a \sin(nt + \epsilon)$ where a, n and ϵ are constants. When $t = 0$, $x = 1$ m and the body is moving so as to increase x. Find the smallest positive value of ϵ, giving your answer in radians.
Find the value of t when the particle is first (a) $\sqrt{2}$ m, (b) $\sqrt{3}$ m away from the centre of oscillation.

21. The diagram shows a particle P moving in a circular path of radius r, with constant speed v. The point Q is the projection of P onto the diameter AB and is obtained by drawing a line from P to meet AB at right angles. O is the centre of the circle and x is the displacement of Q from O at any instant. Show that, as P moves around the circle, the point Q performs S.H.M. along AB with time period $\frac{2\pi r}{v}$.
If $r = 30$ cm and $v = 2$ m/s, find the maximum speed of the point Q as it moves along AB and the time taken for it to move from B to the mid-point of OB.

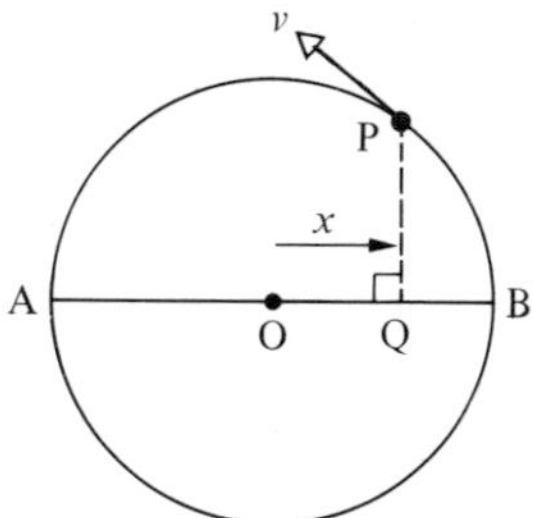

Exercise 16D Examination questions

1. A particle P moves in a straight line so that its distance, s metres, from a fixed point O is given by $s = 12t - t^3$ where t is the time in seconds after leaving O. Calculate
   (i) the distance travelled by P in the fourth second,
   (ii) the value of s when P comes instantaneously to rest,
   (iii) the velocity of P when the acceleration is zero. (Cambridge)

2. A car starts from rest, and its acceleration after t seconds is $\frac{84 + 5t - t^2}{48}$ m/s^2 until the acceleration becomes zero. After this instant it moves with the constant speed it has then reached. Find (i) the time taken to reach its greatest speed; (ii) the greatest speed; (iii) the distance travelled in the first minute. (Oxford)

3. Two particles A and B are moving in horizontal straight lines in the plane of the coordinate axes Ox and Oy, the units of distance and time being the metre and the second respectively. At time t seconds, A has position vector $t^2\mathbf{i} + 2\mathbf{j}$ and B has position vector $t(3\mathbf{i} - 4\mathbf{j})$.
   (a) Calculate the distance between A and B when $t = 3$.
   (b) Show that the velocity of B is constant and calculate its magnitude and direction.
   (c) Calculate the acceleration of A.
   (d) Find the magnitude and the direction of the velocity of B relative to A when $t = 2$.
   (London)

4. Prove that, with the usual notation, $\dfrac{dv}{dt} = v\dfrac{dv}{ds}$.
   A particle moving in a straight line on a smooth horizontal plane is subjected to a single variable resistive force which produces a retardation whose magnitude is proportional to the cube of the speed of the particle at any instant. The initial speed of the particle is 10 m/s and the initial retardation is 2 m/s^2. Show that the speed v m/s of the particle after t seconds is given by $v^2 = \dfrac{500}{2t + 5}$.
   Given that the mass of the particle is 0·2 kg, find, in joules, the change in kinetic energy between the instants when $t = 0$ and when $t = 10$.
   Find also the distance covered by the particle between the instants when $t = 0$ and $t = 10$.
   (A.E.B)

5. (a) A particle travelling in a straight line is subject to a retardation of $(v^2 + 25)$ m s^{-2}, where v m s^{-1} is its speed at t seconds. If its initial speed is 12 m s^{-1}, show that it travels a distance of $\log_e \frac{13}{5}$ metres before coming to rest.
   (b) A particle moving with S.H.M. oscillates between two points A and B which are 10 m apart. When it is 1 m from A, its speed is 16 m s^{-1}. Find (i) the periodic time; (ii) the maximum speed it attains; (iii) the maximum acceleration; (iv) the shortest time it takes to travel from A to C, where AC is $\frac{1}{4}$ AB.
   (S.U.J.B)

6. A particle of mass 5 kilograms executes simple harmonic motion with amplitude 2 metres and period 12 seconds. Find the maximum kinetic energy of the particle, leaving your answer in terms of π.
   Initially the particle is moving with its maximum kinetic energy. Find the time that elapses until the kinetic energy is reduced to one quarter of the maximum value, and show that the distance moved in this time is $\sqrt{3}$ metres.
   [Standard formulae relating to simple harmonic motion may be quoted without proof.]
   (J.M.B)

7. A gauge on the side of a harbour shows the depth of water at any time. The point of the scale at the water level can be regarded as moving up and down with simple harmonic motion. High tide on a particular occasion coincided with midnight when the gauge indicated that the depth was 18 m. At the following low tide at 06 20 the depth was 2 m. The harbour is considered safe for the movement of shipping when the depth of water is at least 5 m. Find the times (to the nearest five minutes) between midnight and mid-day when the harbour is safe for the movement of shipping.
   (Cambridge)

17

Variable force

If a force, which is constant in direction but variable in magnitude, acts upon a body, then the acceleration produced by the force will be variable. Using Newton's Second Law, $F = ma$, one of the differential forms $v\frac{dv}{ds}$ or $\frac{dv}{dt}$ must be used for the acceleration. Which one is used will depend upon the data given and the type of relationship required.

The worked examples in this chapter will, unless stated otherwise, involve a horizontal force F of constant direction and variable magnitude acting upon a body of constant mass m resting on a smooth horizontal surface.
The displacement s of the body from a fixed point O and the velocity v of the body after time t will be in SI units.

Example 1

If $F = 3t + 1$, $m = 4$ kg and the body is initially at rest at a point O, find
(a) v when $t = 2$ s, (b) s when $t = 2$ s.

(a) Given: $F = 3t + 1$ To find: v when $t = 2$ s
$m = 4$ kg
$s = 0$ and $v = 0$ when $t = 0$

Since v and t are involved, use $\frac{dv}{dt}$ for a in Newton's Second Law equation:

$$F = ma$$
$$3t + 1 = 4\frac{dv}{dt}$$
$$\int (3t + 1)dt = \int 4\, dv$$
$$\tfrac{3}{2}t^2 + t = 4v + C$$
$$v = 0 \text{ when } t = 0, \text{ hence } C = 0$$
$$\therefore \quad 4v = \tfrac{3}{2}t^2 + t$$
$$\text{when } t = 2 \text{ s}, \quad 4v = \tfrac{3}{2}(4) + 2$$
$$\therefore \quad v = 2 \text{ m/s}$$

When $t = 2$ s the velocity is 2 m/s.

(b) To find: s when $t = 2$ s

From part (a) $$4v = \tfrac{3}{2}t^2 + t$$

$$\text{thus} \quad 4\frac{ds}{dt} = \tfrac{3}{2}t^2 + t$$

$$\therefore \quad \int 4ds = \int (\tfrac{3}{2}t^2 + t)\,dt$$

$$\therefore \quad 4s = \frac{t^3}{2} + \frac{t^2}{2} + C'$$

$s = 0$ when $t = 0$, hence $C' = 0$

$$\therefore \quad 4s = \frac{t^3}{2} + \frac{t^2}{2}$$

when $t = 2$ s, $4s = \tfrac{8}{2} + \tfrac{4}{2}$

$$\therefore \quad s = 1{\cdot}5 \text{ m}$$

When $t = 2$ s, the displacement is 1·5 m.

Example 2

If $F = 5s + 6$, $m = 1$ kg and the body is initially at rest at a point O, find
(a) v when $s = 4$ m, (b) s when $v = 9$ m/s.

(a) Given: $F = 5s + 6$ To find: v when $s = 4$ m

$m = 1$ kg

$s = 0$ and $v = 0$ when $t = 0$

Since v and s are involved, use $v\dfrac{dv}{ds}$ for a in Newton's Second Law equation:

$$F = ma$$

$$5s + 6 = (1)\,v\frac{dv}{ds}$$

$$\int (5s + 6)ds = \int v dv$$

$$\therefore \quad \tfrac{5}{2}s^2 + 6s = \frac{v^2}{2} + C$$

$v = 0$ when $s = 0$, hence $C = 0$

$$\therefore \quad \frac{v^2}{2} = \frac{5}{2}s^2 + 6s$$

when $s = 4$ m, $\dfrac{v^2}{2} = \dfrac{5}{2}(16) + 6(4)$

$$\therefore \quad v = 8\sqrt{2} \text{ m/s}$$

When $s = 4$ m, the velocity is $8\sqrt{2}$ m/s.

(b) To find: s when $v = 9$ m/s

From part (a) $v^2 = 5s^2 + 12s$

when $v = 9$ m/s, $81 = 5s^2 + 12s$

$\therefore \quad 5s^2 + 12s - 81 = 0$

or $(5s + 27)(s - 3) = 0$

$$\therefore \quad s = -5\tfrac{2}{5} \text{ or } +3$$

Taking the positive answer, when the velocity is 9 m/s, then $s = 3$ m.

Example 3

If $F = \dfrac{3}{2v+1}$, $m = 2$ kg and the body is initially at rest at a point O, find t when $v = 6$ m/s.

Given: $F = \dfrac{3}{2v+1}$ To find: t when $v = 6$ m/s

$m = 2$ kg

$s = 0$ and $v = 0$ when $t = 0$

Since t is required use $\dfrac{dv}{dt}$ for a in $F = ma$:

$$\frac{3}{2v+1} = 2\frac{dv}{dt}$$

$$\int 3dt = \int 2(2v+1)\,dv$$

$$\text{or} \quad 3t = 2(v^2 + v) + C$$

$$v = 0 \text{ when } t = 0, \text{ hence } C = 0$$

$$\therefore \quad 3t = 2(v^2 + v)$$

$$\text{when } v = 6 \text{ m/s}, \quad 3t = 2(36 + 6)$$

$$\therefore \quad t = 28 \text{ s}$$

When the velocity is 6 m/s, $t = 28$ s.

Example 4

A force $\mathbf{F}$ acts on a body of mass 250 g which is initially at rest at a fixed point O. If $\mathbf{F} = [(5t-2)\mathbf{i} + 4t\mathbf{j}]$ N, where t is the time for which the force has been acting on the body, find expressions for (a) the velocity vector of the body at time t, (b) the position vector of the body at time t.

(a) Given: $\mathbf{F} = [(5t-2)\mathbf{i} + 4t\mathbf{j}]$ N To find: $\mathbf{v}$

$m = 250$ g

$\mathbf{s} = 0$ and $\mathbf{v} = 0$ when $t = 0$

Since $\mathbf{v}$ is required, and t is involved, use $\dfrac{d\mathbf{v}}{dt}$ for $\mathbf{a}$ in $\mathbf{F} = m\mathbf{a}$:

$$(5t-2)\mathbf{i} + 4t\mathbf{j} = \tfrac{1}{4}\frac{d\mathbf{v}}{dt}$$

$$\therefore \quad \int [(5t-2)\mathbf{i} + 4t\mathbf{j}]\,dt = \int \tfrac{1}{4}\,d\mathbf{v}$$

$$\therefore \quad (\tfrac{5}{2}t^2 - 2t)\mathbf{i} + 2t^2\mathbf{j} = \frac{\mathbf{v}}{4} + A\mathbf{i} + B\mathbf{j}$$

$$\text{since } \mathbf{v} = 0 \text{ when } t = 0, \; 0\mathbf{i} + 0\mathbf{j} = 0 + A\mathbf{i} + B\mathbf{j}$$

equating coefficients of $\mathbf{i}$ and $\mathbf{j}$, $A = 0$ and $B = 0$

$$\therefore \quad (\tfrac{5}{2}t^2 - 2t)\mathbf{i} + 2t^2\mathbf{j} = \frac{\mathbf{v}}{4}$$

The velocity vector $\mathbf{v} = [(10t^2 - 8t)\mathbf{i} + 8t^2\mathbf{j}]$ m/s.

(b) Using $\frac{ds}{dt}$ for **v** from part (a):

$$\frac{d\mathbf{s}}{dt} = (10t^2 - 8t)\mathbf{i} + 8t^2\mathbf{j}$$

$$\therefore \quad \mathbf{s} = (\tfrac{10}{3}t^3 - 4t^2)\mathbf{i} + \tfrac{8}{3}t^3\mathbf{j} + C\mathbf{i} + D\mathbf{j}$$

$$\mathbf{s} = 0 \text{ when } t = 0, \text{ hence } C = 0 \text{ and } D = 0$$

The position vector $\mathbf{s} = [(\frac{10}{3}t^3 - 4t^2)\mathbf{i} + \frac{8}{3}t^3\mathbf{j}]$ m.

Example 5

A resistance of $\frac{8}{3s+1}$ N acts upon a body, where s is the distance in metres that the body has fallen under gravity from rest. The mass of the body is 2 kg.
Find the speed of the body when it has fallen a distance of 1·2 m. (Take $g = 10$ m/s^2).

Given:

$$\text{resistance } F = \frac{8}{3s+1} \text{ N}$$
$$m = 2 \text{ kg}$$
$$s = 1{\cdot}2 \text{ m}$$

Since speed and displacement are involved, use $v\frac{dv}{ds}$ for a in $F = ma$:

$$2g - \frac{8}{3s+1} = 2v\frac{dv}{ds}$$

$$\therefore \quad \int \left[g - \frac{4}{3s+1}\right] ds = \int v\,dv$$

$$\therefore \quad 10s - \tfrac{4}{3}\ln(3s+1) = \frac{v^2}{2} + A$$

$v = 0$ when $s = 0$, hence $-\frac{4}{3}\ln 1 = A$ or $A = 0$

$$\therefore \quad 10s - \tfrac{4}{3}\ln(3s+1) = \frac{v^2}{2}$$

when $s = 1{\cdot}2$ m, $12 - \frac{4}{3}\ln(4{\cdot}6) = \frac{v^2}{2}$ giving $v = 4{\cdot}464$ m/s

When the body has fallen 1·2 m, its speed is 4·46 m/s.

Exercise 17A

Questions **1** to **10** involve a horizontal force F of constant direction and variable magnitude, acting on a body of mass m initially at rest at a point O on a smooth horizontal surface. After a time t, the displacement of the body from O is s and the body has velocity v (SI units used throughout).

1. If $F = 2t$ and $m = 5$ kg, find v when $t = 4$ s.
2. If $F = 6t - 4$ and $m = 2$ kg, find s when $t = 4$ s.
3. If $F = 2s + 1$ and $m = 1{\cdot}5$ kg, find v when $s = 3$ m.
4. If $F = \frac{1}{4s+5}$ and $m = 250$ g, find v when $s = 8$ m.
5. If $F = \frac{2}{s+5}$ and $m = 470$ g, find s when $v = 2$ m/s.
6. If $F = \frac{1}{v+2}$ and $m = 250$ g, find t when $v = 4$ m/s.

7. If $F = 8 - 3v^2$ and $m = 6$ kg, find s when $v = 1$ m/s.
8. If $F = 5 - 3v$ and $m = 6$ kg, find t when $v = 1$ m/s.
9. If $F = 4 + v^2$ and $m = 4$ kg, find v when $t = \frac{\pi}{2}$ s (hint: $\int \frac{dx}{\alpha^2 + x^2} = \frac{1}{\alpha} \tan^{-1} \frac{x}{\alpha} + c$)
10. If $F = 40 - v$ and $m = 5$ kg, find s when $v = 15$ m/s.
11. A resultant force **F** acts on a body of mass 500 g initially at rest at an origin. $\mathbf{F} = (3t\mathbf{i} + \mathbf{j})$ N where t seconds is the time for which the force has been acting on the body. Find expressions for (a) the velocity vector of the body at time t,
(b) the position vector of the body at time t.
12. A resultant force **F** acts on a body of mass 500 g initially at rest at an origin O. $\mathbf{F} = [(4t - 1)\mathbf{i} + 4\mathbf{j}]$ N where t seconds is the time for which the force has been acting on the body. Find the body's speed and distance from O when $t = 2$.
13. A body of mass 750 g is initially at rest at a point O on a smooth horizontal surface. A horizontal force F acts on the body and causes it to move in a straight line across the surface. The magnitude of F is given by $F = (3t + 1)$ N where t seconds is the time for which the force has been acting on the body. Find the speed of the body when $t = 3$ and its distance from O at that time.
14. A body of mass 1 kg is released from rest and falls under gravity against a resistance of $\frac{7}{s + 1}$ N where s is the distance (in metres) that the body has fallen since release. Find the speed of the particle when it has fallen a distance of 6·4 m. (Take $g = 10$ m/s^2.)
15. A horizontal force F is applied to a body of mass 5 kg, initially at rest at a point O on a rough horizontal surface, coefficient of friction $\frac{2}{7}$. The force causes the body to move in a straight line across the surface. The magnitude of the force is given by $F = (5s + 25)$ N where s is the distance in metres that the body is from O. Find the speed of the body when $s = 10$ m.
If the applied force were $(5s + 12)$ N, would the body move?
16. A body of mass 4 kg is released from rest and falls under gravity against a resistance to motion of $2v$ N where v is the speed of the body in m/s. Find the time taken for the body to attain a speed of 14 m/s.
17. A body of mass 500 g is released from rest and falls under gravity against a resistance of $\frac{v^2}{2}$ N where v is the speed of the body in m/s. Find how far the body falls when its speed increases from 1 m/s to 3 m/s.
18. With its engine working at a constant rate of 10 kW, a car of mass 800 kg accelerates along a level road from rest to a maximum speed v_{max}. Throughout the motion the car experiences a resistance to motion of $20v$ N where v is the speed of the car in m/s. Find (a) v_{max}, (b) the time taken for the car to increase its speed from 10 m/s to 20 m/s.
19. With its engines developing a constant 100 kW of power, a train of mass 90 tonnes accelerates up an incline of 1 in 98. If the air and frictional resistances are a constant 1000 N, find (a) v_{max}, the maximum speed of the train up the incline, (b) the time taken for the train to accelerate up the incline from rest to a speed equal to $\frac{v_{max}}{2}$.
20. A horizontal force F of variable magnitude and constant direction acts on a body of mass m which is initially at rest at a point O on a smooth horizontal surface. The magnitude of F is given by $F = \beta + \alpha t$ where t is the time for which the force has been acting on the body and α and β are positive constants. If s is the distance the body is from O at time t, show that $s = \frac{t^2}{6m}(3\beta + \alpha t)$.

21. A body of mass m is initially at rest at a point O on a smooth horizontal surface. A horizontal force F is applied to the body and causes it to move in a straight line across the surface. The magnitude of F is given by $F = \dfrac{1}{s+\alpha}$ where s is the distance of the body from O and α is a positive constant. If v is the speed of the body at any moment, show that $s = \alpha(e^{\frac{1}{2}mv^2} - 1)$.

22. A body of mass m is released from rest and falls under gravity against a resistance to motion of mkv where v is the speed of the body at a time t after release and k is a positive constant. Show that (a) $kt = \ln\left(\dfrac{g}{g-kv}\right)$, (b) as the motion continues, v approaches a maximum value of $\dfrac{g}{k}$.

Impulse of a variable force

As has already been seen on page 257, the impulse of a constant force F is $F \times t$, where t is the time for which the force is acting.
If the force is variable, the calculus notation must be used, and

$$\begin{aligned}
\text{impulse} &= \int F\,dt \\
&= \int m\frac{dv}{dt} \times dt \qquad \left(\text{using } F = ma = m\frac{dv}{dt}\right) \\
&= \int_u^v m\,dv \\
&= [mv]_u^v \\
&= mv - mu \\
&= \text{change in momentum}
\end{aligned}$$

So the impulse of a variable force is also equal to the change in momentum.

Example 6

A horizontal force F, of variable magnitude and constant direction, pushes a body of mass 4 kg along a smooth horizontal surface. The magnitude of F is given by $F = (2t^3 + 3)$ N, where t seconds is the time for which the force has been acting on the body. When $t = 0$ the body is at rest, and when $t = 2$ s the force ceases to act. Find

(a) the magnitude of the impulse given to the body,
(b) the magnitude of the final momentum of the body,
(c) the final speed of the body.

(a)

$$\text{Impulse} = \int F\,dt$$

$$= \int_0^2 (2t^3 + 3)\,dt = \left[\frac{t^4}{2} + 3t\right]_0^2$$

$$= 14 \text{ N s}$$

The impulse given to the body is 14 N s.

(b)

$$\text{Since} \quad \text{impulse} = \text{change in momentum}$$

$$14 = \text{final momentum} - \text{initial momentum}$$

$$\therefore \quad 14 = \text{final momentum} - 0$$

The final momentum of the momentum is 14 N s.

(c)

$$\text{Final momentum} = 14 \text{ N s}$$

$$\therefore \quad 14 = mv$$

$$\text{but } m = 4 \text{ kg} \quad \text{and} \quad \therefore \quad v = 3{\cdot}5 \text{ m/s}$$

The final speed of the body is 3·5 m/s.

Example 7

A body of mass 2 kg is initially moving with a constant velocity of $(\mathbf{i} + 4\mathbf{j})$ m/s when a force $\mathbf{F}$ commences to act on it. If $\mathbf{F} = (5\mathbf{i} + 4t\mathbf{j})$ N, where t is the time for which the force has been acting on the body, find the speed of the body when $t = 2$ s.

$$\text{Impulse} = \int \mathbf{F}\,dt$$

Impulse given to the body in the first 2 seconds

$$= \int_0^2 (5\mathbf{i} + 4t\mathbf{j})\,dt = [5t\mathbf{i} + 2t^2\mathbf{j}]_0^2$$

$$= (10\mathbf{i} + 8\mathbf{j}) \text{ N s}$$

$$\text{But} \quad \text{impulse} = \text{change in momentum}$$

$$= m\mathbf{v} - m\mathbf{u}$$

$$\therefore \quad 10\mathbf{i} + 8\mathbf{j} = 2\mathbf{v} - 2(\mathbf{i} + 4\mathbf{j})$$

$$\text{or} \quad \mathbf{v} = 6\mathbf{i} + 8\mathbf{j}$$

$$\therefore \quad \text{speed} = \sqrt{(6^2 + 8^2)} = 10 \text{ m/s}$$

When $t = 2$ s the body has a speed of 10 m/s.

Work done by a variable force

The work done by a constant force F is $F \times s$ where s is the distance moved in the direction of the force. If the force is variable, the calculus notation must be used, and

$$\text{work done} = \int F\,ds$$

$$= \int mv\frac{dv}{ds} \times ds \qquad \left(\text{using } F = ma = mv\frac{dv}{ds}\right)$$

$$= \int_u^v mv\,dv = \tfrac{1}{2}mv^2 - \tfrac{1}{2}mu^2 = \text{change in kinetic energy.}$$

Example 8

After release from rest at a point O, a body of mass 1 kg falls under gravity against a resistance of $\frac{24}{25}s$ N, where s metres is the distance the body is below O at any instant.
Find the amount of work done by the body against the resistance, from release until it passes through a point P, 10 m below O, and find the speed of the body at that instant.

$$\text{Work done against resistance} = \int_0^{10} \frac{24}{25}s\,ds$$

$$= \left[\frac{12}{25}s^2\right]_0^{10}$$

$$= 48 \text{ J}$$

Loss in potential energy = gain in kinetic energy + work done against resistance

$$\therefore \quad mgh = \tfrac{1}{2}mv^2 + 48$$

$$\therefore \quad (1)(9{\cdot}8)(10) = \tfrac{1}{2}(1)v^2 + 48$$

$$\text{or} \quad v = 10 \text{ m/s}$$

In travelling from O to P the body does 48 J of work against the resistance and passes through the point P with speed 10 m/s.

Exercise 17B

1. A horizontal force F, of variable magnitude and constant direction, pushes a body of mass 3 kg along a smooth horizontal surface. The magnitude of F is given by $F = (2t + 1)$ N where t seconds is the time for which the force has been acting on the body. When $t = 0$ the body is at rest, and when $t = 3$ the force ceases to act. Find
   (a) the magnitude of the impulse given to the body by the force,
   (b) the magnitude of the final momentum of the body,
   (c) the final speed of the body.
2. A horizontal force F, of variable magnitude and constant direction, pushes a body of mass 9 kg along a smooth horizontal surface. The magnitude of F is given by $F = (3t^2 + 2)$ N where t seconds is the time for which the force has been acting on the body. When $t = 0$ the body is at rest and when $t = 4$ the force ceases to act. Find
   (a) the magnitude of the impulse given to the body by the force,
   (b) the magnitude of the final momentum of the body,
   (c) the final speed of the body.
3. A horizontal force F acts in a constant direction, on a body of mass 2 kg initially at rest on a smooth horizontal surface. $F = (10 - 3t)$ N where t is the time in seconds for which the force has been acting on the body. Find the speed of the body when $t = 4$ s.
4. A body of mass 2 kg is initially moving with a constant velocity of $3\mathbf{i}$ m/s when a force $\mathbf{F}$ acts upon it. If $\mathbf{F} = (2t + 1)\mathbf{i}$ N, where t seconds is the time for which the force has been acting on the body, find the velocity of the body when $t = 4$.
5. A body of mass 2 kg is initially moving with a constant velocity of $-3\mathbf{i}$ m/s when a force $\mathbf{F}$ acts upon it. If $\mathbf{F} = (2t + 1)\mathbf{i}$ N, where t seconds is the time for which the force has been acting on the body, find the velocity of the body when $t = 4$.

6. A body of mass 2 kg is initially moving with a constant velocity of $-2\mathbf{i}$ m/s. A force $\mathbf{F}$ acts upon the body for 3 seconds and at the end of this time the body has velocity $7\mathbf{i}$ m/s. If $\mathbf{F} = at\mathbf{i}$ N, where t seconds is the time for which the force has been acting and a is a constant, find the value of a.
7. A body of mass 1 kg is initially moving with a constant velocity of $4\mathbf{i}$ m/s when a force $\mathbf{F}$ acts upon it. If $\mathbf{F} = (2t\mathbf{i} + 3\mathbf{j})$ N, where t seconds is the time for which the force has been acting on the body, find the velocity and speed of the body when $t = 2$.
8. A horizontal force F acts on a body of mass 6 kg initially at rest on a smooth horizontal surface and causes the body to move across the surface in a straight line. The magnitude of F is given by $F = (2s + 1)$ N, where s metres is the distance of the body from its original position. If the force ceases when $s = 3$, find (a) the work done by the force, (b) the final kinetic energy of the body, (c) the final speed of the body.
9. A horizontal force F acts on a body of mass 250 g initially at rest on a smooth horizontal surface and causes the body to move across the surface in a straight line. The magnitude of F is given by $F = (4s + 5)$ N, where s metres is the distance of the body from its original position. If the force ceases when $s = 2$, find (a) the work done by the force, (b) the final kinetic energy of the body, (c) the final speed of the body.
10. A body of mass 500 g lies at a point O on a horizontal surface. The body is projected from O with a speed of 3 m/s and moves against a resistance of $2s$ N, where s metres is the distance of the body from O. If the body comes to rest at a point B, find (a) the loss in the kinetic energy possessed by the body, (b) the distance OB.
11. A body of mass 2 kg is released from rest at a point O and falls under gravity against a resistance of $\frac{16}{5}s$ N, where s metres is the distance of the body from O at any instant. When the body has fallen 3·5 m, find (a) the loss in potential energy, (b) the work done against the resistance, (c) the gain in the kinetic energy, (d) the speed of the body.
12. A body of mass 5 kg is projected vertically upwards from a point O and comes to instantaneous rest at a point B, 6 m from O. In addition to the gravitational pull slowing the body it experiences a resistance of $\frac{11}{3}s$ N, where s metres is the distance the body is from O. For the motion from O to B, find (a) the gain in potential energy, (b) the work done against the resistance, (c) the speed of projection.
13. A body of mass 4 kg is released from rest at the top of a slope of length 10 m which is inclined at an angle θ above the horizontal where $\sin\theta = \frac{1}{14}$. The body moves down the slope against a resistance of $\frac{2}{5}s$ N, where s metres is the distance of the body from the top. For the motion of the body from the top to the bottom of the slope, find (a) the loss in potential energy, (b) the work done against the resistance, (c) the final speed of the body.
14. Use the methods of this chapter to prove the result obtained in chapter 15 that the work done in stretching an elastic string from its natural length l to a stretched length $l + x$ is $\dfrac{\lambda x^2}{2l}$ where λ is the modulus of the string.

S.H.M. and variable forces

A body moving with simple harmonic motion gives an example of a variable force acting. Since the acceleration $\ddot{x}$ is $-n^2x$, then the force producing this acceleration is $-mn^2x$, where m is the mass of the body.

Example 9

A body of mass 250 g moves horizontally with S.H.M. about a mean position O. Find the time period of the motion, given that, when the body is 40 cm from O the horizontal force acting on the body is of magnitude 2·5 N.

Since the motion is simple harmonic, $\ddot{x} = -n^2 x$

When the body is 0·4 m from O, it has an acceleration of $0{\cdot}4n^2$ towards O

Using force = mass × acceleration

$$2{\cdot}5 = \frac{250}{1000} \times 0{\cdot}4n^2$$

$$\therefore \quad n = 5$$

using $T = \dfrac{2\pi}{n}$ gives $T = \dfrac{2\pi}{5}$

The period of the motion is $\dfrac{2\pi}{5}$ s.

Springs and S.H.M.

Consider a light spring attached to a fixed point A on a smooth horizontal surface and with a body of mass m attached to its other end B. The body is now moved horizontally and held at rest in a position such that the spring is stretched. If the body is now released, the horizontal force acting on it will be the tension in the spring and this will be variable. When the spring returns to its natural length, the body will be moving and the spring will then start being compressed until the body is eventually brought to rest.

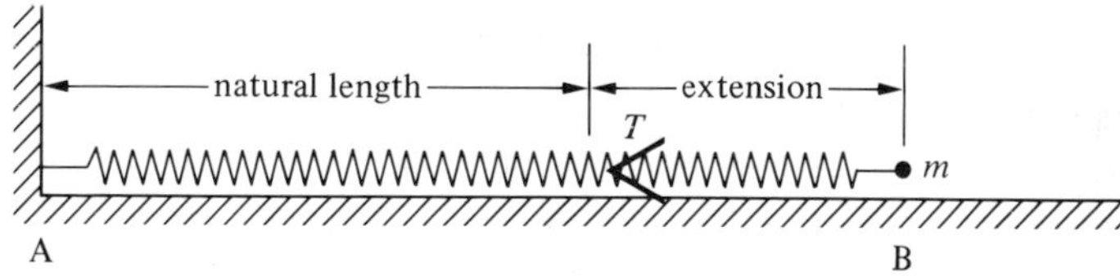

It can be shown that the motion of the body is simple harmonic and therefore the formulae associated with S.H.M. may then be applied.

In Example 9 it was stated that the motion was simple harmonic. Therefore, in addition to using $F = ma$, the S.H.M. equations already obtained in chapter 16 may also be used, i.e.

1. $\ddot{x} = -n^2 x$
2. $v^2 = n^2(a^2 - x^2)$
3. $v_{max} = na$
4. $x = a \sin(nt + \epsilon)$
5. $v = an \cos(nt + \epsilon)$
6. $T = \dfrac{2\pi}{n}$

In some questions, it may *not* be stated that the motion is simple harmonic, and this fact may first have to be proved.

In such cases, it must be first shown that a relationship of the form $\ddot{x} = -n^2 x$ does apply. Having done this and so proved that the motion is simple harmonic, the S.H.M. equations can then be used as previously.

The following points and method of approach should be noted:

1. If the motion is believed to be simple harmonic, consider the force acting on the body when it is at some distance x from what is believed to be the centre of the motion. When force = mass × acceleration is applied, then a relationship of the form $\ddot{x} = -n^2 x$ proves the motion is simple harmonic.
2. If x is not measured from the centre of the motion, an equation of the form
$$\ddot{x} = -n^2 x + c$$
may result. This equation can be rearranged as
$$\ddot{x} = -n^2 (x - \frac{c}{n^2}).$$
Then by the substitution $y = x - \frac{c}{n^2}$ (from which it follows that $\ddot{y} = \ddot{x}$), the equation may then be written as:
$$\ddot{y} = -n^2 y$$
The motion is therefore simple harmonic about the position $y = 0$, i.e. where $x = \frac{c}{n^2}$. The second method of solution of example 12 shows this type of approach.
3. S.H.M. questions often involve springs and elastic strings, in which case Hooke's Law $T = \lambda\frac{x}{l}$ should be applied.
4. When elastic strings are involved, it must be remembered that these go slack under compression and this can then affect the type of motion (see example 12).

Example 10

One end of a light elastic spring of natural length 2 m and modulus 10 N is fixed to a point A on a smooth horizontal surface. A body of mass 200 g is attached to the other end of the spring and is held at rest at a point B on the surface, causing the spring to be extended by 30 cm. Show that, when released, the body will move with S.H.M. and find the amplitude of the motion.

Let O be the end of the spring when it is unstretched.

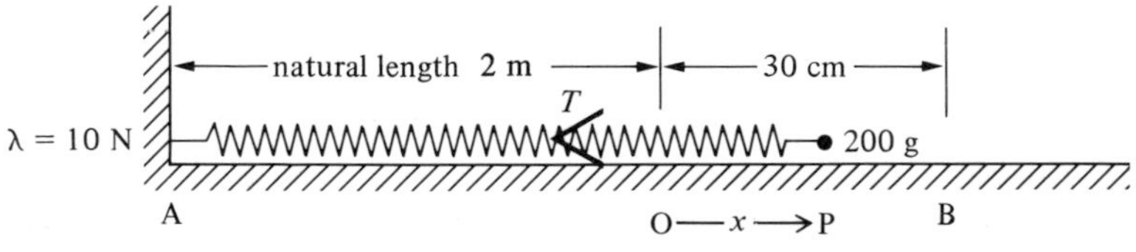

Consider the situation when the body is at a point P and the spring is extended a distance x.

By Hooke's Law $\qquad T = \lambda\frac{x}{l} = 10\frac{x}{2}$

$\therefore \quad T = 5x$

using $F = ma$, $\qquad T = -\frac{200}{1000}\ddot{x}$ (−ve sign since force is in direction of decreasing x)

$$\therefore \quad 5x = -\frac{200}{1000}\ddot{x}$$

$$\text{or} \quad \ddot{x} = -25x$$

This is of the form $\ddot{x} = -n^2x$ and thus the motion is simple harmonic about $x = 0$, i.e. about the point O, and $n = 5$.

During S.H.M. the body is furthest from the mean position when its speed is zero, i.e. at B. Thus OB is the amplitude of the motion.

The amplitude of the motion is 30 cm.

Example 11

A light elastic string of natural length 2·4 m and modulus 15 N is stretched between two points A and B, 3 m apart on a smooth horizontal surface. A body of mass 4 kg attached to the mid-point of the string is pulled 10 cm towards B and then released. Show that the subsequent motion is simple harmonic and find the speed of the body when it is 158 cm from A.

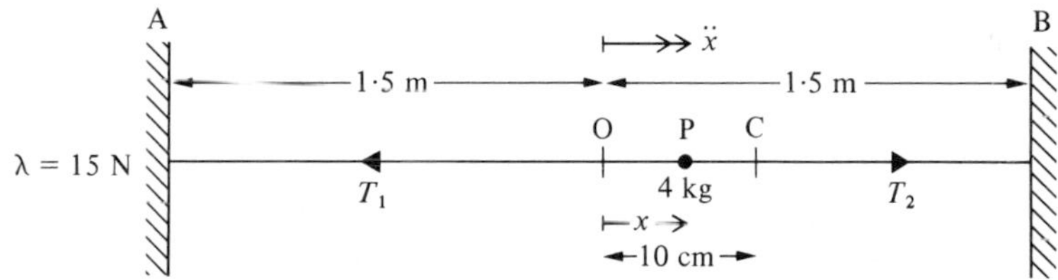

Let O be the centre of AB; C is the point from which the body is released and P is a point such that OP $= x$ metres.
When the body is at P, there are two horizontal forces acting on it: the tensions in the two parts of the string.

Force tending to increase $x = T_2 - T_1$

applying $F = ma$, $\qquad T_2 - T_1 = m\ddot{x} \qquad \ldots [1]$

Considering AP and PB as separate strings of modulus 15 N and natural length 1·2 m, and using Hooke's Law, equation [1] becomes:

$$\lambda\frac{(0{\cdot}3 - x)}{1{\cdot}2} - \lambda\frac{(0{\cdot}3 + x)}{1{\cdot}2} = 4\ddot{x}$$

$$\therefore \quad 0{\cdot}3\lambda - \lambda x - 0{\cdot}3\lambda - \lambda x = 4{\cdot}8\ddot{x}$$

$$\therefore \quad -2\lambda x = 4{\cdot}8\ddot{x}$$

and using $\lambda = 15$ N, $\qquad \ddot{x} = -\frac{25}{4}x$

This is of the form $\ddot{x} = -n^2x$, and thus the motion is simple harmonic about $x = 0$, i.e. about the point O, and $n = \frac{5}{2}$.

Note that, as the body is released from rest at C and it oscillates about a mean position O, then CO is the amplitude of the motion. Thus neither string goes slack at any stage of the motion.

When the body is 158 cm from A, $x = \frac{8}{100}$ m.

Amplitude = CO = 10 cm

Using $v^2 = n^2(a^2 - x^2)$ gives $\quad v^2 = \frac{25}{4}[(\frac{1}{10})^2 - (\frac{8}{100})^2]$

and $\quad v = 0{\cdot}15$ m/s

The motion is simple harmonic, and when the body is 158 cm from A its speed is 0·15 m/s.

Example 12

A light elastic spring, of natural length 50 cm and modulus $20g$ N, hangs vertically with its upper end fixed and a body of mass 6 kg attached to its lower end. The body initially rests in equilibrium and is then pulled down a distance of 25 cm and released.
Show that the ensuing motion will be simple harmonic, and find the period of the motion and the maximum speed of the body.
Would the answers have been the same had an elastic string been used in place of the spring?

Method 1

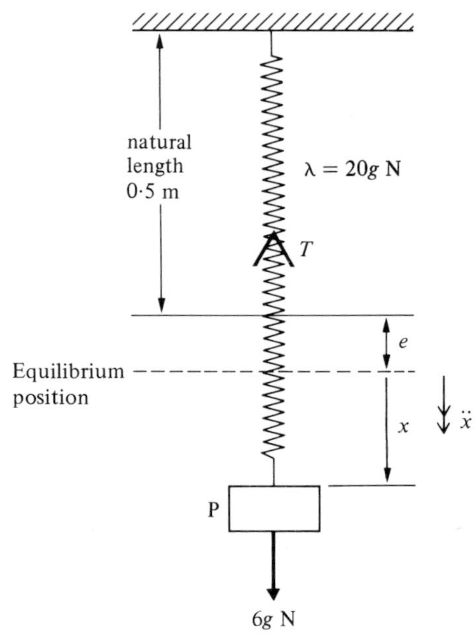

In this method, first find the equilibrium position, which is then expected to be the centre of the oscillation.

Let the extension in the equilibrium position be e

By Hooke's Law $\quad T = \dfrac{\lambda e}{l} \quad \ldots [1]$

In equilibrium $\quad T = 6g \quad \ldots [2]$

from these equations, $\quad \dfrac{\lambda e}{l} = 6g$

substituting $\lambda = 20g$ and $l = 0{\cdot}5$ gives

$e = 15$ cm

Consider the body at a point P, distance x below the equilibrium position.

Applying $F = ma$ gives $\quad 6g - T = 6\ddot{x}$

using Hooke's Law, $\quad 6g - \dfrac{\lambda(x+e)}{l} = 6\ddot{x}$

$$\therefore \quad 6g - 20g\frac{(x + 0{\cdot}15)}{0{\cdot}5} = 6\ddot{x}$$

$$\therefore \quad 6g - 40gx - 6g = 6\ddot{x}$$

or $\quad \ddot{x} = -\frac{20}{3}gx$

Hence the motion is simple harmonic about $x = 0$, the equilibrium position, and

$$n = \sqrt{(\tfrac{20}{3}g)} = \frac{14}{\sqrt{3}}$$

The body is released from a point 25 cm below the equilibrium position, i.e. the body is at rest when 25 cm from the centre of the oscillation; hence the amplitude of the motion is 25 cm.

Using $T = \dfrac{2\pi}{n}$ and $v_{max} = na$ gives $T = \dfrac{\pi\sqrt{3}}{7}$ s and $v_{max} = \dfrac{7}{6}\sqrt{3}$ m/s

Method 2

In this method, take the general position of the body to be when the spring is extended a distance x from its unstretched length.

Applying $F = ma$, $\quad 6g - T = 6\ddot{x}$

using Hooke's Law,

$$6g - \lambda\frac{x}{l} = 6\ddot{x}$$

substituting $\lambda = 20g$ and $l = 0{\cdot}5$ gives

$$\ddot{x} = g - \tfrac{20}{3}gx$$

which can be written $\quad \ddot{x} = -\tfrac{20}{3}g(x - \tfrac{3}{20})$

substituting $y = x - \tfrac{3}{20}$ gives

$$\ddot{y} = -\tfrac{20}{3}gy$$

Hence the motion is simple harmonic about $y = 0$, i.e. about $x = 15$ cm as in method 1.

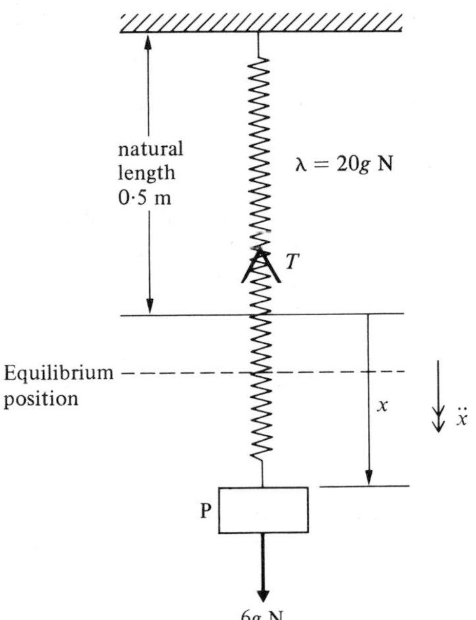

Again $n = \sqrt{(\tfrac{20}{3}g)}$ and the amplitude and time period can be obtained as in method 1.

As will be seen from the diagram on the right, during part of the motion the body will be above the point at which the spring has its natural length, i.e. the spring is being compressed. Thus, were a string to replace the spring, the body would be in free flight for some period of time. The motion would still be periodic but the period of the motion would be different.

The maximum speed occurs when the body passes through the equilibrium position and this would be unchanged if a string replaced the spring.

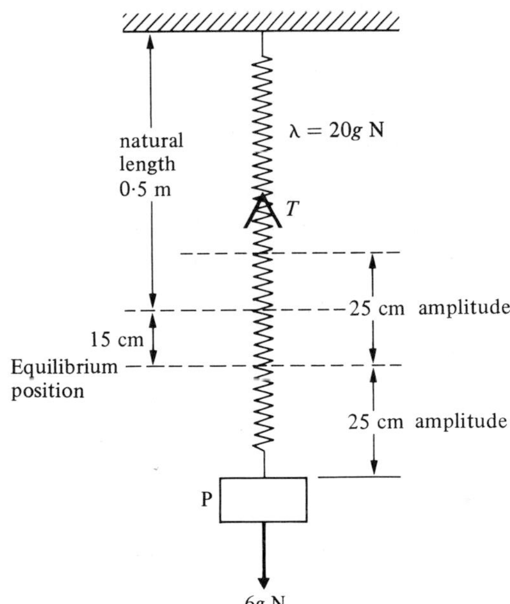

Exercise 17C

1. A body of mass 500 g moves horizontally with S.H.M. of amplitude 1 m and time period $\frac{\pi}{2}$ s. Find the magnitude of the greatest horizontal force experienced by the body during the motion.

2. A body of mass 100 g moves horizontally with S.H.M. about a mean position O. When the body is 50 cm from O the horizontal force on the body is of magnitude 5 N. Find the time period of the motion.
3. A body of mass 5 kg is placed on a rough horizontal surface, coefficient of friction $\frac{11}{49}$. State whether or not the body will slide across the surface when the surface is moved horizontally with simple harmonic motion of amplitude 55 cm and (a) a time period of $\frac{\pi}{2}$ s, (b) a time period of $\frac{3\pi}{2}$ s, (c) a frequency of 15 oscillations per minute, (d) a frequency of 20 oscillations per minute.
4. A horizontal platform is made to move vertically up and down with simple harmonic motion according to the relationship $\frac{d^2x}{dt^2} = -49x$ where x m is the vertical displacement of the platform from its mean position at time t s. Show that any mass placed on the platform will leave the platform if the amplitude of the motion is greater than 20 cm.
5. A light spring is of natural length 1 m and modulus 2 N. One end of the spring is attached to a fixed point A on a smooth horizontal surface and to the other end is attached a body of mass 500 g. The body is held at rest on the surface at a distance of 1·25 m from A. Show that on release the body will move with S.H.M. and find the amplitude and time period of the motion.
6. Two points A and B are 1 m apart on a smooth horizontal table. A light spring, of natural length 75 cm and modulus 54 N, has one end fastened to the table at A and the other end fastened to a body of mass 8 kg which is held at rest at B. Show that when the body is released it moves with S.H.M. and find the maximum speed of the body during the motion.
7. A body of mass 2 kg is fixed to the mid-point of a light elastic string of natural length 1 m and modulus 18 N. The ends of the string are attached to two points A and B 2 m apart on a smooth horizontal surface. The body is pulled a distance y towards A ($y < 50$ cm) and released. Show that the subsequent motion is simple harmonic and find the time period of the motion.
   If the maximum speed of the body is 1·5 m/s, find the value of y.
8. A light elastic string of natural length 1·5 m and modulus 12 N is stretched between two points A and B 2 m apart on a smooth horizontal surface. A body of mass 2 kg is attached to the mid-point of the string, pulled 20 cm towards A and released. Show that the subsequent motion is simple harmonic and find the speed of the body when 88 cm from A.
9. A and B are two fixed points on a smooth horizontal surface with AB = 2 m. A body of mass 4 kg lies on the line AB at a point P and is in equilibrium with a light elastic string of natural length 75 cm and modulus 18 N connecting it to A and a light elastic string of natural length 50 cm and modulus 6 N connecting it to B. Show that P is midway between A and B.
   If the body is then pulled 20 cm towards A and released, show that the subsequent motion is simple harmonic and find the maximum speed of the body during the motion.
10. A light elastic string is of natural length 60 cm and modulus $3mg$ N. The string hangs vertically with its top end fixed and a body of mass m kg fastened to the other end. Find the extension in the string when the body hangs in equilibrium.
    If the body is then pulled vertically downwards a distance of 10 cm and released, show that the ensuing motion will be simple harmonic, and find the time period of the motion and the maximum speed of the body.

11. A body of mass 500 g is attached to end B of a light elastic string AB of natural length 50 cm. The system rests in equilibrium with the string vertical and end A fixed. The body is then pulled vertically downwards a small distance and released. If the ensuing motion is simple harmonic of time period $\frac{1}{5}\pi$ s, find the modulus of the string.
12. A light spring hangs vertically with its top end fixed and a body of mass m kg attached to the other end. The spring is of natural length 1 metre and modulus $5mg$ N and initially the system rests in equilibrium. The body is then pulled vertically downwards a distance of 30 cm and released. Show that the subsequent motion is simple harmonic and find the greatest acceleration experienced by the body. Would the motion have been simple harmonic if an elastic string had been used in place of the spring?
13. A light elastic string of natural length 20 cm and modulus 40 N has one end attached to a fixed point A on a smooth horizontal surface and a body of mass 2 kg attached to the other end. The body is held on the surface at a point which is 40 cm from A, and released. Show that the subsequent motion will be periodic and find the time period of the motion and the speed of the body as it passes through A.
14. A light elastic string of natural length 50 cm hangs vertically with its top end fixed and a body of mass 2 kg attached to the other end. With the body hanging in equilibrium, the string has a total length of 70 cm. Find the modulus of the string.
    If the body is then pulled vertically downwards a distance of 10 cm and released, show that the subsequent motion is simple harmonic and find the speed of the particle when it is 2 cm above the point of release.
15. A light spring of natural length 40 cm and modulus $2g$ N hangs vertically with its upper end fixed and a particle attached to its other end. When the particle hangs in equilibrium the extension of the string is 5 cm. The spring is now replaced by a different spring of natural length 50 cm. The system is again allowed to settle in a position of equilibrium and then the particle is pulled vertically downwards a short distance and released. If the subsequent motion is simple harmonic with time period $\frac{1}{10}\pi$ s find the mass of the particle and the modulus of this second spring.
16. A and B are two points 25 cm apart on a smooth horizontal surface. A particle of mass 500 g lies at A and is connected to B by a light spring of natural length 25 cm and modulus 50 N. If the particle is projected directly towards B with speed 4 m/s, show that the ensuing motion will be simple harmonic and find the time period and amplitude.
17. A light spring of natural length 50 cm and modulus 147 N hangs vertically with its upper end fixed and a body of mass 1·5 kg attached to the lower end. With the system resting in equilibrium, the body is projected vertically downwards with a speed of 1·4 m/s. Show that the resulting motion will be simple harmonic and find the amplitude of the motion.
18. A light elastic string of natural length 1 m and modulus $2mg$ N hangs vertically with its upper end fixed and a body of mass m kg attached to its other end. Find the total length of the string when the mass hangs in equilibrium.
    The body is then pulled vertically downwards a distance d metres and released. Show that the body will move with S.H.M provided $d \leqslant 0{\cdot}5$.
19. A light elastic string hangs vertically with its upper end fixed and a body of mass 1 kg attached to its lower end. The string is of natural length 1 m and modulus 100 N. Find the extension in the string when the body hangs in equilibrium.
    The body is then pulled vertically downwards to a point 35 cm below the equilibrium position and released. Find the speed of the body when the string first goes slack and the greatest height reached by the particle above the equilibrium position (hint: use energy considerations).

20. A light elastic string of natural length 1 m and modulus $10g$ N hangs vertically with its upper end fixed and a body of mass 2 kg attached to its lower end. Show that in equilibrium the string is extended to a length of 1·2 m.
If the body is pulled down a further 25 cm and released, find the speed of the particle at the instant the string first goes slack and the time after release that this occurs.
21. A light elastic string of natural length l and modulus λ has one end attached to a fixed point A on a smooth horizontal surface and a body of mass m attached to the other end. The body is held on the surface at a distance $(l + s)$ from A and released. Show that the subsequent motion will be periodic of time period $\frac{2}{s}(\pi s + 2l)\sqrt{\left(\frac{lm}{\lambda}\right)}$.
22. A, B, C and D are four fixed points on a smooth horizontal surface. ABCD is a straight line with AB = BC = CD = 1 m. A light elastic string of natural length 2 m and modulus 16 N lies on the surface with one end attached to A and the other end attached to B.
A body of mass 3 kg is attached to the mid-point of the string, pulled aside to the point A and released from rest. Find the speed of the body as it reaches points B, C and D (hint: use energy considerations).
Find the time taken for the body to reach the mid-point of AD after its release from A.

Simple pendulum

A simple pendulum of length l makes small oscillations, i.e. it moves through a small angle θ on either side of the vertical through the point of suspension.

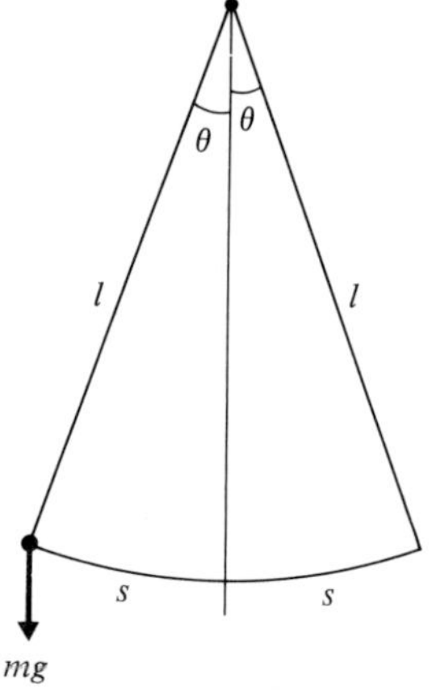

If the length of the arc from one extreme position to the centre of the arc is s,

$$\begin{aligned} \text{then} \quad & s = l\theta \quad \text{where } \theta \text{ is in radians} \\ \therefore \quad & \dot{s} = l\dot{\theta} \\ \text{and} \quad & \ddot{s} = l\ddot{\theta} \end{aligned}$$

The force along the tangent, in the extreme position, is $mg \sin \theta$

Hence, applying $F = ma$,

$$\begin{aligned} & mg \sin \theta = -m\ddot{s} \\ \text{or} \quad & mg \sin \theta = -ml\ddot{\theta} \end{aligned}$$

But for small oscillations, $\sin \theta \approx \theta$

$$\begin{aligned} \therefore \quad & mg\,\theta = -ml\ddot{\theta} \\ \text{or} \quad & \ddot{\theta} = -\frac{g}{l}\theta \qquad \ldots [1] \end{aligned}$$

Hence the motion of a simple pendulum approximates to S.H.M. Compare equation [1] with $\ddot{x} = -n^2 x$ and using $T = \frac{2\pi}{n}$, it is seen that the period T of the pendulum is $T = 2\pi\sqrt{\left(\frac{l}{g}\right)}$.

Exercise 17D

1. (a) If a simple pendulum of length l makes small oscillations, show that the motion approximates to simple harmonic motion of time period $2\pi\sqrt{\left(\frac{l}{g}\right)}$.
   (b) Find the time period if (i) $l = 80$ cm, (ii) $l = 1{\cdot}25$ m, (iii) $l = 50$ cm.
   (c) Find l if the time period is (i) $\frac{2\pi}{7}$ s, (ii) $\frac{\pi}{\sqrt{5}}$ s, (iii) 1 s.
2. Find the time period of a simple pendulum of length 2 m performing small oscillations. What is the increase in the time period when the length of the pendulum is increased by 10%?
3. A simple pendulum performs small oscillations with time period $2T$. By what percentage should the string be shortened for the time period to be T?
4. A simple pendulum performs small oscillations with time period $5T$. By what percentage should the string be shortened for the time period to be $4T$?

Exercise 17E Examination questions

1. At time t the position vector of a particle of mass m is $t^2\mathbf{i} + (\sin t)\mathbf{j}$. Find the resultant force acting on the particle when $t = \pi/2$. (London)

2. The force acting at time t $(0 \leqslant t < 2)$ on a particle of unit mass is $24t^2\mathbf{i} + 6\mathbf{j}$. At time $t = 0$ the particle is at rest at the point with position vector $-2\mathbf{i} + 3\mathbf{j}$. Find the position vector of the particle at time $t = T$ $(0 \leqslant T < 2)$.
   For time $t \geqslant 2$ the force acting on the particle is $6\mathbf{j}$. Find the position vector of the particle at time $t = 3$. (London)

3. A body of mass m is released from rest at a point O. It falls under gravity against a resistance to motion of kv^2, where v is the speed of the body when it is a distance s from O and k is a positive constant. Show that
   (a) $s = \frac{m}{2k}\ln\left(\frac{mg}{mg - kv^2}\right)$, (b) as s increases, v tends to a maximum value and find this value in terms of m, g and k.

4. A particle falls vertically from rest under gravity in a medium exerting a resistance proportional to its speed. Show that the particle has a terminal speed V (i.e. a limiting value to which its speed tends as the time increases). Find the time taken for the particle to reach a speed $\frac{2}{3}V$. (Oxford)

5. A particle of mass m, subject to a resistance mk times the square of its speed, is projected vertically downwards with speed w, where $kw^2 < g$. Find the speed of the particle when it has descended a distance x. (J.M.B)

6. With the usual notation, prove that $\frac{dv}{dt} = v\frac{dv}{ds}$.

A particle P of mass m moves in a straight line and starts from a point O with velocity u. When OP $= x$, where $x \geqslant 0$, the velocity v of P is given by $v = u + \frac{x}{T}$, where T is a positive constant. Show that, at any instant, the force acting on P is proportional to v.
Given that the velocity of P at the point A is $3u$, calculate
(i) the distance OA in terms of u and T,
(ii) the time, in terms of T, taken by P to move from O to A,
(iii) the work done, in terms of m and u, in moving P from O to A. (A.E.B)

18
Jointed rods and frameworks

Jointed rods in equilibrium

When two rods are smoothly jointed together, each rod will exert a force on the other at the joint. If the joint is in equilibrium, with no external forces acting, the forces on the rods must be equal and opposite to each other. These forces may, in general, be in any direction. To find these forces it is convenient to 'separate' the rods at the joints and suppose that these forces have horizontal and vertical components, X and Y:

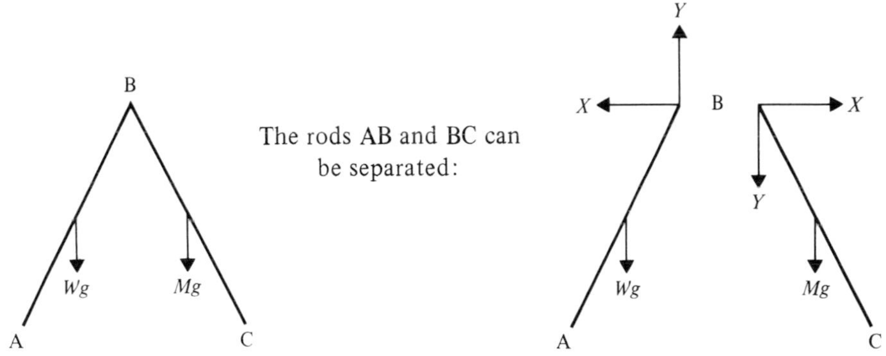

Since the joint B is in equilibrium, the forces exerted on rod AB by BC, must be equal and opposite to those exerted on rod BC by rod AB.

It is sometimes possible to evaluate some of these forces by symmetry considerations. In the above diagram, if the system is symmetrical, i.e. AB = BC in length and mass, then there must still be this symmetry after 'separation'. Thus $Y = 0$.

In addition to considering the equilibrium of each rod separately, the equilibrium of the whole system may also be considered.

Example 1

The diagram shows two uniform rods AB and BC, of equal length, smoothly jointed together at B. The mass of AB is 6 kg and the mass of BC is 8 kg. The end C is freely hinged to a rough horizontal surface and end A rests on the same surface, coefficient of friction μ. The points A, B and C lie in the same vertical plane and angle BAC = 30°.

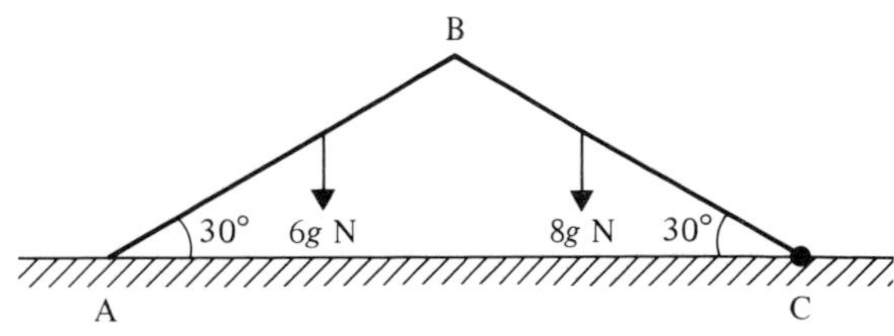

If A is on the point of slipping, find

(a) the horizontal and vertical components of the reaction at B,
(b) the horizontal and vertical components of the reaction at C,
(c) the value of μ.

First, draw the 'separated' diagram:

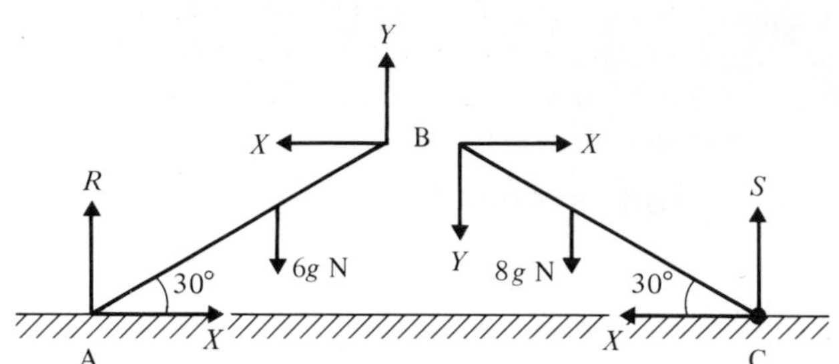

Note that:

(a) the forces on the ends of the rods at B are shown to be equal and opposite, since the joint is in equilibrium,
(b) the horizontal forces at A and C must be X as shown, so as to ensure horizontal equilibrium for each rod considered separately,
(c) as the system is not symmetrical (the masses of the rods are unequal) it cannot be assumed that $Y = 0$.

For the whole system, taking AB = BC = $2l$

$\overset{\frown}{\text{C}}$
$$R(4l \cos 30) = 6g(3l \cos 30) + 8g(l \cos 30)$$
$$\therefore \quad R = \tfrac{13}{2}g \text{ N}$$

Resolving vertically for the whole system

$$R + S = 6g + 8g$$
$$\therefore \quad S = \tfrac{15}{2}g \text{ N}$$

For rod AB

$\overset{\frown}{\text{B}}$
$$R(2l \cos 30) = 6g(l \cos 30) + X(2l \sin 30)$$

substituting for R, $\quad X = \tfrac{7}{2}g\sqrt{3}$ N

Resolving vertically for rod AB

$$Y + R = 6g$$
$$\therefore \quad Y = -\frac{g}{2} \text{ N}$$

(negative sign indicates that the force is in the opposite direction to that shown on diagram).

The horizontal and vertical components at B are $\tfrac{7}{2}g\sqrt{3}$ N and $\frac{g}{2}$ N respectively.

The horizontal and vertical components at C are $\tfrac{7}{2}g\sqrt{3}$ N and $\tfrac{15}{2}g$ N respectively.

If A is on the point of slipping

$$\mu = \frac{X}{R} = \tfrac{7}{13}\sqrt{3}$$

The value of μ is $\tfrac{7}{13}\sqrt{3}$.

Example 2

The diagram shows two identical uniform rods AB and AC each of mass 4 kg. The third uniform rod BC is of mass 6 kg and angle CAB is 90°. The rods are smoothly jointed to form a framework which hangs in equilibrium in a vertical plane, freely suspended by a string attached at A. Find
(a) the tension in the string,
(b) the horizontal and vertical components of the reaction at C.

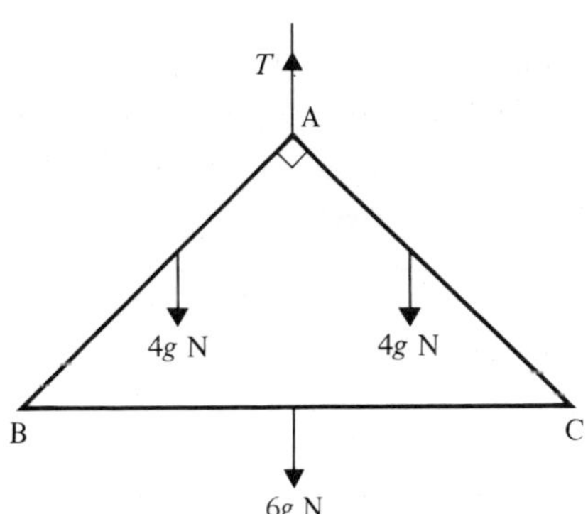

First, draw the 'separated' diagram:

Note that, with the external force T acting at A, the vertical forces of reaction on each rod need not be equal and opposite. Symmetry consideration enables them to be as shown and

$$T = 2Y \qquad \ldots [1]$$

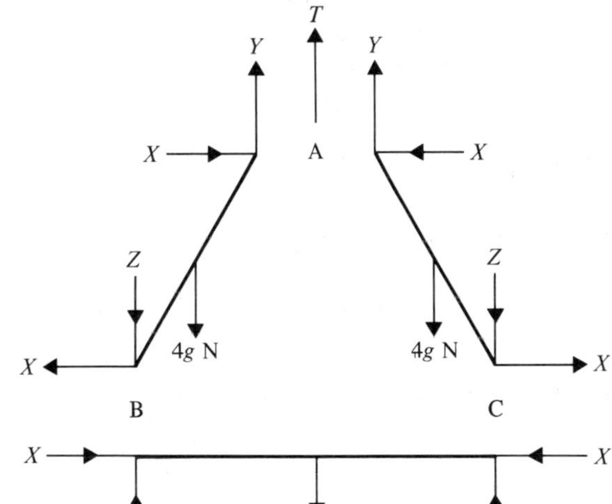

(a) For the whole framework

resolving vertically

$$T = 4g + 4g + 6g$$
$$\therefore \quad T = 14g$$

The tension in the string is $14g$ N.

(b) From equation [1],

$$Y = \frac{T}{2} = 7g \text{ N}$$

For rod AC, taking AC = AB = $2l$

$$\overset{\curvearrowright}{\mathrm{C}} \qquad X(2l \sin 45) + 4g(l \cos 45) = Y(2l \cos 45)$$
$$\text{or} \quad X + 2g = Y$$

substituting for Y,
$$X + 2g = 7g$$
$$\therefore \quad X = 5g \text{ N}$$

resolving vertically for rod AC,
$$Y = 4g + Z$$
$$\text{hence} \quad Z = 3g \text{ N}$$

The horizontal and vertical components of the reaction at C are $5g$ N and $3g$ N respectively.

Exercise 18A

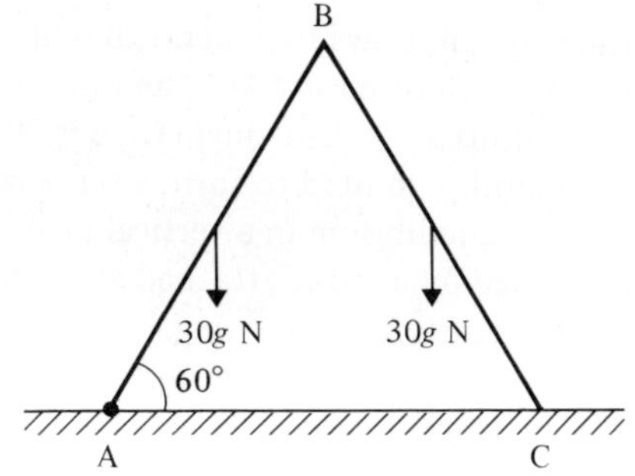

1. The diagram shows two identical uniform rods AB and BC smoothly jointed together at B. End A is freely hinged to a rough horizontal surface and end C stands on the same surface, coefficient of friction μ. A, B and C all lie in the same vertical plane. Each rod is of mass 30 kg and angle BAC = 60°. If C is just on the point of slipping, find

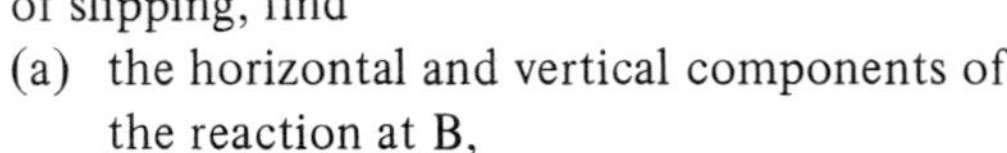

   (a) the horizontal and vertical components of the reaction at B,
   (b) the horizontal and vertical components of the reaction at A,
   (c) the value of μ.

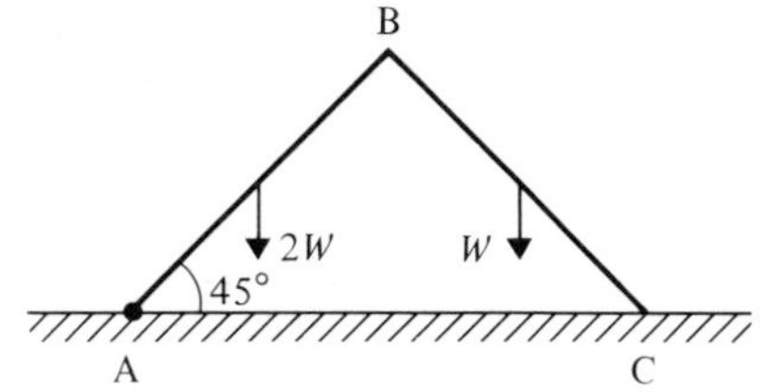

2. The diagram shows two uniform rods AB and BC of equal length, smoothly jointed together at B. End A is freely hinged to a rough horizontal surface and C stands on the same surface, coefficient of friction μ; A, B and C all lie in the same vertical plane. AB and BC have weights of $2W$ and W respectively and angle BAC = 45°. Show that slipping will not occur provided $\mu \geqslant 0{\cdot}6$.
   Find the horizontal and vertical components of the reaction at B.

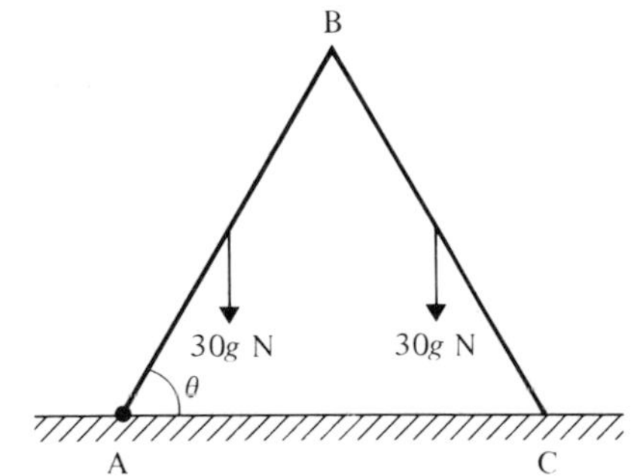

3. The diagram shows two identical uniform rods AB and BC, each of mass 30 kg and lying in the same vertical plane. The rods are smoothly jointed together at B and ends A and C rest on rough horizontal ground, coefficient of friction μ. If angle BAC = θ, where $\tan\theta = \frac{3}{2}$, and the rods are just on the point of slipping, find the value of μ and the magnitude of the reaction at B.

4. Two identical uniform heavy rods, AB and BC, are smoothly jointed together at B and have ends A and C resting on rough horizontal ground, coefficient of friction μ. The points A, B and C lie in the same vertical plane and angle BAC = α. Show that slipping will not occur provided $\mu \geqslant \frac{1}{2}\cot\alpha$.

5. The diagram shows a framework of three rods AB, BC and AC each of weight W, smoothly jointed together at their ends and lying in the same vertical plane. The framework hangs in equilibrium, suspended by a string attached to the mid-point of AB. The rods AB, BC and AC have lengths $l\sqrt{2}$, l and l respectively. Find
   (a) the tension in the string,
   (b) the magnitude of the reaction at C,
   (c) the magnitude of the reaction at B.

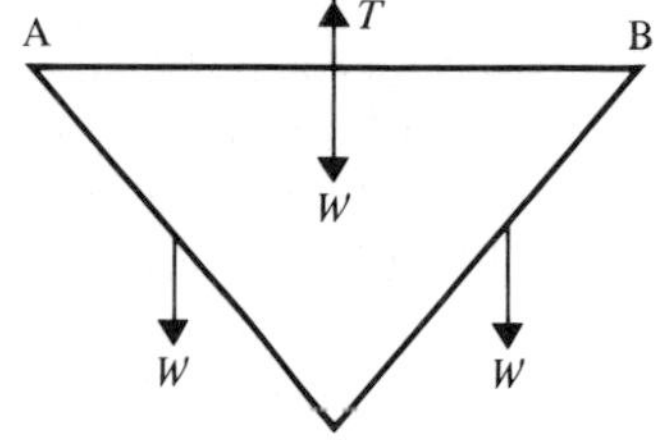

6. The diagram shows three identical uniform rods AB, BC and CA smoothly jointed together at A, B and C and lying in the same vertical plane. The framework is suspended by a string attached to the mid-point of AB. The weight of each rod is W. With the system in equilibrium, find
   (a) the tension in the string,
   (b) the magnitude of the reaction at C,
   (c) the magnitude of the reaction at B.

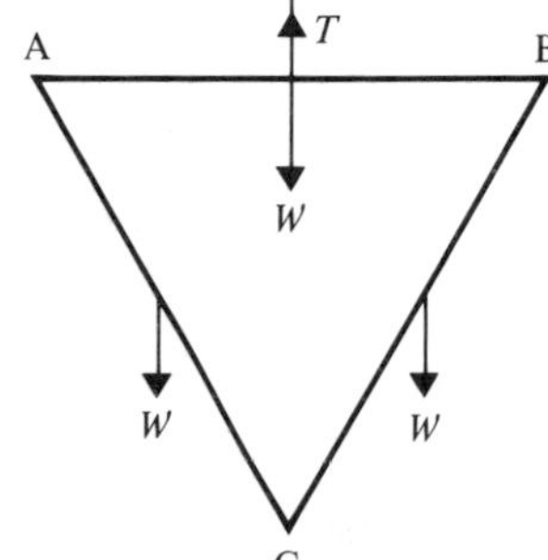

7. Two rods AB and BC, of equal length and weights of $5W$ and W respectively, are smoothly jointed together at B. Ends A and C rest on rough horizontal ground and B is vertically above AC. The coefficient of friction between the ground and the rods is $\frac{3}{8}$ and angle BAC $= \alpha$. Show that neither rod will slip provided $\tan\alpha \geqslant 2$.
   With $\tan\alpha = 2$, find the magnitude of the reaction at B.

8. Two uniform heavy rods AB and BC, each of length l, lie in the same vertical plane and are smoothly jointed together at B. Ends A and C rest on rough horizontal ground and angle BAC $= \alpha$. The rod AB is twice as heavy as rod BC. The coefficients of friction at A and C are μ_1 and μ_2 respectively. Show that neither rod will slip provided

$$\mu_1 \geqslant \frac{3}{7\tan\alpha} \quad \text{and} \quad \mu_2 \geqslant \frac{3}{5\tan\alpha}$$

   Suppose now, with these conditions fulfilled, $\mu_1 = \mu_2$ and α is gradually decreased (both A and C remaining in contact with the ground). Show that slipping will tend to occur at C before it does at A.

9. The diagram shows three identical uniform rods AB, BC and AC, each of weight W. The rods are smoothly jointed at their ends to form a framework which hangs in equilibrium in a vertical plane, freely suspended by a string attached at A. Find
   (a) the tension in the string,
   (b) the horizontal and vertical components of the reaction at B.

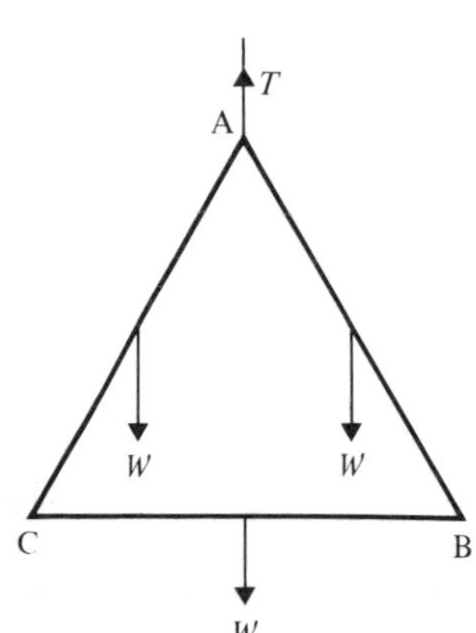

10. The diagram shows four identical uniform rods each of weight W and length l, freely hinged together at their ends. A light inextensible string of length l connects A to C and the framework hangs in equilibrium in a vertical plane, with A freely hinged to a horizontal ceiling. Find the tension T in the string, and the horizontal and vertical components of the reaction at B.

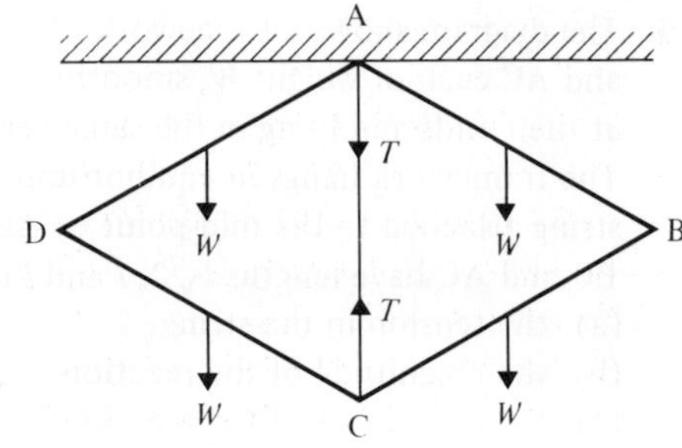

Light rods

A framework may consist of a number of *light* rods, i.e. ones for which the weight may be ignored.

Consider two such rods AB and BC, freely hinged together at B:

We can 'separate' each rod, and show the reaction on each, due to the other, at B as a single force R:

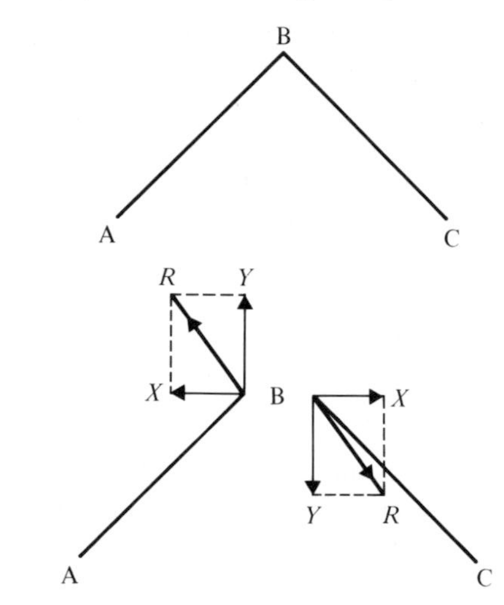

As the rod AB has no weight, it can be seen that for equilibrium R must have no turning effect about A, i.e. R must act along AB. Thus all the reaction forces between light rods will act *along* the rods.

A particular light rod AB may be acting as a *tie* between the points A and B, i.e. preventing A and B from moving apart, and the rod is then in *tension.*

Alternatively, the rod AB may be acting as a *strut*, i.e. preventing points A and B moving closer together, and the rod is then exerting a *thrust.*

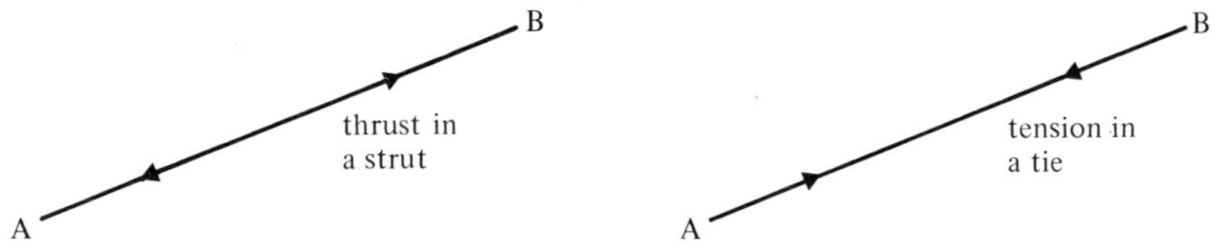

A light rod which is acting as a *tie* could be replaced by a string, but a *strut* could not.

For a framework in equilibrium, equations can be formed by:
(i) taking moments about a convenient point, for the whole system,
(ii) resolving horizontally and vertically for the whole system,
(iii) resolving horizontally and vertically at the ends of each rod.

In addition the forces in some rods may be determined by considering the symmetry of the framework.

Example 3

The diagram shows a framework of four light rods smoothly jointed together, freely hinged at A and B to a vertical wall, and carrying a load of 100 N at D. BC and AD are horizontal, $C\hat{A}D = C\hat{D}A = 30°$ and $CD = l$.

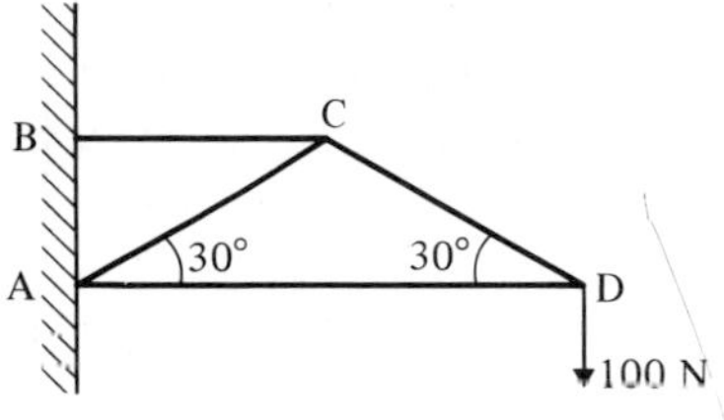

Find (a) the force of reaction exerted on the rods at A,
(b) the force of reaction exerted on the rods at B,
(c) the stresses in each rod stating whether in tension or in thrust.

First draw a diagram showing each rod in tension. If any of these tensions are found to have a negative value, then this will imply there is a thrust in that rod. The horizontal and vertical components of the reactions on the framework at A and B are also shown. Note that (i) the vertical component at B must be zero for vertical equilibrium at B, (ii) the horizontal component at A must be equal and opposite to that at B for horizontal equilibrium of the whole framework.

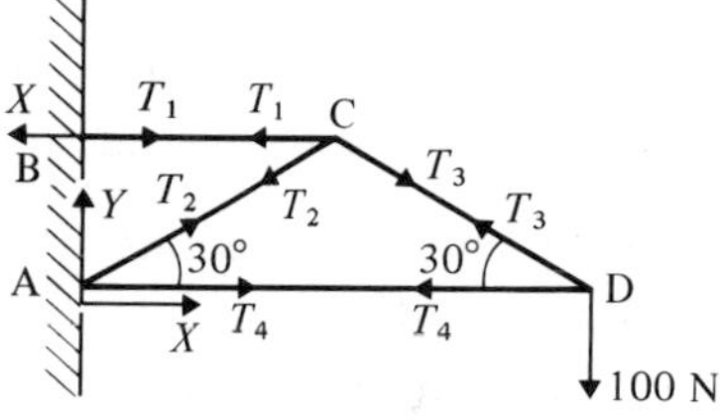

For the whole system: resolving vertically $Y = 100$ N

$$\overset{\frown}{A} \quad Xl \sin 30 = 100 \times 2 \times l \cos 30$$
$$\therefore \quad X = 200\sqrt{3} \text{ N}$$

The reaction at A is $\sqrt{(X^2 + Y^2)}$ at $\tan^{-1} \dfrac{Y}{X}$ to AD i.e. $100\sqrt{13}$ N at $\tan^{-1} \dfrac{\sqrt{3}}{6}$ to AD

At B, resolving horizontally $\quad T_1 = X = 200\sqrt{3}$ N

At A, resolving vertically
$$T_2 \sin 30 + Y = 0$$
$$\therefore \quad T_2 = -200 \text{ N}$$

resolving horizontally
$$T_2 \cos 30 + T_4 + X = 0$$
$$\therefore \quad T_4 = -100\sqrt{3} \text{ N}$$

At D, resolving vertically
$$T_3 \sin 30 = 100$$
$$\therefore \quad T_3 = 200 \text{ N}$$

(a) The force of reaction at A is $100\sqrt{13}$ N at $\tan^{-1} \dfrac{\sqrt{3}}{6}$ to AD

(b) The force of reaction at B is $200\sqrt{3}$ N in direction CB

(c) BC is in tension of $200\sqrt{3}$ N   CD is in tension of 200 N
CA is in thrust of 200 N   AD is in thrust of $100\sqrt{3}$ N

Example 4

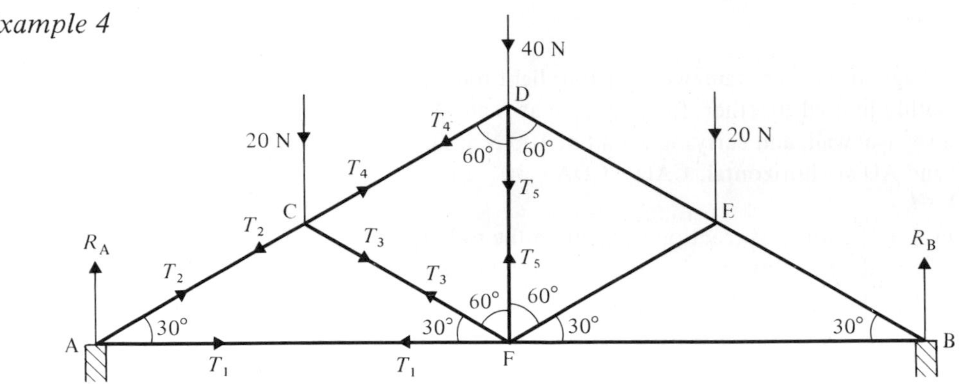

The framework shown is smoothly supported at A and B and carries loads of 20 N, 40 N and 20 N at C, D and E respectively. AB is horizontal and the other angles are as shown. Find the forces of reaction on the rods at A and at B, and the stresses in each rod.

Let the tensions in the rods be as shown (as the framework is symmetrical, the forces in some rods can be determined by symmetry).

For the whole framework,

resolving vertically, $R_A + R_B = 20 + 40 + 20$

by symmetry, $R_A = R_B$

$\therefore\ R_A = R_B = 40$ N

(Alternatively, R_A and R_B are found by taking moments about A or B for the whole system)

At A, resolving horizontally, $T_2 \cos 30 + T_1 = 0$ ... [1]

At A, resolving vertically, $T_2 \cos 60 + R_A = 0$ ... [2]

hence $T_2 = -80$ N and $T_1 = 40\sqrt{3}$ N

At C, resolving horizontally, $T_4 \cos 30 + T_3 \cos 30 = T_2 \cos 30$
or $T_4 + T_3 = T_2$ ... [3]

At C, resolving vertically, $20 + T_3 \cos 60 + T_2 \cos 60 = T_4 \cos 60$
or $40 + T_3 + T_2 = T_4$ ... [4]

solving equations [3] and [4], using the values of T_2 and T_1 gives

$T_3 = -20$ N and $T_4 = -60$ N

At D, resolving vertically, and remembering that by symmetry the stress in CD will be the same as that in DE,

$$40 + T_5 + T_4 \cos 60 + T_4 \cos 60 = 0$$

which by substitution gives $T_5 = 20$ N

The reactions at A and B are both 40 N.

The stresses in the rods are as follows.

In AF and FB a tension of $40\sqrt{3}$ N
In DF a tension of 20 N
In AC and BE a thrust of 80 N
In CD and ED a thrust of 60 N
In CF and EF a thrust of 20 N.

Exercise 18B

Each question of this exercise involves light rods lying in the same vertical plane. In each case find (a) the force of reaction exerted on the rods by the wall (or support) at A,
(b) the force of reaction exerted on the rods by the wall (or support) at B,
(c) the stresses in each rod stating whether in tension or thrust.

1. The diagram shows a framework of three identical light rods smoothly jointed together and resting on smooth supports at A and B.
   AB is horizontal and a load of W N is suspended from C.

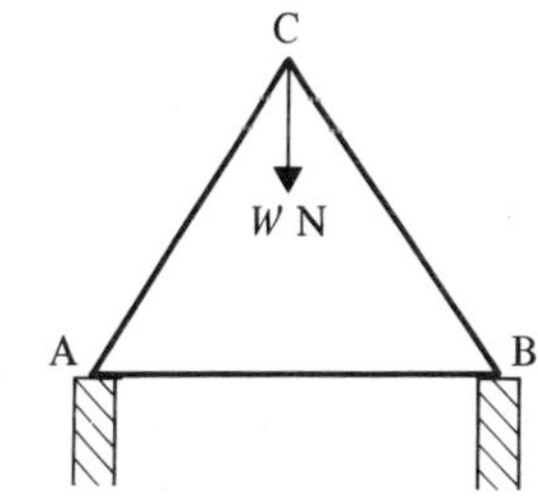

2. The diagram shows a framework of four light rods smoothly jointed together, freely hinged at A and B to a vertical wall, and carrying a load of 100 N at D.
   $C\hat{A}D = C\hat{D}A = 45°$; BC and AD are horizontal.

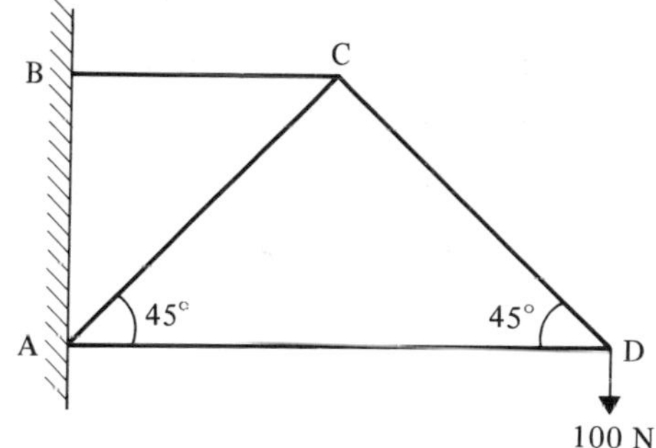

3. The diagram shows a framework of four light rods smoothly jointed together, freely hinged at A and B to a vertical wall, and carrying a load of $500\sqrt{3}$ N at D.
   AC = CD = AD = 2BC and AD and BC are horizontal.

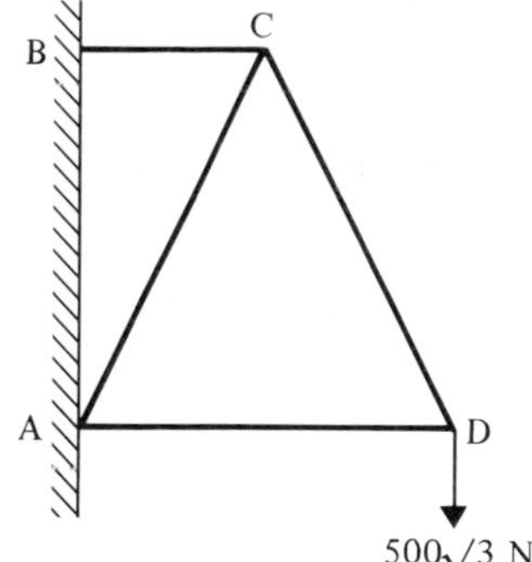

4. The diagram shows a framework of seven identical light rods smoothly jointed together and resting on smooth supports at A and B;
   AB is horizontal and loads of W N, W N and $4W$ N are suspended from D, C and E respectively.

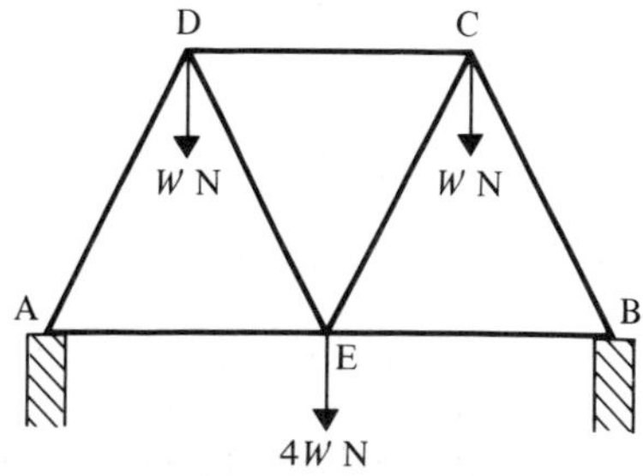

5. The diagram shows a framework of five identical light rods smoothly jointed together. The framework is freely hinged at A to a vertical wall, rests on a smooth support at B, and carries a load of W N at C; BC is horizontal.

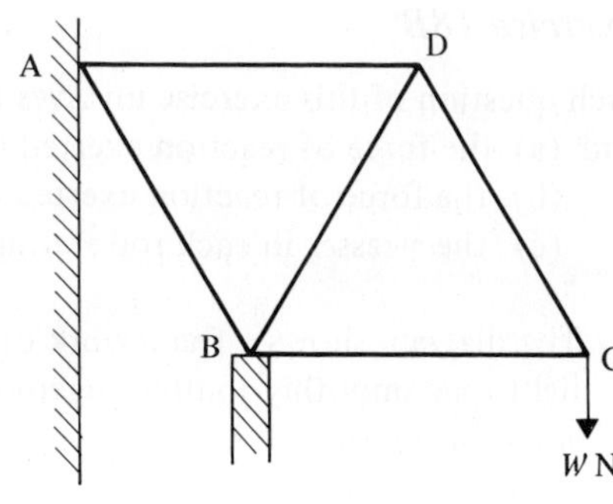

6. The diagram shows a framework of five light rods smoothly jointed together; ABCD is a parallelogram, AB is horizontal and angle DAB $= 60°$. The framework is freely hinged to a vertical wall at A, rests on a smooth support at B, and carries a load of 3 kN at C.

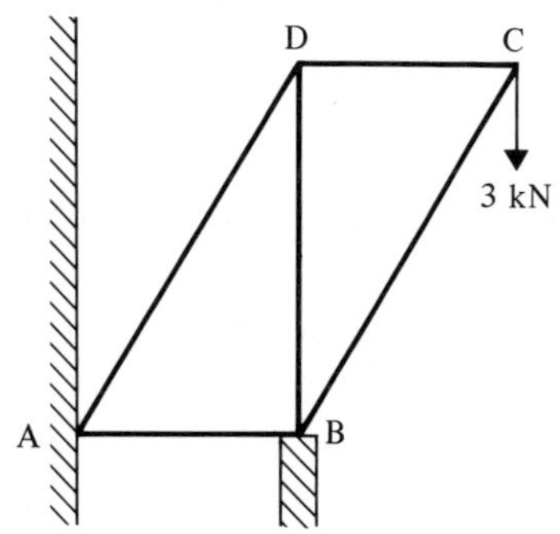

7. The diagram shows a framework of seven identical rods smoothly jointed together and resting on smooth supports at A and B; AB is horizontal and loads of $6W$ N, $\frac{9}{2}W$ N and $3W$ N are suspended from D, E and C respectively.

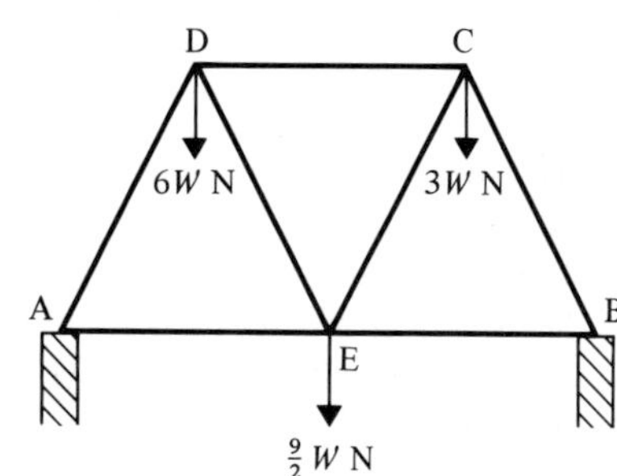

8. The diagram shows a framework of nine identical light rods smoothly jointed together. The framework is freely hinged at A to a vertical wall, rests on a smooth support at B, and carries a load of $2\sqrt{3}$ kN at D; AB is horizontal.

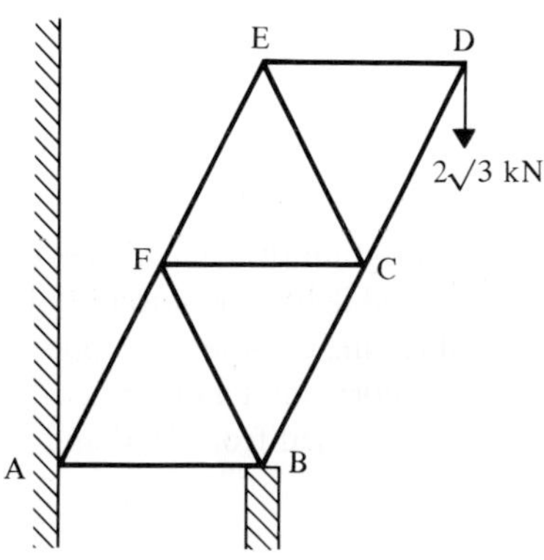

Exercise 18C Examination questions

1. The rigid framework OABC consists of five equal light rods as indicated in Fig. 1. It is freely hinged to a fixed point at O, and a mass of 1 kg is suspended from B. The system is kept in equilibrium in a vertical plane, with the rods OA and BC horizontal, by a light horizontal string connected to the framework at C. Find the tension P in the string and the forces in the rods, in newtons, and prove that the reaction of the hinge on the framework makes an angle $\frac{1}{6}\pi$ with the horizontal. (Take g, the acceleration due to gravity, to be 10 m s^{-2}.) (Oxford)

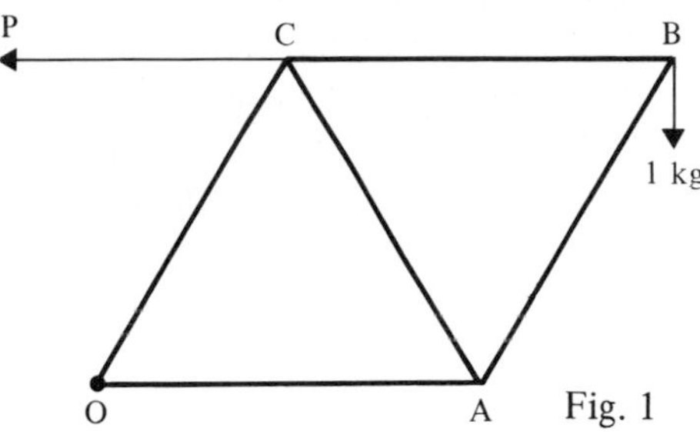

Fig. 1

2. Two straight uniform rods AB and BC, each of length $2a$ and weight W, are smoothly hinged together at B and are in equilibrium with ABC in the same horizontal line. The rod AB is simply supported at the point X in AB, where BX $= x$, and the rod BC is simply supported at the point Y in BC, where BY $= y$.
   (i) By considering the equilibrium of the system, show that the reactions at X and Y are $\dfrac{2yW}{x+y}$ and $\dfrac{2xW}{x+y}$ respectively.
   (ii) By considering the equilibrium of each rod separately, show that
   (a) if $x > a$, then $y < a$, (b) $2xy = a(x+y)$, (c) the mutual reaction between the rods at B has magnitude $\dfrac{W(x-y)}{x+y}$.
   (iii) Find the value of y when $x = 2a$. (A.E.B)

3. Fig. 2 shows a smoothly-jointed framework consisting of 7 light rigid rods; AB = BD = 2 m and BC = 1 m. The rods AB and BC are horizontal and the angles BED and DBC are right angles. The framework is simply supported at A and B and is in equilibrium in a vertical plane when loads of 300 N, 400 N and 100 N are attached at E, D and C respectively.
   (i) Find the magnitudes of the external forces supporting the framework at A and B.

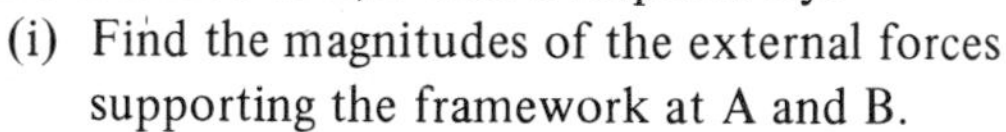

   (ii) Find the magnitude of the force acting in each member of the framework.

   State which rods are in tension.

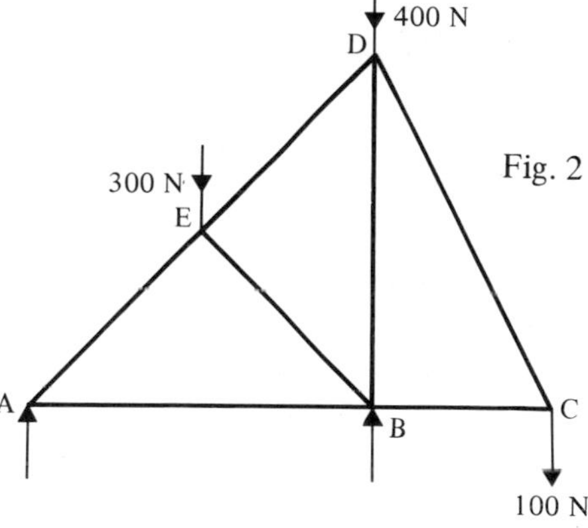

Fig. 2

(A.E.B)

4. Fig. 3 shows a framework of seven freely jointed light rods, AB, BD being horizontal and BE being vertical. The framework is freely hinged at A to a fixed smooth inclined plane of slope 30°. It carries loads 600 N, 150 N at C, D respectively, and is in equilibrium in a vertical plane with E resting on the slope.
Find the magnitude of the reaction at E, and the magnitude and direction of the reaction at A.
Find which rod bears the largest tension, and which bears the largest compression. State the magnitude of each of these two stresses.

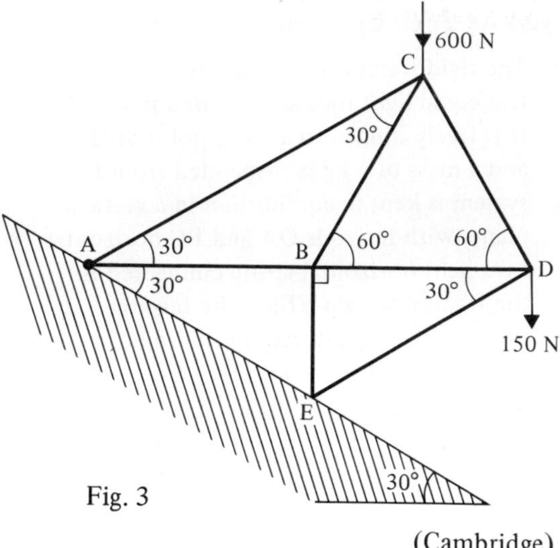

Fig. 3

(Cambridge)

5. Fig. 4 shows eleven equal light rods, freely jointed to form a framework. It is loaded with weights of 16 kN, 12 kN and 8 kN at the points G, F and E. It is placed on supports at A and D.

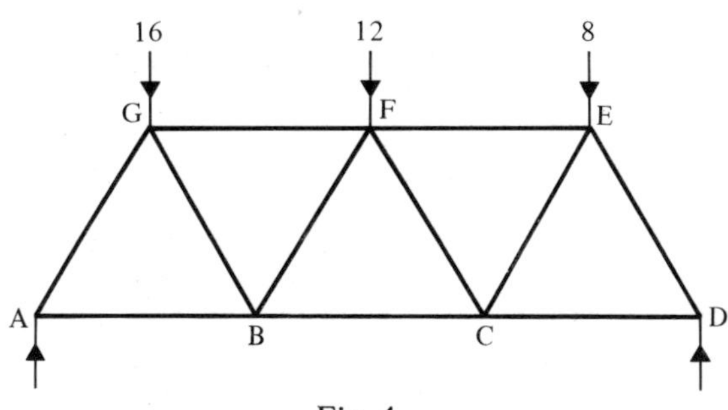

Fig. 4

(i) Find the supporting forces at A and D.
(ii) Find, by calculation or scale drawing, the nature and the magnitude of the stresses in the rods BG, CF and EF.

(S.U.J.B)

Answers

Exercise 1A page 2

1. 2 km west
2. 50 km, S36·9° E
3. 6·5 km, 247·4°
4. 7·81 km, N56·3° E, yes
5. 700 m, 101·8°
6. 65·7 km, 328·2°
7. N 78·7° W, S 11·3° E
8. 6·8 km, N 77° E

Exercise 1B page 5

1. 6·77 units, 083·7°
2. 10·8 units, 249·1°
3. 8·06 units, 100·3°
4. 5·74 units, 299·1°
5. 11·2 units, 083°
6. 4·1 units, 112°
7. (a) $\mathbf{b}$ (b) $-\mathbf{b}$ (c) $\mathbf{a}$ (d) $-\mathbf{a}$ (e) $\mathbf{a}+\mathbf{b}$ (f) $-\mathbf{a}-\mathbf{b}$ (g) $\mathbf{b}-\mathbf{a}$ (h) $\mathbf{a}-\mathbf{b}$
8. (a) $\mathbf{b}-\mathbf{a}$ (b) $\mathbf{a}-\mathbf{b}$ (c) $\frac{1}{2}(\mathbf{b}-\mathbf{a})$ (d) $\frac{1}{2}(\mathbf{a}+\mathbf{b})$
9. (a) $\mathbf{b}$ (b) $\mathbf{a}+\mathbf{b}$ (c) $\mathbf{b}-\mathbf{a}$
10. (a) $\mathbf{b}-\mathbf{a}$ (b) $\frac{2}{3}(\mathbf{b}-\mathbf{a})$ (c) $\frac{1}{3}(\mathbf{a}-\mathbf{b})$ (d) $\frac{1}{3}(\mathbf{a}+2\mathbf{b})$
11. (a) $\frac{1}{3}\mathbf{c}$ (b) $\frac{1}{4}\mathbf{a}$ (c) $\mathbf{a}+\frac{1}{3}\mathbf{c}$ (d) $\mathbf{c}+\frac{1}{4}\mathbf{a}$ (e) $\mathbf{c}-\frac{3}{4}\mathbf{a}$ (f) $\frac{2}{3}\mathbf{c}-\frac{3}{4}\mathbf{a}$
12. (a) $\frac{1}{2}\mathbf{a}$ (b) $\mathbf{b}-\mathbf{a}$ (c) $\frac{3}{4}(\mathbf{b}-\mathbf{a})$ (d) $\frac{1}{4}(\mathbf{a}+3\mathbf{b})$ (e) $\mathbf{b}-\frac{1}{2}\mathbf{a}$ (f) $\frac{1}{4}(3\mathbf{b}-\mathbf{a})$

Exercise 1C page 8

1. (a) (i) $4\mathbf{i}+3\mathbf{j}$ (ii) 5 units (iii) 36·9° (b) (i) $4\mathbf{j}$ (ii) 4 units (iii) 90°
   (c) (i) $\mathbf{i}+4\mathbf{j}$ (ii) $\sqrt{17}$ units (iii) 76·0° (d) (i) $5\mathbf{i}+5\mathbf{j}$ (ii) $5\sqrt{2}$ units (iii) 45°
   (e) (i) $-3\mathbf{i}+3\mathbf{j}$ (ii) $3\sqrt{2}$ units (iii) 135° (f) (i) $-3\mathbf{i}-\mathbf{j}$ (ii) $\sqrt{10}$ units (iii) 198·4°
   (g) (i) $-2\mathbf{i}-3\mathbf{j}$ (ii) $\sqrt{13}$ units (iii) 236·3° (h) (i) $3\mathbf{i}-2\mathbf{j}$ (ii) $\sqrt{13}$ units (iii) 326·3°
   (i) (i) $-3\mathbf{j}$ (ii) 3 units (iii) 270°
2. $\mathbf{a}=4\mathbf{i}$, $\mathbf{b}=-7\mathbf{j}$, $\mathbf{c}=5\mathbf{i}+5\mathbf{j}$, $\mathbf{d}=5\sqrt{3}\mathbf{i}+5\mathbf{j}$, $\mathbf{e}=-3\sqrt{3}\mathbf{i}-3\mathbf{j}$, $\mathbf{f}=-4{\cdot}23\mathbf{i}+9{\cdot}06\mathbf{j}$
3. (a) $10\mathbf{i}+28\mathbf{j}$ (b) $15\mathbf{i}+9\mathbf{j}$ (c) $6\mathbf{i}-14\mathbf{j}$ (d) 5 units
   (e) 25 units (f) 17 units (g) $\sqrt{26}$ units (h) $9\mathbf{i}+12\mathbf{j}$
   (i) $\frac{7}{2}\mathbf{i}+12\mathbf{j}$ (j) $9\mathbf{i}+12\mathbf{j}$ (k) $28\mathbf{i}+96\mathbf{j}$ (l) $-4\sqrt{5}\mathbf{i}+2\sqrt{5}\mathbf{j}$
   (m) $15\mathbf{i}+20\mathbf{j}$ (n) $-2\sqrt{5}\mathbf{i}+\sqrt{5}\mathbf{j}$ (o) $\frac{51}{5}\mathbf{i}+\frac{68}{5}\mathbf{j}$

Exercise 2A page 13

1. 10 m/s
2. 22·5 m/s
3. 126 km/h
4. 79·2 km/h
5. 100 m/s
6. 25 m/s
7. 25 m/s (90 km/h)
8. 16 m
9. 200 m
10. 6 s
11. 12 m, $5\frac{1}{2}$ s
12. (a) 80 000 km (b) 20 000 km (c) $22\frac{2}{9}$ km
13. 40·8 km
14. 8 min 20 s
15. 1·7 km
16. 7·04 m/s
17. (a) 45 min (b) 12 km/h
18. (a) 2 m/s (b) 1 m/s north
19. (a) 80 km/h (b) zero
20. (a) 5 m/s (b) 8 m/s (c) 10 s (d) 3 m/s in direction $\overrightarrow{AB}$

Exercise 2B page 15

1. 25 m
2. (a) 17 m (b) 13 m (c) $(-3\mathbf{i}+3\mathbf{j})$ m (d) $3\sqrt{2}$ m
3. (a) 5 m (b) $8\sqrt{2}$m (c) 7 m (d) $(5\mathbf{i}+12\mathbf{j})$ m
   (e) $(-8\mathbf{i}-15\mathbf{j})$ m (f) $(8\mathbf{i}+15\mathbf{j})$ m (g) 13 m (h) 17 m
4. 10 m/s
5. 25 m/s
6. $\sqrt{17}$ m/s
7. B
8. 4·8
9. −15 or 8
10. (a) $(7\mathbf{i}+7\mathbf{j})$ m (b) $(9\mathbf{i}+11\mathbf{j})$ m
11. (a) $(11\mathbf{i}+\mathbf{j})$ m (b) $(15\mathbf{i}-\mathbf{j})$ m
12. 1, −2
13. 13 m/s, $(16\mathbf{i}-30\mathbf{j})$ m, 34 m.

Exercise 2C page 19

1. 128 m **2.** 15 m/s **3.** 16 m **4.** 40 m **5.** 4 m/s^2
6. 11 m/s **7.** −4 m/s^2 **8.** 5 m/s **9.** 4 s **10.** 28 m
11. 3 m/s^2 **12.** 9·8 m/s^2 **13.** 12 m **14.** 5 s **15.** 2 s
16. 2 m/s **17.** 30 m **18.** 15 m/s **19.** 16 m **20.** 5 m/s
21. 300 m, 20 s **22.** 18 m/s^2 **23.** 2·5 m/s^2 **24.** 31 500 m/s^2, $\frac{1}{150}$ s
25. $6\frac{2}{3}$ m/s^2, 3 s **26.** (a) 8 m (b) 18 m, 10 m
27. 1·6 m/s^2, 100 m **28.** (a) $7\frac{1}{2}$ m (b) $5\frac{1}{2}$ m
29. (a) 21 m (b) 25 m, 5 s (c) 29 m **30.** 10 s, 29 m
31. 2·5 m, 15 m/s **32.** (a) 1·82 km (b) 1 min 44 s (c) 17·5 m/s **33.** 2 m/s^2, 3 m/s
34. 4, 6 **35.** 300 m

Exercise 2D page 22

1. 5·6 m/s **2.** $2\frac{6}{7}$ s **3.** 10 m **4.** (a) 27·6 m (b) 23·6 m/s
5. (a) 16·1 m (b) 22·4 m (c) 18·9 m **6.** (a) 28 m/s (b) 41 m
7. $4\frac{2}{7}$ s **8.** 10 m, $2\frac{6}{7}$ s **9.** 22·5 **10.** (a) 1 s (b) 4 s, 3 s
11. (a) 40 cm (b) 10 cm **12.** 6 s **13.** 10 s
14. (a) 9·1 m (b) 8·4 m, 2·5 m **15.** 180 m

Exercise 2E page 25

1. (a) (i) zero (ii) 6 m/s (iii) 1·5 m/s^2 (iv) 12 m
(b) (i) 2 m/s (ii) 6 m/s (iii) 1 m/s^2 (iv) 16 m
(c) (i) 6 m/s (ii) 3 m/s (iii) −0·75 m/s^2 (iv) 18 m
2. (a) (i) 2 m/s^2 (ii) 3 m/s^2 (iii) 39 m
(b) (i) 4 m/s^2 (ii) $1\frac{1}{3}$ m/s^2 (iii) 20 m
3. (a) $1\frac{1}{2}$ m/s^2 (b) 2 m/s^2 (c) 324 m
4. (a) 6 m/s^2 (b) 20 m/s (c) 2 m/s^2 (d) 240 m
5. 70 s

Exercise 2F page 26

1. (i) 19**i** + 88**j** (ii) −84**i** + 288**j** **2.** (a) 1·12 m/s^2 (b) 31·25 s
3. (a) 45 m/s (b) 2175 m
4. (i) 10·2 m (ii) 8**i** + 6**j** i.e 10 m/s, N 53·1°E **5.** $2\frac{6}{7}$ s, 40 m
6. (i) 4 s (ii) 20 m **7.** 2, 4, 24 m/s, 10 s **8.** 200 s
9. (ii) 3 s (iii) 15 m/s (iv) $1\frac{2}{3}$ m/s^2

Exercise 3A page 30

1. (a) 50 N (b) 20 N, 98 N (c) 90 N (d) 60 N, 50 N
2. (a) 30 N (b) 10 N, 25 N (c) 40 N, 20 N (d) 60 N, 10 N
3. 2 m/s^2 **4.** 6 N **5.** 8 kg **6.** 50 m/s^2
7. (10**i** + 4**j**) N **8.** (6**i** + 6**j**) m/s^2 **9.** 3 m/s^2, 1800 N
10. (a) 6 m/s^2 (b) 4 m
11. (a) 118 N (b) 78 N (c) 49 N, 25 N (d) 49 N, 30 N (e) 49 N, 10 N (f) 240 N
12. 500 N **13.** 100 N **14.** 0·75 m/s^2 **15.** $2\sqrt{10}$ N
16. 8i **17.** 2, $-\frac{1}{2}$ **18.** (a) 1250 N (b) 1600 N
19. 770 N **20.** 1200 m **21.** 4050 m

Exercise 3B page 34

1. 39·2 N **2.** 500 kg **3.** 0·98 N
4. (a) 58 N (b) 138 N (c) 49 N, 25 N (d) 5 kg (e) 15 kg (f) 3 kg
(g) 5 m/s^2 (h) 0·2 m/s^2 (i) 4·8 m/s^2

5. (a) 148 N (b) 48 N (c) 98 N (d) 98 N 6. 1·1 N
7. 500 N 8. (a) 0·4 m/s^2 (b) 10 200 N (c) 9400 N 9. 2575 N
10. 0·2 N 11. 3 N 12. 5810 N 13. (a) 24 N (b) 147 N

Exercise 3C page 38

1. 98 N 2. 1·96 N 3. 39·2 N 4. 49 N, 68·6 N 5. 19·6 N, 78·4 N
6. (a) $F = (m_1 + m_2)a$ (b) $F - T = m_1 a$ (c) $T = m_2 a$, $R_1 = m_1 g$, $R_2 = m_2 g$
7. (a) $m_1 g - T = m_1 a$ (b) $T - m_2 g = m_2 a$
8. (a) $T = m_1 a$ (b) $m_2 g - T = m_2 a$
9. (a) $m_1 g - T_1 - m_1 a$ (b) $T_1 - T_2 = m_2 a$ (c) $T_2 - m_3 g = m_3 a$
10. (a) $T - mg - Mg = (m + M)a$ (b) $T - R - Mg = Ma$ (c) $R - mg = ma$
11. 1·96 m/s^2, 47·04 N 12. 510 N 13. 660 N
14. 1·4 m/s^2, 13·44 N 15. 1·5 m/s^2, 1050 N
16. (a) (i) 4·2 m/s^2 (ii) 28 N (iii) 56 N
(b) (i) zero (ii) 49 N (iii) 98 N
(c) (i) 2·8 m/s^2 (ii) 6·3 N (iii) 12·6 N
17. 4·9 m/s^2, 9·8 m
18. (a) 1·4 m/s^2, 44·8 N, 58·8 N (b) 1·4 m/s^2, 16·8 N, 11·2 N
19. (a) 0·7 m/s^2, (b) 0·0455 N (c) 1·4 m
20. 1080 N, 980 N, 860 N 21. 1300 N, $1\frac{2}{3}$ m/s^2, $\frac{2}{3}$ m/s^2. Thrust of 100 N.

Exercise 3D page 41

3. $\frac{1}{10}g$, $\frac{33}{10}mg$, $\frac{18}{5}mg$ 4. (a) 2100 N (b) 700 N 6. 4200 N, 1050 N
7. 1·96 m/s^2 (a) 1·4 m/s (b) 30 cm 8. $\frac{11}{14}$ 9. 66 cm

Exercise 3E page 45

1. $\frac{4}{11}g$ m/s^2 ↑, $\frac{15}{11}g$ N 2. $\frac{1}{9}g$ m/s^2 ↓, $\frac{8}{3}g$ N
3. (a) 1·4 m/s^2 ↑ (b) 7 m/s^2 ↑, 4·2 m/s^2 ↓, 1·4 m/s^2 ↓ (c) 33·6 N, 16·8 N
4. $\frac{9}{37}g$ m/s^2 ↓, $\frac{224}{37}g$ N, $\frac{112}{37}g$ N 5. 1·4 m/s^2, 42 N 6. $\frac{1}{17}g$ m/s^2 ↓, $\frac{48}{17}mg$, $\frac{24}{17}mg$
8. $\frac{3}{19}g$ m/s^2 ↓, $\frac{48}{19}mg$ 9. 8 kg 10. 1 kg

Exercise 3F page 46

1. 5 m/s^2 at $\tan^{-1}\frac{7}{24}$ to direction of **i**.
2. 2000 N 3. (a) 360 N (b) 860 N 4. 1·4 N
5. (i) 4200 N (ii) 6 m/s (iii) 3400 N
6. (i) 1·96 m/s^2 (ii) 9·8 m/s (iii) 4·9 m (iv) 2 s
7. (i) $\frac{2}{k}$ s, $\frac{5}{k}$ s (ii) 0·7 (iii) 32·5 N, 50 N, 57 N 8. $\frac{1}{2}g$, $\frac{3}{2}mg$

Exercise 4A page 53

1. (a) 7·2 N, 34° (b) 11·6 N, 12° (c) 4·4 N, 37°
2. (a) 5 N, 36·9° (b) 6·71 N, 26·6° (c) 4·47 N, 26·6°
(d) 7·61 N, 28·4° (e) 17·9 N, 18·9° (f) 13·5 N, 45·7°
3. (a) 9·85 N, 15·3° to 8 N force (b) 10·2 N, 13·0° to 8 N force
(c) 5·28 N, 11·2° to 8 N force
4. 1·62 N, 38·3° 5. 5·29 N, 40·9° 6. 70·5° 7. 130·5°
8. 5·29 9. $F\sqrt{7}$ N, 40·9° 10. 8 11. 8·24
12. (a) 17·2 N, 32° (b) 3·8 N, 172° (c) 4·1 N, 52° 13. 7·2 N, N 9° E
14. 5·7 N, 101° 15. 8·2 N, 76° 16. 10 N, 60° 17. 740 N, 025·8°
18. 60 N, 6° 19. 20°

Exercise 4B page 57

1. (a) (i) $4\sqrt{3}$ N (ii) 4 N (b) (i) 0 (ii) 10 N
(c) (i) −7·66 N (ii) 6·43 N (d) (i) $2\sqrt{2}$ N (ii) $-2\sqrt{2}$ N
(e) (i) $\sqrt{3}$ N (ii) 3 N (f) (i) $P\cos\theta$ (ii) $P\sin\theta$

2. (a) $(8\mathbf{i}+8\sqrt{3}\mathbf{j})$ N (b) $(3\mathbf{i}+3\mathbf{j})$ N (c) $(-3\mathbf{i}+3\mathbf{j})$ N (d) $(-5\mathbf{i}-5\sqrt{3}\mathbf{j})$ N
(e) $P\cos\alpha\mathbf{i}+P\sin\alpha\mathbf{j}$ (f) $-Q\cos\phi\mathbf{i}-Q\sin\phi\mathbf{j}$

3. (a) (i) −5 N (ii) $-5\sqrt{3}$ N (b) (i) $5\sqrt{3}$ N (ii) −5 N
(c) (i) $-5\sqrt{2}$ N (ii) $-5\sqrt{2}$ N (d) (i) 3·42 N (ii) −9·40 N
(e) (i) −6·43 N (ii) −7·66 N (f) (i) 1·74 N (ii) −9·85 N

4. (a) (i) $\sqrt{3}$ N (ii) 7 N (b) (i) 12·3 N (ii) −0·071 N
(c) (i) $P\cos\theta+R-Q\sin\phi$ (ii) $P\sin\theta-Q\cos\phi$

5. (a) (i) 1 N (ii) −3·66 N (b) (i) −8·86 N (ii) −0·66 N
(c) (i) $-11\sqrt{2}$ N (ii) $\sqrt{2}$ N (d) (i) 13·4 N (ii) 0·60 N
(e) (i) −17 N (ii) $7\sqrt{3}$ N (f) (i) $R\sin\phi+Q\cos\theta$ (ii) $P+Q\sin\theta-R\cos\phi$

Exercise 4C page 61

1. (a) 5i N (b) (4i − 3j) N (c) 4i N (d) (4i − 6j) N
2. −4, 1 **3.** 4, 2 **4.** 5 N. 36·9° **5.** 4·47 N, 116·6°
6. (a) 5·66 N, 45° (b) 7·07 N, 98·1° (c) 3·61 N, 326·3° (d) 2 N, 180°
7. (a) (13i + 8·66j) N, 15·6 N, 33·7° (b) 5j N, 5 N, 90°
(c) (5·2i − 4·46j) N, 319·3°
8. $\sqrt{101}$ N, 354·3° **9.** −6·59i + 13·49j **10.** 12·1 N, 145·6°
11. 5·66 N, 45° on same side as C **12.** 4 N, 30° on other side from C
13. 2·83 N, 45° with BA on other side from C **14.** 2 N parallel to AB
15. 13·4 N, 26·6° to AB on same side as C **16.** 20 N, 60° with AB on same side as C

Exercise 4D page 62

1. (a) 20·3 N, 042·3° (b) 6·2 N, 143·8°
2. (i) 12·3 N, N 86° W (ii) 6·9 N, N 33° E
3. $2\sqrt{3}$, 4 **4.** $2\sqrt{3}$ N at 30° with BA **5.** 10, 090°, 223·9°, $5\sqrt{13}$ N
6. 25 N **7.** (a) 4 : 1 (b) 6, −24

Exercise 5A page 69

1. (a) 32 N, 151° (b) 6 N, 141° (c) 3 N, 5·2 N
2. (a) 7·4 N, 105° (b) 4·8 N, 35° (c) 6·3 N, 8·9 N
3. (a) 5 N, 126·9° (b) 6·56 N, 67·6° (c) 8·89 N, 77°
4. (a) 5 N, 8·66 N (b) 2·18 N, 4·89 N (c) 2·59 N, 7·07 N
5. 8·89 N, 43° **6.** 36·9°, 36·75 N, 61·25 N **7.** 11·3 N, N 6° E
8. 9·8 N, 17·0 N **9.** 11·3 N, 22·6 N **10.** 35 N, $28\sqrt{2}$ N

Exercise 5B page 75

1. (a) $P\cos 30°=4\sqrt{3}$ (b) $P\sin 30°+Q=6$ (c) $P=8$ N, $Q=2$ N
2. (a) $P\cos 45°=Q$ (b) $P\sin 45°=5\sqrt{2}$ (c) $P=10$ N, $Q=5\sqrt{2}$ N
3. (a) $4=P\cos\theta$ (b) $3=P\sin\theta$ (c) $P=5$ N, $\theta=36{\cdot}9°$
4. (a) $5+P\cos\theta=10\cos 20°+10\sin 20°$ (b) $P\sin\theta+10\sin 20°=10\cos 20°$
(c) $P=9{\cdot}84$ N, $\theta=37{\cdot}4°$
5. (a) $P\cos 30°=Q\cos 60°+10\cos 60°$ (b) $P\sin 30°+Q\sin 60°=10\sin 60°+8$
(c) $P=(5\sqrt{3}+4)$ N, $Q=(5+4\sqrt{3})$ N
6. (a) $P=10\cos 60°$ (b) $Q=10\sin 60°$ (c) $P=5$ N, $Q=5\sqrt{3}$ N
7. (a) $P\sin 30°=Q$ (b) $P\cos 30°=12$ (c) $P=8\sqrt{3}$ N, $Q=4\sqrt{3}$ N

8. (a) $10 \cos 30^\circ + 10 \cos 60^\circ = Q$ (b) $P + 10 \sin 30^\circ = 10 \sin 60^\circ$
(c) $P = (5\sqrt{3} - 5)$ N, $Q = (5\sqrt{3} + 5)$ N **9.** (a) $P \cos\theta = 3 + 5\sqrt{2} \cos 45^\circ$
(b) $P \sin\theta + 1 = 5\sqrt{2} \sin 45^\circ$ (c) $P = 8{\cdot}94$ N, $\theta = 26{\cdot}6^\circ$
10. (a) $-4, 1$ (b) $-8, -5$ (c) $5, 2$ (d) $1, 2$ (e) $7, -14$
12. 58·8 N, 98 N **13.** $\frac{5mg}{13}, \frac{12mg}{13}$ **14.** 3·16 N, N 71·6° E **15.** 49 N, 84·9 N
16. 56·6 N, 113 N **17.** 51·7° **18.** (a) 19·5° (b) 9·8 N (c) 27·7 N
19. 44·4 N, 49 N, 28·3 N **20.** 30°, 41·8° **21.** 29·0°, 75·5°

***Exercise 5C** page 80*

1. (a) $R + 50 \sin 30^\circ = 10g$ (b) $50 \cos 30^\circ = 10a$ (c) $R = 73$ N, $a = \frac{5\sqrt{3}}{2}$ m/s^2
2. (a) $R + P \sin 30^\circ = 10g$ (b) $P \cos 30^\circ = 50\sqrt{3}$ (c) $P = 100$ N, $R = 48$ N
3. (a) $R + 50 \sin 60^\circ = 10g$ (b) $50 \cos 60^\circ - 10 = 10a$ (c) $R = 54{\cdot}7$ N, $a = 1{\cdot}5$ m/s^2
4. (a) $R + 50 \sin 20^\circ = 10g + 50 \sin 20^\circ$ (b) $50 \cos 20^\circ + 50 \cos 20^\circ = 10a$
(c) $R = 98$ N, $a = 9{\cdot}40$ m/s^2
5. (a) $88 + P \sin\theta = 10g$ (b) $P \cos\theta = 10\sqrt{3}$ (c) $P = 20$ N, $\theta = 30^\circ$
6. (a) $R = 10g \cos 30^\circ$ (b) $10g \sin 30^\circ = 10a$ (c) $R = 84{\cdot}9$ N, $a = 4{\cdot}9$ m/s^2
7. (a) $R = 10g \cos 30^\circ$ (b) $P - 10g \sin 30^\circ = 20$ (c) $R = 84{\cdot}9$ N, $P = 69$ N
8. (a) $R + 50 \sin 40^\circ = 10g \cos 40^\circ$ (b) $50 \cos 40^\circ + 10g \sin 40^\circ = 10a$
(c) $R = 42{\cdot}9$ N, $a = 10{\cdot}1$ m/s^2 **9.** (a) $49 = 10g \cos\theta$
(b) $10g \sin\theta - 50 = 10a$ (c) $a = 3{\cdot}49$ m/s^2, $\theta = 60^\circ$
10. 65·3 N, 1·63 m/s^2, 84·9 N **11.** 34·6°, 80·7 N, 25·6 N **12.** 2 m/s^2, 9 m
13. 0·8 m/s^2, $4\sqrt{2}$ N, 45 N **14.** 69·2 N, 2·12 m/s^2, 16·96 m
15. 14 m/s, 14 m/s **16.** 38 N **17.** 28 N **18.** 0·74 s
19. 3·8 m/s^2, 7·6 m **20.** 31 N **21.** 30° **22.** 0·663 s

***Exercise 5D** page 82*

12. 10 m **15.** 2·42 m/s

***Exercise 5E** page 84*

1. (a) $\sqrt{122}$ (b) 11 **2.** 8·83 N, 316·6°; 8·83, 0·24 **3.** 4·3, 17·9
4. (i) $\frac{13}{10}mg$ (ii) $\frac{20}{11}Mg$ **5.** 164 N, 2·64 m/s^2
6. (i) 3 m/s^2 (ii) 5·48 m/s (iii) 1 m/s^2 (iv) 4·47 m/s (v) 2·83 s
7. (a) 1·4 m/s^2 (b) 2·52 N (c) 3 s (d) 4·2 m/s

***Exercise 6A** page 89*

1. (a) 10 N, rest (b) 14 N, rest (c) 14 N, accelerate (d) 10 N, rest
(e) 14 N, rest (f) 18 N, accelerate (g) 10 N, rest (h) 10 N, accelerate
(i) 10 N, accelerate (j) 12 N, accelerate (k) 12·12 N, rest (l) 22 N, accelerate
2. (a) $\frac{2}{7}$ (b) $\frac{1}{2}$ (c) $\frac{1}{4}$ (d) 0·33
(e) 0·348 (f) 0·677 **3.** $\frac{4}{7}$ **4.** $\frac{1}{10}$
5. 49 N, yes, 0·05 m/s^2 **6.** 50 N, no **7.** 0·49 N, yes, 1·02 m/s^2
8. 19·8 N, 10 N **9.** $\frac{1}{4}$ **10.** 1·02 m
11. (a) 0·7 m/s^2 (b) 3·15 N (c) 2·45 m/s^2 (d) 11·25 m
12. (a) 9·8 (b) 8·78 (c) 15·9
13. (a) 3·92 (b) 4·62 (c) 6·93 **14.** 0·58, 0·63 N
15. (a) no sliding (b) no sliding (c) sliding
16. (a) no sliding (b) sliding (c) sliding
17. (a) no sliding (b) no sliding (c) sliding **18.** Yes
19. 9·8 N, no **20.** (a) 1·4 m/s^2 (b) 0·42 N (c) 70 cm
21. (a) 1·09 m/s^2 (b) $2\frac{1}{3}$ m/s^2 (c) 1·48 m/s
23. 2·94 m/s^2 **24.** $\frac{1}{7}$

Exercise 6B page 95

1. (a) 33·5 N, rest (b) 42·4 N, accelerate (c) 37·5 N, accelerate
   (d) 17·0 N, rest (e) 23·7 N, rest (f) 23·0 N, accelerate
2. (a) 1·61 N (b) 10·2 N (c) 18·4 N (d) 19·7 N (e) 18·3 N (f) 23·2 N
3. (a) 19·3 N (b) 23·2 N (c) 27·7 N (d) 33·2 N (e) 27·3 N (f) 22·0 N
4. (a) 0·15 (b) 0·20 (c) 0·24
5. 2·25 N, body will slide
6. 0·202
7. 0·279
8. 0·523
9. (a) 2·78 N (b) 7·02 N (c) 8·52 N
10. (a) $2g$ N (b) $18g$ N (c) $6(8 + 3g)$ N
11. $\frac{1}{2}$
12. yes, $2\frac{1}{3}$ m/s
13. 0·26, 4 s
14. 0·23
15. 3 s, 4·2 m/s
16. (b) yes (c) no (d) 0·577
17. 0·29
18. 5·6 m/s
19. 1·96, 0·1
20. 0·98 m/s^2, 44·1 N

Exercise 6C page 102

1. 11·3°
2. 0·577
3. (a) 51·0 N, 15·9° (b) 52·9 N, 22·2° (c) 44·4 N, 24·1°
   (d) 5·60 N (e) 5·60 N (f) 5·25 N
   (g) 76·6 N (h) 63·4 N (i) 76·6 N
   (j) $\frac{19\sqrt{3}}{3}$ N, 30° (k) 44·1 N, 15·8° (l) 28·2 N, 26·2°
4. 20·2 N, 14·0°
5. 14·9 N, 2·16 m/s^2
6. (b) (i) Yes (ii) no (c) 25°, 0·466
7. 13·5 N
8. 10·5 N
9. 28·0 N
10. 25·9°, Remain at rest
11. (a) 17·0 N (b) 16·8 N (c) 17·8 N
13. (a) 4·12 N (b) 4·09 N (c) 4·17 N
14. (a) 20·8 N (b) 20·7 N (c) 21·4 N
16. (a) 15·2 N (b) 15·0 N (c) 15·2 N
17. (a) 24·5 N (b) 24·9 N (c) 26·1 N

Exercise 6D page 105

1. 0·077
2. 14
3. 0·304
4. (a) 4·5 m/s^2, 42 N, 24 N (b) 2 m/s^2, 42 N, 24 N
5. (i) $\frac{6u}{g}$ (iii) $u\sqrt{5}$ (iv) $0{\cdot}85u$
6. $mg \sin \lambda$
7. $\frac{(4M - 3m)g}{4(M + m)}, \frac{7mMg}{4(M + m)}, M < \frac{1}{4}m$

   (i) Q ascends, $\frac{25}{24}Mg, \frac{5}{2}Mg\sqrt{3}, \frac{5}{4}Mg$ (ii) No motion, $Mg, \frac{3}{2}Mg\sqrt{3}, \frac{1}{2}Mg$
   (iii) Q descends, $\frac{7}{8}Mg, \frac{1}{2}Mg\sqrt{3}, \frac{1}{4}Mg$

Exercise 7A page 111

1. 10 N m clockwise
2. 2 N m anticlockwise
3. 10 N m clockwise
4. 10 N m anticlockwise
5. 24 N m clockwise
6. 16 N m clockwise
7. 8 N m clockwise
8. 3 N m anticlockwise
9. zero
10. 12 N m clockwise
11. 14 N m clockwise
12. 10 N m clockwise
13. 4 N m clockwise
14. 6 N m anticlockwise
15. 6 N m anticlockwise
16. 40 N m clockwise
17. 24 N m anticlockwise
18. 80 N m clockwise
19. zero
20. 25 N m clockwise
21. 34·6 N m clockwise
22. 61·3 N m clockwise
23. 9·85 N m anticlockwise
24. 20 N m anticlockwise
25. 20 N m clockwise
26. 9 N m clockwise
27. 2 N m clockwise
28. 5 N m clockwise
29. 7 N m clockwise
30. 17 N m anticlockwise

Exercise 7B page 115

1. (a) 40 N m clockwise (c) 5 N m anticlockwise (e) 19 N m clockwise
   (f) 17 N m anticlockwise (g) 2 N m clockwise (h) 2 N m clockwise
2. 1·75 N m
3. 21 N m in sense ABCD
4. 15

5. 15 N, 14 N m in sense ABCD
6. 2·4 N m in sense ADCB
7. 12 N m in sense ABCD, 1·5
8. 80 N
9. 10 N m clockwise
10. 7 N m clockwise
11. 13 N m anticlockwise
12. −6, 4, 26 N m clockwise
13. 12 N m clockwise, independent of P.

***Exercise 7C** page 117*

1. (a) 6 N, 1 m (b) 2 N, $7\frac{1}{2}$ m (c) 2 N, $7\frac{1}{2}$ m
2. 5 N, 4 m
3. 20 N, 3 m
4. 2 N, 10 m
5. 4 N, 15 m
6. 6, 2
7. 4, 6
8. 10 N, 1·1 m
9. 1·5 N, $1\frac{2}{3}$ m
10. up, down, up, 3 m

***Exercise 7D** page 120*

1. $P = 30$ N, $x = 1\frac{1}{3}$m
2. $P = 20$ N, $Q = 10$ N
3. $P = 8$ N, $x = 1\frac{1}{2}$m
4. $P = 5g$ N, $Q = 10g$ N
5. $P = 160g$ N, $Q = 100g$ N
6. $P = 25g$ N, $x = 2{\cdot}5$ m
7. $P = 20g$ N, $Q = 30g$ N
8. $P = 190$ N, $Q = 130$ N
9. $30g$ N, $40g$ N
10. 2 m from 10 kg mass
11. 75 cm from other end
12. 24 cm from head
13. $2\frac{3}{4}$ N, $2\frac{1}{4}$ N
14. 1·4 m from A
15. $1\frac{1}{2}$ m from A
16. $7\frac{1}{2}g$ N downwards, $12\frac{1}{2}g$ N upwards
17. $8\frac{4}{7}$, $14g$ N
18. (a) 66 cm from handle (b) $2\frac{1}{4}g$ N, $12\frac{3}{4}g$ N

***Exercise 7E** page 122*

1. (a) 98 N (b) 49 N (c) 44 N
2. (a) 4·9 N (b) 5·9 N (c) 3·57 N
3. (a) 42·4 N (b) 27·7 N (c) 30·1 N
4. 13·7 N
5. 19·6 N
6. $12g$ N
7. 46 cm
9. 31·4 N
10. 614 N

***Exercise 7F** page 127*

1. (a) 13 N, 22·6° to AB. 7 cm from A (b) 13 N, 22·6° to AB. At A
   (c) 25 N, 73·7° to BA. 3 m from A (d) 8·54 N, 324·2° to AB. 8·31 m from A
2. (a) 0·14 N m in sense ADCB (b) 41 N m in sense ADCB
   (c) $10\sqrt{3}$ N m in sense ABC (d) $3\,Pa\sqrt{3}$ in sense ABCDEF
3. 5 N, 53·1° to AB. $12\frac{1}{2}$ cm from A
4. 12·8 N, 2·49 N, 65·9 N m in sense ADCB
5. 4·47 N, 26·6° to AB. $\frac{7}{2}\,a$ from A on BA produced
6. 30 N, 30° to AB. 2 m from A on BA produced
7. 18·3 N, 10·9° to AB. $\frac{3}{2}\,a$ from A on BA produced
8. (4**i** + 6**j**) N, **i**
10. 11·3 N, 16·3° to BA. 89 cm from A on BA produced
11. 8·84 N, 57·6° to AB. 2·32 from A on BA produced
12. 12·5 N, 49·6° to AB. 1 m from A
13. 20·1 N, 15·2° to BA, 3·04 cm from A. Same force but now 6·46 cm from A on BA produced
14. 8·25 N, 76·0° to AB, 25 cm from A. 8·54 N, 69·4° to AB, 2 N m in sense ABCD
15. 16·4 N, 37·6° to AB, 5·5 m from A on AB produced. Same force after couple introduced
16. (4**i** + 3**j**) N, −4**i**, $a = 14$, $b = -4$, $c = -3$

***Exercise 7G** page 129*

1. 2, 8 N
2. 15 units
3. (a) 14 (b) 168 N
4. 2·75 m from A. $\frac{41}{2}g$ N, $\frac{79}{2}g$ N
5. (b) $5P$, at $\tan^{-1}\frac{4}{3}$ below AB, $2a$ from A on AB produced. $5P$, $4P$
6. 8·06 N at $\tan^{-1} 8$ to OA, $y = 8x + 3$, 3 N in direction AB
7. $P\sqrt{14}$, $P(1 + \sqrt{3})$, $P\sqrt{3}$, $P(2 - \sqrt{3})$
8. (a) $25P$, $7x - 24y = 0$
   (b) (i) $-18P$, $15P$ (ii) $72Pa$ (iii) clockwise
9. (i) 2, 6 (iii) $x = 3a\sqrt{3}$; $P\sqrt{31}$, $5Pa\sqrt{3}$

***Exercise 8A** page 134*

1. (a) $(2\frac{1}{2}, 0)$ (b) $(2, 2)$ (c) $(\frac{1}{2}, 2)$ (d) $(1, -1)$ **2.** $(0, 4)$
3. $(2, 1\frac{1}{2})$ **4.** $(-1, 3)$ **5.** $-3, 2$ **6.** 1 cm, $1\frac{1}{2}$ cm **7.** $1\frac{1}{3}$ m, $1\frac{3}{4}$ m
8. $5\mathbf{i} + 3\mathbf{j}$ **9.** $\mathbf{i} - 3\mathbf{j}$ **10.** $2, -4$ **11.** $(2{\cdot}4, -3{\cdot}2)$ **12.** $1, 2{\cdot}5$
13. $(0, -2)$ **14.** $(3, 1)$ **15.** 1 g, 5 cm

***Exercise 8B** page 139*

1. $(2, 2)$ **2.** $(2\frac{1}{2}, 1\frac{1}{2})$ **3.** $(1, 2\frac{1}{2})$ **4.** $(2, 3)$ **5.** $(2, 2)$
6. $(3, 2)$ **7.** $(2, 2)$ **8.** $(1, 2)$ **9.** $(2, 2)$ **10.** $(3, 2)$
11. $(1\frac{1}{3}, 2)$ **12.** $(2\frac{2}{3}, 1)$

***Exercise 8C** page 144*

1. $(2\frac{1}{2}, 1\frac{1}{2})$ **2.** $(1{\cdot}8, 1{\cdot}3)$ **3.** $(2, 1\frac{3}{4})$ **4.** $(2{\cdot}6, 1{\cdot}9)$ **5.** $(2{\cdot}3, 1{\cdot}4)$
6. $(1, 2\frac{1}{3})$ **7.** $(1{\cdot}4, 2{\cdot}2)$ **8.** $(3, 3)$ **9.** $(1\frac{13}{14}, 2)$ **10.** $(2\frac{4}{9}, 3\frac{1}{9})$
11. $(4{\cdot}1, 2{\cdot}95)$ **12.** $(-\frac{2}{15}, 0)$ **13.** $(-\frac{2}{3}, 0)$ **14.** (a) $\frac{1}{3}$ m (b) $1\frac{1}{3}$ m
15. 1·8 m from AD, 1·8 m from AB **16.** 2·18 m **17.** 0·5 m **18.** $1\frac{2}{3}$ m, 3 m
19. $(-\frac{4}{13}, -\frac{15}{26})$ **20.** $1\frac{2}{3}$ cm **21.** On axis of symmetry, $1\frac{6}{7}$ cm above base
22. $\frac{28}{3\pi}$ m **23.** On axis of symmetry, $\frac{28}{5\pi}$ cm into larger semicircle from common diameter
24. On axis of symmetry, $\frac{4}{\pi}$ cm into larger semicircle from common diameter
25. On axis of symmetry, $3\frac{8}{11}$cm above base of cylinder
26. On axis of symmetry, $2\frac{5}{7}$ cm from undrilled end
27. On axis of symmetry, 10·8 cm from tip of cone

***Exercise 8D** page 149*

1. 26·6° **2.** 56·3° **3.** 39·5° **4.** 31·0° **5.** 35°
6. 23·2° **7.** 14·4° **8.** 38·9° **9.** 3·5°

***Exercise 8E** page 153*

7. $(1\frac{1}{2}, 1\frac{1}{5})$ **8.** $(1, 0{\cdot}4)$ **9.** $(1{\cdot}56, 2{\cdot}25)$ **10.** $(5{\cdot}31, 3{\cdot}21)$ **11.** $(1{\cdot}5, 3{\cdot}6)$
12. $(\frac{3}{8}, 2\frac{1}{5})$ **13.** $(2\frac{2}{3}, 0)$ **14.** $(3\frac{1}{9}, 0)$ **15.** $(3{\cdot}39, 0)$ **16.** $(1{\cdot}30, 0)$
17. $(2\frac{2}{5}, 7\frac{5}{7})$, $(2\frac{5}{8}, 0)$

***Exercise 8F** page 156*

1. $11\frac{1}{4}$ cm, 6 cm **2.** 38° **3.** $\frac{a(3a-2s)}{3(2a-s)}$ **4.** $3r$ **5.** $3Wa\sqrt{3}$
6. (a) $2\pi a^3$ (b) $(\frac{2}{3}a, 0)$, $\frac{8\pi a^3}{5}$ **7.** $\frac{6l^2 - h^2}{4h + 12l}$ from interface

***Exercise 9A** page 165*

1. (a) 0, 49 N, 49 N (b) 49 N, 49 N, 69·3 N (c) 49 N, 49 N, 69·3 N
(d) 0, 49 N, 49 N (e) 21·2 N, 61·3 N, 42·4 N (f) 21·2 N, 61·3 N, 42·4 N
2. (a) 28·3 N, 28·3 N, 30° (b) 24·5 N, 42·4 N, 30° (c) 42·4 N, 64·8 N, 40·9°
(d) 42·4 N, 24·5 N, 60° (e) 21·2 N, 32·4 N, 19·1° (f) 50·7 N, 21·6 N, 81·8°
3. (a) 42·4 N, 21·2 N, 61·3 N, 0·346 (b) 57·8 N, 19·8 N, 43·7 N, 0·453
(c) 49 N, 42·4 N, 73·5 N, 0·577 **4.** (a) 84·9 N (b) 53·5 N (c) 25·9 N
5. (a) 56·6 N, 56·6 N, 196 N, 0·289 (b) 98 N, 98 N, 196 N, 0·5
(c) 26·3 N, 26·3 N, 196 N, 0·134 **6.** (a) $\frac{W}{3}$N, (b) $\frac{W}{2}$N, (c) W N

7. (a) 67·4° (b) 54·5° (c) 53·1°
8. (a) 19·4° (b) 36·1° (c) 49·1°
9. 49 N, 49 N at 60° to wall
10. 39·9 N, 45·2 N
11. 11·3 N at 58° to AB, 2·08 m
12. 60°, 5 N, 8·66 N
13. 35·4 N
14. 157 N, 547 N
15. 141 N, 245 N
16. 0·346
18. 27·7 N
19. 245 N, 32·8 N, 0·134
20. 3·05 N
21. 0·269
23. 6 m
30. 11·4 N, 158·4 N
31. 12·1 N, 138 N
32. 3·8 m, 514·5 N
33. 1176 N, 467 N, 0·505
34. 6 m, $\frac{8W}{17}$
35. 9 m, $\frac{7W}{11}$
36. 8 m, 126 N

***Exercise 9B** page 172*

2. 5g N along AB, 5g N
3. $\frac{15}{2}g\sqrt{2}$ N, $\frac{15}{2}g\sqrt{2}$ N
4. $\frac{1}{6}a$
5. (i) $\frac{25}{37}W$, $\frac{20}{37}W$ (ii) 0·325
6. $\frac{4}{15}W$, $\frac{7}{15}W$, $\frac{1}{7}W$
7. $\frac{mgl}{2a}$

***Exercise 10A** page 178*

1. (a) (9**i** − **j**) m/s (b) (7**i** + 9**j**) m/s (c) (−**i** − 6**j**) m/s
 (d) (54**i** + 27**j**) km/h (e) 63**i** km/h
2. (a) 25 m/s N 16° E (b) 8·7 km/h N 30° E (c) 6·2 m/s N 76° E
 (d) 16·9 m/s N 48° E (e) 82 km/h N 73° E
3. (a) 15 m/s N 36·9° W (b) 4·25 m/s N 48·3° W (c) 31·2 km/h N 12·2° E
 (d) 15·7 km/h S 77·9° W (e) 32·8 m/s S 80° E
4. 4, −5
5. 7, 3
6. 10 km/h N 30° E
7. 9·54 m/s S 63° E
8. 11·5 m/s S 34·2° E
9. 53·1°, 25 s
10. 70·5° to the bank, 44·2 s
11. 25 s, 75 m
12. 1 min, 30 m
13. N 73·7° E, 50 min
14. 054·3°, 2 h 8 min
15. 330·8°, 1 h 19 min
16. (a) $3\frac{3}{4}$ min (b) 3 min
17. upstream at 45·6° to bank or upstream at 24·4° to bank, 28 s, 48 s
18. N 8·3° E, 1 h 50 min, 1 h 33 min
19. 3 h 15 min
20. $\frac{v+u}{2}$

***Exercise 10B** page 184*

1. 10 km/h north
2. 50 m/s north
3. 30 km/h east, 45 km
4. 15 m/s east, 900 m
5. (8**i** + 2**j**) m/s
6. (2**i** − 7**j**) m/s
7. (100**i** + 600**j**) km/h
8. 12**j** km/h, −12**j** km/h
9. 5 km/h N 36·9° W
10. 17 km/h S 61·9° E
11. 10·3 km/h N 46·9° E
12. 25·2 m/s from 018·1°
13. 292 km/h S 77·9° E
14. 20 m/s N 36·9° W, 20 m/s S 36·9° E
15. (**i** + 3**j**) km/h
16. (300**i** + 240**j**) km/h
17. 6 m/s from S 30° W
18. 520 km/h west
19. 7·81 km/h N 26·3° E
20. 5·32 km/h from N 21·5° E
21. 15·4 km/h 238·7°
22. 11·4 m/s 262·4°
23. 196 km/h N 36·5° E
24. 9·64 km/h from 024·8°
25. 7·57 km/h from N 72·5° W
26. (250**i** − 100**j**) m/s, (300**i** + 70**j**) m/s
27. (5**i** + 7**j**) km/h
28. 12·2 km/h from S 34·7° E
29. 10·6 m/s from S 69·9° W
30. 26·8 km/h from N 33·4° W

***Exercise 10C** page 189*

1. (4**i** + 3**j**) m/s, 50 s
2. (5**i** − 12**j**) m/s, 3 s
3. 12.30 p.m. (4**i** + 8**j**) km
4. 12.20 p.m., (10**i** + $5\frac{1}{3}$**j**) km
5. 1.15 p.m., (22**i** + 5**j**) km
6. 12.06 p.m., (80**i** + 460**j**) km
7. (a) 2.20 p.m., (8**i** + 7**j**) km (b) (16**i** + **j**) km, 10 km (c) 3.00 p.m.
8. 1.10 p.m., (21**i** + 27**j**) km, 1.30 p.m.
9. A and C at 12.45 p.m., (11**i** − 8**j**) km, 1.09 p.m.
10. B and C at 8.48 a.m., (18**i** − 6**j**) km, (15**i** − 20**j**) km/h, 9·00 a.m.

Exercise 10D *page 191*

1. 12.35 p.m.
2. 11.47 p.m.
3. N 41·4° E, 1.45 a.m.
4. S 23·6° W, $26\frac{1}{2}$ min past 9 p.m.
5. 126·8°, 12.15 p.m.
6. 345·0°, 92 s
7. N 15·5° W, 9 min 7 s
8. N 24·7° E, 2·4 s

Exercise 10E *page 193*

1. (a) 25 km (b) 15 km (c) 1.20 p.m., (−5**i** + 22**j**) km, (4**i** + 10**j**) km
2. 1·81 km, 9.47 a.m.
3. 3·08 km, 8.04 p.m.
4. 6·32 km, 1340 hours
5. (a) (9**i** + 18**j**) km (b) 13·4 km (c) 1548 hours
6. 2·81 km, 12.57 p.m.
7. (a) (5**i** + 4**j**) km (b) 5·83 km (c) 1·70 km (d) 1.21 p.m.
8. 4·47 km, 10 min
9. 1·41 km, 12 min

Exercise 10F *page 195*

1. 60 km, 58 min
2. 3 km, $13\frac{1}{3}$ s
3. 800 m, 5 min 18 s
4. 3·6 km, 7.42 a.m.
5. 330 m, 112 s, 198 m north, 264 m west
6. 571 m, 5·8 s
7. 6·58 km, 1518 hours, 22 min
8. 5·46 km, 12 min 27 s

Exercise 10G *page 196*

1. N 53·1° E, 360 m, 80 s
2. 2 km, 8.57 and 36 s
3. 108·2°, 5·4 min
4. N 16·4° E, 6·89 km, 45 min
5. N 30° W, 12.24 p.m., 21 min, 14 min
6. 8·27 m, 2·61 s

Exercise 10H *page 197*

1. 7 min
2. (a) (**i** − 2**j**) (b) $\sqrt{17}$
3. 061·3°, 5·4 min
4. (i) $(10 - 2t)$**i** + $(5 + 4t)$**j** (ii) 2 p.m. (iii) 8·94 km
5. (i) 500 km/h N 36·9° W, 12 km, S 53·1° W (ii) N 48·6° E, 12.04 and 32 seconds
6. (i) −2**i** + 2**j**, 9·90 km, 0·5 h (ii) 9·91 km
7. $2 - \cot\alpha$
8. N 7·2° W, S 47·5° W, 28·3 m/s
9. 9·27 km/h from N 35·4° W

Exercise 11A *page 202*

1. 4 s, 80 m
2. 60 m, 40 m below, 72·1 m
3. 14 m/s
4. (a) (5**i** + 15·1**j**) m (b) (10**i** + 0·4**j**) m
5. (11**i** + 5·4**j**) m
6. 60 m
7. 28 m/s
8. 2 m
9. 20 cm
10. 1·1 m
11. 630 m
12. 1·44
13. 20 m/s, 19·6 m/s
14. 175 m/s, 16·3° below horizontal
15. $1\frac{3}{7}$, 21
16. (a) 35 m (b) 35 m
17. 2·4 m, 4 m
18. 2 m
20. 40 m
21. 60 m

Exercise 11B *page 208*

1. (a) $2\frac{6}{7}$ s (b) 40 m (c) $5\frac{5}{7}$ s (d) 277 m
2. (a) 3 s (b) 45 m (c) 6 s (d) 312 m
3. 120 m
4. 180 m, 6 s
5. 5·4 s
6. 2·25 km
7. Free kick
8. 1·6 m
9. (8**i** + 6**j**) m, 10 m
10. (30**i** + 15**j**) m, (10**i** − 10**j**) m/s
11. 21·7 m/s, 12 m/s down, 24·8 m/s 29° below horizontal
12. 60 m, 40·4 m, 72·3 m
13. 2 s, 2
14. 75 m, 3 s
15. 2 s, 2 m
16. 5 s, 24·5 m
17. 50 m/s elevation 36·9°
18. 30°
19. (15**i** + 29**j**) m/s
20. 5 m/s, 17 m/s
21. 15·6 m/s at 45° above horizontal
22. 3 s
23. 4 s
24. 3 s, 21
25. 5 s, 60 m
26. 90 m
27. 20 m
28. 15°, 75°
29. 14·7°, 75·3°
30. 11·8°, 78·2°
31. 14·8 m/s, 58° above horizontal
32. 31·3 m/s
33. 42 m, 36·4 m, 7, 53·1°
34. 3 s, 60 m
35. 15·7°, 24·7°, 65·3°, 74·3°
36. 20·4° → 24·5° or 65·5° → 69·6°

Exercise 11C page 212

1. (d) 45° 2. 250 m 3. 171 m, 500 m, 45°
4. 10 s, 5 s 5. 140 m/s 6. 5·8° or 84·2°
7. 63·4° 9. 25 m/s 10. 45°, 71·6°
11. 63·4°, 76·0°

Exercise 11D page 215

1. (a) $1\frac{3}{7}$ s (b) 10 m 2. (a) 2·42 s (b) 53·8 m
3. (a) 3 s (b) 44·1 m 4. (a) $1\frac{3}{7}$ s, 19·3 m 5. 47·3 m/s, 173 m
6. 12 m/s, 26·8 m 7. 44·4° 8. 30°, 90° 9. $2\theta + \phi = 90°$

Exercise 11E page 216

1. 30 m 2. 10 s
3. (i) 25 m/s at 16·3° (ii) 4 s (iii) 96 m (iv) 40·8 m/s
4. (i) 50 m/s, 20 m/s (ii) 20 m (iii) 100 m (iv) 38·7°
5. (i) 5·66 m/s (ii) 59 cm (iii) 8·5 m/s, 62° to the horizontal
6. 26·6°, 56·3°, 1·61 : 1 7. 2 tan α

Exercise 12A page 221

1. 200π rad/min 2. 3π rad/s 3. $\frac{300}{\pi}$ rev/min
4. (a) 6 rev/min (b) 12π rad/min (c) $\frac{\pi}{5}$ rad/s
5. (a) $\frac{3\pi}{2}$ rad/s (b) $\frac{10\pi}{9}$ rad/s 6. (a) 10 m/s (b) 50π m/s
7. (a) 5 m/s (b) 20π m/s 8. (a) 24 rad/s (b) $\frac{720}{\pi}$ rev/min
9. (a) 1 rad/s (b) $\frac{30}{\pi}$ rev/min 10. $\frac{2\pi}{5}$ m/s, $\frac{3\pi}{5}$ m/s
11. (a) 2 rad/s (b) 10π s (c) 3·5 m/s

Exercise 12B page 226

1. 6 m/s² 2. 18 m/s² 3. 1·11 m/s², 2·22 m/s², 4·44 m/s²
4. 5 N 5. 8 N 6. 592 N 7. 0·152 N
8. (a) 5 N (b) 20 N (c) 5·48 N
9. (a) 1 N (b) 25 N (c) 100 N 10. 6 rad/s
11. 4 m/s, 7 m/s 12. (a) 0·98 N down (b) 0·98 N up (c) 4·5 N away from O (d) 4·5 N towards O
13. $\mu \geqslant 0·4$ 14. 0·417
15. 12·6 m/s 16. 12·5 cm, 12·5 cm 17. 0·151 18. 360 m
19. 30 000 N 20. 50 cm 21. 1·75 m/s

Exercise 12C page 233

1. (a) 122·5 N, 3·5 rad/s (b) 196 N, 7 rad/s (c) 245 N, 66·4°
2. 21·8° 3. 17·5 m/s 4. 14·0° 5. 16·4 m/s
6. 11·8° 7. 25 N 8. 1·25 m
9. (a) 9800 N (b) 37·5 mm 10. (a) 4900 N (b) 15 mm
11. 22 m/s 12. 0·5 13. 0·4 14. 33·7°
18. 0·364, 22·2 m/s 19. 42 m/s, 14 m/s 20. (a) 4 N (b) 2·9 N
21. 16·9, 4·9 N 22. 14 rad/s 23. 1·3 N, 6·37 N

Exercise 12D page 236

2. (i) 4·9 N (ii) 1·48 3. 12 N, 20 m/s 4. 0·392, 21·4°, 35·5 m/s
5. $\frac{1}{9}mg, \frac{2}{3}mg$ 6. (a) 9·9 cm (b) 9·12 cm
7. (i) $\frac{10}{3}mg, \frac{5}{3}mg$ (iii) $2\pi\sqrt{\frac{l}{5g}}$ 8. $mg \sin\alpha + \frac{mv^2}{l}\cot\alpha, \frac{mv^2}{l} - mg\cos\alpha$

Exercise 13A *page 240*

1. 98 J **2.** 29·4 J **3.** 19 600 J **4.** 5880 J **5.** 98 J
6. 3·92 J **7.** 800 J **8.** 56 J **9.** 98 J **10.** 168 J
11. 0·125 **12.** 735 J **13.** (a) 1470 J (b) 250 J
14. (a) 336 N (b) 16 800 J (c) 24 500 J **15.** (a) 1700 J (b) 9800 J
16. 6370 J

Exercise 13B *page 243*

1. (a) 40 J (b) 9 J (c) 40 000 J (d) 20 J (e) 800 J
2. (a) 490 J (b) 2940 J (c) 196 000 J
3. (a) 392 J (b) 7840 J (c) 98 000 J **4.** (a) 5500 J (b) 125 J
5. (a) 3 J (b) 10 000 J **6.** 4 m/s **7.** 5 m/s **8.** 45 000 J
9. 147 J

Exercise 13C *page 247*

1. (a) 16 J (b) 16 J (c) 4 N **2.** (a) 78 J (b) 78 J (c) 13 N
3. (a) 27 J (b) 27 J (c) 3 m/s **4.** (a) 40 J (b) 40 J (c) 6 m/s
5. (a) 2 J (b) 2 J (c) 1 N **6.** (a) 90 J (b) 90 J (c) 6 m/s
7. (a) 70 J (b) 70 J (c) $\frac{2}{7}$ **8.** (a) 196 J (b) 196 J (c) 14 m/s
9. (a) 147 J (b) 147 J (c) 171 J **10.** (a) 98 J (b) 98 J (c) 2·5 m
11. (a) 21 J (b) 21 J (c) $4\frac{2}{7}$ m
12. (a) 735 m (b) 375 J (c) 360 J (d) 12 m/s **13.** 17 **14.** 16·25 m
15. (a) 1200 J (b) 1200 J (c) 4800 N **16.** 20 cm
17. (a) 98 J (b) 98 J (c) $4\frac{1}{6}$ m **18.** 7 m/s **19.** 26 m
20. (a) 49 J (b) 4 J (c) 45 J (d) 9 **21.** (a) 1340 J (b) 0·342
22. 12·5 m **23.** (a) 8·4 J (b) $4\sqrt{7}$ m/s (c) 16·8 J (d) 6·69 m/s

Exercise 13D *page 250*

1. 700 W **2.** 14·7 kW **3.** 40 W **4.** 3 W
5. 196 W **6.** 49 W **7.** 1·05 kW **8.** 5·17 kW
9. 6·4 kW **10.** (a) 80 kg (b) 10·4 kW **11.** 446 W
12. (a) 8 m/s (b) 3·24 kW (c) 4·32 kW
13. (a) 25 cm^2 (b) 2 m **14.** 10 m/s, 5 cm^2

Exercise 13E *page 253*

1. (a) 10 m/s (b) 15 m/s (c) 22 m/s
2. (a) 1·4 m/s^2 (b) 0·6 m/s^2 (c) 40 m/s
3. (a) 3 m/s^2 (b) 1 m/s^2 (c) 40 m/s **4.** 400 N
5. 400 W **6.** 200 W **7.** 2000 N, 0·02 m/s^2 **8.** 10 m/s, 3 m/s
9. (a) 1·25 m/s^2 (b) 30 m/s **10.** 5000 N **11.** 16·8 kW, 20 m/s
12. 49 N, 4 m/s **13.** (a) 800 N (b) 600 N (c) 10 kW
14. (a) 350 N (b) 35 m/s **15.** 420 N, 17·5 m/s

Exercise 13F *page 254*

1. 680 J **2.** (i) 343 000 J (ii) 3500 J, 339 500 J **3.** 2 W
4. 52 J (i) $25d$ (ii) $40d$, 80 cm **5.** (ii) 0·5 m/s^2 (iii) $18\frac{2}{3}$ m/s
6. (i) 18 (ii) 1250 kg (iii) 16·16
7. $3mg\cos\theta + \dfrac{mv^2}{a} - 2mg$, when particle is level with B
8. 60° to upward vertical, $\frac{3}{4}a$ **9.** (ii) $\frac{7}{2}ag$, $\sqrt{\dfrac{ag}{2}}$

Exercise 14A *page 258*

1. (a) 14 000 N s (b) 3 N s (c) 300 N s (d) 3 600 000 N s (e) 6 N s
2. (a) 15 N s (b) 50 N s (c) 55 N s (d) 70 N s
3. (a) 18 N s (b) 18 N s (c) 9 m/s
4. 35 N s
5. (a) 60**i** N s (b) 7**i** m/s
6. (a) −20**i** N s (b) 4**i** m/s
7. (a) −20**i** N s (b) −2**i** m/s
8. (a) 8**i** N s (b) 48**i** N s (c) 8
9. −8
10. 3
11. 4
12. (2·5**i** − 6**j**) m/s, 6·5 m/s
13. (4**i** − 3**j**) m/s, 5 m/s
14. 1, −2
15. (a) (i) −12**i** N s (ii) 12**i** N s (b) (i) −4**i** N s (ii) 4**i** N s (c) (i) −12**i** N s (ii) 12**i** N s

Exercise 14B *page 261*

1. 50 N 2. 6400 N 3. 3840 N 4. 0·064 N 5. 75 N 6. 0·0412 N

Exercise 14C *page 265*

1. 2·5 m/s 2. 5 m/s 3. 5 m/s 4. 6 m/s 5. 3 m/s 6. 1 m/s
7. 4 m/s 8. 4 m/s 9. (a) 4 m/s (b) 3·6 J (c) 2·4 J (d) 1·2 J
10. 10, 3J 11. 3**i** m/s 12. 3 m/s, 0·9 J 13. 4 m/s, 2520 J
14. 1 m/s, 401 000 J 15. 6 m/s, 20 cm 16. (a) 5 m/s, 5 m/s (b) 2 m/s, 3 m/s
17. 1 m/s, 9 J 19. $m_3 > \dfrac{2m_2}{5}$

Exercise 14D *page 267*

1. (a) 3 m/s (b) 0·525 J
2. 0·005 kg
3. (i) 10 cm/s, 18 cm/s (ii) 0·045 N s
4. $\frac{1}{4}$
5. 2000 N, $10\sqrt{2}$ m/s
6. 18 m, 32 400 J
7. (ii) $\frac{12}{5}m\sqrt{gc}$ (iii) $\frac{33}{25}c$ (iv) $\frac{12}{5}mgc$

Exercise 15A *page 272*

1. 4 N 2. 2 N 3. 15 N 4. 25 N 5. 20 cm 6. 25 cm
7. 12 cm 8. 25 cm 9. 49 N 10. 40 cm, 1·6 kg 11. 3 m/s^2
12. 0·4 m/s^2 13. (a) 20 cm (b) 20 cm (c) 30 cm (d) 10 cm 14. 1 m
15. 5·88 m/s^2, 3·92 m/s^2 16. 19·6 N, 9·8 17. $1\frac{7}{8}$ 18. 1 m 19. 14 N

Exercise 15B *page 276*

1. 2·5 J 2. 0·9 J 3. 12·5 J 4. 1 J 5. 45 J 6. 9 J
7. (a) 10 cm (b) 1·47 J
8. (a) 0 (b) 2 J (c) 2 J
9. (a) 1 m (b) 2·45 J (c) 9·8 J (d) 2·45 J
10. (a) 1·2 m/s (b) 12 cm

Exercise 15C *page 281*

1. (a) $\frac{2}{3}$ (b) $\frac{1}{2}$ (c) $\frac{2}{3}$ (d) 0, 2 m/s
(e) 2 m/s, 3 m/s (f) 1 m/s, 3 m/s (g) 3 m/s, 4 m/s (h) 2 m/s, 1 m/s
2. 4 m/s, $\frac{3}{4}$ 3. −**i** m/s, **i** m/s 4. $\frac{3}{5}$, 0·75 J 5. mu^2
6. (a) 4 m/s, 10 m/s, 5 m/s, 8 m/s (b) 3 m/s, 3 m/s, 2 m/s, 3 m/s
7. 1 m/s, 1·5 m/s, 3 m/s 8. −**i** m/s, 4**i** m/s, 5**i** m/s 9. $\dfrac{u}{3}(5 - e), \dfrac{u}{3}(5 + 2e)$
10. $4mu^2\,(1 - e^2)$

Exercise 15D *page 286*

1. (a) $\frac{1}{2}$ (b) $\frac{3}{4}$ (c) 6 m/s (d) 1·5 m/s (e) 3 m/s (f) 15 m/s
2. −4·5**i** m/s 3. $\frac{4}{5}$ 4. 4·5 s 5. 250 g, 5 m/s, 1 m/s
6. 3·6 m/s, 2 m/s 9. (b) 1·5 s (c) 4·5 m (d) 2·5 m/s, 0·25 m/s

***Exercise 15E** page 288*

1. 100 J **3.** (i) $\frac{5}{4}c$ (ii) $2c$

4. (i) $\frac{1}{2}$ (ii) $\frac{1}{4}mu^2$ (iii) $\frac{2}{3}$ (iv) u **5.** $\frac{169}{384}u$

7. $\frac{1}{10}u$, $\frac{3}{5}u$, $\frac{9}{5}mu$, $\frac{9}{20}mu^2$, $\frac{6}{5}mv$, $\frac{2}{5}v$

***Exercise 16A** page 292*

1. 40 m **2.** 28 m/s **3.** 4 m/s^2 **4.** 24 m/s **5.** 12 m/s^2

6. 21 m **7.** 30 m/s^2 **8.** 48 m/s **9.** 34 m

10. (a) 1 m (b) $2t$ (c) 4 m/s (d) 4 s (e) 13 m

11. (a) 45 m (b) 2 s, 5 s (c) 60 m/s (d) $12t - 42$ (e) -42 m/s^2

12. (a) 4 m/s (b) $8 - 6t$ (c) -10 m/s^2 (d) $4t^2 - t^3$ (e) 9 m

13. (a) 8 m/s (b) $\frac{2}{3}$ s, 4 s (c) 4 m/s^2 (d) 2 m **14.** 9 m/s, 9 m

15. (6i + 4j) m/s^2 **16.** 15 m/s **17.** 12i m/s^2 **18.** 20 m **19.** (24i − 9j) m/s

20. (30i + 5j) m **21.** 29 m/s, 24 m **22.** 3 s **23.** $2\frac{1}{4}$, 3 **24.** $7\frac{1}{2}$ s, $29\frac{1}{4}$ m

25. 8 s, $10\frac{2}{3}$ m

***Exercise 16B** page 298*

1. 2 m/s **2.** 10 m/s^2 **3.** 3·5 s **4.** 2·5 m **5.** 5 m

6. 0·549 s **7.** 0·2 m/s^2 **8.** 2 m/s **9.** 2 m **10.** 4·16 m

11. (a) $v = s + 2$ (b) $t = \ln\left(\frac{s+2}{2}\right)$ **12.** (a) 7 m/s (b) 0·973 s

13. 1 m/s^2 **14.** 9 m **15.** 32 s **16.** 1·95 s **17.** 8·96 m

18. 1·13 m **19.** 8 s **22.** 1 **24.** $\frac{5}{2}(1 - e^{-2t})$, 2·5 m/s

***Exercise 16C** page 304*

1. (a) 2π s (b) π s (c) $\frac{2\pi}{3}$ s **2.** 16 m/s^2 **3.** π s, 0·8 m/s^2

4. 12 m/s^2 **5.** 8 m/s **6.** 1·25 m/s, 1 m/s **7.** 60 cm

8. (a) 3 m/s (b) 50 cm (c) $\frac{\pi}{3}$ s **9.** 1 m, π s

10. 40 cm, 3 s, 0·84 m/s **11.** 2·5 m, 5 m/s

12. $\sqrt{2}$ m (a) 2 (b) 10 (c) 14

13. $\frac{\sqrt{3}}{2}$ m (a) 2 (b) 10

14. (a) $\frac{1}{3}\pi$ m/s (b) $\frac{1}{6}\pi$ m/s (c) $\frac{1}{6}\pi$ m/s (d) $\frac{\sqrt{3}}{4}$ m

15. 0·5 s, 2 m **16.** 15 cm **17.** 60 cm

18. 0·4 s, $\frac{5}{12}\pi\sqrt{3}$ m/s, 0·6 s, $\frac{5}{12}\pi\sqrt{2}$ m/s **19.** 0·43 s **20.** $\frac{1}{6}\pi$, (a) $\frac{1}{2}$ s, (b) 1 s

21. 2 m/s, $\frac{1}{20}\pi$ s

***Exercise 16D** page 306*

1. (i) 25 m (ii) 16 (iii) 12 m/s

2. (i) 12 s (ii) 16·5 m/s (iii) 912 m

3. (a) 14 m (b) (3i − 4j) i.e. 5 m/s at 53·1° below x-axis (c) 2i m/s^2

(d) (−i − 4j), i.e. $\sqrt{17}$ m/s, 104° below x-axis **4.** 8 J, 61·8 m

5. (b) (i) $\frac{3}{8}\pi$ s (ii) $26\frac{2}{3}$ m/s (iii) $142\frac{2}{9}$ m/s^2 (iv) $\frac{1}{16}\pi$ s **6.** $\frac{5}{18}\pi^2$, 2 s

7. midnight → 04.30, 08.10 → mid-day

***Exercise 17A** page 311*

1. 3·2 m/s **2.** 16 m **3.** 4 m/s **4.** 2 m/s **5.** 3 m

6. 4 s **7.** 0·47 m **8.** 1·83 s **9.** 2 m/s **10.** 19 m

11. (a) $(3t^2\mathbf{i} + 2t\mathbf{j})$ m/s (b) $(t^3\mathbf{i} + t^2\mathbf{j})$ m
12. 20 m/s, $17\frac{1}{3}$ m
13. 22 m/s, 24 m
14. 10 m/s
15. 12 m/s, No
16. 2·51 s
17. 1·20 m
18. (a) 22·4 m/s, (b) 27·7 s
19. (a) 10 m/s, (b) 17·4 s

Exercise 17B *page 315*

1. (a) 12 N s (b) 12 N s (c) 4 m/s
2. (a) 72 N s (b) 72 N s (c) 8 m/s
3. 8 m/s
4. 13**i** m/s
5. 7**i** m/s
6. 4
7. (8**i** + 6**j**) m/s, 10 m/s
8. (a) 12 J (b) 12 J (c) 2 m/s
9. (a) 18 J (b) 18 J (c) 12 m/s
10. (a) $2\frac{1}{4}$ J (b) $1\frac{1}{2}$ m
11. (a) 68·6 J (b) 19·6 J (c) 49 J (d) 7 m/s
12. (a) 294 J (b) 66 J (c) 12 m/s
13. (a) 28 J (b) 20 J (c) 2 m/s

Exercise 17C *page 321*

1. 8 N
2. $\frac{\pi}{5}$ s
3. (a) Yes (b) No (c) No (d) Yes
5. 25 cm, π s
6. 0·75 m/s
7. $\frac{\pi}{3}$ s, 25 cm
8. 0·64 m/s
9. 0·6 m/s
10. 20 cm, $\frac{2\pi}{7}$ s, 0·7 m/s
11. 25 N
12. 14·7 m/s^2, No
13. $\frac{1}{5}(2 + \pi)$ s, 2 m/s
14. 49 N, 0·42 m/s
15. 250 g, 50 N
16. $\frac{\pi}{10}$ s, 20 cm
17. 10 cm
18. 1·5 m
19. 9·8 cm, 3·36 m/s, 67·4 cm
20. 1·05 m/s, 0·357 s
22. 4 m/s, 4 m/s, 0, 0·572 s

Exercise 17D *page 325*

1. (b) (i) $\frac{4}{7}\pi$ s (ii) $\frac{5}{7}\pi$ s (iii) $\frac{\pi\sqrt{10}}{7}$ s
(c) (i) 20 cm (ii) 49 cm (iii) 24·8 cm
2. 2·84 s, 0·14 s
3. 75%
4. 36%

Exercise 17E *page 325*

1. $2m\mathbf{i} - m\mathbf{j}$
2. $(2T^4 - 2)\mathbf{i} + (3T^2 + 3)\mathbf{j}$, $94\mathbf{i} + 30\mathbf{j}$
3. (b) $\sqrt{\frac{mg}{k}}$
4. $\frac{V}{g}\ln 3$
5. $\sqrt{\frac{g}{k}(1 - e^{-2xk}) + w^2e^{-2xk}}$
6. (i) $2uT$ (ii) $T\ln 3$ (iii) $4u^2m$

Exercise 18A *page 330*

1. (a) $49\sqrt{3}$ N, zero (b) $49\sqrt{3}$ N, 294 N (c) $\frac{\sqrt{3}}{6}$
2. $\frac{3W}{4}, \frac{W}{4}$
3. $\frac{1}{3}$, 98 N
5. (a) $3W$ (b) $\frac{W}{2}$ (c) $\frac{W}{2}\sqrt{5}$
6. (a) $3W$ (b) $\frac{W\sqrt{3}}{6}$ (c) $\frac{W\sqrt{39}}{6}$
7. $\frac{5W}{4}$
9. (a) $3W$ (b) $\frac{W\sqrt{3}}{3}, \frac{W}{2}$
10. $2W$, $\frac{W\sqrt{3}}{2}$, zero

Exercise 18B *page 335*

1. (a) $\frac{1}{2}W$ N (b) $\frac{1}{2}W$ N (c) Thrust of $\frac{W\sqrt{3}}{3}$ N in AC and CB. Tension of $\frac{W\sqrt{3}}{6}$ N in AB
2. (a) $100\sqrt{5}$ N at $\tan^{-1}\frac{1}{2}$ to AD (b) 200 N in direction CB (c) Tension of 200 N in BC. Thrust of $100\sqrt{2}$ N in AC. Thrust of 100 N in AD. Tension of $100\sqrt{2}$ N in CD
3. (a) $500\sqrt{7}$ N at $\tan^{-1}\frac{\sqrt{3}}{2}$ with AD (b) 1000 N in direction CB
   (c) Thrust of 500 N in AD. Thrust of 1000 N in AC. Tension of 1000 N in CD. Tension of 1000 N in BC
4. (a) $3W$ (b) $3W$ (c) Thrust of $2W\sqrt{3}$ N in AD and CB. Tension of $W\sqrt{3}$ N in AE and EB. Tension of $\frac{4W}{3}\sqrt{3}$ N in DE and EC. Thrust of $\frac{5W}{3}\sqrt{3}$ N in DC
5. (a) $2W$ N vertically downwards (b) $3W$ N
   (c) Thrust of $\frac{4W\sqrt{3}}{3}$ N in AB. Thrust of $\frac{W\sqrt{3}}{3}$ N in BC
   Tension of $\frac{2W\sqrt{3}}{3}$ N in CD. Thrust of $\frac{2W\sqrt{3}}{3}$ N in DB. Tension of $\frac{2W\sqrt{3}}{3}$ N in DA
6. (a) 3 kN vertically downwards (b) 6 kN
   (c) Thrust of $\sqrt{3}$ kN in AB. Thrust of $2\sqrt{3}$ kN in BC
   Tension of $\sqrt{3}$ kN in CD. Thrust of 3 kN in DB. Tension of $2\sqrt{3}$ kN in DA
7. (a) $\frac{15W}{2}$ N (b) $6W$ N (c) Thrust of $5W\sqrt{3}$ N in AD.
   Tension of $\frac{5W\sqrt{3}}{2}$ N in AE. Tension of $W\sqrt{3}$ N in DE. Thrust of $3W\sqrt{3}$ N in DC.
   Tension of $2W\sqrt{3}$ N in CE. Tension of $2W\sqrt{3}$ N in EB. Thrust of $4W\sqrt{3}$ N in CB.
8. (a) $2\sqrt{3}$ kN vertically downwards (b) $4\sqrt{3}$ kN
   (c) Thrust of 2 kN in AB Thrust of 6 kN in BC Thrust of 4 kN in CD
   Tension of 2 kN in ED Tension of 2 kN in EF Thrust of 2 kN in EC
   Tension of 2 kN in FC Thrust of 2 kN in FB Tension of 4 kN in AF

Exercise 18C *page 337*

1. 17·3 N, OA thrust 11·5 N, AB thrust 11·5 N, AC tension 11·5 N, OC thrust 11·5 N, CB tension 5·77 N
2. (iii) $\frac{2}{3}a$
3. (i) 100 N, 700 N (ii) AB 100 N, BC 50 N, CD $50\sqrt{5}$ N, DE $50\sqrt{2}$ N, EB $150\sqrt{2}$ N, EA $100\sqrt{2}$ N, BD 550 N; AB, CD, DE.
4. $600\sqrt{3}$ N, $150\sqrt{13}$ N, 16·1° below DA; BD tension $350\sqrt{3}$ N, BC thrust $400\sqrt{3}$ N
5. (i) $20\frac{2}{3}$ kN, $15\frac{1}{3}$ N (ii) $\frac{28}{9}\sqrt{3}$ kN tension, $\frac{44}{9}\sqrt{3}$ kN thrust, $\frac{68}{9}\sqrt{3}$ kN thrust

Index